Iron-Sulfur Proteins

VOLUME III

Structure and Metabolic Mechanisms

MOLECULAR BIOLOGY

An International Series of Monographs and Textbooks

Editors: BERNARD HORECKER, NATHAN O. KAPLAN, JULIUS MARMUR, AND HAROLD A. SCHERAGA

A complete list of titles in this series appears at the end of this volume.

Iron-Sulfur Proteins

VOLUME III
Structure and Metabolic Mechanisms

Edited by Walter Lovenberg

SECTION ON BIOCHEMICAL PHARMACOLOGY
EXPERIMENTAL THERAPEUTICS BRANCH
NATIONAL HEART AND LUNG INSTITUTE
NATIONAL INSTITUTES OF HEALTH
BETHESDA, MARYLAND

ACADEMIC PRESS   New York   San Francisco   London   1977
A Subsidiary of Harcourt Brace Jovanovich, Publishers

Copyright © 1977, by Academic Press, Inc.
ALL RIGHTS RESERVED.
NO PART OF THIS PUBLICATION MAY BE REPRODUCED OR
TRANSMITTED IN ANY FORM OR BY ANY MEANS, ELECTRONIC
OR MECHANICAL, INCLUDING PHOTOCOPY, RECORDING, OR ANY
INFORMATION STORAGE AND RETRIEVAL SYSTEM, WITHOUT
PERMISSION IN WRITING FROM THE PUBLISHER.

ACADEMIC PRESS, INC.
111 Fifth Avenue, New York, New York 10003

United Kingdom Edition published by
ACADEMIC PRESS, INC. (LONDON) LTD.
24/28 Oval Road, London NW1

Library of Congress Cataloging in Publication Data

Lovenberg, Walter.
　　Iron-sulfur proteins.

　　(Molecular biology ; an international series of
monographs and textbooks)
　　Includes bibliographies.
　　CONTENTS: v. 1.　　Biological properties.–v. 2.
Molecular properties.
　　1.　Iron sulfur proteins.　　I.　Title.　　II.　Series.
QP552.I7L69　　　　574.1'9245　　　　72-13613
ISBN 0–12–456003–2

PRINTED IN THE UNITED STATES OF AMERICA

Contents

List of Contributors	ix
Preface	xi
Contents of Other Volumes	xiii

1. Nitrogenase-Derepressed Mutants of *Klebsiella pneumoniae*

K. T. Shanmugam, C. Morandi, and R. C. Valentine

I.	Introduction	1
II.	Glutamine Synthetase as Regulatory Protein	2
III.	Nitrogenase-Derepressed Mutants	4
IV.	Discussion and Summary	11
	References	13

2. Current Topics and Problems in the Enzymology of Nitrogenase

W. H. Orme-Johnson and L. C. Davis

I.	Introduction	16
II.	Composition and Structure of Nitrogenase Protein Components	17
III.	Interaction of MgATP and the Fe Protein	34
IV.	Sequence of Electron Transfers between Nitrogenase Components	43
V.	Structural and Mechanistic Hypotheses	55
	References	58

3. Iron–Sulfur Centers of the Mitochondrial Electron Transfer System—Recent Developments

Helmut Beinert

I.	Introduction	61
II.	Iron–Sulfur Centers of the Mitochondrial Electron Transfer System	66
III.	Relation of Iron–Sulfur Centers to Energy Conservation	89
IV.	Outlook	93
	References	97

4. Biosynthesis of Iron–Sulfur Proteins

James W. Brodrick and Jesse C. Rabinowitz

I.	Introduction	101
II.	Bacterial Protein Synthesis	103
III.	Regulation of Synthesis of Ferredoxin	106
IV.	Conversion of Apoferredoxin to Holo- or Native Ferredoxin	107
V.	Synthesis of Specific Proteins *in Vitro*	110
VI.	Immunological Studies with Clostridial Ferredoxin	111
VII.	Enzymatic Assays for Ferredoxin	112
VIII.	Synthesis of Clostridial Ferredoxin or Apoferredoxin *in Vitro*	113
	References	118

5. Role of Iron–Sulfur Proteins in Formate Metabolism

Rudolf K. Thauer, Georg Fuchs, and Kurt Jungermann

I.	Introduction	121
II.	Role of Formate in Metabolism	122
III.	Ferredoxin and Formate Metabolism	129
IV.	Formate Dehydrogenases and Formate Metabolism	143
V.	Concluding Remarks	150
	References	151

6. X-Ray Analysis of High-Potential Iron–Sulfur Proteins and Ferredoxins

Charles W. Carter, Jr.

I.	Introduction	158
II.	$Fe_4S_4^*$ Active Centers and the Three-State Hypothesis	160
III.	Comparison of the Protein-Bound and Synthetic $Fe_4S_4^*$ Clusters	165
IV.	Characteristic Features of the Cluster Binding Cavities	172
V.	Analysis of Reduced versus Oxidized HiPIP	187
VI.	Conclusions and Review	196
	References	202

7. Synthetic Analogues of the Active Sites of Iron–Sulfur Proteins

R. H. Holm and James A. Ibers

I.	Introduction	206
II.	Synthesis and Structures of 1-Fe, 2-Fe, and 4-Fe Active-Site Analogues	217
III.	Physical Properties of Analogues	228

IV.	Chemical Reactivity of Analogues	256
V.	Iron–Sulfur Units as Redox Centers	265
VI.	Perspectives and Conclusions	271
	References	272

8. Evidence from Mössbauer Spectroscopy and Magnetic Resonance on the Active Centers of the Iron–Sulfur Proteins

R. Cammack, D. P. E. Dickson, and C. E. Johnson

I.	Introduction	283
II.	Spectroscopy of the Active Center	289
III.	Proteins with 1 Fe Centers	298
IV.	Proteins with 2 Fe–2 S* Centers	301
V.	Proteins with 4 Fe–4 S* Centers	305
VI.	Conclusions	326
	References	327

9. Redox Mechanisms of Iron–Sulfur Proteins

Larry E. Bennett

I.	Introduction	331
II.	Recent Development in Simple Redox Chemistry	332
III.	Redox Dynamics of Structurally Characterized Iron–Sulfur Proteins	352
IV.	Survey of the Redox Behavior of Structurally Less Characterized Iron–Sulfur Proteins	372
	References	375

10. Recent Mössbauer Results of Some Iron–Sulfur Proteins and Model Complexes

P. G. Debrunner, E. Münck, L. Que, and Charles E. Schulz

I.	Introduction	381
II.	Rubredoxin	388
III.	The MoFe Protein of Nitrogenase	400
IV.	Model Compounds for Rubredoxin and 4 Fe–4 S* Clusters	411
	References	415

Author Index	419
Subject Index	436

List of Contributors

Numbers in parentheses indicate the pages on which the authors' contributions begin.

HELMUT BEINERT (61), *Institute for Enzyme Research, The University of Wisconsin, Madison, Wisconsin*

LARRY E. BENNETT (331), *Department of Chemistry, San Diego State University, San Diego, California*

JAMES W. BRODRICK (101), *Enzymology Research Laboratory, Martinez Veterans Administration Hospital, Martinez, California*

R. CAMMACK (283), *Department of Plant Sciences, King's College, University of London, London, England*

CHARLES W. CARTER, JR. (157), *Departments of Biochemistry and Anatomy, University of North Carolina, School of Medicine, Chapel Hill, North Carolina*

L. C. DAVIS* (15), *Department of Biochemistry and the Center for the Study of Nitrogen Fixaton, The College of Agricultural and Life Sciences, The University of Wisconsin, Madison, Wisconsin*

P. G. DEBRUNNER (381), *Physics Department, University of Illinois, Urbana, Illinois*

D. P. E. DICKSON (283), *Department of Physics, University of Liverpool, Liverpool, England*

GEORG FUCHS (121), *Lehrstuhl für Biochemie der Pflanzen, Abteilung für Biologie, Ruhr-Universität Bochum, Bochum, Germany*

R. H. HOLM (205), *Department of Chemistry, Stanford University, Stanford, California*

JAMES A. IBERS (205), *Department of Chemistry, Northwestern University, Evanston, Illinois*

* Present address: Department of Biochemistry, Kansas State University, Manhattan, Kansas.

C. E. JOHNSON (283), *Department of Physics, University of Liverpool, Liverpool, England*

KURT JUNGERMANN (121), *Biochemisches Institut, Universtät Freiburg, Freiburg, Germany*

C. MORANDI* (1), *The Plant Growth Laboratory and Department of Agronomy and Range Science, University of California, Davis, California*

E. MUNCK (381), *Freshwater Biological Institute, University of Minnesota, Navarre, Minnesota*

W. H. ORME-JOHNSON (15), *Department of Biochemistry and the Center for the Study of Nitrogen Fixation, The College of Agricultural and Life Sciences, The University of Wisconsin, Madison, Wisconsin*

L. QUE (381), *Freshwater Biological Institute, University of Minnesota, Navarre, Minnesota*

JESSE C. RABINOWITZ (101), *Department of Biochemistry, University of California, Berkeley, California*

CHARLES E. SCHULZ (381), *Physics Department, University of Illinois, Urbana, Illinois*

K. T. SHANMUGAM (1), *The Plant Growth Laboratory, and Department of Agronomy and Range Science, University of California, Davis, California*

RUDOLF K. THAUER† (121), *Lehrstuhl für Biochemie der Pflanzen, Abteilung für Biologie, Ruhr-Universität Bochum, Bochum, Germany*

R. C. VALENTINE (1), *The Plant Growth Laboratory and Department of Agronomy and Range Science, University of California, Davis, California*

* Present address: Universitá degli Studi di Milano, Cattedra di Chimica Macromolecolore, Milan, Italy.

† Present address: Mikrobiologie, Fachbereich Biology, Phillipps-Universität, Lahnberge, Marburg, Germany.

Preface

In the few years that have elapsed since the first two volumes of "Iron–Sulfur Proteins" were published, this important class of proteins has continued to occupy the attention of many scientists around the world. As a result, numerous advances have been made in our understanding of the vital role these proteins play in biological processes, and we have gained an insight into their detailed and unique chemical structure. The objective of this third volume is to collate some of these important advances and, in essence, to update the original volumes. Of the ten chapters comprising this work, five present recent advances in biochemical areas and five are devoted to some of the elegant physical studies of the past several years.

A total understanding of biological nitrogen fixation is of special importance when one considers the worldwide need for fertilizer and food supplies. Two chapters are devoted to this important subject, one of which deals with the nitrogenase gene and the other with the molecular mechanism of this complex enzyme. The role of iron–sulfur proteins in mammalian mitochrondrial function is covered in another chapter. While exciting and important advances have been made in this area, the complexity of these organelles precludes a total understanding of the role of these proteins in mitochondrial function. As for the biochemical aspects, two chapters concern bacterial ferredoxin. These include a discussion of the mechanism of biosynthesis and the function of iron–sulfur proteins in formate metabolism.

Our recognition of iron–sulfur complexes as mediator of electron-transfer reactions is relatively recent. However, our understanding of their detailed physical and structural properties has been one of the most rapid and dramatic scientific achievements of our time. In this volume two different approaches that have yielded far-reaching and complementary advances in our understanding of the iron–sulfur clusters are presented. Dr. Holm and Dr. Ivers discuss their work on the syntheses of model complexes corresponding to the natural active centers of these proteins, and Dr. Carter presents recent X-ray crystallographic findings. Iron–sulfur proteins have also presented a unique opportunity to utilize Mössbauer spectroscopy in probing biological material, and two chapters deal largely with this technique. The theoretical aspects of the redox properties of iron–sulfur proteins have been dealt with in depth. This is particularly important since the primary role of iron–sulfur clusters is to mediate redox reactions.

As in the first two volumes, the style of presentation of each author has been maintained with the hope that the excitement and philosophy of each of the contributing laboratories can be transmitted to the reader.

WALTER LOVENBERG

Contents of Other Volumes

Volume I

BIOLOGICAL PROPERTIES

Iron–Sulfur Proteins: Development of the Field and Nomenclature
  Helmut Beinert

Bacterial Ferredoxins and/or Iron–Sulfur Proteins as Electron Carriers
  Leonard E. Mortenson and George Nakos

Comparative Biochemistry of Iron–Sulfur Proteins and Dinitrogen Fixation
  R. W. F. Hardy and R. C. Burns

Iron–Sulfur Proteins in Photosynthesis
  Charles F. Yocum, James N. Siedow, and Anthony San Pietro

Ferredoxin and Carbon Assimilation
  Bob B. Buchanan

Structure and Reactions of a Microbial Monoxygenase: The Role of Putidaredoxin
  I. C. Gunsalus and J. D. Lipscomb

Role of Rubredoxin in Fatty Acid and Hydrocarbon Hydroxylation Reactions
  Eglis T. Lode and Minor J. Coon

Adrenodoxin: An Iron–Sulfur Protein of Adrenal Cortex Mitochondria
  Ronald W. Estabrook, Koji Suzuki, J. Ian Mason, Jeffrey Baron, Wayne E. Taylor, Evan R. Simpson, John Purvis, and John McCarthy

Iron–Sulfur Flavoprotein Dehydrogenases
  Thomas P. Singer, M. Gutman, and Vincent Massey

Iron–Sulfur Flavoprotein Hydroxylases
  Vincent Massey

Author Index–Subject Index

Volume II

MOLECULAR PROPERTIES

The Chemical Properties of Ferredoxins
  Richard Malkin

The Types, Distribution in Nature, Structure-Function, and Evolutionary Data of the Iron–Sulfur Proteins
  Kerry T. Yasunobu and Masaru Tanaka

The Iron–Sulfur Complex in Rubredoxin
  William A. Eaton and Walter Lovenberg

Crystal and Molecular Structure of Rubredoxin from *Clostridium pasteurianum*
  L. H. Jensen

Probing Iron–Sulfur Proteins with EPR and ENDOR Spectroscopy
  W. H. Orme-Johnson and R. H. Sands

Mössbauer Spectroscopy of Iron–Sulfur Proteins
  Alan J. Bearden and W. R. Dunhan

NMR Spectroscopy of the Iron–Sulfur Proteins
  W. D. Phillips and Martin Poe

Current Insights into the Active Center of Spinach Ferredoxin and Other Iron–Sulfur Proteins
  Graham Palmer

Author Index—Subject Index

CHAPTER 1

Nitrogenase-Derepressed Mutants of *Klebsiella pneumoniae*

K. T. SHANMUGAM, C. MORANDI, and R. C. VALENTINE

I. Introduction.. 1
II. Glutamine Synthetase as Regulatory Protein..................... 2
III. Nitrogenase-Derepressed Mutants................................ 4
    A. Isolation... 4
    B. Biochemical Properties...................................... 5
    C. Mapping of Nitrogenase Regulatory Genes.................... 8
IV. Discussion and Summary... 11
    A. Gln (AC)$^-$.. 12
    B. Gln C$^-$... 12
    C. Gln$^-$ Nif C$^-$... 13
    References... 13

I. INTRODUCTION

Nitrogenase is a crucial iron–sulfur- (and molybdenum-) containing enzyme that catalyzes the reductant- and ATP-dependent interconversion of gaseous nitrogen to ammonium ion. The adjective "crucial" is used because this is the only known enzyme that bridges the gap between what amounts to a huge reservoir of atmospheric nitrogen and ammonium ion, needed as raw material for building cellular proteins and other nitrogen-containing compounds.

Just as the 1960's were an active period for exploring the biochemistry of nitrogenase and its ancillary reactions, the 1970's promise to be a time when the genes that govern nitrogen fixation activity are unraveled. Following the first paper on nitrogen fixation (Nif) genetics in 1971 (Streicher *et al.*, 1971), a number of significant findings have been made in more or less chronological order as follows: (i) mapping the nitrogen fixation genes on the chromosome of *Klebsiella* (Streicher et al., 1971; Shanmugam et al., 1974; St. John et al., 1975); (ii) transfer of nitrogen fixation genes to *Escherichia coli*, creating nitrogen-fixing hybrids (Dixon and Postgate, 1972); (iii) construction of infectious (F′ and P′)-Nif episomes capable of transferring nitrogen fixation genes to a variety of gram-negative bacteria (Cannon et al. 1976; R. A. Dixon, personal communication); (iv) derepression of nitrogen fixation genes (nitrogenase activity no longer repressed by NH_4^+) (Streicher et al., 1974; Tubb, 1974; Shanmugam et al., 1975); (v) construction of strains that export large quantities of NH_4^+ (Shanmugam and Valentine, 1975a). Since the first three findings have been amply reviewed elsewhere (Streicher et al., 1972; Shanmugam and Valentine, 1975b), we will concentrate here mainly on the fourth topic of regulation of nitrogen fixation. A historical perspective has recently been provided by Brill (1975). In a broader context, it is safe to say that regulation of iron–sulfur proteins is one of the most poorly understood aspects of this growing field. Perhaps some general concepts regarding the regulation of iron–sulfur proteins may emerge from the intensive studies of regulation of nitrogen fixation now under way in several laboratories. The underlying issue we shall deal with throughout this short review is: What are the signals that switch the nitrogen fixation genes on and off?

II. GLUTAMINE SYNTHETASE AS REGULATORY PROTEIN

There is now mounting evidence, recently summarized in a timely review by Goldberger (1974), that enzymes may often have dual biological functions, behaving both as enzymes and genetic regulators that control gene expression. A great deal of attention has recently been focused on glutamine synthetase as such a genetic control protein.

Glutamine synthetase is a protein containing twelve identical subunits (with a subunit molecular weight of 50,000). Experiments conducted at the laboratory of Earl Stadtman in the United States and Holzer in Germany (Ginsburg and Stadtman, 1973; Wohlhueter et al., 1973) revealed the existence of an enzyme system that modifies glutamine synthetase (adenylylation enzyme cascade). Within 10–20 seconds after addition of NH_4^+ to a culture, a surge of NH_4^+ enters the cell, causing a marked perturbation of the intracellular levels of glutamine, glutamate, and related metabolites (Wohlhueter et al., 1973). The sudden raising or lowering of the concentration of these metabolites apparently triggers the activation of the enzymes of the adenylylation cascade system that catalyze the covalent modification of glutamine synthetase by attachment (or removal) of adenylyl moieties on twelve specific tyrosine residues, one on each of the twelve identical subunits (see Ginsburg and Stadtman, 1973; Senior, 1975). The fully modified protein has greatly reduced glutamine synthetase biosynthetic activity in comparison to the nonadenylylated enzyme; in addition, a variety of catalytic parameters of the enzyme are influenced by modification of the tyrosine residues.

Glutamine synthetase, in the unadenylylated form, has been shown to act as a genetic activator for the genes responsible for histidine utilization (*hut*) of *Klebsiella aerogenes*, a non-nitrogen-fixing organism (Tyler et al., 1974). Since so much of the interpretation of the mechanism of nitrogen fixation genes regulation hinges on the *hut* system, this system will be briefly described.

The histidine degradation (*hut*) pathway of *Klebsiella* consists of four enzymes capable of converting histidine to glutamate, ammonia, and formamide:

$$\text{Histidine} \rightarrow NH_4^+ + \text{uroconate} \rightarrow \rightarrow \text{glutamate} + \text{formamide}$$

During a continuing study of over two decades, Magasanik and co-workers (1974) have detailed how NH_4^+ represses the synthesis of these enzymes. Highlights of the study carried out with whole cells can be summarized as follows.

  i. Histidase, the first enzyme of the pathway, is repressed by NH_4^+ in a medium containing glucose, histidine, and ammonium.

  ii. Mutants constitutive for the synthesis of glutamine synthetase have high levels of histidase even when grown with an excess of ammonium. Of particular interest is the case of a glutamine-requiring mutant that synthesizes enzymatically inactive glutamine synthetase antigen (CRM) constitutively. In this mutant, the level of histidase is elevated when the cells are grown in a medium containing an excess of ammonium.

  iii. Introduction of an episome carrying the *E. coli* glutamine synthetase gene into the CRM$^+$, and glutamine-requiring mutant results in a repression of the glutamine synthetase antigen and normal control of histidase.

  iv. Certain glutamine-requiring mutants missing glutamine synthetase (CRM$^-$) do not induce histidase even in conditions of low NH_4^+. Thus, Magasanik and co-workers (1974) conclude that glutamine synthetase appears to be directly involved in the regulation of histidase.

The strongest evidence for the role of glutamine synthetase in *hut* regulation comes from *in vitro* experiments in which DNA specific for *hut* was used as template for production of *hut* messenger RNA (Tyler et al., 1974). In these studies, which utilize purified RNA polymerase of *E. coli*, *in vitro* transcription of the *hut* message was found to be stimulated by the nonadenylylated form of glutamine synthetase. These results are in excellent accord with the observations cited above using whole cells; in addition, they prove that glutamine synthetase acts by stimulating the transcription of the *hut*-specific DNA. In essence, activation by glutamine synthetase requires the conversion of this protein to the active state by the removal of the attached adenylyl groups and by an increase in its cellular concentrations—an elegant example of a cascade effect. In this manner, glutamine synthetase appears to be an example of a protein that, in addition to its enzymatic function, can act as a genetic regulator protein.

III. NITROGENASE-DEREPRESSED MUTANTS

A. Isolation

Based on the studies on the regulation of histidase biosynthesis in *K. aerogenes*, we have devised a procedure for isolation of nitrogenase derepressed mutants of *Klebsiella pneumoniae* which is as follows: *Klebsiella pneumoniae* is capable of utilizing one of two different *enzyme* systems involving a total of three enzymes for assimilation of NH_4^+ into glutamate (depending on the source of nitrogen in the medium).

$$\alpha\text{-Ketoglutarate} + NH_4^+ + NADPH \rightarrow \text{L-glutamate} + NADP^+ \quad (1)$$

$$\text{L-Glutamate} + NH_4^+ + ATP \rightarrow \text{L-glutamine} + ADP + P_i \quad (2)$$

$$\text{L-Glutamine} + \alpha\text{-ketoglutarate} + NADPH \rightarrow 2\ \text{L-glutamate} + NADP^+ \quad (3)$$

Reaction (1), catalyzed by glutamate dehydrogenase, is the major source of glutamate for the cell when the concentration of NH_4^+ in the medium is higher than 1 mM (Nagatani et al., 1971). Under nitrogen-fixing conditions (in the absence of added NH_4^+, under nitrogen), NH_4^+ produced by nitrogenase is assimilated into glutamate through the Reactions (2) and (3), catalyzed by glutamine synthetase and glutamate synthase, respectively (Nagatani et al., 1971).

It has been shown previously that Gln C$^-$ mutants producing constitutive levels of glutamine synthetase (Reaction 2) are simultaneously derepressed for nitrogenase activity (Streicher et al., 1974). These mutants have no detectable glutamate dehydrogenase activity (Reaction 1). Brenchley et al. (1973) have isolated Gln C$^-$ mutants as revertants from

Asm⁻ mutants [which lack glutamate synthase activity, Eq. (3)] of K. aerogenes. These mutants are found to be glutamate auxotrophs. Using this rationale, we have isolated glutamate auxotrophs from Asm⁻ mutants of *K. pneumoniae*. About 50% of these mutants were found to be derepressed for nitrogenase and glutamine synthetase biosynthesis (Nif C⁻, Gln C⁻). In addition to the Nif C⁻, Gln C⁻ phenotype, two additional classes of nitrogenase derepressed mutants were also obtained as glutamine auxotrophs, lacking glutamine synthetase activity, from Asm⁻ mutants. In the case of Asm-1, about 5–10% of the auxotrophic colonies were found to be glutamate or glutamine auxotrophs, with about half of these (3/106 auxotrophs) being derepressed for nitrogenase biosynthesis. Nitrogenase-derepressed mutants were also isolated, starting with strain Asm-3 (strain SK-28) and strain Asm-24 (strains SK-29, 54–57, 59, 60).

B. Biochemical Properties

Specific activities of nitrogenase levels in the mutants and their parental strains are presented in Table I. Nitrogenase activities, in the presence of NH_4^+, in the mutants (strains SK-24–29, 54–57, 59, 60) ranged from 40 to 100% of the value compared to those observed in the absence of NH_4^+. Under the same conditions, the parental strains produced no nitrogenase activity in the presence of NH_4^+. Nitrogenase activity was also determined in these mutants as a function of NH_4^+ concentration (data not presented). The derepressed mutants induced about 2–4 units of nitrogenase activity in the concentration range of 0–15 mM of NH_4^+. The parental strains (M5A1 and Asm⁻ strains) were completely repressed by concentrations of NH_4^+ as low as 1.5 mM. Because of the previously observed correlation between high glutamine synthetase activity and nitrogenase derepression (Streicher *et al.*, 1974), the specific activities of glutamine synthetase and other ammonia assimilatory enzymes, such as glutamate synthase and glutamate dehydrogenase of nitrogenase-derepressed mutants and their parental strains, were determined (see Table I).

Note that glutamine synthetase activity is fully derepressed (in the presence of NH_4^+) in strains SK-24, 28, 29, 57, and 60, which require glutamate for growth. Glutamate dehydrogenase activity is not detected in these strains, a property shared with other glutamine synthetase constitutive mutants [see Magasanik *et al.* (1974) for a discussion of the possible role of glutamine synthetase in the repression of glutamate dehydrogenase]. However, in the other glutamate-requiring nitrogenase-derepressed strains (strains SK-54 and 59), glutamine synthetase activity is not derepressed. Only small amounts of glutamine synthetase (100–200 units) were detected, even in the absence of NH_4^+ (as compared to 600–

TABLE I
Specific Activities of Nitrogenase, Glutamine Synthetase, Glutamate Synthase, and Glutamate Dehydrogenase in Various Nitrogenase-Derepressed Strains of *K. pneumoniae*[a]

Strain	Nitrogenase (μmoles/hour/mg protein)		Glutamine synthetase (nmoles/minute/mg protein)		Glutamate dehydrogenase (nmoles/minute/mg protein)		Glutamate synthase (nmoles/minute/mg protein)	
	$-NH_4^+$	$+NH_4^+$	$-NH_4^+$	$+NH_4^+$	$-NH_4^+$	$+NH_4^+$	$-NH_4^+$	$+NH_4^+$
M5A1	4.04	0.00	878	229	<5	126	37	53
Asm-1	3.14	0.00	653	147	18	116	<5	<5
Asm-3	4.74	0.00	963	171	20	156	<5	<5
Asm-24	4.24	0.00	643	165	20	150	<5	<5
SK-24	2.80	2.42	684	866	<5	<5	<5	<5
SK-28	3.27	3.05	1006	1055	<5	<5	<5	<5
SK-29	2.82	2.01	739	627	<5	<5	<5	<5
SK-54	3.77	2.12	193	162	39	35	<5	<5
SK-57	3.25	2.90	490	933	10	15	<5	<5
SK-59	2.85	1.68	124	222	35	34	<5	<5
SK-60	2.60	2.21	566	880	<5	<5	<5	<5
SK-25	4.47	3.70	0	0	12	16	<5	<5
SK-26	3.82	2.91	0	0	17	18	<5	<5
SK-27	2.88	1.81	0	0	17	15	<5	<5
SK-37	2.45	0.95	0	0	147	88	<5	<5
SK-55	2.53	1.92	6	0	20	20	<5	<5
SK-56	3.19	2.29	14	21	22	24	<5	<5

[a] Nitrogenase activity was detected in whole cells using the acetylene reduction technique as described in Shanmugam et al. (1974). Other enzyme activities were determined in crude extracts of cultures grown in sucrose-minimal medium with either glutamate (100 μg/ml) or glutamine (100 μg/ml) as described before (Shanmugam et al., 1975). Strains SK-25, 26, 27, 37, 55, and 56 were grown in the medium containing glutamine, while all other strains were cultured in a glutamate-supplemented minimal medium. The concentration of $(NH_4)_2SO_4$ was 1 mg/ml. See Shanmugam et al. (1975) for the properties of the various strains. Strains SK-54, 57, 59, and 60 were isolated as glutamate auxotrophs after mutagenesis of strain Asm-24 with N-methyl-N'-nitro-N-nitrosoguanidine (NTG). Strains SK-55 and 56 were obtained after mutagenesis of strain Asm-24 with NTG as glutamine auxotrophs.

1000 units for the wild type). These strains also produced about 40 units of glutamate dehydrogenase even in the absence of NH_4^+ (compared to undetectable levels in the parent). Addition of NH_4^+ (15 mM) to the culture medium had no effect on either of the two enzyme activities. Strain SK-57, although derepressed for glutamine synthetase, synthesizes small amounts of glutamate dehydrogenase. Glutamate synthase activity is not observed with any of the nitrogenase-derepressed mutants, a property of the parental strains.

In contrast to the nitrogenase-derepressed mutants such as strain SK-24, glutamine-requiring mutants (strains SK-25–27, 37, 55, 56) produce no detectable glutamine synthetase activity or produce very low levels. Low but significant levels of glutamate dehydrogenase activities were detected in all strains except in SK-37, which produced high levels even in the absence of NH_4^+. The specific activities of the four enzymes in the parental strains are given for comparison. To test the possibility that glutamine added to the medium as a growth supplement might have acted as a repressor of glutamine synthetase in the nitrogenase-derepressed mutants requiring glutamine, glutamine synthetase activity was determined in the parent, Asm-1, in the presence and absence of glutamine. Addition of glutamine to the growth medium did not repress glutamine synthetase activity.

In order to rule out the possibility that catalytically active glutamine synthetase was produced during a specific stage of growth in the glutamine-requiring strains (strains SK-25–27, 37, 55, 56), the levels of glutamine synthetase were monitored throughout the growth cycle in SK-25. No glutamine synthetase activity was detected at any time during the growth period. The nitrogenase induction pattern was similar to the parent strain. Low levels of glutamate dehydrogenase were observed late in the growth period.

During their studies of the regulation of the histidine utilization (*hut*) operon, Magasanik *et al.* (1974) described glutamine-requiring mutants of *K. aerogenes*, a non-nitrogen-fixing organism, which produced catalytically inactive glutamine synthetase protein that was detected immunologically. This raised the possibility that the nitrogenase-derepressed, glutamine-requiring mutants of *K. pneumoniae* (strains SK-25–27, 37, 55, 56) may also produce glutamine synthetase protein lacking catalytic activity. Extracts prepared from the nitrogenase-derepressed mutants lacking glutamine synthetase activity were tested for the presence of immunologically active glutamine synthetase protein [antigenic cross-reacting material (CRM)]. A precipitin band corresponding to glutamine synthetase protein (CRM) was observed with extracts of strains SK-27, 37, 55, and 56 (Fig. 1). No detectable precipitin band was observed with the extracts of strain SK-25 or strain SK-26. Extracts of strains M5A1, Asm-1, and SK-27, 37, 55, and 56 gave one band each, corresponding to the presence of a homologous antigen in all extracts. In the extract from strain SK-56, an additional fast-migrating band was also observed. We do not have an explanation for this second band.

A second procedure involving the use of the antibody neutralization by glutamine synthetase protein was also used to confirm the absence

Fig. 1. Immunodiffusion analysis for the presence of glutamine synthetase protein (CRM) in extracts of nitrogenase-derepressed mutants. Ouchterlony-type plates (Hyland Laboratories, Costa Mesa, California) were equilibrated with 0.01 M imidazole buffer, pH 7.0, containing 0.15 M NaCl for 1 hour. Ten microliters of serum was placed in the center well. Ten microliters of crude extracts (about 350–400 µg protein) obtained from various nitrogenase-derepressed mutants grown under nitrogen-fixing conditions were placed in the respective outer wells. The presence or absence of precipitin band(s) corresponding to glutamine synthetase antigen–antibody complex was scored after 16–18 hours of incubation at room temperature.

of the antigen in the extracts of strains SK-25 and 26. In this experiment, extracts of strain SK-25 or strain SK-26 added to diluted samples of the glutamine synthetase immune serum were found not to decrease the effectiveness of the antibody for inactivation of glutamine synthetase catalytic activity. Under the same conditions, extracts of strain SK-27 completely neutralized the antibody.

C. Mapping of Nitrogenase Regulatory Genes

Because of the relatedness between *E. coli* and *K. pneumoniae*, it is possible to localize Gln⁻ or Glu⁻ mutations on the chromosome of *K. pneumoniae* through complementation by *E. coli* F′ genetic elements. Using this technique, we have observed that glutamine-independent colonies can be obtained from Gln⁻ strains of *K. pneumoniae* by constructing hybrid clones carrying *E. coli* F′133 (Streicher et al., 1974), which harbors the chromosomal markers between 72 and 77.5 minutes of the genetic map (Low, 1972). F′ complementation analysis was carried out with the glutamate- or glutamine-dependent mutants described above in order to map the nitrogenase regulatory mutations on the *K. pneumoniae* genome. As seen from Table II, strains lacking glutamine synthetase activity (strains SK-25–27, 37, 56) were complemented by the episome, F′133, suggesting a location of nitrogenase regulatory genes in or near the glutamine synthetase genes on the chromosome. Hybrids derived from strains such as strain SK-27 are readily cured of the Gln⁺ property if they are tested immediately after their initial selection. Pro-

TABLE II
F' COMPLEMENTATION ANALYSIS OF NITROGENASE-DEREPRESSED MUTANTS[a]

Recipient	Relevant phenotype (Gln)	Parent	Number of prototrophic colonies with F'133[b] (nitrogen source)	
			NH_4^+	Glutamate[c]
SK-24	Gln C−	Asm-1	0	—
SK-28	Gln C−	Asm-3	0	—
SK-29	Gln C−	Asm-24	0	—
SK-25	Gln−	Asm-1	670	N.D.
SK-26	Gln−	Asm-1	620	N.D.
SK-27	Gln−	Asm-1	960	N.D.
SK-37	Gln−	SK-24	720	N.D.
SK-56	Gln−	Asm-24	2.0×10^5	N.D.
SK-512	Gln−	SK-24	0	2.5×10^3
SK-513	Gln−	SK-24	0	5.0×10^3
SK-514	Gln−	SK-24	0	4.0×10^3

[a] Conjugation experiments were conducted as described before (Shanmugam et al., 1975).

[b] Glutamate- or glutamine-independent colonies obtained per milliliter in sucrose-minimal medium containing either NH_4^+ (15 mM) or glutamate (1 mg/ml) as the sole source of nitrogen.

[c] N.D. = not determined; — = not applicable.

longed storage in a selective medium (in order to maintain the episome) leads to an eventual loss of curability, even in the presence of acridine orange.

It was not possible using these techniques to map the nitrogenase regulatory loci of strains SK-24, 28, and 29, because of the inability to obtain any Glu+ prototrophic colonies with *E. coli* episome F'133.

The inability to obtain any glutamate-independent prototrophic colonies in the crosses between Gln C− strains and F'133 could be due to the inability of these recipient strains to inherit F'133. To rule out this possibility, strains carrying Gln− mutations were constructed from strain SK-24 (strains SK-511–514) (see Table III). Upon sexual transfer of F'133, prototrophic colonies for glutamine independence were obtained (see Table II). None of these was found to be glutamate independent.

Biochemical properties of *K. pneumoniae*/F'133 hybrids are presented in Table III. In contrast to the parental strains (see Table I), all hybrid strains constructed from strains SK-25–27, 37, and 56 produced glutamine synthetase (presumably the *E. coli* enzyme) as well as glutamate dehydrogenase. It should be emphasized that addition of NH_4^+ to

TABLE III

Nitrogenase, Glutamine Synthetase, and Glutamate Dehydrogenase Activities of Nitrogenase-Derepressed Strains of *K. pneumoniae* Carrying *E. coli* F'133 (F' Glutamine Synthetase)[a]

Strain	Recipient strain[b]	Nitrogenase (μmoles/hour/mg protein)		Glutamine synthetase (nmoles/minute/mg protein)		Glutamate dehydrogenase (nmoles/minute/mg protein)	
		$-NH_4^+$	$+NH_4^+$	$-NH_4^+$	$+NH_4^+$	$-NH_4^+$	$+NH_4^+$
SK-31	SK-37	2.83	0.04	130	101	181	128
SK-32	SK-25	3.32	0.00	128	109	77	72
SK-33	SK-25	2.79	0.07	121	94	191	72
SK-34	SK-26	3.63	0.09	163	172	233	89
SK-35	SK-26	4.19	0.00	111	120	269	100
SK-36	SK-27	3.94	0.12	349	72	235	118
SK-535	SK-512	2.69	1.97	63	53	<5	<5
SK-536	SK-513	1.86	0.63	179	242	<5	<5
SK-537	SK-514	1.86	1.59	183	189	<5	<5
SK-61	SK-56	2.90	0.33	48	226	196	300
SK-511	—	0.00	0.00	72	91	<5	<5
SK-512	—	2.45	2.18	10	10	<5	<5
SK-513	—	2.39	2.01	0	0	<5	<5
SK-514	—	2.17	1.81	0	0	<5	<5

[a] The conditions were the same as for Table I.
[b] The parental Gln⁻ *K. pneumoniae* strain used as the recipient in the crosses.

the sucrose-minimal medium completely repressed nitrogenase induction (Nif C⁺ phenotype) in the hybrid strains SK-32 and 35. Strains SK-31, 33, 34, and 36 were found to induce traces of nitrogenase activity in the presence of NH_4^+ (about 1–2% of the derepressed value). Under these same conditions, strain SK-61 produced about 11% of nitrogenase in the presence of NH_4^+ compared to the derepressed levels. F'133 hybrid colonies obtained with strains SK-512–514 retained the ability to derepress nitrogenase activity. These strains, while producing low levels of glutamine synthetase, a property similar to other *K. pneumoniae/E. coli* hybrids, synthesized no detectable glutamate dehydrogenase activity. Nitrogenase and glutamate dehydrogenase expression in these strains resemble their parental recipient strains (strains SK-512–514) and their parent (strain SK-24). Enzyme activities of the recipient strains (SK-512–514) are presented in Table III for comparison. Strains SK-512–514 have similar enzymatic pattern, both among themselves and with the parent strain SK-24, except for the absence of glutamine synthetase activity. Transfer of the *E. coli* episome F'133 to strains SK-512–514 led to their synthesis of glutamine synthetase. However, due to the absence of both glutamate dehydrogenase and glutamate synthetase, the hybrid clones carrying F'133 failed to grow on NH_4^+ as sole source of nitrogen and required glutamate for growth.

Glutamate dehydrogenase was detected in all other hybrid strains carrying F'133 (strains SK-31–36, 61), even in the absence of NH_4^+. In wild type *E. coli*, glutamate dehydrogenase is produced irrespective of the presence or absence of NH_4^+ in the medium (see Senior, 1975), a property that is quite different from that observed in *K. pneumoniae* (see Table I). In *E. coli*, addition of NH_4^+ to the medium decreased the levels of glutamate dehydrogenase activity. In most of the *K. pneumoniae/E. coli* F'133 hybrids (strains SK-33–35), addition of NH_4^+ decreased the levels of glutamate dehydrogenase to about 37% (as compared to NH_4^+-free medium). This suggests that in these hybrids, *E. coli* glutamine synthetase is being produced and that the *Klebsiella* glutamate dehydrogenase may be under the influence of the *E. coli* regulatory system. Under these conditions, in which the *E. coli* regulatory system apparently dominates, nitrogenase is repressed by NH_4^+. These observations, combined with the production of nitrogenase, even in the presence of small amounts of glutamine synthetase, suggest that the *E. coli* glutamine synthetase may be structurally different to elicit different regulatory properties. In *K. pneumoniae/E. coli* F'133 hybrids, nitrogenase activity is detected, even when the glutamine synthetase activity (produced under the influence of *E. coli*, *gln* genes) is considerably lower than 350 units, in contrast to previous observations with *K. pneumoniae*, where no nitrogenase activity was detected when the glutamine synthetase levels decreased below 350 units (Shanmugam *et al.*, 1976).

Derepressed synthesis of nitrogenase was again observed in the segregant Gln⁻ clones that had been cured of the F'133. Similar genetic and biochemical results were obtained when the *E. coli* episome F'105 was used instead of F'133 for F' complementation analysis.

IV. DISCUSSION AND SUMMARY

Nitrogenase-derepressed mutants described above can be obtained as a result of one of three different classes of mutations, all of which affect glutamine synthetase: (i) mutations that lead to production of catalytically inactive glutamine synthetase protein [Gln (AC)⁻] (strains SK-27, 37, 55, 56); (ii) strains that produce constitutive levels of glutamine synthetase [Gln C⁻] (strains SK-24, 28, 29, 57, 60); (iii) strains producing no detectable levels of either glutamine synthetase activity or protein [Gln⁻, Nif C⁻] (strains SK-25, 26). We will summarize the properties of each of these classes in this section.

A. Gln (AC)⁻

This class of nitrogenase-derepressed mutants (strains SK-27, 37, 55, 56) produces catalytically inactive glutamine synthetase protein that is detected immunologically. Glutamine synthetase–CRM was detected in amounts comparable to the derepressed levels of glutamine synthetase in a Gln C⁻ mutant, irrespective of the NH_4^+ concentration in the medium. The Gln (AC)⁻ mutation in strains SK-27, 55, and 56, which abolishes catalytic activity of glutamine synthetase and simultaneously leads to derepressed synthesis of nitrogenase, is presumably a point mutation as judged by its reversion frequency (4.7×10^{-10} for strain SK-27).

Using F′ analysis, the mutations of the strains SK-27, 37, 55, and 56 were mapped in a region of the *K. pneumoniae* chromosome corresponding to 75.5–77.5 minutes of the *E. coli* genome of the Taylor map (Low, 1972). The fact that active glutamine synthetase was produced following introduction of episomes carrying glutamine synthetase genes is strong evidence that the mutation maps in this region. This evidence and the fact that the mutation leads to the production of a catalytically inactive glutamine synthetase protein make it highly probable that the mutation is in the structural gene for glutamine synthetase (see also De Leo and Magasanik, 1975). Also, nitrogenase biosynthesis is repressed by NH_4^+, following the transfer of *E. coli* F′Gln into the nitrogenase-derepressed strains, SK-27, 37, and 56. This suggests that the Nif C⁺ property (repression by NH_4^+) is dominant over the Nif C⁻ phenotype in the Gln (AC)⁻ strains.

B. Gln C⁻

Mutants belonging to this class (strains SK-24, 28, 29, 57, and 60) produce derepressed levels of glutamine synthetase and nitrogenase. These strains, which are fully derepressed for nitrogenase biosynthesis, differ from the previously reported Gln C⁻ strains (Streicher *et al.*, 1974), which are only partially derepressed. The nitrogenase derepressed strain, described by Streicher *et al.* (1974) (strain KP5069), was constructed by transferring the Gln C⁻ gene from *K. aerogenes* to *gln*⁻ strain of *K. pneumoniae* using the transducing phage P1. The low levels of nitrogenase derepression (about 30%) observed with KP5069 could be due to some structural differences between the glutamine synthetase of *K. aerogenes* and *K. pneumoniae* that is manifested in its regulatory properties.

The Gln C⁻ phenotypes described in this communication have not been mapped. However, Streicher *et al.* (1975), using P1 transductional analysis, mapped the *gln* C⁻ locus in the glutamine synthetase (*gln* A) region of the *K. aerogenes* chromosome. It is probable that the mutations producing the Gln C⁻ phenotype isolated from *K. pneumoniae* may also map in the same *gln* A region. Additional fine-structure mapping is required to determine this point.

C. Gln⁻ Nif C⁻

Mutants SK-25 and 26 synthesize nitrogenase in the presence of NH_4^+ under conditions in which no glutamine synthetase catalytic activity or protein (CRM) can be detected. This finding may have at least three explanations: (i) that catalytically and antigenically inactive glutamine synthetase is present in these strains (see De Leo and Magasanik, 1975, for further discussion), (ii) there is a substitute for glutamine synthetase that regulates nitrogenase biosynthesis in its absence; and (iii) glutamine synthetase does not directly regulate nitrogenase expression. Obviously, additional experiments are necessary to distinguish among the three possibilities. A similar mutant of *K. aerogenes* (MK 9021) that maps in the *gln* A region and produces no detectable glutamine synthetase antigen has been described by De Leo and Magasanik (1975). They suggested that this mutant may be producing a molecule with regulatory properties but without the enzymatic or antigenic properties of glutamine synthetase. The *K. pneumoniae* mutants do share certain properties with the Gln (AC)⁻ strains, i.e., (i) lack of glutamate dehydrogenase even in the presence of NH_4^+, (ii) as determined by F' complementation analysis, the *K. pneumoniae* mutants map in the same chromosomal region in which the Gln (AC)⁻ strains map.

Following the introduction of F' Gln of *E. coli*, nitrogenase regulation returned to that of Nif C⁺ phenotype, indicating that the Nif C⁺ phenotype from *E. coli* is dominant over the Nif C⁻ phenotype of *K. pneumoniae*. Catalytically active glutamine synthetase is also produced by these hybrids.

The major conclusion from these studies is that mutations affecting the structure of glutamine synthetase leading to the production of catalytically inactive glutamine synthetase protein (strains SK-27, 55, 56) result in the derepression of nitrogenase biosynthesis.

REFERENCES

Brenchley, J. E., Prival, M. J., and Magasanik, B. (1973). *J. Biol. Chem.* **248,** 6122.

Brill, W. J. (1975). *Annu. Rev. Microbiol.* **29,** 109.
Cannon, F. C., Kennedy, C. K., Postgate, J. R., Tubb, R. S., and Dixon, R. A. (1976). *In* "Symposium on Dinitrogen Fixation" (W. E. Newton and C. J. Nyman, eds.), Vol. 2, p. 320. Washington State Univ. Press, Pullman.
De Leo, A. B., and Magasanik, B. (1975). *J. Bacteriol.* **121,** 313.
Dixon, R. A., and Postgate, J. R. (1972). *Nature (London)* **237,** 102.
Ginsburg, A., and Stadtman, E. R. (1973). *In* "The Enzymes of Glutamine Metabolism" (E. R. Stadtman and S. Prusiner, eds.), p. 9. Academic Press, New York.
Goldberger, R. F. (1974). *Science* **183,** 810.
Low, K. B. (1972). *Bacteriol. Rev.* **36,** 587.
Magasanik, B., Prival, M. J., Brenchley, J. E., Tyler, B. M., De Leo, A. B., Streicher, S. L., Bender, R. A., and Paris, C. G. (1974). *Curr. Top. Cell. Regul.* **8,** 119.
Nagatani, H., Shimizu, M., and Valentine, R. C. (1971). *Arch. Mikrobiol.* **79,** 164.
Senior, P. J. (1975). *J. Bacteriol.* **123,** 407.
Shanmugam, K. T., and Valentine, R. C. (1975a). *Proc. Natl. Acad. Sci. U.S.A.* **72,** 136.
Shanmugam, K. T., and Valentine, R. C. (1975b). *Science* **187,** 919.
Shanmugam, K. T., Loo, A., and Valentine, R. C. (1974). *Biochim. Biophys. Acta* **338,** 545.
Shanmugam, K. T., Chan, I., and Morandi, C. (1975). *Biochim. Biophys. Acta* **408,** 101.
Shanmugam, K. T., Streicher, S. L., Morandi, C., Ausubel, F., Goldberg, R. B., and Valentine, R. C. (1976). *In* "Symposium on Dinitrogen Fixation" (W. E. Newton and C. J. Nyman, eds.), Vol. 2, p. 313. Washington State Univ. Press, Pullman.
St. John, R. T., Johnston, H. M., Seidman, C., Garfinkel, D., Gordon, J. K., Shah, V. K., and Brill, W. J. (1975). *J. Bacteriol.* **121,** 759.
Streicher, S. L., Gurney, E., and Valentine, R. C. (1971). *Proc. Natl. Acad. Sci. U.S.A.* **68,** 1174.
Streicher, S. L., Gurney, E., and Valentine, R. C. (1972). *Nature (London)* **239,** 495.
Streicher, S. L., Shanmugam, K. T., Ausubel, F., Morandi, C., and Goldberg, R. B. (1974). *J. Bacteriol.* **120,** 815.
Streicher, S. L., Bender, R. A., and Magasanik, B. (1975). *J. Bacteriol.* **121,** 320.
Tubb, R. S. (1974). *Nature (London)* **251,** 481.
Tyler, B. M., De Leo, A. B., and Magasanik, B. (1974). *Proc. Natl. Acad. Sci. U.S.A.* **71,** 225.
Wohlhueter, R. M., Schutt, H., and Holzer, H. (1973). *In* "The Enzymes of Glutamine Metabolism" (E. R. Stadtman and S. Prusiner, eds.), p. 45. Academic Press, New York.

CHAPTER 2

Current Topics and Problems in the Enzymology of Nitrogenase*

W. H. ORME-JOHNSON and L. C. DAVIS

I. Introduction	16
II. Composition and Properties of Nitrogenase Protein Components	17
A. Subunit Structure	17
B. Metal Composition	19
C. Visible and Ultraviolet Absorption Spectra	25
D. Displacement and Identification of Iron–Sulfur Centers	26
E. EPR Properties of the Protein Components	27
F. Mössbauer Spectra and Interpretations	32
III. Interaction of MgATP and the Fe Protein	34
A. Binding Studies and K_m in the Overall Reaction	34
B. Effects of MgATP Binding on EPR Spectrum and Redox Potential	36
C. Control of Overall Reaction Rate by [MgATP]/[MgADP]	40
IV. Sequence of Electron Transfers between Nitrogenase Components	43
A. Steady-State Evidence	43
B. Energy of Activation for Nitrogenase Activity and Complex Formation	46
C. EPR and Absorbance Studies of the Functioning Enzyme	49
V. Structural and Mechanistic Hypotheses	55
A. The Fe Protein	55
B. The MoFe Protein and the Enzyme	55
References	58

Abbreviations used in this chapter: Fe protein—the iron-containing MgATP-binding protein of nitrogenase; MoFe protein—the molybdenum-containing protein component of nitrogenase. We also employ the notation of Eady et al. (1972), namely, Cp1, Av1, Kp1, Rj1, etc., the respective MoFe proteins of *Clostridium pasterianum, Azotobacter vinelandii, Klebsiella pneumoniae, Rhizobium japonicum,* etc.; Cp2, Av2, Kp2, Rj2, etc., the Fe proteins of the nitrogenase found in these organisms.

I. INTRODUCTION

Biological nitrogen fixation as a topic for study by molecular biologists has come of age. This can be discerned from the number of reviews and books on the subject. We will focus our attention on what we see to be some of the aspects of the enzymology of nitrogenase that are of most current interest and importance and discuss what is known of a few of the problems that we expect will be at the center of research in the next few years. To supplement this chapter, Chapter 1 should be consulted, as well as an earlier account of this subject matter by Hardy and Burns (Chapter 3, Volume I). For a fuller discussion of the wider topic of nitrogen fixation, see the books edited by Postgate (1971) and Quispel (1974). For earlier reviews on the nitrogenase enzyme system, see Dalton and Mortenson (1972), Eady and Postgate (1974), Burris and Orme-Johnson (1974), Hardy and Burns (1968), and Zumft and Mortenson (1975).

Currently studied aspects of nitrogenase may be divided into several broad areas—prerequisites [see the chapter of that title in Quispel (1974)], mechanism (see reviews cited above), integration of nitrogenase into the cell economy (Dalton and Mortenson, 1972; Hardy and Havelka, 1975), and genetics (Shanmugam and Valentine, 1975; Streicher and Valentine, 1973; Brill, 1975). The prerequisites for nitrogen fixation and the integration of nitrogenase into the cell economy are primarily a direct consequence of its mechanism of action and physicochemical properties; the occurrence and physiology of nitrogen-fixing organisms reflect in an evolutionary sense the integration of nitrogenase into the economy of the cell and the larger ecosystem.

The nitrogenase of free-living heterotrophs has been the most intensively studied, primarily because of stability and availability (Zumft and Mortenson, 1975; Burris and Orme-Johnson, 1974). Under optimal conditions, nitrogenase may constitute 15% of the soluble protein in *Clostridium pasteurianum*; yields of 500 mg purified MoFe protein component from 30 liters of cultured clostridia (85 gm cell paste) have been obtained in this laboratory. This contrasts with the situation in the rhizobial–legume association where $7\frac{1}{2}$ kg of nodules would yield the same amount of protein (Israel et al., 1974). Fortunately, the nitrogenase from different sources seems to be rather similar in most of its requirements, so that results obtained with the heterotrophs are probably of general validity. Exceptions to this statement will no doubt be found on more intensive investigation.

In the organisms thus far examined in detail, the nitrogenase consists of two protein components, designated the MoFe protein and the Fe protein.

The former contains approximately 15–30 Fe, comparable amounts of acid-labile sulfur, and 1–2 Mo per 200,000–300,000 MW. The latter contains approximately 4 Fe and 4 sulfur per 60,000 MW.

Hydrolysis of ATP is an obligatory part of the nitrogenase reaction, although the precise role(s) of ATP in the mechanism has yet to be elucidated. The ATP requirement with purified enzyme preparations is variable (Ljones and Burris, 1972) and of uncertain relationship to the ATP requirement *in vivo*. Values from 3 ATP per pair of electrons transferred to more than 20 have been reported (Zumft and Mortenson, 1975; Ljones and Burris, 1972; Dilworth, 1974).

All known reactions of nitrogenase require that protons be activated and transferred to a reducible substrate. In the absence of a sufficient concentration of a reducible substrate, the enzyme activates protons and yields hydrogen. If this were to occur *in vivo*, it would be a most wasteful metabolic process, since ATP is required in large quantities. This phenomenon is under control at least for *Azotobacter*. As far as we are aware, there are no reports of this aerobe evolving hydrogen in the absence of nitrogen in active cultures.

Protection from oxygen is essential, since the nitrogenase protein components are rapidly destroyed in the presence of molecular oxygen. Half-lives range from a few seconds to several minutes, depending on the organism or component protein tested.

Among the free-living heterotrophs, growth and nitrogen fixation occurs neither at very low nor at very high temperatures (see discussion by Postgate, 1971). However, among the blue-green, algae there appears to be a much wider tolerance of temperatures with reports of fixation in hot springs and in the Antarctic [see the article by Stewart in Quispel (1974)].

In what follows, we will examine the present status of knowledge about the structural and catalytic properties of nitrogenase and its component proteins.

II. COMPOSITION AND PROPERTIES OF NITROGENASE PROTEIN COMPONENTS

A. Subunit Structure

We will only deal with recent studies in which purified proteins were used. For discussions of older work, see Zumft and Mortenson (1975) and Tso (1974). In discussing nitrogenase from different organisms, we will adopt the shorthand notation proposed by the Sussex group (see footnote regarding abbreviations at the beginning of this chapter).

A detailed analysis of the molecular weight of a nitrogenase component was carried out by Israel et al. (1974), who examined Rj1 by the methods of high- and low-speed sedimentation equilibrium and sedimentation velocity and obtained 200,000 MW. A single type of subunit was found on sodium dodecyl sulfate (SDS) gel electrophoresis, with an apparent molecular weight of 55,000. Centrifugation in 6 M guanidine hydrochloride gave a subunit molecular weight of 50,300.

Eady et al. (1972) determined the molecular weights of the Kp1 and Kp2 components and their subunits by several techniques. Sedimentation velocity gave an apparent molecular weight for Kp1 of 200,400, while gel electrophoresis as a function of gel concentration indicated a value of 217,000. Thin-layer gel filtration yielded a value of 220,000, compared to 221,800 for the sum of the molecular weights of the two nonidentical subunits observed in SDS gel electrophoresis. The molecular weight used for calculations cited later was 218,000. For Kp2, Eady et al. found a molecular weight of 68,200 by sedimentation velocity and 62,000 by gel filtration. The subunit composition in SDS gel electrophoresis showed one band of 34,600 MW, indicating that the protein is a dimer.

Although Mortenson and co-workers (Zumft and Mortenson, 1975) have done extensive studies on Cp1, there have been relatively few studies reported for pure highly active preparations. Huang et al. (1973) showed that carboxymethylated subunits chromatographed in 6 M guanidine hydrochloride had apparent molecular weights of 52,500 and 60,000. A rather remarkable dependence of sedimentation coefficient on protein concentration made the determination of molecular weight of the native protein by this method unreliable. However, if the Cp1 is an associating system over the concentration range tested, there should be comparable problems in gel filtration experiments (Zimmerman and Ackers, 1971, 1974). The predicted skewing and displacement of peaks have not been reported, suggesting either that this was overlooked or that the anomalies in the sedimentation velocity pattern had another cause.

Tso (1974) examined the molecular weight of Cp1 by gel filtration, obtaining a value of 210,000. Glycerol gradient centrifugation indicated a molecular weight of 201,000. For Cp2, she found a molecular weight by gel filtration of 56,000. Electrophoresis in SDS gels yielded two subunits of Cp1 with weights of 51,000 and 60,000. For Cp2, the subunit molecular weights determined similarly were 27,500.

Burns et al. (1970) reported that the molecular weight of Av1 (Archibald method of approach to equilibrium) was 270,000. More recently, Kleiner and Chen (1974) have used both sucrose gradient centrifugation and gel filtration and reported values of 215,000 and 213,000 for the two techniques. SDS gel electrophoresis showed only a single type of subunit with an apparent molecular weight of 58,000–58,900 (or 55,000–56,000).

2. ENZYMOLOGY OF NITROGENASE

The Av2 component gave a single subunit type of molecular weight about 33,000, and gel filtration indicated a dimer of 64,000 for the native protein.

In summary, we find that in the concentration range used for the measurements and in the presence of moderate ionic strength buffers, the apparent molecular weights of the MoFe protein of four different organisms is in the range of 200,000–230,000. The apparent molecular weight of the Fe protein from three organisms is in the range of 55,000–65,000. For the Fe protein, there appears to be general agreement that the native protein is constituted of two identical subunits. For the MoFe protein, however, there is a divergence of results with Rj1 and Av1 showing a single type and Kp1 and Cp1 showing two nonidentical subunits of differing molecular weight. These divergent findings may result from the type of SDS used in different laboratories. R. Dixon, C. Kennedy, and R. R. Eady (personal communication) have found that the brand of SDS determines the degree of resolution of subunits from Kp1. The origin of these variations appears to be in variable amounts of contaminants, such as dodecanol, in the SDS. These workers have observed that careful adjustment of the level of dodecanol in the SDS used is required to optimize the resolution of subunits from a given MoFe protein. The finding of two subunits in Kp1 is not an artifact in that the isolated bands have different amino acid analyses and yield different tryptic peptide maps. Another point to be considered is that rather drastic treatments are required to remove all of the iron and molybdenum from the protein (see Huang *et al.*, 1973). Also, if there is a molybdenum cofactor as postulated by Nason and co-workers (Ketchum *et al.*, 1970), this may have a sufficient molecular weight to introduce an observable discrepancy between the sum of the subunit weights and the molecular weight of the holoprotein. At this point, however, the experimental uncertainties do not allow the settling of this question.

B. Metal Composition

In general, the metal analyses relevant to the present discussion were obtained by the same authors as the molecular weights just discussed. At this point, one has the option of assuming either that the molecular weight determinations are reliable and that metal content should be expressed as a mole fraction, or that all enzymes may have a comparable molecular weight, so that metal content should be expressed per unit protein. By either method, a nonunitary stoichiometry for molybdenum content has generally been observed. These difficulties have four origins:

1. Incorrect metal analyses—Iron analyses in particular are plagued by nonspecific contamination effects, and the growth of organisms in molybdenum-enriched media may introduce a similar problem with that

metal. One can also be deceived by artifacts introduced by incomplete decomposition of the protein (Beinert et al., 1970).

2. Incorrect molecular weight determination introduced by concentration-dependent effects, extreme molecular shapes, and unusual partial specific volumes, among other phenomena.

3. Incorrect determination of the amount of protein in a solution utilized for metal analysis. Perhaps the only totally reliable methods are careful dry weight determinations with allowance for the counterions or, ultimately, analyses based on the correct amino acid sequence and cofactor content. Colorimetric methods, which are commonly employed, must be used with particular care to avoid artifacts associated with the metal-containing chromophores and with buffer salts.

4. Mixed populations of molecules in the preparation analyzed. For example, Mortenson and his co-workers (Zumft and Mortenson, 1973; Huang et al., 1973) have discovered a protein contaminant in their Cp1 preparation. This protein has the same molecular weight and subunit composition but is deficient in (but not devoid of) molybdenum, iron, and labile sulfur. The amino acid analysis is reported to differ somewhat. This protein might be derived from Cp1 by loss of metals and a small peptide component, or alternatively be a precursor or be an entirely irrelevant contaminant.

It should be emphasized that the oxygen lability of nitrogenase proteins makes it likely that most preparations will contain varying quantities of inactivated species. Whether such preparations will exhibit low metal analyses probably depends strongly upon whether the proteins have been subject to purification steps subsequent to inactivation. It is quite possible that preparations of low specific activity might still be homogeneous in the sense of protein sequence and gross metal content. However, we do not yet know what the ultimate specific activity, the attainment of which would guarantee homogeneity, might be. We are presently in the position of having to persevere in the simplification and improvement of preparative procedures, and in this regard, the protracted saga of xanthine oxidase (Bray, 1975) may hint at the difficulties ahead. In that case, the presence of demolybdo species and later a desulfo species had to be detected and dealt with during the nearly twenty years following the first preparation of a respectably active crystalline homogenous enzyme. Nitrogenase components may yet have to be subjected to much further refinement, including affinity chromatographic methods, before definitive compositions are available.

We conclude that none of the reported studies, including our own, is entirely satisfactory with respect to the above comments. Nonetheless, it is useful to consider the range of values found to date in order to set the stage for further endeavor and discussion. For convenience of discussion, we have tabulated the data of interest (Table I). Where they are available, the molecular weight determinations are indicated and the content of both iron and molybdenum expressed relative to either molecular weight of the protein or per milligram of protein. The preparations shown varied widely in reported specific activities and the methods used for their determination. It is not clear that the reported activities in fact represent maximal activities (see Section IV,A). However, they all represent proteins that meet at least two criteria of homogeneity.

The highest iron content per milligram was found by Israel et al. (1974), the lowest by Tso (1974) and Eady et al. (1972). The molar iron content thus varied from as little as 12 to as much as 29 moles/mole. The molybdenum content varied from 4.7 nmoles/mg (Tso, 1974) to 9.0 nmoles/mg (Huang et al., 1973) without a comparable difference in reported specific activities. There can be no simple explanation of discrepancies of this magnitude.

Unfortunately, it is not the usual procedure to indicate the number of replicates for each analysis nor the number of different preparations analyzed for metals. The results of Huang et al. (1973) are derived from a single graph, while those cited from the work of Tso represent the lower end of the range of values for which no mean and standard deviation were indicated. Burns et al. (1970) provide analytical data for steps in the recrystallization procedure for Av1. From these, one can derive a standard deviation of $\pm 7\%$ for the molybdenum and $\pm 5\%$ for the iron analyses. This appears to be quite reasonable analytical precision but allows a considerable range in the ratio of molybdenum/iron and in the apparent molar metal content of the Av1.

Values for the iron content of the Fe protein from several organisms show a similar scattering in the analytical data. In this case, the result is somewhat less surprising, since the Fe protein is considerably more labile to oxygen, and preparations of high specific activity are harder to obtain and maintain than those of the MoFe protein. All of the results are in agreement in indicating that the Fe protein contains 4 or fewer Fe per mole.

Several authors have recently assumed that the MoFe protein should have 2 Mo per mole protein. A result equally in accord with the data in hand is that in fact it contains 1 mole Mo per mole protein plus a variable amount of adventitiously bound metal. Significant amounts of other metals have been found in MoFe protein preparations. For instance, Eady et al. (1972) indicated that their preparation of Kp1, after passage over Chelex 100 resin to remove weakly bound metals, still contained more

TABLE I

MOLECULAR WEIGHTS (HOLOPROTEIN AND SUBUNITS) AND METAL AND SULFIDE COMPOSITION OF NITROGENASE PROTEINS

Organism and reference	Molecular weight		Mo (nmole/mg)	Fe (nmole/mg)	Ratio (Fe/Mo)	Mo (g atom/mole)	Fe (g atom/mole)	S^{2-} (g atom/mole)
	Holoprotein	Subunits						
MoFe protein								
R. japonicum								
Israel et al., 1974	200,000	50,000	6.5	144	22	1.3 ± 0.15	28.8 ± 0.53	26.2 ± 0
Bergerson and Turner, 1973	180,000	—	5.2	49	9.4			
K. pneumoniae								
Eady et al., 1972	218,000	60,000	4.8	80	17	1.04 ± 0.1	17.5 ± 0.7	16.7 ± 1
		51,000						
C. pasteurianum								
Tso, 1974	210,000	60,000	4.7–6.5	57–105	11–18	1–1.5	12–18	8–15
		51,000						
Huang et al., 1973	220,000	60,000	9.0	—	—	1.95	ca. 24	ca. 24
		51,000						
A. vinelandii								
Kleiner and Chen, 1974	216,000	56,000	7.1	111	16	1.54	24	—
Burns et al., 1970	270,000	—	6.5–8.3	116–131	14.5–19	2	32	24
						$(1.4–1.8)^a$	$(26–29)^a$	
A. vinelandii (^{57}Fe)								
Münck et al., 1975	—	—	6.0	69	11.5	1.5^a	17^a	19^a
Chromatium								
Evans et al., 1973	—	—	5.6	84	15	—	—	—
Calculated^b	—	—	9.1	109		—	—	—
Fe protein								
K. pneumoniae								
Eady et al., 1972	66,800	34,600	—	60	—	—	4	3.85
C. pasteurianum								
Tso et al., 1972	56,000	27,500	—	53–73	—	—	3–4	—
A. vinelandii								
Kleiner and Chen, 1974	64,000	33,000	—	54	—	—	3.45	2.85

^{a} Calculated assuming the molecular weight of the holoprotein to be 220,000.
^{b} Calculated assuming 2Mo, 24Fe, and 24S^{2-} atoms per mole of holoprotein (molecular weight 220,000).

than 1 mole of Cu, Mg, and Ca and nearly a mole of Zn per mole of protein. Dalton et al.(1971), with a less pure preparation of Cp1, found 1.7 Ca and 0.5 Mg per Mo.

If one assumes that all of the activity in cells is due to MoFe proteins with 2 Mo per molecule and that the purified protein fractions are characteristic of all of the MoFe protein going through the fractionation procedure, a suggestive calculation can be made. In the preparations of Av1 and Cp1, with which we have personal acquaintance, the recovery is about 65% of the initial activity in the finally purified fraction. Much of the remaining activity can be accounted for in the side fractions of chromatography procedures and, in the case of Av1, in the crystallization mother liquor. If we make the not unreasonable assumption that 80% of all activity can be accounted for, the Mo content of purified MoFe proteins is too low if one had to believe that 2 Mo were present in each functioning molecule. Even if one assumed the presence of two independent catalytic sites per molecule so that the loss of 1 Mo atom would halve the specific activity, then the recovery of purified protein with less than 1.6 Mo per molecule (80%) is difficult to understand. We feel that one or more of the molecular weight, the Mo analyses, the recovery of activity, and the maximal specific activity (in the case of Az1 for a recrystallized protein) is incorrect.

In regard to the Mo content of functioning MoFe protein, another interesting finding was made during the purification of Cp1 described by Huang et al. and Zumft and Mortenson. The Mo content of the starting Cp1 (reportedly half active) was 4.0 nmoles Mo/mg, while the most active material had 9 nmoles Mo/mg, and the inactive material had 2.0–4.3 nmoles Mo/mg. The Fe content of the starting material was not stated, but the active material had about 100 nmoles Fe/mg, while the inactive material varied from 47 to 80 nmoles Fe/mg. The authors concluded that functioning Cp1 contains 2 Mo per molecule, but they did not comment on the fact that the metal content of the inactive species appears to lie within the typical range of other highly purified preparations in which there is not a large loss of activity in the purification process itself.

We have recently obtained preparations of Cp1 of high specific activity and high yield using cells that had been stored in liquid nitrogen prior to purification rather than being dried. These Cp1 preparations show none of the inactive species described by Huang et al. (1973) and Zumft and Mortenson, according to EPR spectroscopy (Fig. 1), yet the metal analysis indicates 5.25–6.25 nmoles Mo/mg and 97–122 nmoles Fe/mg (which corresponds to about 1.3 Mo/mole and 23 Fe/mole protein) (Table I). Our preparations and those reported by Mortenson and co-workers con-

Fig. 1

tain similar amounts of Fe but differ by a factor of 1.5 in Mo content. If Mo content is limiting, their preparation should have a specific activity 1.5 times that of ours, which is not the case. Certainly, present data do not establish whether one Mo atom or two are essential to functioning Cp1, nor is it known that any Mo, in fact, is required to be present in functional Cp1 (as opposed to the long-ago demonstrated nutritional requirement for Mo). We wish only to point out that much remains to be done in this most elementary area of nitrogenase research, rather than try to sweep aside our misgivings on the compositions of the MoFe proteins.

C. Visible and Ultraviolet Absorption Spectra

The spectrum of the MoFe protein is relatively featureless in the recently described preparations. There appears to be only a steadily declining absorption from about 350 to 650 nm. Features at shorter wavelengths are frequently obscured by dithionite present in the preparations under reduced conditions. Baseline matching is usually done at 600–650 nm, and the true intensity of protein absorption is hard to determine, since it trails off to longer wavelengths.

Eady et al. (1972) have reported data for which baseline corrections were made, giving absorption coefficients for *Klebsiella* MoFe protein at selected wavelengths. The protein in the presence of dithionite yields a value of 35 liters mmole^{-1} cm^{-1} at 420 nm, while the oxidation of the protein with thionine or ferricyanide increased the millimolar absorbancy

---

Fig. 1. EPR spectra obtained from nitrogenase fractions from *C. pasteurianum* at various stages of purification by a modification of the method of Tso et al. (1972). (A) Crude extract obtained by centrifugation of lysozyme-treated cell paste. (B) MoFe protein fraction after chromatography on DEAE cellulose. (C) MoFe protein after chromatography on Agarose 6B. The protein is homogenous, according to gel electrophoresis, at this point. The inset at the right of the main spectrum is taken at lower power and higher amplifier gain to show the absence of the impurity studied by Zumft et al. (1974). (D) Fe protein fraction after DEAE chromatography. (E) Fe protein after chromatography on Sephadex G-100. The protein is homogenous, according to gel chromatography, at this point. Conditions of EPR spectroscopy: A Varian E-9 spectrometer operating at 9.19 GHz was used with a quartz insert dewar and helium gas flow system maintaining the temperature of the sample at 13°K. Modulation frequency, 100 kHz; modulation amplitude, 10 G; sweep rate, 1000 G min^{1}; time constant 0.1 second; microwave power, 3 mW. The relative final amplifier gains are given next to the spectra, which are displayed as the first derivative of the microwave absorption (ordinate) as a function of field (abscissa). The field increases linearly to the right, and selected values of the frequency to field ratio (*g* value) are given on the abscissa. From M. T. Henzl, L. C. Davis, and W. H. Orme-Johnson, unpublished.

to 50. Ljones (1973) has developed a very useful technique in which traces of the MoFe protein plus a MgATP generator system are used to enzymatically oxidize the Fe protein. The reduced form of the Fe protein from *C. pasteurianum* had a millimolar absorbancy of 5.6, and the oxidized form 10. In both these studies, the absorbance increased during oxidation, which is a feature in common with ferredoxins (Orme-Johnson, 1973). However, the absorbancies per iron atom at 430 nm were: MoFe protein, 2.3 (oxidized) and 2 (reduced); Fe protein 2.5 (oxidized) and 1.4 (reduced), which compare with values of about 4 (oxidized) and about 2 (reduced) for ferredoxins. In the case of Cp2 (see next section), we are now reasonably sure that the chromophore is an $Fe_4S_4(RS)_4$ cluster similar to that in bacterial ferredoxins, so that the discrepancy in that case may, as Ljones (1973) observes, be due to the presence of inactive protein.

Given our present conceptions of the oxidation state of the MoFe protein as isolated in the presence of dithionite, namely that it is an oxidized state relative to the form that reduces nitrogen (see Secion IV), what the results of Eady *et al.* seem to mean is that if the iron–sulfur centers in the MoFe protein are like those known in simpler iron–sulfur proteins, then the majority (but not all) of the iron–sulfur centers in the MoFe protein are relatively reduced in this state of the protein, which, however, is not further oxidized by the natural substrates N_2 and H^+. If further oxidized states are produced transiently during enzyme turnover, then they must be rereducible by dithionite and, therefore, do not require the critical reductant, the MgATP complex of the Fe protein, which is needed to produce the catalytically active species of nitrogenase. M. N. Walker and Mortenson (1974), working with Cp1, have recently compared the spectrum of the MoFe protein as isolated with that from MoFe protein in the presence of the Fe protein plus a MgATP generator and reported no increase in A_{440}, but rather a 2–3% decrease. They also found that the dye-oxidized form of the MoFe protein was not reducible by *C. pasteurianum* ferredoxin in the presence of hydrogenase, according to absorbance measurements. They concluded that the dye-oxidized form of the MoFe protein is, in fact, unlikely to be an intermediate in the reduction of N_2 by nitrogenase. Thorneley has recently made use of the difference spectrum of the Fe protein to monitor nitrogenase function in the millisecond time range, as will be discussed below.

D. Displacement and Identification of Iron–Sulfur Centers

In Chapter 7, the lability of the thiol ligands in synthetic iron–sulfur complexes is well documented. When taken with the observation of

McDonald et al. (1973) that ferredoxins can reversably unfold in the presence of high concentrations of dimethyl sulfoxide (DMSO), this offers a useful method of removing and subsequently identifying the iron–sulfur centers from iron–sulfur enzymes. Que et al. (1975) have shown that ferredoxins from *C. pasteurianum* and *Spirrullina* yield the expected low molecular weight analogs when treated with simple thiols in the presence of 80% DMSO, and Bale et al. (1976) have shown that not only is this true for a range of simple iron–sulfur proteins, but that the iron–sulfur centers can actually be transferred intact between proteins by brief exposure of mixtures of holo and apo proteins to denaturing solvents in the presence of low molecular weight thiols, which apparently act as carrier ligands for the exchange. This should allow the transfer of iron–sulfur centers from complex enzymes to air-stable simple ferredoxins, which may be separated and quantitated somewhat more easily than can the air-sensitive synthetic analogs.

For enzymes that contain only one type of center, simply unfolding the protein in the presence of thiols and subsequent recording of the absorbance spectrum may suffice. Erbes et al. (1975) were able to displace the centers from *C. pasteurianum* hydrogenase with thiophenol in the presence of 80% hexamethylphosphoramide. Comparison of the absorbance spectrum of the resulting mixture with those obtained from *C. pasteurianum* ferredoxin (a source of four-iron clusters) and bovine adrenodoxin (a source of two-iron clusters) revealed the displacement of $Fe_4S_4(\phi S)_4$ from hydrogenase in quantitative yield. Thus, hydrogenase appears to contain the same basic iron–sulfur cluster as the bacterial ferredoxins. This procedure was also applied to the Fe protein of *C. pasteurianum* (Averill et al., 1975). The Fe protein yielded four-iron–four-sulfur clusters as is depicted in Fig. 2. When the same procedure was applied to the MoFe protein, a complex mixture resulted whose spectrum could not be analyzed as the sum of two- and four-iron cluster spectra. This matter is under further investigation, but other spectroscopic evidence (Section IV) suggests the presence of familiar iron–sulfur clusters. It is clear, however, that the basic $Fe_4S_4(RS)_4$ system is utilized in active sites that are responsible for both ligand activation (hydrogenase) and energy transduction (nitrogenase Fe protein). The study of the means by which the protein environment utilizes iron–sulfur centers for these tasks must be an active field of endeavor in the near future.

E. EPR Properties of the Protein Components

The earliest attempt to identify the EPR of nitrogenase was the work of Nicholas et al. (1962). The first correct identification of the EPR

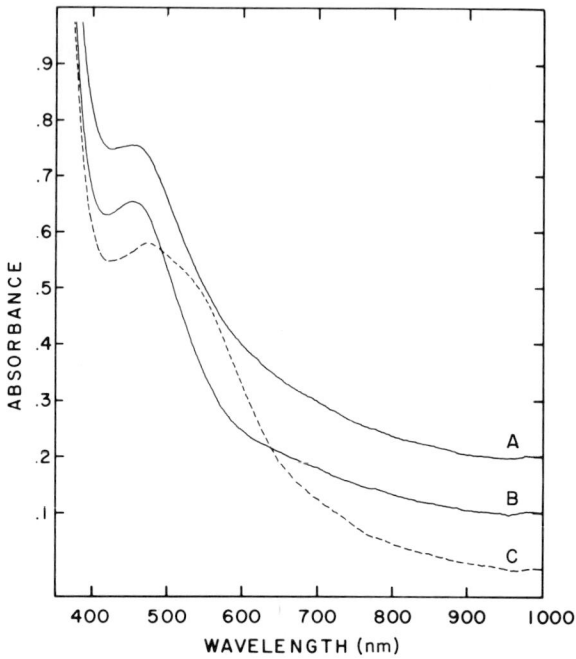

Fig. 2. Displacement of FeS center from the Fe protein of *C. pasteurianum*. In a final volume of 1.5 ml, iron sulfur proteins (120 nmoles Fe) were dissolved anaerobically in 80% hexamethyl phorphoramide: 20% 100 mM tris · Cl, pH 8.5, containing 10 mM thiophenol. Absorbance spectra (1 cm path length) were examined after 8 minutes of incubation at 23°C in a Cary 14 spectrometer. (A) Fe protein of *C. pasteurianum*; (B) *C. pasteurianum* ferredoxin [source of Fe$_4$S$_4$(ϕS)$_4$ clusters]; (C) spectrum, bovine adrenal iron–sulfur protein [source of Fe$_2$S$_2$(ϕS)$_4$ clusters]. From Averill *et al.* (1976).

features of the MoFe protein at g values of 4.3, 3.6, and 2.01 (Fig. 2) as due to a single anisotropic center appears to be in the work of Davis *et al.* (1972). The exact positions of the g values differ with the organism examined (Table II) and for Kp1 with the pH of the system (Smith *et al.*, 1973), but highly purified preparations of the native enzyme as obtained under reducing conditions contain only these resonances. The impurity in MoFe protein preparations, detected by a resonance at $g = 1.94$ (Davis *et al.*, 1972) and subsequently separated and studied by Huang *et al.* (1973; see also Zumft and Mortenson, 1975), is evident in most earlier published EPR spectra of this protein, but apparently is not required for functional nitrogenase.

The EPR observable in the MoFe protein as isolated has been pointed out by Palmer *et al.* (1972), Smith and Lang (1974), and Münck *et al.*

TABLE II
EPR OF NITROGENASE PROTEINS

Organism	MoFe protein observed g values			Fe protein observed g values[a]		
	g_x	g_y	g_z	g_x	g_y	g_z
A. vinelandii	4.32	3.65	2.01	2.05	1.94	1.88
C. pasteurianum	4.29	3.77	2.01	2.05	1.94	1.88
B. polymyxa	4.37	3.53	2.01			
Chromatium D[b]	4.3	3.7	2.02			
R. rubrum	4.34	3.65	2.01			
R. japonicum	4.17	3.73	2.03			
K. pneumoniae (low pH)[c]	4.32	3.63	2.01	2.05	1.94	1.87
Calculated values[d]	4.32	3.67	2.01 $(m = \pm\frac{1}{2})$			
	0.32	0.34	6 $(m = \pm\frac{3}{2})$			

[a] From Eady et al. (1972).
[b] From Evans and Albrecht (1974).
[c] From Smith et al. (1973).
[d] From the spin Hamiltonian for $S = \frac{3}{2}$ with $\lambda = 0.055$, $g_\perp = 2.0$, $g_\parallel = 2.03$.

(1975) to be that expected from the $m = \pm\frac{1}{2}$ Kramers doublet in a spin = $\frac{3}{2}$ system (see Fig. 3). Münck et al. commented that the observed g values are generated for such a system if the rhombicity parameter E/D was 0.055 compared to a maximum possible value of $\frac{1}{3}$, i.e., the g values show a slight distortion from the axial case. They further fitted

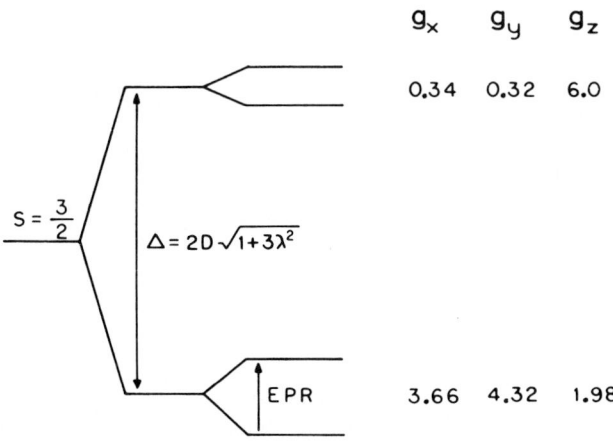

Fig. 3. Energy level diagram for the $S = \frac{3}{2}$ Kramers manifold with $D > 0$. The effective g values are given for $\lambda = 0.055$. From Münck et al. (1975).

spin concentrations from integrated EPR spectra of Av1 to a Boltzman function (Fig. 4). The EPR emanates from a ground state that is depopulated as the temperature rises, and the separation of the ground and excited states, given by $\Delta = 2D[1 + 3(E/D)^2]^{\frac{1}{2}}$ (Fig. 3), appears to be $\approx 15°K$. The fit to the Boltzman function was reasonable for data points at 15°K and below; above this temperature, the signal broadens rapidly and could not be integrated with precision. This rapid broadening with increase in temperature between 10° and 30°K is seen with all species of MoFe protein examined so far. The effect is striking with Av1 and much less so with Cp1. Thus, there is a temperature window for the successful observation of the MoFe protein in the state as isolated. The limits of this window, which express the relaxation properties and therefore, in some way as yet not understood, the environment of the center, differ rather dramatically between, for example, Cp1 and Av1.

Münck et al. utilized the fit of the low-temperature points for Av1 to the Boltzman curve to estimate the spins per Mo evidenced in the ground

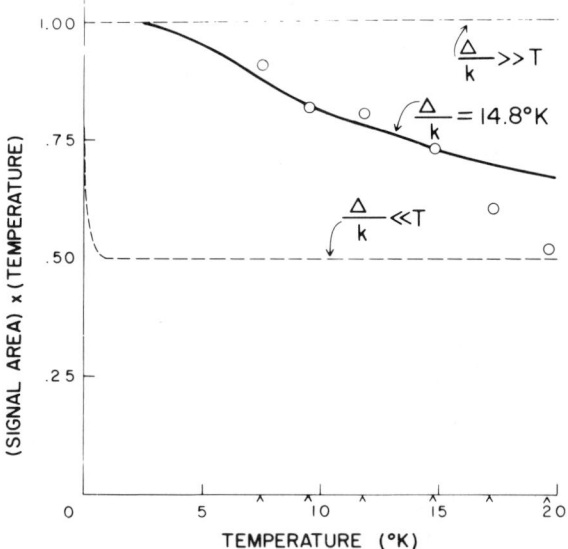

Fig. 4. Temperature dependence of the integrated EPR signal of ^{57}Fe-substituted MoFe protein of *A. vanelandii*. Spectra were obtained under the following conditions: microwave frequence 9.21 GHz, microwave power 1 mW, modulation frequency 100 KHz, modulation amplitude 10 G, magnetic field sweep rate 500 G min^{-1}, time constant 0.25 second. The temperature was varied from 7° to 20°K by varying the helium flow rate. The temperature of the sample was monitored with a gold–iron thermocouple, and integrations of signal area were performed using 2 mM CuEDTA solution as a standard. Above 15°K, the signal could not be integrated accurately. From Münck et al. (1975).

state at 0°K, where the population of the state should be unity. They estimated the integrated spectrum to correspond to 0.9 spins per Mo, by comparison to a CuEDTA standard using a transition probability correction, due to Aasa and Vanngard (1970), proportional to g^2. R. Aasa and T. Vanngard (personal communication) have recently suggested that for field-swept spectrometry, the correction should rather be proportional to g. When this is taken into account, as well as the fact that the MoFe protein used by Münck et al. contained 1.5 Mo per molecule of 220,000 MW, there appear to be 2.1 spins present per molecule; i.e., about two spin = $\frac{3}{2}$ centers per MoFe protein.

The EPR spectra of Av1 that had been prepared from organisms grown on media enriched in ^{57}Fe or ^{95}Mo ($>90\%$) were carefully examined at 12°K, and no evidence was found for broadening due to the interaction of the electron spin with the nuclear spin in these isotopes, except for the feature at $g = 2.01$ in the case of ^{57}Fe–MoFe protein. A small (ca. 7 G) reproducible broadening indicated that the spin $\frac{3}{2}$ centers contain iron. The lack of broadening for the ^{95}Mo specimen constitutes no evidence for or against its presence in these centers, since Mo might be present in a case where spin was not sufficiently localized on the metal for observation of the electron nuclear interaction.

The postulated excited-state EPR ($g = 6, 0.32, 0.32$) has not been observed. This is possibly due to the low transition probability predicted (Münck et al., 1975) for the $g = 6$ resonance; the other resonances (at 20 KG for X-band) have not yet been searched for.

In summary then, the MoFe protein as isolated appears to possess two iron-containing spin $\frac{3}{2}$ centers per molecule. Subsequent discussion will take up the question of the state of this and other iron-containing prosthetic groups during the catalytic cycle of nitrogenase.

EPR spectra of the Fe protein were reported by Palmer et al. (1972) for Cp2, by Orme-Johnson et al. (1972) for Cp2 and Av2, and by Eady et al. (1972) for Kp2. In each case, the protein shows a rhombic spectrum with g values about 2.05, 1.94, and 1.86. Upon addition of ATP in the presence of Mg at a pH near 8, there is a change of the shape of the spectrum to a simpler axial form concomitant with a loss of signal intensity. The native protein also shows a variable amount of a signal near $g = 4.3$, which may be relevant to the structure of the active site (Orme-Johnson et al., 1972). Physiologically oxidized Fe protein shows no EPR signal. As mentioned, the center present in this protein appears to be a four iron, four sulfide cluster, and the EPR of the Fe protein as isolated is that expected of an reduced bacterial-type ferredoxin containing a single iron–sulfur cluster (Stombaugh et al., 1973), although the middle resonance feature is rather narrower than those seen in the smaller ferredoxin (Fig. 1). The EPR appears to arise from a ground state doublet (Orme-Johnson and Sands, 1973). The form of the protein obtained by oxidation with the MoFe protein in the presence of MgATP displays no EPR, and these results are consistent with the notion that the Fe protein operates between the −2 (diamagnetic) and −3 (paramagnetic) net charge states of a $Fe_4S_4(RS)_4$ center (Holm and Ibers, Chapter 7). Thus the protein should transfer $1e^-$ during an oxidation–reduction event.

F. Mössbauer Spectra and Interpretations

Smith and Lang (1974) have examined the Mössbauer effect from Kp1 and Kp2, while Münck et al. (1975) have studied Av1. Both groups utilized proteins isolated from organisms grown in the presence of ^{57}Fe-enriched media. Debrunner, Münck, Que, and Schulz discuss the Mössbauer effect of this and other systems in Chapter 10, so only the highlights relevant to our argument are recapitulated here.

Both sets of authors identified four species of iron in the MoFe protein as isolated. These are listed in Table III. The Mössbauer parameters of these species of iron are closely similar in the two MoFe proteins, and their quantities are similar. The Mössbauer spectrum of the magnetically split species decreases in magnitude as the MoFe protein is brought into the nitrogen-fixing steady state in the presence of reducing agents, the Fe protein, and a MgATP generator. At the same time, as will be dwelt

TABLE III
MÖSSBAUER RESULTS[a] ON MoFe PROTEINS FROM
A. vinelandii[b] AND K. pneumoniae[c]

	A. vinelandii				K. pneumoniae		
Spectral species	ΔE_Q (mm/ second)	δ (mm/ second)	% of total absorption[d]	Spectral species	ΔE_Q (mm/ second)	δ (mm/ second)	% of total absorption[d]
Fe^{2+}	3.02	0.69	14	M4	3.02	0.66	13
D	0.81	0.64	42.5	M5	0.83	0.61	47
M	0.76	0.40	38.5	M6	0.71	0.37	40
S	1.4	0.6	≈5	(Species corresponding to S of A. vinelandii present in spectra but not evaluated)			

[a] ΔE_Q = quadruple splitting, δ = isomer shift relative to Fe metal.
[b] Münck et al. (1975). Data taken at 30°K.
[c] Smith and Lang (1974). Data taken at 77°K.
[d] Assuming equal recoilless fractions for all species.

upon below, the EPR signal of the MoFe protein decreases correspondingly. The magnetically split Mössbauer species, called M by Münck et al. and M6 by Smith and Lang, is therefore identified with the center detected by EPR in the MoFe protein as isolated. In the nitrogen-fixing steady state, this species exhibits a quadrupole doublet spectrum. In the resting state of the protein, the magnetically split Mössbauer spectrum of this species also collapses to a quadrupole doublet at 30°K, a temperature at which the electron spin relaxation time has grown very short (see the previous section). The electron magnetic field at the ^{57}Fe nuclei is time-averaged to zero at these higher temperatures. Comparison of the areas under the quadrupole doublets seen at 30°K, corresponding to the different Mössbauer species (assuming that the recoilless fraction or Debye–Waller factor is the same for each iron atom), yields the result that the iron atoms in the EPR-active center comprise roughly 40% of the total iron in the Av1 preparation studied. This amounts to about 6–7 Fe per molecule of 220,000 (or 8–10 Fe atoms per Mo found by analysis). Münck et al. proposed that the two spins associated with the $g =$ 4.3, 3.6, 2.0 EPR center emanate from a pair of four-iron clusters in the protein.

Smith and Lang made the interesting suggestion that the Mössbauer spectrum of the EPR active center is reminiscent of that from the high potential iron–sulfur protein of chromatium (Evans et al., 1970). R. H. Holm (personal communication) has suggested to us that in theory the oxidized model compound Fe_4S_4(thiophenolate)$_4^{-1}$ should exhibit a spin = $\frac{3}{2}$ ground state. A successful experimental test of this suggestion would lend much credence to the speculation of Smith and Lang. The uncertainties in the above calculations are such that they are only approximate at this point, but clearly about 60% of the iron atoms or 10–11 Fe per molecule are not associated with the EPR-active center. These remaining iron atoms in the MoFe protein are disposed in three classes with relative populations of 2–3, 7–8, and about 1 Fe per molecule. The first of these species seems to be formally Fe^{2+}, according to the Mössbauer parameters. The approximate ratio of 1:3 of the first two of these species led Münck et al. to speculate that they might be associated in tetrameric clusters in the protein. The presence of two to three such clusters is suggested by the analytical data. The origin of the remaining iron atom is now known, but it is a differentiable entity in spectra obtained from both Kp1 and Av1. Smith and Lang also reported Mössbauer studies on Kp2, which yielded spectra much like those of bacterial (four-iron) ferredoxins.

In summary, the Mössbauer evidence shows us that the MoFe protein contains iron atoms of four differentiable types. It should not be inferred from this evidence alone that there are four distinct iron-containing centers in the protein, as two types of Mössbauer atoms may be present in a single complex (Evans et al., 1970). What the Mössbauer results do establish is that only a portion of the iron atoms appear to be involved in the paramagnetic center studied by EPR spectroscopy. This suggests the presence of at least two distinct iron-containing electron carriers in the MoFe molecule.

III. INTERACTION OF MgATP AND THE Fe PROTEIN

A. Binding Studies and K_m in the Overall Reaction

Bui and Mortenson (1968) showed semiquantitatively, by the technique of gel filtration (Hummel and Dreyer, 1962), that MgATP binds Cp2. More recently, Tso and Burris (1973) have examined in detail the binding of MgATP and MgADP to highly purified Cp2, using a simple gel equilibration method (Hirose and Kano, 1971). The partitioning of radioactive nucleotide between the internal and external phases of a G-50 Sephadex was measured as a function of the concentration of Cp2 and other reactants. Binding of nucleotide to Cp2 results in an increase in concentration of nucleotide in the external phase, since the Cp2 is relatively excluded from the gel. Binding constants for ATP or ADP in the presence of excess Mg were 16.7 and 5.2 μM, respectively. Two molecules of MgATP or one molecule of MgADP were found to bind each molecule of Cp2. MgADP was a nonlinear inhibitor of MgATP binding, indicating that in the presence of one MgADP molecule, MgATP was bound at a second site with approximately twofold greater affinity. A point not discussed is the observation that significant binding occurs in a mixture of Fe and MoFe proteins in the absence of added Mg, indicating either that the protein can bind nucleotides in the absence of Mg or that there is a metal ion present in the preparation at a sufficiently high concentration to stimulate the binding. If a metal chelate is involved in this type of binding, it is not one that is readily hydrolyzed to ADP by the action of nitrogenase.

The association constant for MgATP determined by Tso and Burris should be clearly distinguished from the K_m for the MgATP determined in the enzymatic reaction. As discussed by Davis and Orme-Johnson (1976), many of the published values of K_m ATP are of doubtful validity because of inappropriate assay conditions. The inhibitory nature of MgADP makes the properties of the MgATP regenerating system of critical concern in kinetic studies. Some representative values will be cited with the awareness that the final interpretation of the role of MgATP in substrate reductions must await further experiments under more ideal conditions.

A significant paper for the understanding of ATP function is that of Silverstein and Bulen (1970). They used a complex of Av1 and Av2 of unspecified stoichiometry (probably near 1:1 molar, Bulen and Lecomte, 1972). Systematic variation in the concentration of complex and MgATP resulted in a change of the relative efficiency of nitrogen reduction vs hydrogen evolution. Computer simulation of results suggested that the active nitrogenase enzyme consists of a complex of the MoFe and Fe proteins, with MgATP altering the relative amount of the state of the enzyme responsible for either H_2 evolution of N_2 reduction. At relatively low MgATP concentrations, the enzyme exists in a state E_1 that evolves H_2, and at high MgATP concentration, it exists primarily as a state E_3 that reduces N_2. MgATP concentration versus activity plots show sigmoidal patterns, indicating that MgATP is involved in more than one step, or at multiple sites, or both.

Silverstein and Bulen observed that there was a slight sigmoidicity to the plots of rate of H_2 evolution versus ATP concentration and a much larger effect in similar plots for N_2 reduction. The levels of MgATP used were so low (down to 25 μM), that effects due to the properties of the MgATP regenerating system (Davis and Orme-Johnson, 1976) or the finite binding constant of MgATP with the Fe protein (Tso and Burris, 1973) cannot be ruled out as the cause of sigmoidicity in the H_2 evolution profiles. However, since it is clearly established that the rate of electron utilization is independent of whether N_2 is present or not, this difference between H_2 evolution and N_2 reduction in rate versus ATP concentration profiles must be indicative of an additional step in the process leading to N_2 reduction. Silverstein and Bulen conclude that this additional process is the conversion of enzyme E to a particular form E_3, which can reduce N_2. This concept is further discussed below in Section IV,A.

This type of study has been extended (Davis et al., 1975) to the effect of ATP concentration and component ratio on the relative efficiency of reduction of the alternative substrates acetylene and dinitrogen. The results and interpretations of Silverstein and Bulen were in general confirmed, indicating that for substrate reductions, the observed K_m for ATP utilization is affected by component ratio and enzyme concentrations.

Bergersen and Turner (1973), using Rj1 and 2 in varying ratio, found that the kinetic patterns for ATP, S_2O_4, C_2H_2, and N_2 were all complex functions. As found by Silverstein and Bulen, Hill plots of activity versus ATP had slopes greater than 1. They interpreted this in terms of an allosteric effector function for the Fe protein in the presence of ATP (Bergersen and Turner, 1973). From the data in their paper, the ATP concentration for half maximal velocity in either C_2H_2 or N_2 reduction appears to be several millimolar. In the studies of Silverstein and Bulen, the level of ATP for half maximal velocity is about 0.1 mM. This is the same value cited by Eady et al. for Kp1 and 2 in 1:1 molar ratio.

Ljones used the continuous method of dithionite oxidation to assay for the MgATP requirement (Ljones, 1973). He made an unfortunate choice of MgATP regenerating systems, namely, pyruvate kinase plus phosphoenol pyruvate. The product pyruvate also oxidizes dithionite, or at least diminishes its absorbance, resulting in spurious results (Davis and Orme-Johnson, 1976). We have done similar dithionite oxidation assays with the usual creatine kinase regenerating system. For dithionite oxidation by the *Azotobacter vinelandii* system, the K_m ATP is about 0.3 mM for a wide range of component ratios and enzyme concentrations, although in agreement with Silverstein and Bulen, nonlinear kinetics are obtained with dilute enzyme. For Cp proteins, we have found similarly a K_m near 0.5 mM. The results of all these studies suggest that the observed K_m is substantially larger than the equilibrium binding constant, which means that kinetic steps subsequent to binding are rate limiting under the conditions used for typical enzyme assays.

B. Effects of MgATP Binding on EPR Spectrum and Redox Potential

The binding of MgATP to Fe can be detected directly by changes in the EPR signals (Fig. 5) of the component (Orme-Johnson et al., 1972). Binding as observed in this way, need not show the same concentration dependence as the binding of [^{14}C]ATP described by Tso and Burris (1973), since it may detect weaker binding sites than observable with the radiochemical method.

Zumft et al. (1973, 1974) stated that an EPR signal change from rhombic to axial occurred when Cp2 was treated with two moles of MgATP^{2-}. They used a protein concentration of 18.6 mg/ml, which corresponds to approximately 0.3 mM, in the presence of 0.25 M NaCl, 0.05 M tris Cl buffer pH 7.5 (at room temperature). The shape of the EPR signal changed as MgATP was added, until 2 moles of MgATP had been added per mole of Cp2, after which there were no further changes observed. The apparent linearity of the change in signal shape suggests that the binding of MgATP must be relatively tight. Other nucleotides were tested and there was a response to MgGTP and βCH$_2$ATP, $\beta\alpha$CH$_2$ATP and MgADP. The latter appeared to give the same type of response as MgATP when present at 10 mM. AMP, UTP, CTP, and NAD gave no detectable change at 10 mM.

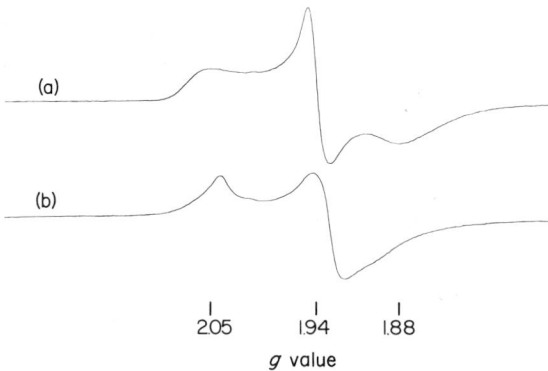

Fig. 5. EPR of the Fe protein of *C. pasteurianum* nitrogenase, with and without MgATP. The samples contained 8.8 mg of Fe protein per milliliter (160 μM based on a MW 55,000), as well as 2 mM sodium dithionite and 25 mM tris · HCl (pH 8.0). (a) Contains the described mixture; (b) also contains 1 mM ATP and 2 mM MgCl$_2$. The conditions of EPR spectroscopy were: microwave frequency, 9.18 GHz; microwave power, 9 mW; modulation frequency, 100 kHz; modulation amplitude, 12 G; magnetic field sweep rate, 400 G min^{-1}; time constant, 0.25 second; sample temperature, 13°K. The g values indicated on the abscissa were calculated from magnetic field positions derived from a proton NMR probe next to the cavity and klystron frequencies obtained with a counter. The ordinates are an arbitrary function of the first derivative of the microwave absorption with respect to the magnetic field. From Orme-Johnson et al. (1972).

Tso (1974) presents data for a titration of Cp2 with MgATP at much lower concentrations. Using 30 μM Cp2 and varying MgATP from 0 to 300 μM, she found that the change of EPR signal shape was in agreement with the expected fractional conversion to the MgATP form, assuming a binding constant of 20 μM for MgATP. This compares well with the value of 16.7 μM reported in her earlier study (Tso and Burris, 1973).

Smith et al. (1973) present data for the titration of Kp2 by ATP in the presence of an excess of Mg. They found rather similar results to those for Cp2, although they calculated the fractional binding from simulation of the spectra of intermediate forms rather than by simply measuring amplitude changes as done by Tso and by Zumft et al. Unfortunately, the experiments with Kp2 were done with a relatively high concentration (0.1–0.2 mM protein) of relatively low specific activity in the presence of 50 mM Mg (Mg$_2$ATP as well as MgATP present), so that an accurate derivation of a binding constant is difficult. Smith et al. calculated a K_{diss} of 0.4 mM for the complex, based on change of the EPR signal shape under their conditions of observation. This is twentyfold higher than the value determined by Tso for Cp2 using direct measurement of ATP binding, although the method she used would not accurately detect such a weak binding. The EPR data she shows for Cp2 are in agreement with the binding constant she reported, so there could be a difference between organisms or between conditions of determination of the binding constants, including higher ionic strength and the presence of $(Mg)_2ATP$ (Thorneley and Willison, 1974).

The pH at which EPR studies of the MgATP binding phenomenon have been done (7.5–8) is far from the optimum for activity of at least the Cp nitrogenase. When we made a study of the EPR signal changes on binding as a function of pH, we found that the shape of the EPR signal did not change on addition of MgATP at pH 7, although we know from the studies of Tso and Burris (1973) that MgATP is bound tightly at this pH. Apparently, either the Fe protein selectively binds $MgATP^-$ (as opposed to $MgATP^{2-}$) with no conformational change at low pH, or the Fe protein possess a pK in this region. Furthermore, we have observed a consistent loss of 30% of the integrated EPR intensity when Cp2 binds MgATP at any pH between 6.5 and 9.0. These two curious phenomena are not understood at present, but their exisence does inject a note of caution into the interpretation of binding studies as monitored by EPR.

The effect of MgATP on the redox potential of the Fe protein may be simply demonstrated. If Cp2 is incubated under 1 atm of H_2 with catalytic amounts of ferredoxin and hydrogenase at pH 7 so that the potential of the system is maintained at -420 mV, the EPR signal seen at equilibrium is much diminished in the presence of MgATP, compared to the signal found in the absence of the nucleotide (Fig. 6) (T. Ljones, M.-Y. W. Tso, and W. H. Orme-Johnson, unpublished). Thus, the binding of MgATP causes a lowering of the potential of the Fe protein. Zumft et al. (1974) made an extended series of electrometric measurements of the potential of Cp2 at pH 7.5. They were careful to use a low enough level of electrode mediators (redox dyes) so that dye-complexed protein species probably do not complicate the interpretation of their measurements. They found that Cp2 has a potential of -294 ± 20 mV, which is shifted to -400 ± 20 mV upon saturation of the protein with MgATP. The potential measurements were done at room temperature, and the EPR signals were evaluated on subsequently frozen samples. The data were fitted by the Nernst equation for which $N = 1$. The potentials may not be precise, because the authors used N-tris(hydroxymethyl)-methyl-2-aminomethanesulfonic acid buffers, which may shift in pH by 0.5 units during slow cooling and freezing. However, the results are probably approximately correct and suggest that the oxidized form of Cp2 binds MgATP one to two orders of magnitude more tightly than does the reduced form. When the rate constants for the binding of MgATP to Cp2 in various oxidation states become available, it may be possible to utilize the findings of Zumft et al. in describing the order of events

2. ENZYMOLOGY OF NITROGENASE

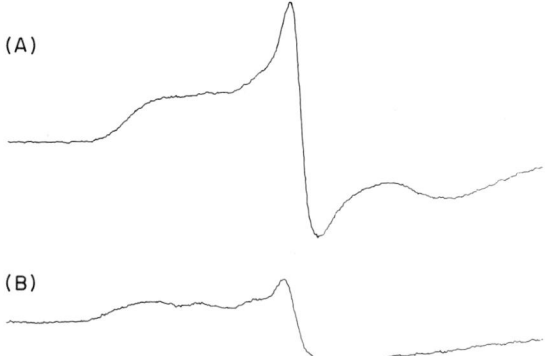

Fig. 6. The Fe protein of *C. pasteurianum* nitrogenase in the presence of hydrogenase and ferredoxin. The samples and conditions were similar to those of Fig. 5, except that the buffer was 0.05 M potassium phosphate, pH 7.0, and 0.5 mg/ml hydrogenase and 0.05 mg/ferredoxin (both from *C. pasteurianum*) were also present. The samples were incubated under 1 atm of H_2 for 1 hour at 25°C, at which time they were at equilibrium. The spectra are presented as they appeared after the signals due to the hydrogenase and ferredoxin were removed by computer subtraction of spectra from samples from which the Fe protein was omitted. (A) Fe protein; (B) Fe protein plus 1 mM MgATP. T. Ljones, M.-Y.W. Tso, and W. H. Orme-Johnson, unpublished results.

(MgATP binding and reduction) during this portion of the nitrogenase mechanism. Since this lowering of the redox potential is not coupled to MgATP hydrolysis, the observed phenomenon is not the production of the "superreductant" that is proposed to be a characteristic feature of nitrogenase action.

Zumft *et al.* did not comment on the question of whether Cp2 transfers one or more electrons during oxidation. The finding of $N = 1$ in the Nernst equation implies independent transfer of electrons, but says nothing about the number available from a single molecule (cf. the eight-iron ferredoxins; see Stombaugh *et al.*, 1976). M. N. Walker and Mortenson (1973) found that Cp2 had an apparent potential of -240 mV and yielded two electrons upon equilibration with redox dyes. However, this interpretation is likely to be in error for the reasons explained in Section II,D and E. Cp2 yields a single 4Fe–4S cluster in displacement experiments and appears to utilize the -2 and -3 net charge states of the clusters in a single electron transfer step.

Thus, we have the current picture of the Fe protein of nitrogenase as a single electron carrier utilizing a $Fe_4S_4(RS)_4$ cluster as a prosthetic group that can bind 2 MgATP or 1 MgATP and 1 MgADP. The meaning of these two modes of binding will become apparent in the next section, but it should be noted that the finding of several workers that the limiting

(MgATP hydrolyzed/(2e⁻transferred) ratio is 4 has a satisfactory explanation in the hypothesis that in combination with the MoFe protein, the Fe protein becomes the ATPase of nitrogenase. The two MgATP binding sites would then represent the incipient locales of energy transduction. Whether this simple view is sufficient remains to be shown.

C. Control of Overall Reaction Rate by [MgATP]/[MgADP]

One of the earliest studies on the influence of ATP and ADP on the functioning of nitrogenase is that of Moustafa and Mortenson (1967). They showed that reduction of acetylene by Cp nitrogenase (crude fractions) was dependent on the ratio of ATP/ADP in the presence of a limiting amount of Mg in the absence of an ATP regenerating system. Kennedy et al. (1968) extended this work to N_2 reduction and H_2 evolution and also examined the influence of Mg on the inhibition. They made the observation that as the Mg concentration was increased, both the initial rate of reaction promoted by ATP and the inhibition of ADP was increased. Use of substrate level ATP, rather than an ATP generating system, limited the time of reaction that could be monitored and made difficult any reliable quantitation of the phenomenon of ADP inhibition.

As discussed in Davis and Orme-Johnson (1976), many more recent studies of nitrogenase kinetics have suffered from similar difficulties of interpretation because authors were unaware of the importance of Mg in the function of ATP during the nitrogenase reaction. A second, more severe deficiency for any quantitative work is the apparent disregard of the capabilities of the commonly used creatine phosphate plus creatine kinase ATP generating system. Nitrogenase and creatine kinase share several properties—their true substrate is the Mg chelate of the nucleotide, the product of the reaction is inhibitory, and the uncomplexed nucleotide at high concentration is inhibitory.

Some of the apparent behavior of nitrogenase in assays where creatine kinase serves to regenerate ATP is explicable as a consequence of the properties of the kinase. For instance, at high pH (8 or above), where many EPR experiments have been carried out, the kinase loses its affinity for creatine phosphate so that the enzymatic activity rapidly diminishes during the course of an assay or EPR experiment. At low pH values (below pH 7), the protonation of ADP results in its being an ineffective substrate for the kinase (since it fails to bind Mg) so that the level of ADP rises even as the MgADP level is kept relatively low by the action of kinase (Morrison and James, 1965).

Since free ADP is a relatively weak inhibitor of nitrogenase, there is a range of pH values and creatine kinase concentrations over which the net activity of nitrogenase is high. However, at both the low and high ends of the pH range, relatively more kinase is needed to keep the concentration of either ADP or MgADP below any given level considered acceptable. When using high concentrations of nitrogenase, as in EPR experiments such as those discussed below, quite large amounts of the kinase are needed to maintain the steady-state level of ADP below detectable inhibition of the nitrogenase. At the high pH end, a steady state can easily be established in which the rate-limiting step of the overall reaction is regeneration of MgATP, and the net rate is controlled by the MgADP/MgATP ratio.

An additional, though usually lesser problem, is that some buffers, notably phosphate, bind Mg and lower the level of the true substrate for nitrogenase and the kinase. Creatine phosphate, the high-energy donor for the ATP regenerating system, also binds Mg, and the by-product inorganic phosphate also binds Mg, making Mg limitation common in coupled assays (see references cited in Davis and Orme-Johnson, 1976). On the other hand, as pointed out by Thorneley and Willison (1974), excessive Mg may inhibit by formation of dimagnesium chelates of the nucleotides.

Taking into account the variables just discussed, we determined the steady-state levels of ADP and ATP present in dithionite oxidation assays of nitrogenase, where the rate of reaction was limited by the concentration of creatine kinase present in the assay (Fig. 7). As initially suggested by Moustafa and Mortenson, the rate of the nitrogenase reaction is indeed quite sensitive to the ratio of nucleotide, although it is the Mg chelate forms that are of relevance.

Haaker et al. (1974) recently suggested that A. vinelandii nitrogenase is limited in vivo by the ATP/ADP ratio and not by (reduced pyridine nucleotides)/(oxidized pyridine nucleotides). They measured these ratios and acetylene-reducing capacity of cells under several oxygen input rates and with several inhibitors of oxidative phosphorylation. They concluded that the source of electrons is probably "energized membranes," i.e., not pyridine nucleotides per se, since the ratio of reduced to oxidized nucleotides did not correlate with nitrogenase activity by the acetylene test. They calculated the levels of ATP to be about 1–2 mM and the levels of ADP to be about half that of the ATP in actively respiring cells. The sensitivity of nitrogenase activity to ATP/ADP in vivo seems comparable to that in vitro and as can be seen by comparison to Fig. 7, nitrogenase would appear to be normally controlled in vivo by this ratio.

Appleby et al. (1975), using bacteroids from Rhizobium, have made similar studies of the effect of nucleotide ratio on nitrogenase activity. They reached similar conclusions concerning the effect of (ADP)/(ATP).

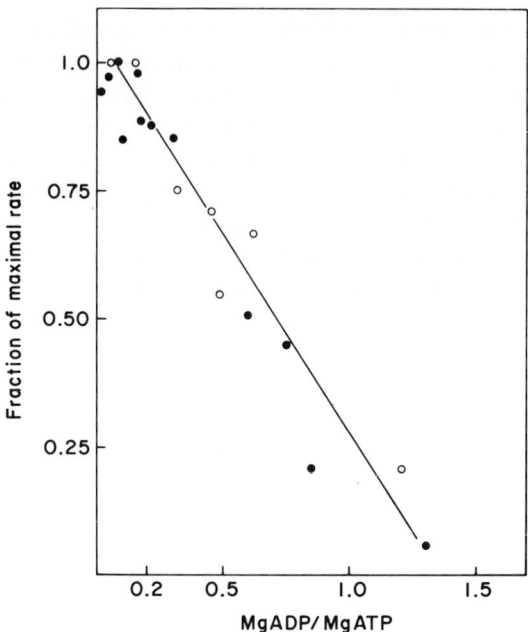

Fig. 7. Dithionite oxidation by *C. pasteurianum* nitrogenase as a function of the relative concentration of MgADP and MgATP. The oxidation of dithionite was monitored spectrophotometrically, and at intervals, portions of the reaction mixture were removed from the cuvette for determination of the concentration of ^{14}C-labeled nucleotides by the method of Morrison and Cleland (1966). Differing ratios of the nucleotides were maintained by variations in the level of creatine kinase present. Solid circles, excess Fe protein; open circles, limiting Fe protein. From Davis and Orme-Johnson (1976).

They also observed that the rate of oxidative phosphorylation was the determinant of ATP production.

Under conditions that exist *in vivo*, the combination of MgATP with the enzyme might be rate controlling, since the enzyme is present at approximately 100 μM and the MgATP is present at a few millimolar with somewhat lower levels of MgADP. The location of the rate-controlling step *in vivo* depends on the relative on and off constants for ATP and ADP in nitrogenase and in the regenerating system and on the steady state concentration of each of these under physiological conditions. Nitrogenase produces a large perturbation in such steady state concentrations. The underlying assumptions of initial velocity kinetics cannot be demonstrated to hold at these high concentrations of enzyme, substrate, and product using steady state kinetic techniques. However, when the results of *in vivo* and *in vitro* kinetic studies are considered together

2. ENZYMOLOGY OF NITROGENASE

with the results of the binding and other physical studies, this level of metabolic control of nitrogenase has a plausible but not established explanation. It would appear that the form of the Fe protein to which two MgATP molecules are bound is competent to transfer electrons to the MoFe protein and, consequently, forms an active nitrogenase complex. On the other hand, the data are consistent with the idea that the forms of the Fe protein to which MgADP are bound are incompetent reductants and, therefore, form inhibited complexes with the MoFe protein. This seemingly simple hypothesis will no doubt receive further scrutiny.

IV. SEQUENCE OF ELECTRON TRANSFERS BETWEEN NITROGENASE COMPONENTS

A. Steady-State Evidence

At least as early as the 1968 review by Hardy and Burns, the role of ATP was postulated to be production of a superreductant in the nitrogenase enzyme system. Silverstein and Bulen (1970) presented evidence that was consistent with more than one role for ATP in the overall reaction and more than one pathway for hydrolysis of ATP (see discussion in Section III,A). On the basis of the concentration dependence for the reaction, they postulated multiple levels of reduction of the enzyme. The work of Vandecasteele and Burris (1970) made clear the critical importance of component ratio in the function of nitrogenase. They showed that as the concentration of Fe protein is raised while the MoFe protein is held constant, the activity rises to a plateau, while in the converse case, where the Fe protein is constant and the MoFe protein concentration is raised, the activity rises and then falls again. They also observed that the classic ATP-independent hydrogenase is an inhibitor of nitrogen reduction but not of ATP-driven hydrogen evolution by nitrogenase.

With the results of Vandecasteele and Burris and of Silverstein and Bulen in mind, we undertook a study of the effects of component ratio and concentration on some of the activities of nitrogenase from *Azotobacter*. Some of the effects observed under standard assay conditions are discussed in Shah *et al.* (1972). There appears to be no system present in *Azotobacter* that will evolve H_2 in the absence of MgATP (Hyndman *et al.*, 1953). Thus, the utilization of electrons from dithionite to form H_2 is dependent on the presence of nitrogenase and MgATP even in crude extracts.

We observed that there was a sigmoidal dependence of rate of dithionite oxidation on MgATP concentration at levels of MgATP below approximately 0.1 mM. The shape of the MgATP saturation function was dependent on the ratio of components present in the assay and can be interpreted in terms of the formation of an active enzyme complex whose concentration is dependent on both MgATP levels and enzyme concentration. Determination of acetylene reduction in the dithionite oxidation assay cuvettes allowed comparison of the relative rates of electron utilization and substrate reduction as a function of ATP concentration and component ratio. It was found that the relative efficiency of substrate reduction was much more markedly decreased by either low ATP concentration or enzyme dilution than was the rate of dithionite oxidation (Davis et al., 1976, and unpublished observations).

Since our results using the dithionite oxidation assay confirmed the observations of Silverstein and Bulen, who utilized product formation assays, we further studied the effect of varying component ratios in product formation assays. The stoichiometry of the nitrogenase complex used by Silverstein and Bulen is probably near 1:1 molar MoFe:Fe protein, as previously mentioned. We used enzyme preparations as varying stages of purification and found consistent results whether crude extracts prepared as described by Shah et al. (1972) or highly purified fractions (Shah and Brill, 1973) were used. That is, the relative activities of a crude extract for reduction of acetylene and N_2 are those expected for a comparable ratio of purified components. The precise ratio of activities depends on method of preparation and storage of the extract and will change if there is selective loss of the Fe protein (Shah et al., 1975), thus changing the component ratio.

As mentioned above, the K_m for ATP is dependent on the component ratio as well as absolute enzyme concentration. Furthermore, the relative efficiency of reduction of C_2H_2 versus N_2 is a function of component ratios (Davis et al., 1975). C_2H_2 is a more potent inhibitor of N_2 reduction when Av2 is limiting in the reaction; H_2 is a more potent inhibitor of N_2 reduction and enhances C_2H_2 reduction when Av2 is limiting. Lowering the level of ATP in the assay has an effect similar to that of reducing the concentration of Av2.

Knowing that component II binds ATP (Section III), the results are most simply interpreted by assuming that component II, in the presence of ATP, supplies electrons to the substrate reducing center on component I. The rate of electron supply versus the rate of leakage of protons determines the average redox state in which the enzyme operates and hence its ability to reduce $C_2H_2(2e)$ versus $N_2(6e)$.

Bergersen and Turner (1973), using soybean bacteroid components, came to similar conclusions on the relative role of the two components, i.e., that component I has the substrate reduction site and component II is an "effector" in the presence of ATP. They found alterations of apparent Michaelis constants and maximal velocities upon changing the component ratio. Unfortunately, they do not present sufficient data on the C_2H_2 reduction versus C_2H_2 concentration to ensure that the apparent K_m effect is not simply the superposition of a single K_m and K_i whose relative importance varies as the ratio of components. This ratio may well determine the distribution of enzyme forms capable of reacting productively and nonproductively with C_2H_2. For *A. vinelandii*, Shah et al. (1975) have shown that there is an inhibition by C_2H_2 that is dependent on the ratio of components in the assay. Data of V. K. Shah (personal communication) indicates that a single K_m and K_i will fit the data for a wide range of component ratios. This gives an apparently changing K_m if an insufficient number of C_2H_2 concentrations are used in derivation of the K_m and K_i values.

The importance of component ratios can further be seen from the following. Using an extract of *A. vinelandii* in which activity was limited by the Fe protein, we added back a purified Fe protein fraction and determined the relative inhibition of N_2 reduction by C_2H_2. Addition of a saturating amount of Av2 increased the activity from 210 to 430 nmoles N_2 reduced per minute per milliliter of extract and raised the apparent K_i of C_2H_2 from 0.015 atm to 0.034 atm ($1/v$ versus I plots at 0.95 atm N_2). We conclude that it would be advisable to interpret the route of electron flow, based on steady state kinetics, from studies in which the variable of component ratio in addition to the variable of ATP reductant and reducible substrates is included.

It is not within the scope of intent of this chapter to make a detailed analysis of the kinetic complexities of nitrogenase with respect to the wide range of possible inhibitors and reducible substrates (see Chapter 3, Volume I). The most extensive recent work on this subject is that of Hwang et al. (1973), Hwang and Burris (1972), and Rivera-Ortiz and Burris (1975). We will limit our discussion to the most commonly used substrate, acetylene, and the inhibitor carbon monoxide. The former is used in most assays of nitrogenase activity at the present time, while the latter is of particular interest for the EPR studies to be discussed below.

Acetylene is a very efficient substrate for nitrogenase and will completely suppress hydrogen evolution activity (Rivera-Ortiz and Burris, 1975). It is a potent noncompetitive inhibitor of N_2 reduction and is also an inhibitor of its own reduction in *A. vinelandii*, as discussed above. Somewhat surprisingly, the reduction of acetylene by *C. pasteurianum* shows behavior very different from that in *A. vinelandii*. In *C. pasteurianum*, the nitrogenase exists in two forms, or states, whose relative concentration in steady-state assays varies as a function of component ratios and pH of the assay. A low K_m form is present in a larger amount when the Fe protein is limiting, while the high K_m form is more abundant when the MoFe protein is limiting. Analysis of the rate data by a 2/1 function (Cleland, 1970) indicates an apparent K_m of 0.23 atm for the high K_m form and approximately 0.002 atm for the low K_m form.

The inhibition of acetylene reduction by carbon monoxide was examined for both forms of *C. pasteurianum* nitrogenase with $1/v$ versus I plots. For the low K_m form, a value of 5.5×10^{-4} atm was obtained; for the high K_m form, a value of 7×10^{-4} atm was obtained. Studies of CO inhibition in *A. vinelandii* indicate that with excess Fe protein, the K_i is about 3×10^{-4} atm, while with limiting Fe protein it is markedly elevated. Analysis of the limiting Fe protein case is complicated by substrate inhibition by the acetylene.

The conclusions that can be drawn from steady-state studies such as these are that the nitrogenase enzyme exists in several different states as a function of component ratio and that the relative distribution and reactivity of these states differs significantly between organisms. The exact nature of these states is undefined, but the concept provides at least a qualitative explanation for some of the kinetic phenomena, including the apparent noncompetitive nature of the mutual inhibition of reducible substrates. This concept will be developed further after consideration of thermochemical and spectroscopic evidence.

B. Energy of Activation for Nitrogenase Activity and Complex Formation

The unusual nature of the temperature dependence of nitrogenase reactions has been recognized for many years (see Hardy and Burns, 1968), but there are relatively few published studies of the activation energy of the different reactions (Hadfield and Bulen, 1969; Burns, 1969; Thorneley et al., 1975). Hadfield and Bulen examined the ATP:2e ratio for *Azotobacter* complex and found that it varied systematically with temperature. More ATP was hydrolyzed per pair of electrons transferred as the temperature was increased above 20°C. This was interpreted in terms of two reaction pathways for ATP hydrolysis—one productive, the other nonproductive. Silverstein and Bulen (1970) made a more detailed correlation of the temperature dependence of ATP hydrolysis with the relative rates of the two ATP hydrolysis processes. Reanalyzing the data of Hadfield and Bulen in terms of the Arrhenius equation, they derived the difference enthalpies and entropies for the productive versus nonproductive reactions. A negative entropic term (-15.9/cal/degree) was interpreted as indicating a requirement for a "more ordered structure in a transition state preceding H_2 formation."

2. ENZYMOLOGY OF NITROGENASE

Burns (1969) carried out a study of the activation energy of the nitrogenase complex over a wide temperature range and found that the apparent K_m for ATP did not vary above and below 21°C, but that there was a sharp break in the measured activation energy at about this temperature.

These studies were done with a limited number of substrates. The work of Silverstein and Bulen clearly established that the responses of different substrates to enzyme concentration is very different. We have shown (Shah et al., 1975; Davis et al., 1975) that the responses of different substrates to component ratio are also quite different. One might thus expect that the activation energy for different substrate reductions might be different. This has not been systematically explored using several substrates under comparable conditions, but we have done some experiments comparing temperature effect on the apparent K_m for ATP crude extracts of *A. vinelandii* for either dithionite oxidation or acetylene reduction (unpublished observations of L. C. Davis, V. K. Shah, and W. J. Brill). In agreement with the results of Burns (1969), the shape of the ATP saturation curve is not altered significantly by temperature of reaction for either dithionite oxidation or acetylene reduction. However, the relative rate of acetylene reduction at ATP levels below 0.8 mM changes approximately seventyfold on going from 11° to 24.5°C, while the rate of dithionite oxidation changes only fourfold. Obviously the formation of the state of enzyme that reduces acetylene is much more temperature dependent than that which oxidizes dithionite, so that observed energies of activation are extremely dependent on such variables as choice of reducible substrate and component ratio. Studies at low temperature are confounded by the slow cold inactivation of the nitrogenase in a reversible manner. We found that enzyme assayed at 11°C lost half its activity for dithionite oxidation in 1 hour, but that the activity was restored on incubation at 30°C.

The concept of nitrogenase as an associating system of dissimilar subunits was clearly established by Silverstein and Bulen (1970) and by Vandecasteele and Burris (1970). The relationship of this association phenomenon to the "dilution effect" in practical terms was pointed out by Hardy and Burns (1968) and systematically studied by Shah et al. (1972). This dilution effect is an expression of the mass action law for formation of an active enzyme complex. The operation of this mass action law leads to the paradoxical behavior of nitrogenase in which titration of the MoFe protein with the Fe protein shows a sigmoidal pattern with a plateau (see Tso et al., 1972; Israel et al., 1974; Ljones and Burris, 1972), while the titration of Fe protein with MoFe protein rises to a maximum and then decreases (Shah et al., 1972; Ljones and Burris, 1972; Eady et al., 1972; Zumft and Mortenson, 1975).

If one carries out a full titration in both dimensions by varying the concentration of both components over a suitable range, one should be able to derive the stoichiometry of complex formation for any particular reaction of the nitrogenase. It must be remembered that the maximum specific activity of either component will be obtained when that component is utilized at its maximal velocity, while the most active nitrogenase enzyme is obtained at the point of highest obtainable activity of both components simultaneously. This point is emphasized in Shah et al. (1975) by comparison of N_2 reduction to acetylene reduction. The lowest ratio of acetylene to nitrogen reduction activity is found when the components are present at about 1:4 molar ratio, whereas the maximal acetylene or nitrogen reduction activity is obtained with considerably more MoFe protein (about 1:1 molar).

Eady et al. (1972) and Eady (1973) suggested that the active enzyme was a 1:1 complex on the basis of the following observations. As mentioned above, the maximal specific activity of the Fe protein is obtained with an approximately 1:1 molar mixture of the two components. Second, there is observed in the ultracentrifuge an approximately 1:1 molar complex formed between Kp1 and Kp2 under conditions of low dithionite concentration.

Thorneley et al. (1975) have extended the studies of Eady et al. by measuring the acetylene reducing activity of K. pneumoniae and A. chroococcum nitrogenase over a range of concentrations with varying temperature and component ratio. There was a rather sharp break at $\sim 17°C$ in the Arrhenius plot of the association constant with temperature such that below 17°C there was a very high energy (ΔH) of activation for complex formation (418 kJ/mole), while above 17°C the ΔH was zero. For the acetylene reduction reaction a ΔH^* of 80 kJ/mole was obtained by difference over the entire range. At 30°C, the association constants for K. pneumoniae and A. chroococcum components and mixtures from the two sources lay in the range 10^7 to 5×10^7 M^{-1}.

Thorneley et al. (1975) generated their data by preparing a 1:1 molar mixture that was diluted to different concentrations and assayed. Analysis of the resulting activity versus concentration curves was done with a mass-action expression that allowed the determination of the binding constant. They used a form applicable to a 1:1 mixture of components, but their data, with nonunitary component ratios suggests the formation of higher complexes. When an excess of Fe protein was used, a very rapid increase in activity with increase in total protein concentration was observed, whereas when an excess of MoFe protein was used, there was an initial increase and then a plateau in activity. The analysis of the effect of dilution on the observed acetylene reduction activity was predicated on

2. ENZYMOLOGY OF NITROGENASE

the theory that 1:1 complexes predominate, which was taken from ultracentrifuge experiments that showed that the Svedberg coefficient of the most rapidly sedimenting species did not increase as the ratio of Kp2 to Kp1 was increased beyond 1:1. The ultracentrifuge experiment may not detect complexes relevant to the functioning enzyme, as Thorneley et al. observed. The departure of the activity measurements made in the presence of the full nitrogenase assay system (i.e., MgATP and large dithionite concentrations) from the behavior expected from the mass-action law for 1:1 complexes suggested that in fact other complexes are present in the active enzyme solution. In particular, the decreases in activity observed when the MoFe protein is in excess suggest that 1:2 (MoFe:Fe) or higher complexes may be of enhanced activity. The combination of a great deal more precise activity and centrifugation data with an extensive computer simulation approach for both phenomena might in the future bring to fruition the promising start made by the Sussex workers.

C. EPR and Absorbance Studies of the Functioning Enzyme

During the period 1971–1972, the purification of nitrogenase components, as well as the development of anaerobic sample handling techniques, had developed to the state that the nitrogenase workers at Wisconsin, Purdue, and Sussex simultaneously and independently realized that the components possessed reproducible EPR features that were altered during the functioning of the enzyme. These physical studies have carried us past the black box approach of steady-state kinetics and onto new and, as yet, incompletely explored ground. Though subsequent experimentation has not resolved all points of interest, due to the considerable experimental problems involved, we will try to glean the seemingly solid facts that add a great deal to our present concept of the enzyme.

Orme-Johnson et al. (1972) reported spectra from 1:4 (molar ratio) mixtures of Av1 and Av2 or Cp1 and Cp2 that were mixed anaerobically with a MgATP generating system in the presence of sodium dithionite. The $g = 3.6$ signal complex of the MoFe protein declined by 90% in samples frozen while an excess of reductant was present. A small quantity of methyl viologen was included in the mixture as a redox indicator. When the blue cast of the viologen cation radical had disappeared, indicating exhaustion of the sodium dithionite, subsequently frozen samples showed the reappearance of the $g = 3.6$ signal. A similar experiment performed with a 1:16 ratio of Cp1 and Cp2 further showed that when the return of the MoFe protein signal was complete, no trace of the previously observable Fe protein (MgATP form) signal could be seen in EPR at 13°K

(Fig. 8). The conclusion was drawn from these experiments that even as isolated in the presence of dithionite, MoFe proteins are relatively oxidized and that they accept electrons from the MgATP·Fe protein complex. The final disappearance of the Fe protein signal, combined with the change in EPR signal shape in the presence of MgATP, suggested that the Fe protein acts as a reductant analogous to the simple ferredoxins, i.e., yielding an EPR signal when reduced, and becoming EPR silent when oxidized by the MoFe protein. The fact that increasing the molar ratio of the Fe protein from 4:1 to 16:1 decreased the steady-state level of residual MoFe protein strengthened the conclusion that the electron flow was in the direction and order indicated.

Smith et al. (1972, 1973) performed similar experiments and reached similar mechanistic conclusions about the Kp proteins. In addition, they examined mixtures prepared by freeze quenching at 10 msecond and longer after mixing. They saw a rapid (half-life 10 msecond at pH 8.1, 23°C) decline of both the Kp1 and Kp2 signals. However, they observed only a 25–50% decrease of the MoFe protein signal in the steady state, whereas in longer term experiments (manual mixing) like those of Orme-Johnson et al., they had observed a 75% decline of the MoFe signal using the same molar ratio of nitrogenase components. The discrepancy no doubt arises from the following subtle artifact; they used an ATP gener-

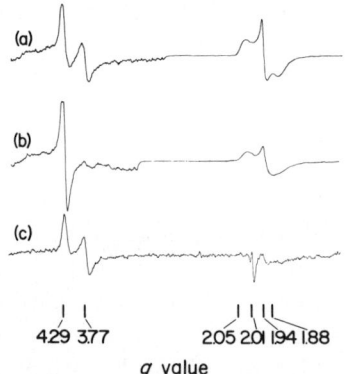

Fig. 8. EPR of components of *C. pasteurianum* nitrogenase mixed at a ratio of 1:16 (MoFe protein:Fe protein). Conditions of EPR spectroscopy were as in Fig. 1. (a) 2.5 mg of MoFe protein/ml, 10 mg of Fe protein/ml, and 5 mM Na$_2$S$_2$O$_4$; no ATP-generating system was added. (b) Same as (a), but including the ATP-generating system (5 mM MgATP); the sample was frozen 45 seconds after mixing. (c) Same as (b), but the Na$_2$S$_2$O$_4$ concentration was decreased to 0.5 mM and 40 μM methylviologen was present. The sample was incubated for 15 minutes before quick-freezing. Note that the final amplifier gain of the left halves of (a) and (b) and all of (c) were increased by tenfold. From Orme-Johnson et al. (1972).

ator that was incapable of coping with the demands of the nitrogenase, and thus the reduction of the MoFe protein was inhibited by the MgADP present (Davis and Orme-Johnson, 1976). Also, as Thorneley and Willison (1974) point out, the presence of a large excess of Mg in their system probably led to inhibition by Mg_2ATP. Thus, the results of Smith et al. cannot be regarded as fully definitive.

Mortenson et al. (1973) performed similar experiments to those of Orme-Johnson et al. and Smith et al., but they proposed that the decline in the EPR of the MoFe protein during turnover was due to a MgATP induced oxidation of the MoFe protein. Mortenson et al. proposed that electron transfer to the MoFe protein was rate limiting, as opposed to the proposal of the other groups that substrate reduction is rate limiting. They did not deal with the fact (which they also observed) that a large excess of Fe protein will ultimately become oxidized during turnover. They adduced in support of this theory their observation that dye-oxidized MoFe protein absorbs light more strongly than the form originally isolated, and stated that during turnover the MoFe protein similarly becomes more strongly absorbing, i.e., more oxidized. G. A. Walker and Mortenson (1974) subsequently reported that, in fact, *no* change in the absorbance of Cp1 occurs during turnover, and they accepted the view of Orme-Johnson et al. and Smith et al. as to the sequence of events. Zumft et al. (1974) then reported a re-examination of the pre-steady-state EPR studies, in which they were careful to insure that the MgATP/MgADP ratio was maintained. They made the important observation that even at extremely low ratios of Cp2 to Cp1 (0.06:1), the MoFe signal declined by at least 90%, though with a half-time of 19 seconds as opposed to a half-time of 70 mseconds for a 2:1 ratio mixture. Under the low Fe protein conditions, the system has hardly any detectable activity and is wildly prodigal of MgATP (MgATP hydrolyzed/$2e^-$ transferred to substrates >20, see Ljones and Burris, 1972). Therefore, the reduction of the $g = 3.6$ signal center of the MoFe is perhaps a necessary but not sufficient condition for substrate reduction by nitrogenase.

Both Smith et al. (1973) and Mortenson et al. (1973) reported that nitrogenase components may be oxidized by dyes to EPR-silent states. Fe protein oxidized by one electron from the paramagnetic state appears to be a state occurring during the action of nitrogenase, but the EPR-silent oxidized state of the MoFe protein is probably irrelevant to catalysis (G. A. Walker and Mortenson, 1974).

In all studies reported to date, the Fe protein becomes somewhat oxidized in the steady state, as evidenced by a decline in the EPR signal size. Since oxidation of this protein is accompanied by an increase in absorbance near 440 nm (Ljones, 1973), Thorneley (1975) was able to follow the pre-steady-state kinetics of Fe protein oxidation with a specially adapted anaerobic stopped-flow spectrophotometer (Thorneley, 1974a). He observed a single first-order absorbance increase at 425 nm, with a time constant of 8 mseconds (23°C) independent of component ratios or order of reactant mixing. Using a ΔA_{425} of 2000 M^{-1} cm^{-1}, he calculated that up to three molecules of Kp2 were rapidly oxidized per Kp1 present. Ljones (1973) calculated a ΔA_{435} of 4500 M^{-1} cm^{-1} for partially active Cp2 and suggested a value of 7000 for fully active material (see also Section II,E). Thorneley's estimate may have to be revised downward by a factor of three; i.e., the initial rapid event may involve only one Fe protein molecule, transferring a single electron to the MoFe protein.

Thorneley also obtained evidence for a tight complex between Kp1 and Kp2, $K_D < 0.5$ μM with a rate of formation $k > 1 \times 10^7$ M^{-1} sec^{-1}. This fit well with the binding behavior deduced by Thorneley et al. (1975) from the dependence of activity upon component concentrations (Section IV,B).

The dependence of the oxidation of Kp2 on MgATP concentration was also studied. As the MgATP was decreased below 300 μM (5 μM components), the half-time increased steadily. Though an ATP-regenerating system was present, it seems possible that what is being measured is the effect of increasing MgADP/MgATP ratio. Barring this objection, the combination of MgATP with either the Fe protein or the Fe protein–MoFe protein complex (the measurements require electron flow and thus the two cases cannot be distinguished) was deduced to have rate constant $k > 2.5 \times 10^6$ M^{-1} sec^{-1}. Electron transfer was a unimolecular process with $k = 2 \times 10^2$ sec^{-1}, and therefore, in general agreement with the earlier EPR studies, neither combination of the Fe and Mo proteins nor combination of nitrogenase with MgATP appears to be rate limiting under normal conditions where the turnover rate of the system is about 3 sec^{-1} for nitrogen reduction. The finding of the deep decline of the $g = 3.65$ MoFe protein signal in the steady state suggests further that the initial electron transfer event is not normally rate limiting, so that either a subsequent movement of electrons within the nitrogenase complex or the ultimate reduction of substrate seems to be the slow electron transfer event in the action of nitrogenase.

We know from Mössbauer studies (Section II,F) that about one-third of the iron atoms in the MoFe protein are associated with the $g = 3.6$ paramagnetic center. The other iron atoms and the molybdenum may, of course, be involved in electron storage and substrate reduction, but they have not manifested themselves spectroscopically during the change to steady-state conditions. Other signals have been noticed during the irreversible oxidative denaturation of Cp1 (see Zumft and Mortenson, 1975), but they have not been related to species present in the functioning enzyme. We sought to elicit EPR from these missing metals by employing the noncompetitive inhibitor carbon monoxide on the theory that if electron flow were interdicted within the nitrogenase complex, then paramagnetic species might become visible as prothetic groups gone into abnormal oxidation states. The following results are seen (Davis et al., 1976): CO does not alter the decline of the $g = 3.6$ signal of the MoFe protein, nor does it affect the rate of electron flow (100% flux to H^+ in the CO inhibited state). Thus, CO has no effect on the initial reduction step nor on the electron pathway leading to hydrogen production. The inhibitor does compete with nitrogen or acetylene for the substrate-reducing form of the enzyme and may in fact interact at a locale other than the substrate-reduction site. The inhibition is reversible.

What occurs is that with a time constant of about 10 seconds (at 30°C), new signals arise from nitrogenase functioning in the presence of CO. When [CO] ≈ [MoFe protein], an EPR signal with g values of 2.08, 1.97, and 1.93 appears. When [CO] > [MoFe protein], a signal with g values of 2.17 and 2.05 replaces the earlier signal (Fig. 9). The process is strongly temperature sensitive, i.e., the new signals do not appear, depending on the species of nitrogenase used, when the temperature of incubation is below 10°–20°C. (EPR is performed at 12°K after freezing-quenching the samples in isopentane at 130°K). The actual inhibition of the enzyme (i.e., inhibition of flux to acetylene reduction) occurs in about one turnover of nitrogenase, so the slow onset of the signals is a consequence of CO inhibition of electron flow, i.e., a slow transition of a previously undetected center to an EPR-active state. Exhaustion of reductants reverses the process, and so the new centers can deliver electrons to the H^+ reducing site. The use of Fe and MoFe proteins selectively labeled with the isotopes ^{57}Fe and ^{95}Mo, and the use of ^{13}CO, allowed us to deduce that the signals arise from the Fe atoms of the MoFe protein and not from Mo or the CO. The shape of the signals (Fig. 9) suggests strongly that the low CO form is a ferredoxin-like FeS center in the −3 net charge state, while the high CO form is a high-potential iron–sulfur proteinlike −1 net charge state. Since we detected the effect of ^{13}CO binding to the $Fe_4S_4RS_4$ center of hydrogenase, we feel that it is unlikely that the centers detected in the MoFe protein are themselves the binding sites for CO, since no effect of ^{13}CO was noted. The simplest explanation appears to be that the new FeS center is connected by an irreversible redox pathway to the H_2 evolution site and that at low [CO], electron flux from the new center to the N_2 reduction site is cut off, leading to the reduction of the new site from the −2 (EPR-silent) to the −3 net charge state.

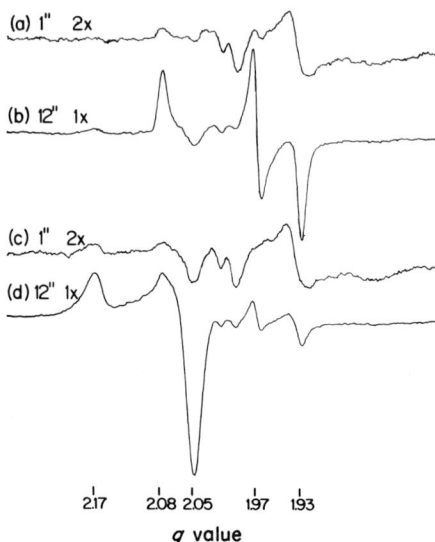

Fig. 9. EPR of nitrogenase in the presence of carbon monoxide. The experiment was performed similarly to that of Fig. 8, except that the ratio of components was 1:1 on a molar basis. (a) and (b) CO was present at a concentration equal to that of the MoFe protein. (c) and (d) CO was present at about 5 times the concentration of the MoFe protein. The samples were incubated for the times indicated at 30°C and then frozen by injection into isopentane at 130°K (Bray, 1961). Spectra were obtained as in Fig. 8. From Davis et al. (1975).

Similarly, if the second CO binding site, occupied at high [CO], lay between the $g = 3.65$ center and the new FeS center, then the irreversible connection to the H_2-evolving site would produce a more oxidized FeS center, i.e., one in the -1 net charge state. This interpretation accords with the facts, and a key point is that the production of the high CO form, surely arising from an oxidized FeS center, requires the presence of a reductant, i.e., the complex of MgATP and the Fe protein. It is not seen during dye oxidations of the MoFe protein. This may be hinting at a second function of MgATP, i.e., to make the pathway from a center in the MoFe protein to H^+ irreversible, presumably by yielding its hydrolysis energy in a transduction event.

Clearly, we are not in a position to state conclusively the character and utility of all the prosthetic groups in the components of nitrogenase, but the $Fe_4S_4(RS)_4$ center of the Fe protein, the $g = 3.65$ center, and the CO-elicited center of the MoFe protein clearly are three parts of a larger set of metal centers that form an electron transport apparatus capable of reductive dephosphorylation and versatile substrate reduction.

V. STRUCTURAL AND MECHANISTIC HYPOTHESES

A. The Fe Protein

Even though the Fe proteins of all species examined are composed of two subunits, the present evidence is that the Fe protein contains a single $Fe_4S_4(RS)_4$ cluster that operates on a single electron transfer basis between the -2 and -3 net charge states. Thus, either two identical protein chains become inequivalent during association and form one FeS cluster site, or the FeS cluster, which has a minimum of fur chain attachment positions (the RS groups), bridges the two subunits. Further relevant features of this protein are its extreme O_2 lability, its possession (for *C. pasteurianum*) of two MgATP and one MgADP sites (competitive with one of the MgATP sites), its participation in an enegry-transducing ATPase, and the conformational changes induced by MgATP. This latter is mimicked by the action of the denaturant area (Zumft et al., 1973) and is accompanied by an increased reactivity of the thiol groups of the protein (Thorneley and Eady, 1973), as well as increased reactivity of the iron atoms to α,α-bipyridyl (G. A. Walker and Mortenson, 1973, 1974). The picture of two protein subunits joined by a bridging FeS cluster, with sites for MgATP in the crevice, action at which generates mechanical stress on the FeS center to accomplish the energy transduction, is attractive but premature. In favor of this sort of picture, however, is the observation of Holm and colleagues (Chapter 7) that relatively nude $Fe_4S_4(RS)_4$ clusters have redox potentials 0.5–1 V more negative than the ferredoxins, and that wrapping the clusters in synthetic peptide ligands raises the potentials dramatically. The reverse process ought to depress the potential of the FeS center, and so the action of MgATP hydrolysis may consist in just such an environmental change around the FeS center to produce the low potential required for N_2 reduction. The expression of this property appears to require the presence of the MoFe protein, which is useful in preventing the hydrolysis of MgATP by the Fe protein alone.

B. The MoFe Protein and the Enzyme

Our picture of the MoFe protein includes an FeS center (about one-third of the Fe) reduced early in turnover (the $g = 3.6$ center) plus one or two centers elicited during CO inhibition. The latter appear to be FeS centers of the normal ferredoxin type. Nothing is known of the nature of the H^+ or N_2 reducing sites, although kinetic complexities suggest up to five sites for ligand interaction on the molecule (Hwang et al., 1973;

Rivera-Ortiz and Burris, 1975). The presence of four subunits has led to the proposal that there are two or more N_2 reduction sites (Zumft and Mortenson, 1975), but the difficulties with the metal composition (Section II,B) make this idea uncertain. No evidence for Mo function have been found, though there is an absolute nutritional requirement for the element in N_2-fixing organisms.

The combining ratio of Fe and MoFe proteins present in the fixing system (i.e., when MgATP and reductant are present) is not known, though evidence for 1:1 complexes has been adduced in nonfixing systems. The fixing system may be a 1:1 complex (Thorneley et al., 1975), but the paradoxical effect of component ratio on the activity (Section IV,B) suggests either that $(MoFe)_2(Fe)_1$ protein complexes are inhibited or that, in fact, the $(MoFe)_1(Fe)_1$ complex is of low activity and the active nitrogenase forms are $(MoFe)_1(Fe)_2$ or higher complexes.

The uncoupling phenomenon or high ATP hydrolysis rates with little electron transfer to substrates may well be a property of $(MoFe)_1(Fe)_1$ complexes because since the Fe protein transfers one electron per two MgATP hydrolyzed, and since the Fe protein binds two MgATP molecules in a 1:1 complex, the following event may take place:

$$2MgATP \cdot Fe(red) \cdot MoFe(ox) \xrightarrow{1e^- \text{ step}} 2MgADP \cdot Fe(ox) \cdot MoFe(red)$$
$$(1) \qquad\qquad\qquad\qquad\qquad\qquad (2)$$

The state of the MoFe protein reduced by $1e^-$ cannot reduce any of the substrates, so that if the MgADP molecules were not replaced by MgATP and the Fe protein were rereduced at a rapid rate, complex (2) might decay into complex (3):

$$2MgADP \cdot Fe(red) \cdot MoFe(ox)$$
$$(3)$$

Replacement of the ADP in (3) by ATP would return the system to form (1), with a net hydrolysis of MgATP and no substrate reduction. When higher levels of the Fe protein are present, then the probability of the activated complex (2) being further reduced to a two electron containing form (4)

$$MgATP \cdot Fe(red) \cdot MoFe(red\ [1e^-]) \rightarrow MgADP \cdot Fe(ox) \cdot MoFe(red\ [2e^-])$$
$$(4) \qquad\qquad\qquad\qquad\qquad\qquad (5)$$

would be enhanced. Species (5) would be capable of $2e^-$ substrate reductions. Alternatively, the substrate reducing form of the enzyme might be a $(Fe)_2(MoFe)_1$ complex, which would have a lower probability of entering the nonproductive cycle, i.e., (1) → (2) → (3) → (1), and a

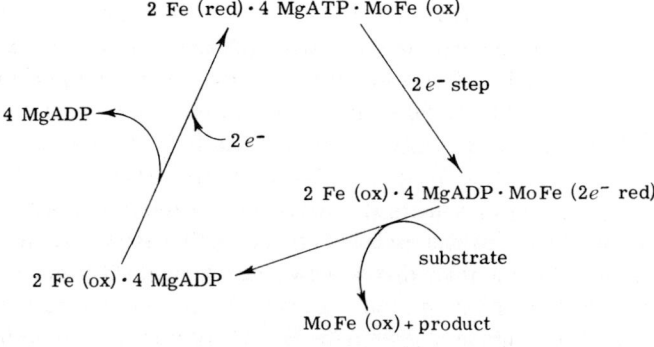

Scheme 1

higher probability of a productive cycle (Scheme 1). This productive cycle would have to go through one turn to produce the form of the enzyme reducing $2e^-$ substrates (i.e., H^+ or acetylene) and three turns to produce the $6e^-$ reductant (i.e., for N_2). If the $(MoFe)_1(Fe)_1$ complex were the active form, the increase in efficiency of utilization of MgATP would be due to a rapid electron transfer from separate MgATP · Fe(red) molecules to the nitrogenase complex. In that case, a productive cycle would turn n times for an n-electron substrate. These ideas are related to those of Silverstein and Bulen (1970), Davis et al. (1975), and Rivera-Ortiz and Burris (1975), who suggested that the differential change in efficiency for ATP utilization during the reduction of two- and six-electron substrates was due to the different reduction levels to which nitrogenase must be pumped in order to be effective. Such a concept would account for, for example, the fact that C_2H_2 can compete N_2 reduction rates down to zero, while the converse is not true.

In summary then, nitrogenase may usefully be thought of as a reductive dephosphorylation system in which both productive and futile cycles are possible. The nature of the Fe protein as a $1e^-$ transfer agent gives rise to a finite probability of $1e^-$-reduced nitrogenase that may decay back to an oxidized state with wastage of the MgATP hydrolysis energy. Presumably, cellular metabolism has evolved to minimize this futile process by having present an excess of the $1e^-$ transfer agent. The productive cycles involve $2e^-$, $4e^-$, or $6e^-$ reduced states of nitrogenase, and their concentration is enhanced in cells both by high Fe protein/MoFe protein ratios and by high MgATP/MgADP ratios. Experiments aimed at the successful introduction of nitrogen fixation into nonfixing organisms by genetic means should be designed with these requirements in mind, as well as attending to the more complicated problems of control at the transcriptional level and of the O_2 sensitivity of nitrogenase.

ACKNOWLEDGMENTS

The authors' research was supported by the NIH through GM17,170, by the College of Agricultural and Life Sciences, and the Graduate Research Committee of the University of Wisconsin, Madison. We thank our collaborators in Madison and Drs. R. N. F. Thorneley, R. R. Eady, B. E. Smith, C. Kennedy, R. Dixon, and R. H. Holm for permission to quote material in advance of publication. N. R. Orme-Johnson is thanked for a critical reading of the text.

REFERENCES

Aasa, R., and Vanngard, T. (1970). *J. Chem. Phys.* **52**, 1612.
Appleby, C. A., Turner, G. L., and Macnicol, P. K. (1975). *Biochim. Biophys. Acta* **387**, 461.
Averill, B. A., Erbes, D. L., Davis, L. C., Henzl, M. T., and Orme-Johnson, W. H. (1976). Submitted for publication.
Bale, J. R., Averill, B. A., and Orme-Johnson, W. H. (1976). Submitted for publication.
Beinert, H., Hartzell, C. R., Van Gelder, B. F., Ganapathy, K., Mason, H. S., and Wharton, D. C. (1970). *J. Biol. Chem.* **245**, 225.
Bergersen, F. J., and Turner, G. L. (1973). *Biochem. J.* **131**, 61.
Bray, R. C. (1961). *Biochem. J.* **81**, 189.
Bray, R. C. (1975). In "The Enzymes" 3rd ed. (P. D. Boyer, ed.), pp. 299–419, Vol. 12. Academic Press, New York.
Brill, W. J. (1975). *Ann. Rev. Microbiol.* **29**, 109.
Bui, P. T., and Mortenson, L. E. (1968). *Proc. Natl. Acad. Sci. U.S.A.* **61**, 1012.
Bulen, W. A., and Lecomte, J. R. (1972). In "Methods in Enzymology" (A. San Pietro, ed.), Vol. 24, Part B, p. 456. Academic Press, New York.
Burns, R. C. (1969). *Biochim. Biophys. Acta* **171**, 253.
Burns, R. C., Holsten, R. D., and Hardy, R. W. F. (1970). *Biochem. Biophys. Res. Commun.* **39**, 90.
Burris, R. H., and Orme-Johnson, W. H. (1974). In "Microbial Iron Metabolism" (J. Neilands, ed.), p. 187. Academic Press, New York.
Cleland, W. W. (1970). In "The Enzymes" 3rd ed. (P. D. Boyer, ed.), Vol. 2, p. 1. Academic Press, New York.
Dalton, H., and Mortenson, L. E. (1972). *Bacteriol. Rev.* **36**, 231.
Dalton, H., Morris, J. A., Ward, M. A., and Mortenson, L. E. (1971). *Biochemistry* **10**, 2066.
Davis, L. C., and Orme-Johnson, W. H. (1976). (in press).
Davis, L. C., Shah, V. K., Brill, W. J., and Orme-Johnson, W. H. (1972). *Biochim. Biophys. Acta* **256**, 512.
Davis, L. C., Shah, V. K., and Brill, W. J. (1975). *Biochim. Biophys. Acta* **384**, 353
Davis, L. C., Henzl, M. T., Burris, R. H., and Orme-Johnson, W. H. (1976). Submitted for publication.
Dilworth, M. J. (1974). *Am. Rev. Plant Physiol.* **28**, 81.
Eady, R. R. (1973). *Biochem. J.* **135**, 531.
Eady, R. R., and Postgate, J. R. (1974). *Nature (London)* **249**, 805.

Eady, R. R., Smith, B. E., Gook, K. A., and Postage, J. R. (1972). *Biochem. J.* **128**, 655.
Erbes, D. L., Burris, R. H., and Orme-Johnson, W. H. (1975). *Proc. Natl. Acad. Sci. U.S.A.* **72**, 4795.
Evans, M. C. W., and Albrecht, S. L. (1974). *Biochem. Biophys. Res. Commun.* **61**, 1187.
Evans, M. C. W., Hall, D. O., and Johnson, C. E. (1970). *Biochem. J.* **119**, 289.
Evans, M. C. W., Telfer, A., and Smith, R. V. (1963). *Biochim. Biophys. Acta* **310**, 344.
Haaker, H., DeKok, A., and Veeger, C. (1974). *Biochim. Biophys. Acta* **357**, 344.
Hadfield, K. L., and Bulen, W. A. (1969). *Biochemistry* **8**, 5103.
Hardy, R. W. F., and Burns, R. C. (1968). *Biochemistry* **37**, 331.
Hardy, R. W. F., and Havelka, M. D. (1975). *Science* **188**, 633.
Hardy, R. W. F., Holsten, R. D., Jackson, E. K., and Boras, R. C. (1971). *Plant Physiol.* **43**, 1185.
Hirose, M., and Kano, Y. (1971). *Biochim. Biophys. Acta* **251**, 376.
Huang, T. C., Zumft, W. G., and Mortenson, L. E. (1973). *J. Bacteriol.* **113**, 884.
Hummel, J. P., and Dreyer, W. J. (1962). *Biochim. Biophys. Acta* **65**, 530.
Hwang, J. C., and Burris, R. H. (1972). *Biochim. Biophys. Acta* **283**, 339.
Hwang, J. C., Chen, C. H., and Burris, R. H. (1973). *Biochim. Biophys. Acta* **292**, 256.
Hyndman, L. A., Burris, R. H., and Wilson, P. W. (1953). *J. Bacteriol.* **65**, 522.
Israel, D. W., Howard, R. L., Evans, H. J., and Russell, A. S. (1974). *J. Biol. Chem.* **249**, 500.
Kennedy, I. R., Morris, J. A., and Mortenson, L. E. (1968). *Biochim. Biophys. Acta* **153**, 777.
Ketchum, P. A., Cambier, H. W., Frazier, W. A., Madanski, C., and Nason, A. (1970). *Proc. Natl. Acad. Sci. U.S.A.* **66**, 1016.
Kleiner, D., and Chen, C. H. (1974). *Arch. Microbiol.* **98**, 93.
Ljones, T. (1973). *Biochim. Biophys. Acta* **321**, 103.
Ljones, T., and Burris, R. H. (1972). *Biochim. Biophys. Acta* **275**, 93.
McDonald, C. C., Phillips, W. D., Lovenberg, W., and Holm, R. H. (1973). *Ann. N.Y. Acad. Sci.* **222**, 789.
Morrison, J. F., and Cleland, W. W. (1966). *J. Biol. Chem.* **241**, 673.
Morrison, J. F., and James, E. (1965). *Biochem. J.* **97**, 37.
Mortenson, L. E., Zumft, W. G., and Palmer, G. (1973). *Biochim. Biophys. Acta* **292**, 422.
Moustafa, E., and Mortenson, L. E. (1967). *Nature (London)* **216**, 1241.
Münck, E., Rhodes, H., Orme-Johnson, W. H., Davis, L. C., Brill, W. J., and Shah, V. K. (1975). *Biochim. Biophys. Acta* **400**, 32.
Nicholas, D. J. D., Wilson, P. W., Heinen, W., Palmer, G., and Beinert, H. (1962). *Nature (London)* **196**, 433.
Orme-Johnson, W. H. (1973). *Annu. Rev. Biochem.* **42**, 159.
Orme-Johnson, W. H., and Sands, R. H. (1973). *In* "Iron–Sulfur Proteins" (W. Lovenberg, ed.), Vol. 2, p. 195. Academic Press, New York.
Orme-Johnson, W. H., Hamilton, W. D., Ljones, T., Tso, M.-Y. W., Burris, R. H., Shah, V. K., and Brill, W. J. (1972). *Proc. Natl. Acad. Sci. U.S.A.* **69**, 3142.
Palmer, G., Multani, J. S., Cretney, W. C., Zumft, W. G., and Mortenson, L. E. (1972). *Arch. Biochem. Biophys.* **153**, 325.
Postgate, J. R. (1971). *In* "Chemistry and Biochemistry of Nitrogen Fixation" (J. R. Postgate, ed.), p. 161. Plenum, New York.

Que, L., Jr., Holm, R. H., and Mortenson, L. E. (1975). *J. Am. Chem. Soc.* **97**, 463.
Quispel, A. (1974). *In* "Biology of Nitrogen Fixation" (A. Quispel, ed.), p. 1. North-Holland Publ., Amsterdam.
Rivera-Ortiz, J., and Burris, R. H. (1975). *J. Bacteriol.* **123**, 537.
Shah, V. K., and Brill, W. J. (1973). *Biochim. Biophys. Acta* **305**, 445.
Shah, V. K., Davis, L. C., and Brill, W. J. (1972). *Biochim. Biophys. Acta* **256**, 498.
Shah, V. K., Davis, L. C., and Brill, W. J. (1975). *Biochim. Biophys. Acta* **384**, 353.
Shanmugam, K. T., and Valentine, R. C. (1975). *Science* **187**, 919.
Silverstein, R., and Bulen, W. A. (1970). *Biochemistry* **9**, 3809.
Smith, B. E., and Lang, G. (1974). *Biochem. J.* **137**, 169.
Smith, B. E., Lowe, D. J., and Bray, R. C. (1972). *Biochem. J.* **130**, 641.
Smith, B. E., Lowe, D. J., and Bray, R. C. (1973). *Biochem. J.* **135**, 331.
Stombaugh, N. A., Burris, R. H., and Orme-Johnson, W. H. (1973). *J. Biol. Chem.* **248**, 7951.
Stombaugh, N. A., Sundquist, J. E., Burris, R. H., and Orme-Johnson, W. H. (1976). *Biochemistry* **15**, 2633.
Streicher, S. L., and Valentine, R. C. (1973). *Annu. Rev. Biochem.* **42**, 279.
Thorneley, R. N. F. (1974a). *Biochim. Biophys. Acta* **333**, 487.
Thorneley, R. N. F. (1974b). *Biochim. Biophys. Acta* **358**, 247.
Thorneley, R. N. F. (1975). *Biochem. J.* **145**, 391.
Thorneley, R. N. F., and Eady, R. R. (1973). *Biochem. J.* **133**, 405.
Thorneley, R. N. F., and Willison, K. R. (1974). *Biochem. J.* **139**, 211.
Thorneley, R. N. F., Eady, R. R., and Yates, M. G. (1975). *Biochim. Biophys. Acta* **403**, 269.
Tso, M.-Y. W. (1974). *Arch. Microbiol.* **99**, 71.
Tso, M.-Y. W., and Burris, R. H. (1973). *Biochim. Biophys. Acta* **309**, 263.
Tso, M.-Y. W., Ljones, T., and Burris, R. H. (1972). *Biochim. Biophys. Acta* **267**, 600.
Vandecasteele, J. F., and Burris, R. H. (1970). *J. Bacteriol.* **107**, 794.
Walker, G. A., and Mortenson, L. E. (1973). *Biochem. Biophys. Res. Commun.* **53**, 904.
Walker, G. A., and Mortenson, L. E. (1974). *Biochemistry* **13**, 2382.
Walker, M. N., and Mortenson, L. E. (1973). *Biochem. Biophys. Res. Commun.* **54**, 669.
Walker, M. N., and Mortenson, L. E. (1974). *J. Biol. Chem.* **249**, 6356.
Zimmerman, J. K., and Ackers, G. K. (1971). *J. Biol. Chem.* **246**, 1078.
Zimmerman, J. K., and Ackers, G. K. (1974). *Anal. Biochem.* **57**, 578.
Zumft, W. G., and Mortenson, L. E. (1973). *Eur. J. Biochem.* **35**, 401.
Zumft, W. G., and Mortenson, L. E. (1975). *Biochim. Biophys. Acta* **416**, 1.
Zumft, W. G., Palmer, G., and Mortenson, L. E. (1973). *Biochim. Biophys. Acta* **292**, 413.
Zumft, W. G., Mortenson, L. E., and Palmer, G. (1974). *Eur. J. Biochem.* **46**, 525.

CHAPTER 3

Iron–Sulfur Centers of the Mitochondrial Electron Transfer System—Recent Developments

HELMUT BEINERT

I. Introduction.. 61
II. Iron–Sulfur Centers of the Mitochondrial Electron Transfer System 66
    A. The NADH Dehydrogenase System........................... 66
    B. The Succinate Dehydrogenase System........................ 73
    C. The Ubiquinol–Cytochrome c Reductase System.............. 80
    D. Unassigned Fe–S Centers..................................... 82
    E. Related Resonances... 88
III. Relation of Iron–Sulfur Centers to Energy Conservation........... 89
IV. Outlook... 93
    References... 97

I. INTRODUCTION

For decades, the classic spectrophotometric studies on cytochromes by Keilin (1966) have formed the basis for our conception of the intracellular "respiratory chain." Traditionally, in our mind, the cytochromes have been the mainstays of this system. It is therefore hard to accept that they should be outnumbered by other—visually at least—less conspicuous constituents that were unknown just a few years ago. In retrospect, then, it seems indeed surprising that a doubtlessly useful and plausible picture of a respiratory chain could be constructed years ago with more than half the components missing—if one considers types—and with by far the major portion missing—if one considers capacity for electron uptake. The latter situation arises from the proportion of ubiquinone (about 10:1) to the electron carriers of the cytochrome type. In these comparisons pyridine nucleotides were not considered electron carriers in the same sense as ubiquinone or cytochromes. One might even go as far as imagining that with our present knowledge of constituents of the electron transfer system other than the cytochromes, one could construct a cytochromeless respiratory chain that would turn out not to differ in essence from the classic one arrived at with knowledge of cytochromes and flavoproteins only. While such thoughts might be looked upon as heresy or a useless mental exercise, it is nevertheless worthwhile to consider the lesson implicit in this comparison; we may find ourselves in a similar situation concerning other aspects of biochemistry right at this moment, without having the benefit of hindsight that we have in the present example.

This chapter is concerned with those developments of the last five years that have made the cytochromes minority constituents of the mitochondrial electron transfer system, namely the discovery of the unexpected multiplicity and diversity of the mitochondrial iron–sulfur (Fe–S) centers.

Without detailed knowledge of the system it is preferable to designate the structures in question as Fe–S *centers* or *clusters*, as this is what we observe by EPR spectroscopy or what we measure by analysis for iron and labile S.† It is hazardous and possibly even incorrect to designate these structures as Fe–S proteins, because they may occur as intimate parts—maybe on different subunits—of proteins bearing other functional groups as well, proteins which we are used to consider as flavoproteins or hemeproteins.

As pointed out in Volume I of this treatise, the fact that Fe–S proteins were discovered only recently is due partly to their lack of unique optical features. EPR spectroscopy remains the method of choice for their detection and assessment of function. According to recent insights (see Chapter 7, this volume), Fe–S structures may occur in several oxidation states, of which three have thus far been observed in nature, namely the state of oxidized, high potential Fe–S protein (HiPIP), $[n\text{Fe}-n\text{S}]^-$, that of reduced HiPIP, which is equivalent to that of oxidized ferredoxins $[n\text{Fe}-n\text{S}]^{2-}$, and that of reduced ferredoxins $[n\text{Fe}-n\text{S}]^{3-}$. In principle, Fe–S structures may be detected by their paramagnetism in the two limiting oxidation states, namely, $[n\text{Fe}-n\text{S}]^-$ and $[n\text{Fe}-n\text{S}]^{3-}$. It is consequently possible that the same Fe–S center may appear in one or the other state depending on conditions, so that in a reductive titration, for

---

† The terms "iron–sulfur center" and "iron–sulfur cluster" will be used interchangeably in this review, although there may be a preference for the term "center" when specific associations with a dehydrogenase are stressed. These terms refer to the minimal independent structural unit, i.e., a [2Fe–2S] or a [4Fe–4S] type of structure. The term "iron–sulfur group" refers to the single iron-labile sulfide pair, which is not a structural unit but merely one useful in calculation and tabulation.

instance, it may be observed consecutively in both states. It has indeed been found with soluble Fe–S enzymes that both extreme oxidation states were accessible (Erbes et al., 1975), although a 1:1 conversion of Fe–S centers from one state to the other has not yet been found with proteins in a biological milieu. While one would thus expect to detect all Fe–S centers by examining oxidized as well as reduced preparations, this is at best true for isolated Fe–S centers.

As is known for soluble Fe–S flavoproteins, such as xanthine oxidase, Fe–S flavoproteins of the electron transfer system also contain multiple Fe–S centers. It is possible that these centers occur in sufficiently close vicinity, such as those of the 2[4Fe–4S] clostridial ferredoxins, so that interactions between centers can occur (Mathews et al., 1974). Depending on the strength of these interactions, the observed EPR signals may show unusual features that in complex materials are not always readily interpreted, or these signals may become undetectable because of excessive linewidth. There may be some chance of observing signals from interacting centers at very low temperatures ($<4.2°K$), again depending on the type and strength of interaction.

Up to the time that biochemists started practicing EPR spectroscopy at temperatures in the range of $\leq 25°K$, the number of known mitochondrial Fe–S centers was three: those of the NADH and succinate dehydrogenases (Sands and Beinert, 1960; Beinert and Sands, 1960) and that of the cytochrome b–c_1 complex (Complex III), also known as Rieske's Fe–S protein (Rieske et al., 1964). The increased spectroscopic resolution realized at temperatures $\leq 25°K$ has made it clear that not only are there several additional components with Fe–S centers present in mitochondria, but also that the dehydrogenases do contain a number of distinct Fe–S centers (Orme-Johnson et al., 1971a, 1974a; Ohnishi et al., 1970, 1971, 1973; Ohnishi, 1975). While on the whole the characterization and assignment of the multiple centers of the dehydrogenases has been relatively straightforward and conclusive, an assessment of the function and identification of the Fe–S centers of unknown association remains a difficult task.

Figure 1 shows mitochondria or whole heart tissue in three different oxidation states. The preparations and oxidation states were chosen so that the most prominent Fe–S centers can be clearly recognized in at least one of the three spectra. These spectra are supposed to serve only as a preliminary visual introduction to the features to be discussed below. For a more detailed description and discussion of the Fe–S centers of various regions or fragments of the electron transfer system, the subsequent exposition should be consulted. Spectrum (A) shows the EPR spectrum of blowfly mitochondria, which can be obtained with low endogenous sub-

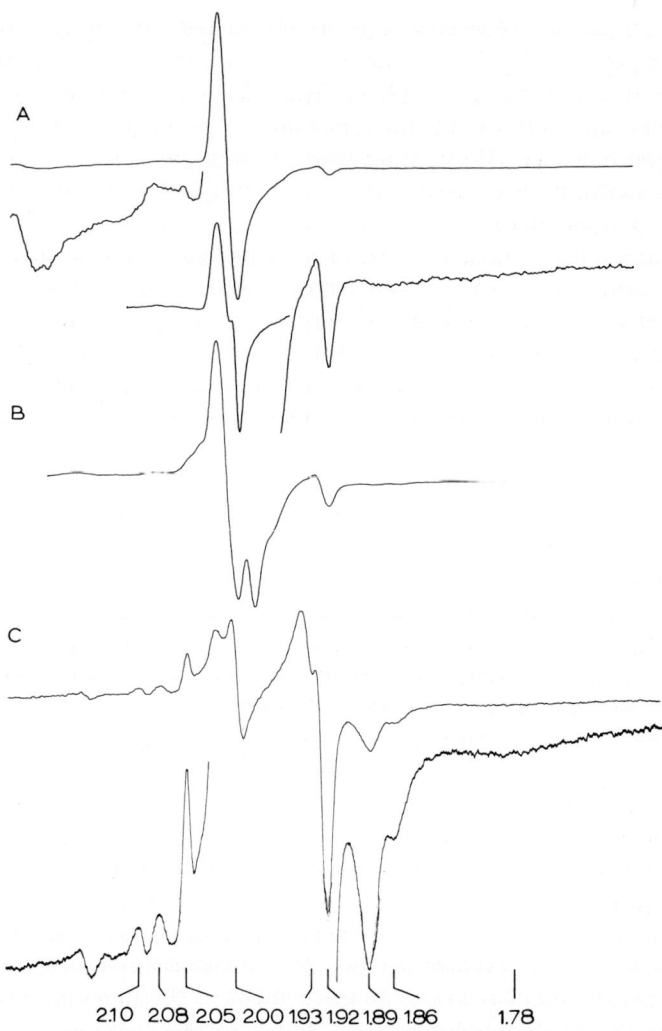

Fig. 1. EPR spectra of mitochondria or whole tissue at different states of reduction. (A) Blowfly mitochondria, approximately 40 mg protein per milliliter, as prepared in 0.15 M KCl, 10 mM tris (pH 7.3), 1 mM EDTA, and 0.5% bovine serum albumin (kindly provided by Dr. B. Sacktor). (B) Beef heart mitochondria, approximately 50 mg/ml, as prepared in 0.25 M sucrose, 0.02 M tris, pH 7.4. (C) Whole pigeon heart frozen immediately after removal of the heart. All EPR spectra shown in this chapter represent the first derivative of the absorption (ordinate) with linearly increasing magnetic field. Unless otherwise mentioned, prominent peaks or shoulders are indicated on a g-factor scale, although this does not mean that the values given are true g values. In many instances they will be very close to actual g values. *Conditions of EPR spectroscopy:* (A) microwave power, 2.7 mW; modulation amplitude, 6.3 G; scanning rate, 400 G/minute; temperature,

3. MITOCHONDRIAL ELECTRON TRANSFER SYSTEM

strate present. Thus, the signal of the high potential type of Fe–S center(s) (Ruzicka and Beinert, 1974; Beinert et al., 1974), centered at $g = 2.01$, is well developed. There is some reduction of center 2 of NADH dehydrogenase ($g = 1.92$) (Orme-Johnson et al., 1971a, 1974a; Ohnishi 1975). This center is rarely seen completely oxidized in mitochondria or submitochondrial particles. Spectrum (B) is that of beef heart mitochondria as isolated. In mitochondria from this source, there appears to be sufficient endogenous substrate that, at the concentration required for EPR spectroscopy, features of partial reduction clearly appear in the spectrum. In this spectrum we can still recognize the signal of the membrane-bound HiPIP Fe–S component. This signal is flanked on either side by a pair of lines (3309 and 3227 G at 9.2 GHz) stemming from the interaction of ubisemiquinone with a paramagnetic species (Ruzicka et al., 1975). The high-field line is more pronounced. Fe–S center 2 of NADH dehydrogenase is more reduced than in (A), and the line typical for reduction of the Fe–S center of Complex III (Rieske's Fe–S center) can just be seen at $g = 1.89$.

Spectrum (C) is that of whole pigeon heart in the reduced state. In this spectrum, the signals of HiPIP and the interacting species of ubisemiquinone are absent, whereas now the resonances of Fe–S centers 1–4 of NADH dehydrogenase ($g = 2.10, 2.05, 2.02, 1.94, 1.92, 1.89, 1.86$), Rieske's Fe–S center ($g = 2.025, 1.89, 1.78$), and of so-called center 5 ($g = 2.08$) (Ohnishi et al., 1972) stand out. Pigeon heart was chosen, because the signal of center 5 is usually very pronounced in preparations from this species.

In the following, the Fe–S centers that could be attributed to known segments of the electron transfer system will be treated first and thereafter those of unknown association or function. Since most, if not all Fe–S centers of mitochondria are bound to membranes, assignment of a certain center to, e.g., a dehydrogenase, may be ambiguous, it may just be the electron acceptor group for this dehydrogenase, and it may then depend on our definition whether we want to consider it as a part of that dehydrogenase or not. Therefore, we will, in the following, speak of associations in broader terms. Unless we have detailed structural information indicating otherwise, we could thus consider the Fe–S center in the

---

$13.3°K$; enlarged wings left and right, modulation amplitude 8 G, 10× amplified; (center inset) 0.09 mW; modulation of 6.3 G; 2.5× amplified. This inset recorded at low power shows that a radical signal is superimposed on the HiPIP signal. (B) Conditions as the main spectrum of A. (C) Conditions as A and B, except microwave power 0.27 mW and scanning rate 200 G/minute. The enlarged wings are 5× amplified. Unless stated otherwise, the modulation frequency was 100 kHz and the microwave frequency 9.2 GHz for all figures of this chapter.

example introduced above as part of that particular dehydrogenase system without specifying to what operational or hypothetical substructure of the system it is attached.

II. IRON–SULFUR CENTERS OF THE MITOCHONDRIAL ELECTRON TRANSFER SYSTEM

A. The NADH Dehydrogenase System

1. Fe–S Centers

Since the studies of Mahler and Elowe (1953, 1954) over twenty years ago, it has been known that preparations of NADH dehydrogenase contained iron. The number of iron atoms per flavin varied with the type of preparation (Lusty *et al.*, 1965; King and Howard, 1967). When the association of labile sulfur with nonheme iron was discovered in the ferredoxins (Fry and San Pietro, 1962), NADH dehydrogenase preparations were also found to contain labile sulfur approximately equivalent in quantity to the iron present (Lusty *et al.*, 1965; King and Howard, 1967). NADH-ubiquinone reductase (Complex I) contains 16–18 Fe-S groups per flavin (Hatefi *et al.*, 1974), whereas in a recently reported soluble preparation of large molecular weight, 28 Fe–S groups per flavin were found (Baugh and King, 1972). Depending on the type of Fe–S centers present, one would then expect to find from 4 to 7 if they were [4Fe–4S] centers or twice as many different Fe–S centers in these types of preparations if they were of the [2Fe–2S] type. No particular significance attaches to the numbering system now in use for Fe–S centers, since they were given numbers in the order in which they were discovered.

By EPR spectroscopy in the temperature range of liquid nitrogen, only center 1 is detectable, so that in older work on the behavior of the "nonheme iron" in NADH dehydrogenase (Beinert *et al.*, 1965), reference is exclusively to this center, which is the one with the lowest apparent oxidation–reduction midpoint potential. In 1970, Ohnishi *et al.* presented spectra of submitochondrial particles from *Candida utilis* in which the signal of center 2, the center of highest midpoint potential, was clearly seen. Although it is a general observation that signals only observable at very low temperatures are not as readily saturated with microwave power as signals detectable also at higher temperatures, this does not apply to the signals of centers 1 and 2. The signal of center 2 is more readily saturated than that of center 1, so that at very low temperatures and high microwave powers the signal of center 1 dominates, as it does at any power at higher temperatures. Whether this is due to heterogeneity, namely, the presence of two species with different saturation behavior, is not clear at this time. There are several lines of independent evidence that the signal of center 1 is composed of approximately equal quantities of two species that differ little in appearance (Albracht and Dooijewaard, 1975) but significantly in their apparent midpoint potentials, as measured potentiometrically in the presence of mediators (Ohnishi, 1975), and in their sensitivity toward prolonged exposure to NADH (Albracht, 1974). Ohnishi (1975) designates the low-potential species as center 1a and the other as 1b; resolution of their potentials was reported for Complex I from beef heart and pigeon heart submitochondrial particles. The question whether there is expected to be in every molecule of NADH dehydrogenase one of each, centers 1a and 1b, has, however, not been answered.

The interpretation of the additional signals observed in reduced NADH dehydrogenase is more ambiguous than that of the signals of centers 1 and 2. With centers 1 and 2, at least, the resonances stand out clearly and qualify as those of Fe–S centers. This cannot be said without qualifications for the additional proposed Fe–S centers. There are definitely spectral features at low ($g = 2.1$) and high ($g = 1.89, 1.86$) field that rapidly respond to NADH addition and are related. To date, the best interpretation remains that they represent two additional Fe–S centers (Orme-Johnson et al., 1974a; Ohnishi, 1975; Albracht and Dooijewaard, 1975), namely, center 3 with relatively high and center 4 with relatively low midpoint potential.† On the basis of this difference, their signals can be separated to some extent by oxidoreductive titration, but only with preparations that exhibit a good signal-to-noise ratio.

At temperatures close to that of liquid helium and at high microwave intensities, additional resonances responding to NADH and apparently related can be observed with Complex I (Fig. 2) or submitochondrial particles (Ohnishi, 1975; Albracht, 1974). They have been attributed to two additional Fe–S centers of low midpoint potential (Ohnishi, 1975). Although these signals may well represent either entities or features separate from those of centers 1–4 described above, the question of whether they are genuine components of the NADH dehydrogenase system or only contaminants or breakdown products is not satisfactorily answered. The fact that these signals are observed in submitochondrial particles makes it likely that they are not due to artifacts of purification, but it gives no clue to their natural association. The failure to observe these signals at 1.5°K at power levels, where the signals of centers 1–4 are well developed (Beinert and Ruzicka, 1975), suggests that

† This applies only to submitochondrial fractions, in mitochondria, no difference was observed (Ohnishi, 1975).

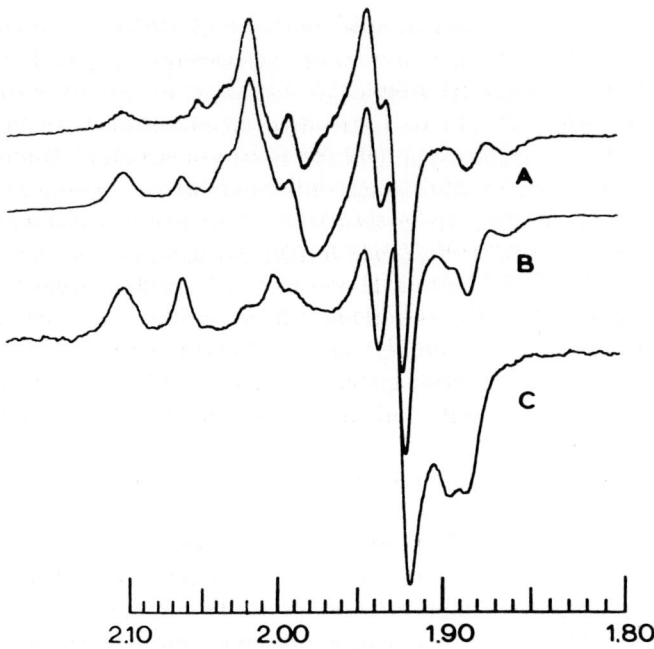

Fig. 2. EPR spectra of NADH-ubiquinone reductase, reduced by NADH at different microwave powers recorded at 4.2°K. Complex I, corresponding to 0.6 nmole of FMN per milligram of protein, was dissolved in 0.66 M sucrose, 0.05 M tris (pH 8), and 1 mM histidine and mixed with 7 mM NADH for 45 seconds at 0°C and then frozen. *Conditions of EPR spectroscopy:* microwave power 0.02 (A), 2 (B) and 207 mW, modulation amplitude 12.5 G; scanning rate 250 G/minute, and amplification ratio 2.5:1.6:1 for A, B and C. The microwave frequency was 9.1 GHz. (With permission from Albracht, 1974.)

these additional signals observed only at highly saturating microwave powers are either instrumental artifacts (see Albracht and Dooijewaard, 1975) or the respective components are not present in quantities sufficient so that there could be any reasonable stoichiometric relationship to the flavin component and the other Fe–S centers of NADH dehydrogenase. In order to accommodate such minority components, one may then have to invoke heterogeneity of the preparations as the explanation. Such an explanation, of course, raises more problems than it answers. Similar considerations apply to the observation that what has been considered as center 1 may in fact be two separate components, called 1a and 1b (Ohnishi, 1975). The midpoint potential of center 1a lies at a value sufficiently low that it is at best partly reduced by NADH, and one may raise questions about its physiological significance.

Despite the call for caution in accepting all of these features as characteristics of equivalent, independent components of the NADH dehydroge-

3. MITOCHONDRIAL ELECTRON TRANSFER SYSTEM

nase system, we must keep an open mind and must prepare ourselves for ever increasing complexity. Up to this point, there is sufficient iron and labile sulfur present to accommodate the number of Fe–S centers postulated even in the most generous estimates, particularly since at this stage there is no information as to whether we are dealing with [2Fe–2S] or [4Fe–4S] centers.

2. Thermodynamic Features

Two types of approaches have been used to assess the redox behavior of the components of complex oxidoreductive systems. Oxidoreductive titration in the presence of potential mediators, in which case the development of optical or EPR features is correlated with electrode potentials (Dutton and Wilson, 1974), and oxidoreductive titration with substrate or chemical reagents, in which case the development of optical or EPR signals is correlated with the quantity of reductant used (Orme-Johnson et al., 1974a,b). The first approach has the advantage that exact midpoint potentials can be derived, but the disadvantage that the total capacity for electron uptake of the system remains unknown. However, the measured potentials must be considered as "apparent" potentials as long as the electrochemical behavior of the system is not understood in detail. The question as to whether at the time of measurement the components of the electron transfer system have reached equilibrium among themselves, with the added mediators and with the electrode, is not readily answered for such a complex system. It is known that on addition of substrates, some components of the electron transfer system come to equilibrium rather slowly, and there is little information on the specificity and kinetics of interaction of the commonly used mediators with the redox carriers and on the influence on such interactions of a membrane environment. It is beyond the scope of this review to consider these aspects in more detail, and the interested reader is referred to recent reviews dealing with these problems (Dutton and Wilson, 1974; Wikström, 1973).

The second approach, namely, direct titration with substrate or chemical reductants or oxidants, has the disadvantage that only the relative sequence of midpoint potentials can be readily determined. But it has the advantage that an estimate of the quantity of the material titrated is obtained, which, in the most unfavorable case, is a maximal value.

A technique that combines advantages of both approaches just discussed is likely to find wider use, namely, the coulometric titration with electrogenerated reductant or oxidant, as it has been successfully applied to purified cytochrome c and cytochrome c oxidase (Mackey et al., 1973).

The apparent midpoint potentials reported for the components of the NADH dehydrogenase system are assembled in Table I. The interpreta-

TABLE I
APPARENT MIDPOINT OXIDATION–REDUCTION POTENTIALS (E_m) OF FE–S CENTERS OF THE MITOCHONDRIAL ELECTRON TRANSFER SYSTEM

Type of preparation	Association	Fe–S center number or name	E_m	pH	n	Reference to E_m values	g factors [peaks]	Reference to position of resonances or g values
Pigeon heart SMP[a]	NADH dehydrogenase	1a	-380 ± 20	7.2	1	Ohnishi, 1975	2.03, 1.94 2.02, 1.94	Ohnishi, 1975 Albracht and Dooijewaard 1975
		1b	-240 ± 20	7.2	1	Ohnishi, 1975	2.03, 1.94, 1.91 2.02, 1.93	Ohnishi, 1975 Albracht and Dooijewaard, 1975
		2	-20 ± 20	7.2	1	Ohnishi, 1975	2.05, 1.93	Ohnishi, 1975; Orme-Johnson et al., 1974a
		3	-240 ± 20	7.2	1	Ohnishi, 1975	[2.10, 1.89, 1.86] 2.10, 1.94, 1.88[b]	Orme-Johnson et al., 1974a Albracht and Dooijewaard, 1975
		4	-410 ± 20	7.2	1	Ohnishi, 1975	[2.11, (?), 1.86] 2.04, 1.93, 1.86[b]	Orme-Johnson et al., 1974a Albracht and Dooijewaard, 1975
		5, 6	-260 ± 20	7.2	1	Ohnishi, 1975	[2.11, 2.06, 2.03] [1.93, 1.90, 1.88]	Ohnishi, 1975 Ohnishi, 1975

Preparation	Enzyme/Complex	Center	E_m (mV)	pH	n	Reference (E_m)	g-values	Reference (g)
Pigeon heart SMP	Succinate dehydrogenase	1	+30	7.2	1	Ohnishi et al., 1972	2.03, 1.93, 1.91	Beinert et al., 1975; Ohnish et al., 1974c
		HiPIP	−60[c]	7.2	≪1	Ohnishi et al., 1972	2.01	Ohnishi et al., 1975
Succinate-Q or cyt. c reductase	Succinate dehydrogenase	1	0 ± 10	7.4	1	Ohnishi et al., 1972	2.03, 1.93, 1.91	Beinert et al., 1975; Ohnish et al., 1974c
		2	−260 ± 15	7.4	1	Ohnishi et al., 1974b	2.03, 1.93, 1.91	Beinert et al., 1975; Ohnish et al., 1974c
		HiPIP	+60 ± 15	7.4	1			
Beef heart SMP	Succinate dehydrogenase	HiPIP	+120	6.0–7.0		Ingledew and Ohnishi, 1975	2.01	
		HiPIP	+80	8.5				
Pigeon heart SMP	b–c_1 complex	Rieske	+280	7.0–7.4	1		2.025, 1.89, 1.81 1.78	Orme-Johnson et al., 1974b
Succinate cyt. c reductase			+280	7.0–7.4	1	Leigh and Erecinska, 1975		
Pigeon heart mitochondria	?	HiPIP (soluble)	>+150	7.2		Ohnishi et al., 1975	2.01	Ruzicka and Beinert, 1974
						Ohnishi et al., 1972	2.02	Ohnishi et al., 1975
Pigeon heart SMP	?	ETF dehydrogenase (center 5)	+40	7.2	1	Ohnishi et al., 1972	2.086 1.94, 1.89	Ruzicka and Beinert, 1975a
Succinate cyt. c reductase			+35	7.4	1	Leigh and Erecinska, 1975		

[a] SMP = submitochondrial particles.
[b] These values obtained in simulations of spectra are considered tentative by the authors.
[c] See, however, Ohnishi et al (1976).

tion and use of these values is complicated, however, by the observation that the midpoint potentials may vary with the type and state of the preparations used. Thus it is necessary always to refer to the preparation used when statements as to the absolute values or relative sequence of potentials are made. Thus, Ohnishi (Ohnishi et al., 1974a) reports a value of $E_{m8.0} = -80$ mV for center 2 in beef heart submitochondrial particles, whereas this value drops to -135 mV in purified Complex I and further to -210 mV for NADH dehydrogenase extracted from Complex I. While this dehydrogenase loses its ability to reconstitute with Complex III an antimycin A-sensitive NADH cytochrome c reductase, the midpoint potential decreases further to -265 mV. With centers 1, 3, and 4, no changes were observed under these conditions. However, when submitochondrial particles are prepared from pigeon heart mitochondria, the potential of center 4, originally equal to that of center 3 is reported to drop to -410 mV (Ohnishi, 1975). This variability of midpoint potentials with the state of preparations explains to some extent discrepancies in values reported from different laboratories before the possibility of changes of this magnitude was appreciated. Thus, with Complex I, centers 2 and 3 were titrated with reductant almost simultaneously, indicating that their midpoint potentials are not widely separated (Orme-Johnson et al., 1974a), whereas, according to Ohnishi (1975), with particles from pigeon heart, the midpoint potentials of these centers at pH 7.2 differ by as much as 220 mV.

3. KINETIC OBSERVATIONS

Reduction of all Fe–S centers in Complex I with NADH is very rapid, so that even at 0°C, rates could not be resolved by the rapid freezing technique (Orme-Johnson et al., 1974a). With acetylpyridine NADH, the rates of appearance of the signals were in the order: center $2 > 3 + 4 > 1$. No clear resolution of centers 3 and 4 could be achieved, but it is very likely that center 4 is not significantly reduced with this substrate of midpoint potential $E_{m7.0} = -248$ mV.

One might have expected, in analogy to what is observed with the mitochondrial electron transfer system as a whole, that the component of lowest midpoint potential is closest to the entry of electrons from the substrate and consequently is reduced first and that of highest potential last. On the contrary, the sequence observed is exactly that found in titrations, suggesting that what we observe is in fact merely a rapid titration. This implies that equilibration between the various Fe–S centers of the NADH dehydrogenase system is very rapid once an electron has entered the molecule, presumably via flavin. Similarly, on reoxidation of reduced Complex I with ubiquinone-1 (Beinert and Ruzicka, 1975), where rates were more readily resolved, the sequence of reoxidation is the reverse, namely, center $1 > 3 + 4 > 2$, supporting the conclusion just drawn.

It had been reported that only the signals of reduced centers 2 and 3 appeared, but not that of center 1 (and presumably also not that of center 4) when Complex I was reduced with NADPH free of significant amounts of NAD or NADH (Hatefi and Hanstein, 1973). This result was interpreted to indicate that center 1 was only involved in the dehydrogenation of NADH but not of NADPH. This observation has additional significance, since phosphorylation accompanies oxidation of NADPH in particles, with an efficiency equal to that of NADH oxidation. Therefore, if center 1 did not participate in the dehydrogenation of NADPH, this center of lowest midpoint potential would not be expected to be involved in the energy conservation mechanism of the NADH dehydrogenase system. Logically, in considerations of energy conservation mechanisms, attention was focussed on the centers of lowest (center 1) and highest (center 2) midpoint potentials (Ohnishi, 1973; Gutman et al., 1975). More recent work (Hatefi and Bearden, 1976), however, has shown that partial reduction of center 1 of Complex I can be observed with NADPH as substrate, indicating that the pathway of electrons from NADPH is not different from that found with NADH.

B. The Succinate Dehydrogenase System

1. Fe–S Centers

The developments concerning the Fe–S character of succinate dehydrogenase were similar to those described for NADH dehydrogenase (Zeylemaker et al., 1965; King, 1964), and as in that case, the number of Fe–S groups present varies with the type of preparation and to a more limited extent also with individual preparations. Approximately four to eight atoms each of iron and labile sulfur may be present per enzyme-bound flavin. However, experience in the analysis of such preparations teaches that the ratios of flavin to iron of 1:4 and 1:8, which are generally held to be the ideal ones, are not always those found in reality.

The situation with succinate dehydrogenase is somewhat simplified in that Fe–S center 1, namely, that which had been observed fifteen years ago by EPR spectroscopy at liquid nitrogen temperature (Beinert and Sands, 1960) is also the one unambiguously involved in enzyme activity. It had long been noticed that reduction by dithionite led to an EPR signal shape of the Fe–S component(s) different from that seen after addition of succinate and to a somewhat increased intensity (Fig. 3). Although this might have been explained by a slight change of the Fe–S

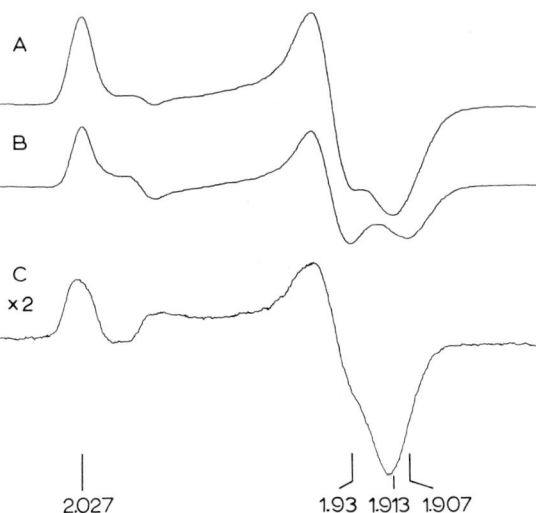

Fig. 3. EPR spectra of succinate–ubiquinone reductase (Complex II) reduced with an excess of dithionite (A) or succinate (B), respectively, and the difference spectrum (C) amplified 2×. Complex II, 5 mg protein per 0.5 ml of 50 mM Hepes buffer of pH 7.8 was anaerobically reduced with 170 nanoequivalents of dithionite or with 62 µmoles of succinate. *Conditions of EPR spectroscopy:* microwave power, 9 µW; modulation amplitude 8 G; scanning rate 200 G/minute, temperature 13.3°K. (With permission from Beinert *et al.*, 1975.)

structure caused by dithionate, there is now strong evidence that an additional Fe–S center is reduced by dithionite over and above that reduced by succinate:

1. With limited amounts of dithionite, initially only the "succinate type" of signal (center 1) is seen (Beinert *et al.*, 1974, 1975), indicating that dithionite does not modify center 1 but that, indeed, a second center is reduced.

2. In line with this are the oxidation–reduction midpoint potentials that were measured, namely, $E_{m7.4} = 0$ mV for center 1 (succinate reducible) and −260 mV for center 2 of Complex II (Ohnishi *et al.*, 1973).

As center 2 can only be observed at oxidation states of the enzyme when center 1 is completely reduced, one can experimentally only obtain a quantitative estimate of center 1 and center 1 + 2, never of center 2 by itself. If the quantity of center 2 present is derived by subtraction of the value found for center 1 from that observed with center 1 + 2, the concentrations of center 2 invariably fall significantly below those for center 1, indicating either that this approach is not valid or that only a fraction of molecules contain this second center. This was observed with a number of different preparations (Beinert *et al.*, 1974, 1975). It must be considered that interaction between centers 1 and 2 may diminish the combined signal intensity when both centers are reduced and are therefore paramagnetic. Signs of an interaction have in fact been reported by Ohnishi et al. (1974c), but those are only apparent at particular states of certain types of preparations, whereas the apparently low concentration of center 2 has been seen in preparations, such as Complex II, which do not show the reported effect (namely, extra lines) for interaction between different centers. Furthermore, since center 2 is not reduced by succinate, its catalytic significance is questionable. The status of center 2 is, therefore, at present poorly defined, and considerations arising in the subsequent discussion on yet a third Fe–S type signal in succinate dehydrogenase only add to the uncertainty as to the significance of center 2.

Ever since preparations of the electron transfer system of mitochondria and also of similar systems in lower organisms were studied by EPR spectroscopy at temperatures in the range of $\leq 20°K$, strong, relatively symmetrical, and sharp ($\Delta H \sim 25$ G) signals at $g = 2.01$ have been observed in the oxidized state (Orme-Johnson et al., 1971a; Ohnishi et al., 1970). On fractionation of mitochondria into subfractions material with this type of signal was mainly recovered in the soluble supernatant fraction (see below) and in Complex II (Ruzicka and Beinert, 1974; Beinert et al., 1974). As will be discussed in more detail below, it could be shown with the soluble material after purification that this type of signal originated from a Fe–S protein. Involvement of iron in the structure producing this type of signal was independently shown by incorporation of ^{57}Fe and ensuing line broadening by unresolved hyperfine structure (DerVartanian et al., 1974). It is known that Fe–S proteins can be paramagnetic in their oxidized forms, analogous to bacterial HiPIP (Dus et al., 1967), when their Fe–S cluster has lost an additional electron as compared to the oxidation state of an oxidized ferredoxin (Carter et al., 1972). By analogy to the situation with the soluble HiPIP-type protein from mitochondria, it became likely that Complex II contains, in addition to ferredoxin type Fe–S centers ($g = 1.94$, when reduced), a HiPIP-type center. Quantitative estimates, assuming an effective spin of $\frac{1}{2}$ for the HiPIP signal, gave values close to those found for center 1 and the bound flavin of Complex II. On titration with dithionite, the HiPIP-type center in Complex II is reduced concomitant with the ferredoxin type center 1. According to the apparent midpoint potentials reported for the HiPIP component, namely, $E_{m7.4} = +60$ mV in a particulate preparation of succinate–cytochrome c reductase (Ohnishi et al., 1973) and $E_{m6-7} = +120$ mV in submitochondrial particles from beef heart (Ohnishi et al., 1972), one would have expected to observe a clear separation of the reduction of center 1 and

HiPIP. Relationships of the HiPIP-type center to enzymatic activities will be discussed below.

Independent evidence for the association of the HiPIP component with the succinate dehydrogenase system was obtained from experiments with mutants of *Saccharomyces cerevisiae* deficient in succinate dehydrogenase (De Kok *et al.*, 1975). The signal of the HiPIP component in submitochondrial particles from two deficient mutants had only 10–25% of the intensity observed in particles from the wild type.

Since the oxidation state of a HiPIP-type center is two electron equivalents above that for a reduced ferredoxin, which is apparent from its EPR signal at $g = 1.94$, it must be considered whether the HiPIP center of Complex II might not be identical to one of the ferredoxin-type centers, simply representing this center in a different oxidation state. The fact, however, that on rapid reduction by succinate, center 1 of Complex II may become $\sim 50\%$ reduced while the HiPIP center changes very little in intensity seems to rule out the possibility that reduced center 1 is the $2e^-$ reduction product of the HiPIP-type center (Beinert *et al.*, 1974, 1975). There is no evidence to date, however, that would eliminate this possibility for center 2. It is not clear, therefore, whether Complex II has two or three Fe–S centers, and the designation of the HiPIP-type center as center 3 (Ohnishi *et al.*, 1974b) must remain tentative. Because of this complication and the uncertainty concerning the stoichiometry of center 2 versus bound flavin, it is also not possible to draw any conclusions as to whether Complex II or other succinate dehydrogenase preparations have Fe–S centers made up of [2Fe–2S] or [4Fe–4S] clusters. Although it might be expected that preparations containing approximately four Fe atoms per bound flavin would allow a decision to be made, the analytical data available on such succinate dehydrogenase preparations are not of the required accuracy.

Note Added in Proof: Experimental evidence is accumulating that Fe–S center 2 of succinate dehydrogenase, the center only detectable after reduction with dithionite, does indeed occur at a higher concentration than indicated by the spectral difference of dithionite and succinate reduced enzyme. The reason is that there is interaction between centers 1 and 2, when both are reduced, which interferes with the detectability of both centers. It is, therefore, not valid to calculate the concentration of center 2 by substracting the integrated intensity of center 1 from that of center $1 + 2$, as indicated in the text. For the same reason, the difference spectrum of Fig. 3C cannot be considered as the true spectrum of center 2. If on this basis we have reason to assume that center 2 occurs in amounts approximately stoichiometric to the bound flavin of succinate dehydrogenase, the fact that both centers 1 and 2 are observed in preparations containing 4 to 6 Fe–S groups per flavin can now be taken as evidence that the HIPIP type center is indeed a separate entity, i.e., not the superoxidized form of center 2, and that it is the HIPIP type center in any oxidation state which is absent from preparations of low Fe–S content.

2. Thermodynamic Features

Values for the midpoint potentials measured in various preparations for centers 1 and 2 and the HiPIP center are incorporated in Table I. Again, the potential depends on the type and state of the preparation studied. In submitochondrial particles from *S. cerevisiae*, where signals from other Fe–S centers do not seriously interfere with the measurement of center 2, an $E_{m7.2}$ value of -245 ± 15 mV was found (Ohnishi *et al.*, 1973) similar to that reported for Complex II from beef heart. However, in soluble succinate dehydrogenase preparations from beef heart, this value decreased to -430 mV (Ohnishi *et al.*, 1973). No changes in the midpoint potential of center 1 of succinate dehydrogenase were observed under similar conditions. Figure 4 shows titrations of three different succinate dehydrogenase preparations with dithionite monitored by EPR spectroscopy.

3. Observations Relating to Enzymatic Activity

The assay for enzymatic activity of the originally membrane bound dehydrogenases of the mitochondrial electron transfer system has posed numerous difficulties and caused controversies in the past. Unfortunately, many of these problems are unresolved and may be insoluble. They are a consequence of the fact that although we know the substrate molecule for the respective enzymes, we have removed the natural acceptor and are, in most instances, not even certain what the acceptor was. Assays with artificial electron acceptors are fairly unambiguous if the enzyme assayed has one active electron-carrying constituent. It is obvious, however, that with the multitude of components shown to be present in such enzymes as NADH or succinate dehydrogenase, a definition as to what reaction we are actually measuring with an artificial electron acceptor becomes increasingly difficult if not impossible.

At this time, succinate dehydrogenase probably represents the most complex situation encountered. It may therefore be useful to discuss this example in more detail. The difficulties which are superimposed in this case are:

1. Turnover is measured with artificial acceptors such as phenazine methosulfate or ferricyanide, and it is not certain whether electrons are transferred to these acceptors from a single or from multiple sites.

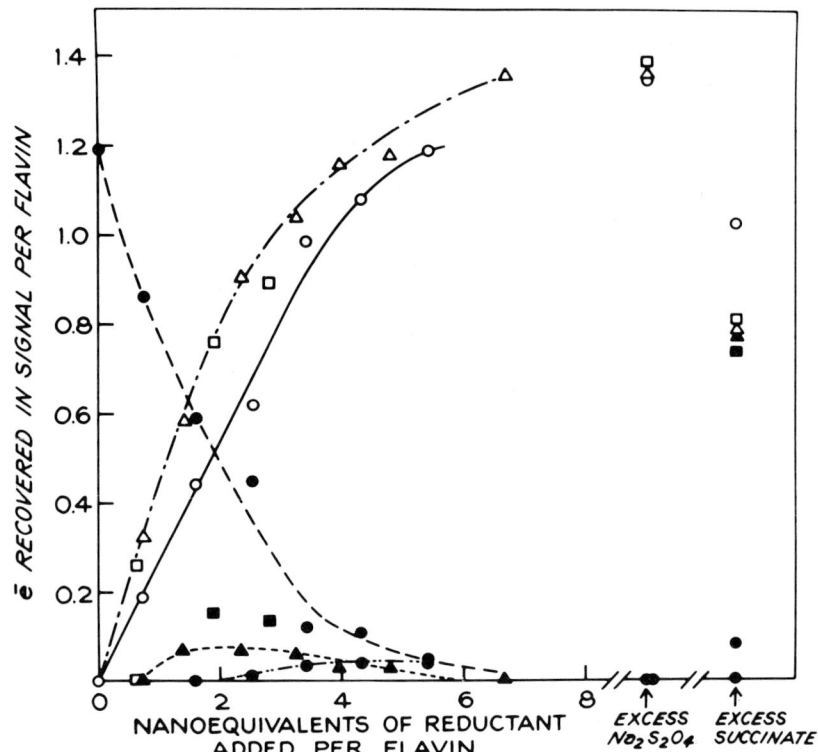

Fig. 4. Anaerobic titration of succinate dehydrogenase preparations with dithionite. The ordinate shows the number of unpaired electrons recovered in the respective signals, and the abscissa shows the number of reducing equivalents added, both in relation to bound flavin. The preparations used were Complex II (Baginsky and Hatefi, 1971) and soluble succinate dehydrogenases of Coles et al. (1972) and Davis and Hatefi (1971). The symbols represent the following signals of these preparations in the above order: $g = 1.94$ (○, □, △); radical (--●--, ■, ▲); and HiPIP —●—, which was only present in Complex II. (With permission from Beinert et al., 1975.)

2. On solubilization and purification, the enzyme is obtained in a state of at least partial "deactivation," so that assays on such material directly provide at best a minimal value for activity. Originally, the enzyme could only be activated by substrate and inhibitors such as malonate (Kearney, 1957). The presence of these substances, however, makes treated preparations unsuitable for EPR spectroscopic activity assays. It is now known that the enzyme can also be activated by univalent ions such as bromide or nitrate (Kearney et al., 1974; Ackrell et al., 1974). "Activation" of soluble preparations by any activators, however, refers to the activity measured by reduction of artificial ac-

ceptors mentioned in 1, above, and we do not know what the manipulations involved in this "activation" may do to the Fe–S centers, whose reactivity we measure by EPR spectroscopy. This is a legitimate question to ask, since preparations activated to between 80 and 100% showed only partial reactivity, in many instances considerably below 80%, of their Fe–S components. It should be added, though, that with particulate preparations, such as submitochondrial particles, activation is also necessary for full succinate dehydrogenase activity, and in this instance, oxygen uptake can be used as the ultimate measure of the entire succinate oxidase system.

3. In order to circumvent the ambiguities arising from the use of artificial electron acceptors, the tactic has been introduced to extract or otherwise eliminate the activity that is to be assayed for from the electron transfer system and then to reincorporate the solubilized enzyme into the deficient electron transfer system and assay the restored activity (Yu et al., 1974). In the case of succinate dehydrogenase, succinate oxidase, or succinate cytochrome c reductase activity would then be assayed for. Thus, it is hoped that an activity more akin to that originally innate in succinate dehydrogenase is being measured. Although this approach probably eliminates objections and uncertainties introduced by the use of artificial electron acceptors, it creates new problems (Singer, 1966) in that the activity now being measured may include additional requirements, possibly artificially created by the process of extraction and reincorporation and not necessarily part of the activity to be measured. In addition, in every instance where such assays were feasible, the activity proved to be very unstable.

It is obvious that, faced with the complications just discussed, it is at present impossible to arrive at a simple and unambiguous picture concerning the involvement of the various components of succinate dehydrogenase in its catalytic function. Rapid reaction studies by EPR spectroscopy showed that even in fully "activated" preparations, only a fraction of center 1 was rapidly reduced by succinate (Beinert et al., 1974, 1975) or rapidly reoxidized by a ubiquinone analogue, well within the overall turnover rate of the enzyme (Singer et al., 1975). The remainder, however, reacted at rates 2–4 orders of magnitude slower. The same applies to the HiPIP component. Apparently then, although all molecules are activated according to the usual criteria and show the usual catalytic activity with artificial acceptors, they are not all active with respect to their Fe–S components. On the other hand, activity in the succinoxidase system after reconstitution was reported to go parallel with the presence (after oxidation) of the EPR signal of the oxidized HiPIP component in the soluble preparation used for reconstitution. This has been interpreted to mean that this component is a necessary ingredient in this reconstitutional activity (Ohnishi et al., 1974b). This is, however, only one possibility, and it could equally well be that the detectability of HiPIP in preparations active in this assay may at best be an indication that they are in such condition as to be suitable for reconstitution without implying that an EPR-detectable HiPIP component is required for this. Thus, one emerges out of these considerations with the conclusions (1) that most, if not all succinate dehydrogenase preparations are not homogeneous with respect to their Fe–S centers, and (2) that there are strong indications that the HiPIP Fe–S center is a catalytically active component of succinate dehydrogenase itself, although this is not expressed in the activity measured by artificial electron acceptors.

C. The Ubiquinol–Cytochrome c Reductase System

Fe–S Centers

The only type of Fe–S center that with certainty is part of this system is that generally known as Rieske's Fe–S protein (Rieske et al., 1964), with g values at 2.025, 1.89, and 1.81 or 1.78. In pigeon heart mitochondria and submitochondrial particles, a midpoint potential $E_{m7.0} = 280$ mV was measured (Leigh and Erecínska, 1975). In a sense, this protein is therefore a high-potential Fe–S protein, although it only exhibits an EPR signal in its reduced state, typical of reduced ferredoxins. According to the reported analyses for iron and labile sulfur, there is one [2Fe–2S] center per cytochrome c_1 and per every two molecules of b cytochromes. The signal of this center is relatively broad, but it is readily detected at 80°K. A change in signal shape has been observed during reduction of Complex III (Orme-Johnson et al., 1971b, 1974b). Without a significant concomitant change in intensity, the signal broadens on extensive reduction of Complex III. This is particularly evident from a shift of the high-field line from $g = 1.81$ to $g = 1.78$. This shift has also been observed in whole tissue (Orme-Johnson et al., 1974b). Judging from the events during reductive titrations, it appears that this shift occurs on reduction of cytochrome b_{562}, indicating that the Fe–S center in some way senses the oxidation state of the associated cytochrome. A protein containing this center has been solubilized by succinylation (Rieske, 1967). No activity has been reported for this solubilized material.

Kinetically, this Fe–S center responds to reductants in a fashion very similar to that of its associated cytochromes (Orme-Johnson et al., 1974b). Its reduction is inhibited by antimycin A, not its reoxidation

(Rieske et al., 1964; Lee and Slater, 1974). An interesting observation is that the signal of this Fe–S center disappears very rapidly on exposure of reduced mitochondria to oxygen pulses (Leigh and Chance, 1974). It has been concluded from studies on kinetics and stoichiometry of fast proton translocation linked to aerobic oxidation of the electron transfer system that electron flow from ubiquinol to cytochrome c is coupled to vectorial proton translocation and that the oxidation–reduction of at least one respiratory carrier is linked to proton translocation across the membrane (Papa et al., 1975). It has been suggested that this carrier may be the Rieske Fe–S center. If this were indeed the case, one would expect that the midpoint potential of this Fe–S center show a dependence on pH. However, in a recent investigation of the pH dependence of the Rieske-type Fe–S center in pigeon heart mitochondria and chromatophores of *Rhodopseudomonas spheroides*, no significant pH dependence was found (Prince et al., 1975). [*Note Added in Proof:* See, however, Prince and Dutton (1976).]

There are a number of other Fe–S centers that have been found associated with Complex III. In some instances, these are likely to be contaminants originating from other complexes, and in others, their quantitative significance or functional relationship to Complex III have remained obscure. A HiPIP type of signal of the same shape as that occurring in Complex II is almost always associated with Complex III, as are signals at $g = 1.94$ after reduction. Since succinate dehydrogenase activity is often found in Complex III and since these signals respond to the addition of succinate, it is likely that at least the components represented in these signals are not genuine constituents of Complex III. A quantitative determination of the components represented in these signals also shows that they are present at a concentration $<10\%$ of those of the recognized components of Complex III. Occasionally, a resonance at $g = 2.08$ is seen in preparations of Complex III. The component represented in this signal [called center 5 (Ohnishi et al., 1972)] has a midpoint potential of $E_{m7.2} = 40$ mV and would thus be expected to operate in the same region of electron transport as does Complex III. This component will be considered in more detail below. Since it is not present in most preparations of Complex III, it becomes difficult to decide whether this material is a contaminant of Complex III or a component closely associated with Complex III *in situ* that may be lost on purification of Complex III.

There has been a report of an additional Fe–S center of Complex III that can be detected at $g = 1.867$ at $<10°$K after reduction by succinate plus ascorbate in the presence of phenazine ethosulfate (Lee and Slater, 1974). It undergoes oxidation and reduction parallel with the Rieske Fe–S center, and its reduction is also inhibited by antimycin. After extraction of ubiquinone, the signal is not observed. Again, the question of the quantitative relationships must be considered here. As mentioned above, there is sufficient iron and labile sulfur in Complex III to allow the presence of one [2Fe–2S] center per cytochrome c_1. If one believes in the existence of any stoichiometric relationships among the electron acceptors of the individual complexes, there is thus little room for additional Fe–S centers, unless we are faced once more with heterogeneity of the preparations.

D. Unassigned Fe–S Centers

In this section we will discuss Fe–S centers that are well recognized as belonging to this class of compounds, but the precise associations or function of which is not certain. To date there are three such centers, namely, the so-called center 5 (Ohnishi et al., 1972), the soluble mitochondrial HiPIP center (Ruzicka and Beinert, 1974), and the ferredoxin-type center apparently located in the outer mitochondrial membrane (Bäckström et al., 1973).

1. ETF Dehydrogenase (Center 5)

This ferredoxin-type center is particularly evident through its low-field resonance at $g = 2.086$, which, in reduced preparations of whole heart mitochondria or sonicated particles (see Fig. 1C), is almost invariably seen side by side with the $g = 2.1$ resonance of centers 3 and 4 of NADH dehydrogenase. It is particularly strong in preparations from pigeon heart and weak but detectable in beef heart preparations. The designation "center 5" was given at a time, when it was thought to be part of NADH dehydrogenase (Ohnishi et al., 1972). It is, however, neither part of the NADH (DerVartanian et al., 1973; Ohnishi et al., 1974a) nor of the succinate dehydrogenase pathway as had been surmised in some schemes (Ohnishi, 1973). The designation "center 5" is therefore meaningless and should be abandoned.

A protein that after reduction with dithionite exhibits the resonances of this center, as they are recognized in particles ($g = 2.086$, $g = 1.89$, Fig. 5), has been purified and identified as an Fe–S flavoprotein with one Fe–S center per FAD (Ruzicka and Beinert, 1975a). Although analyses available at present seem to favor a [4Fe–4S] center, further purification is required to establish this. The protein did not react with any of a number of substrates of low molecular weight, including glycerol 3-phosphate, dihydroorotate, and choline. It was, however, reduced within 100 msec by the reduced electron-transferring flavoprotein (ETF) of the fatty acyl-CoA dehydrogenation pathway (Crane and Beinert, 1956). Fatty

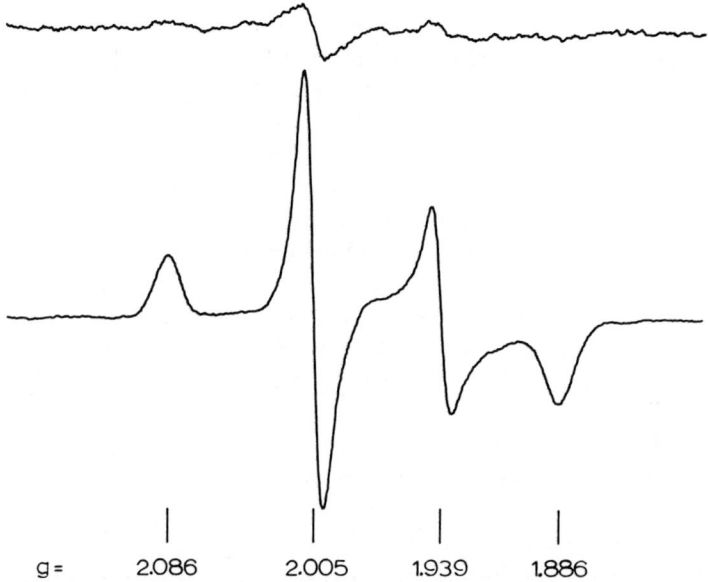

Fig. 5. EPR spectrum of Fe–S flavoprotein from beef heart mitochondria. The protein, 5 mg/ml, was dissolved in 10 mM tris chloride (pH 8). Upper trace, protein as isolated; lower trace, reduced with solid dithionite. *Conditions of EPR spectroscopy:* microwave power, 0.27 mW; modulation amplitude, 8 G; scanning rate, 400 G/minute; temperature, 13.3°K. The radical signal at $g = 2$ is strongly saturated under these conditions. (From Ruzicka and Beinert, 1975a.)

acyl-CoA alone or acyl-CoA plus acyl-CoA dehydrogenase in the absence of ETF did not reduce the Fe–S flavoprotein. In order to learn whether reduced ETF was able to reduce any number of subfragments of the electron transfer system, experiments were done in which the effect of reduced ETF on Complexes I, II, or III was tested. Complexes I and II were not reduced, and Complex III only within seconds. Thus, it appears that the rapid reduction of the Fe–S flavoprotein by reduced ETF is not an unspecific reaction and that the new Fe–S flavoprotein may indeed represent the entrance port of the fatty acyl-CoA dehydrogenation pathway into the electron transfer system. This suggestion for the role of the new Fe–S flavoprotein was supported by EPR studies on brown adipose tissue from guinea pigs. This tissue preferentially utilizes fatty acids as substrate. When compared to the signals of NADH dehydrogenase, the resonance at $g = 2.086$, characteristic of the Fe–S flavoprotein, was four fold higher in brown adipose tissue than in heart (Flatmark et al., 1976). One would have thought that the link to the electron transport system might be accomplished by ETF alone; however, only low rates of reduction of cytochromes in submitochondrial particles were observed when reduced ETF was generated by fatty acyl-CoA and fatty acyl-CoA dehydrogenase (Beinert, 1963). One may, therefore, speculate that the Fe–S feature is required, in addition to flavin, for efficient interaction with the electron transfer system; we may recall here that the two well-documented substrate-linked dehydrogenases of this system are also Fe–S flavoproteins. The new Fe–S flavoprotein may therefore be called more meaningfully "ETF dehydrogenase" rather than "center 5."

2. SOLUBLE HIGH-POTENTIAL FE–S PROTEIN

As pointed out above, EPR spectroscopy at $<20°K$ showed that temperature sensitive signals centered at $g = 2.01$ were rather ubiquitous in biological materials of mammalian and microbial origin. They were generally associated with membranes, and because of the presence of other Fe–S compounds, it was not possible to ascertain whether these signals originated from iron compounds. Evidence indicating that this type of signal was indeed due to iron compounds was presented by DerVartanian et al. (1974), who found broadening of the signal in *Azotobacter* particles after incorporation of ^{57}Fe. It was observed, however, that a signal of the same type was present in the supernatant fraction obtained on sound treatment of beef heart mitochondria (Ruzicka and Beinert, 1974). This made possible purification of a protein, free of ferredoxin-type Fe–S centers, that exhibited the signal observed in the supernatant fraction. This protein contained equivalent quantities of iron and labile sulfur. Although it was of high purity according to criteria of protein chemistry, it apparently did contain approximately equivalent quantities of two closely related species according to properties of its EPR signal (Fig. 6). Also, the ratios of gram atoms of iron per mole of protein (2:1) of molecular weight of 97,000 and of iron atoms per unpaired spins represented in the signal (3:1) are not readily interpretable at this time, except on the assumption of heterogeneity of metal centers.

It should be emphasizd that it is only on the basis of the analytical work on this soluble protein that the Fe–S character of the structures represented in other similar resonances—such as that in Complex II—has been inferred by analogy. It is not that there is much doubt about such inference—after all, similar inferences concerning complex proteins with ferredoxin-type signals were made on the basis of our knowledge of soluble ferredoxins—but it is nevertheless important at this stage to keep this fact in mind.

Obviously, the question was raised whether there might be any relationship between the soluble HiPIP species and the membrane-bound one, which is part of the succinate dehydrogenase system. One might have

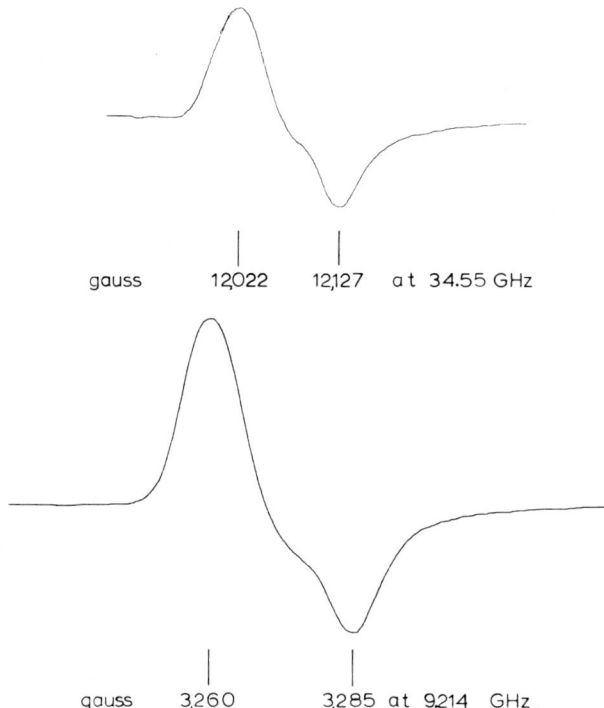

Fig. 6. EPR spectra of the oxidized form of the purified, soluble HiPIP from beef heart mitochondria at 34.6 and 9.2 GHz. Protein, 15 mg/ml, was dissolved in 0.05 M tris chloride (pH 7.4). *Conditions of EPR spectroscopy* at 34.55 and 9.214 GHz, respectively: microwave power, 10 mW and 10 μW; modulation amplitude, 3.2 and 4 G; scanning rate, 400 and 100 G/minute; temperature 13°K.

thought that the soluble protein arose from the membrane-bound one on sonication of mitochondria. However, the molecular weight of the soluble protein as compared to the components of Complex II made this very unlikely from the beginning (Ruzicka and Beinert, 1974). So did the finding that the HiPIP component of succinate dehydrogenase occurred at an approximate 1:1 stoichiometry to the bound flavin (Beinert et al., 1974, 1975). There was thus little room for HiPIP lost on fractionation of mitochondria. Additional evidence for the existence of two separate HiPIP species was presented recently derived from EPR spectroscopy of whole mitochondria, where the signals of both the soluble as well as the membrane-bound HiPIP appear to be present (Ohnishi et al., 1975, 1976).

No activity of the soluble HiPIP has yet been found. It is not reduced by a number of low molecular weight substrates, including glycerol 3-phosphate, and it is not reduced by reduced ETF. No ferredoxin-type

EPR signals were detected when the soluble HiPIP as reduced with dithionite in the presence of methylviologen, indicating that the $2e^-$ reduced state of this structure is not readily accessible. The characterization work on this protein as well as tests for a possible function have been hampered by its instability and apparent heterogeneity, which may be a consequence of this instability. While a fraction of the protein is definitely readily autoxidizable, part of it is not that readily oxidized, so that the protein is often obtained in a partially reduced state when protective agents such as dithiothreitol have been used in the preparation. One may then have to resort to oxidation by ferricyanide, but with the risk that losses of the EPR detectable form of the protein may occur.

3. Fe–S Protein of the Outer Mitochondrial Membrane

The origin of this protein, first observed in a rat kidney cortex microsomal fraction by its characteristic EPR spectrum (g = 2.01, 1.94, 1.89),

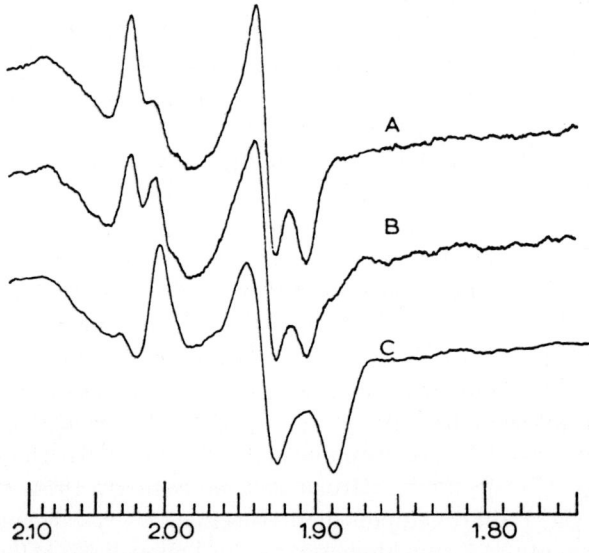

Fig. 7. Comparison of EPR spectra of mixtures of inner and outer membranes from beef heart mitochondria. The preparations were suspended in 0.25 M sucrose and 20 mM tris chloride (pH 7.5). (A) Succinate (25 mM) was added to a preparation of particles partially purified by free-flowing electrophoresis. The suspension was frozen after 3 minutes at 20°C. (B) The tube used for A was thawed, 0.15 mM phenazine methosulfate was added, and the suspension was frozen after 3 minutes at 20°C. (C) A crude preparation of mitochondrial outer membranes was treated as under B. *Conditions of EPR spectroscopy:* microwave power, 130 mW; modulation amplitude, 12.5 G; scanning rate, 250 G/minute; temperature 83°K; microwave frequency, 9.1 GHz. The amplification for A and B is the same. (With permission from Albracht and Heidrich, 1975.)

could be traced to the outer mitochondrial membrane of rat liver mitochondria (Bäckström et al., 1973). In particles, it is reduced by NADH or NADPH, and this reduction is not sensitive to rotenone. According to the number of unpaired spins represented in the signal and analyses for iron and labile sulfur, the protein has a [2Fe–2S] center. The concentration in the outer membrane of rat liver mitochondria is given at 0.5 nmoles per milligram of protein. What apparently is the same component was also found in submitochondrial particles from beef heart (Albracht, 1974). A careful comparison of the spectra observed on reduction with succinate in the presence or absence of phenazine methosulfate (PMS) with that seen after addition of NADH indicated that in addition to the known Fe–S centers of NADH and succinate dehydrogenases, a component was reduced by NADH or succinate plus PMS that had properties of the outer membrane protein previously described by Bäckström et al. (1973). That this component was indeed the outer membrane protein was then conclusively shown by separation of outer and inner membrane fragments by free-flowing electrophoresis (Albracht and Heidrich, 1975). The spectra obtained from various fractions are shown in Figs. 7 and 8.

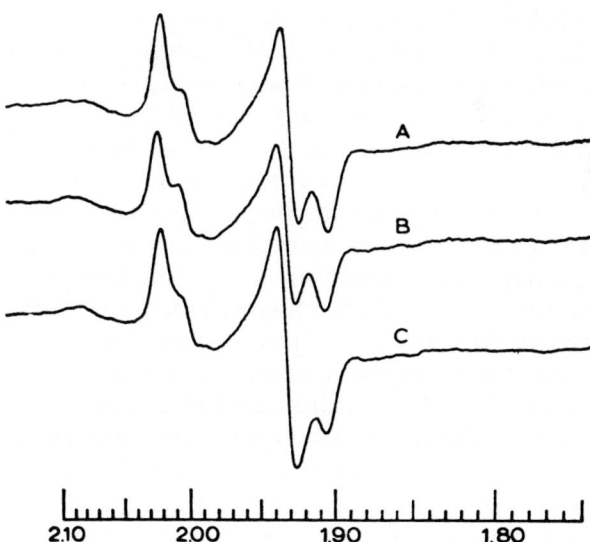

Fig. 8. EPR spectra of pure particles separated by free-flowing electrophoresis. The particles were suspended as for Fig. 7. Substrate was added, and the mixtures were frozen after 3 minutes at 20°C, (A) with 25 mM succinate, (B) with 25 mM succinate plus 0.15 mM phenazine methosulfate, (C) with 6 mM NADH. The conditions of EPR spectroscopy were those of Fig. 7, and the amplification was unchanged throughout. The absence of the Fe–S center of the outer membrane in the purified preparation can be clearly seen in comparison to Fig. 7. (With permission from Albracht and Heidrich, 1975.)

Fig. 9. EPR spectrum of partly purified Fe–S protein from the outer membrane. The protein, 30 mg/ml, was suspended in 0.02 M tris chloride (pH 8) containing 0.5% cholate and was reduced by solid dithionite. *Conditions of EPR spectroscopy:* microwave power, 0.09 mW; modulation amplitude, 8 G; scanning rate, 400 G/minute; temperature 25°K. (The protein was kindly provided by Dr. F. J. Ruzicka.)

The concentration of the outer membrane Fe–S protein in submitochondrial particles from beef heart is ~0.3 nmoles per milligram of protein. The authors also compared the EPR spectra of particles prepared from beef heart mitochondria by the method of Löw and Vallin (1963) and of mitochondria themselves, both after reduction by NADH in the presence of glutamate and malate. The somewhat unexpected result was that the particles, which are generally thought to contain largely inner membrane fragments, contained as much of the outer membrane Fe–S protein as did whole mitochondria. The interpretation of many EPR results obtained on particles through the last years may therefore require revision. It is true, however, that the EPR signal of the outer membrane protein is relatively readily saturated, so that the error introduced when measurements were made <20°K is less serious. A spectrum of a purified fraction containing the protein is shown in Fig. 9. There is at this time no information on the function of this Fe–S protein, other than that in particles it is reduced by pyridine nucleotides, a property that is, however, lost on further purification (D. Bäckström, personal communication).

E. Related Resonances

1. INTERACTION OF UBISEMIQUINONE WITH A
    PARAMAGNETIC COMPONENT

Although not entirely appropriate in the context of unassigned Fe–S centers, this feature deserves mention in a review of the newly detected electron acceptors of the mitochondrial electron transfer system. As shown in Fig. 1B, this interaction contributes a prominent resonance in low-temperature (<20°K) spectra. This resonance shows the behavior expected of a resonance originating from a two-electron acceptor, with the half-reduced state being paramagnetic (Orme-Johnson et al., 1971a, 1974a). If it were not for its unusual shape and temperature dependence, it would have long been obvious that it is due to semiquinone. The signal is not detectable in particles depleted of ubiquinone and is readily seen again after reincorporation of ubiquinone (Ruzicka and Beinert, 1975b; Ruzicka et al., 1975). The signal shape, which cannot be that of an isolated semiquinone species, can be understood if spin–spin interaction of two paramagnetic species is assumed. The extraction experiments indicate that one of these is ubisemiquinone, but the identity of the interaction partner is to date unknown. It could be another molecule of ubisemiquinone; it could also be flavin semiquinone or a paramagnetic Fe–S species. Since the signal has always been observed in association with the membrane-bound HiPIP type of Fe–S center, it was thought that HiPIP may be the interaction partner (Ruzicka and Beinert, 1975b). However, there are arguments against this from considerations of relative signal intensities (Ingledew and Ohnishi, 1975), from simulations of the EPR spectra of the interacting species (Ruzicka et al., 1975), and from structural considerations. The simulations, based on a dipolar spin–spin interaction model indicated a distance of ≤8 Å. From known structural data (Freer et al., 1975), it is not easy to visualize approach of ubisemiquinone to a HiPIP-type Fe–S cluster to within that distance. Concerning interaction with a flavin semiquinone, one may question the probability of this interaction occurring between ubisemiquinone in the lipid phase and flavin in the aqueous protein environment. Also, the oxidation—reduction midpoint potentials seem to be too different for the two redox systems to make flavin and ubisemiquinone interaction very likely. Thus, at this point, interaction between two ubisemiquinone molecules seems to have the highest probability. Ubisemiquinone–ubisemiquinone interaction is by no means trivial, as one would have to visualize a local clustering of these semiquinones in order to produce a signal of the magnitude observed.

III. RELATION OF IRON–SULFUR CENTERS TO ENERGY CONSERVATION

Since the NADH dehydrogenase system includes energy conservation site I, naturally there has been curiosity as to whether the detailed knowledge of the electron transfer components involved in this system might not lend itself to a closer definition of this site. Principally, there have been two approaches taken to this problem. Stimulated by observations made on the b and a cytochromes (Lindsay and Wilson, 1972; Grimmelikhuijzen and Slater, 1972; Chance et al., 1970; Slater et al., 1970; cf. Wikström, 1973; cf. Dutton and Wilson, 1974), there was a search for changes in the oxidation state of Fe–S centers on addition of ATP. The second approach was an attempt to correlate phosphorylation at site I with the presence or development of NADH dehydrogenase in yeast under a number of imposed conditions, such as growth with limiting substrate, iron, or sulfur (Volume I, Chapter 1).

The first approach is experimentally less demanding than the second one. Attention is readily focussed on those components that show changes in oxidation state, i.e., shifts in the apparent midpoint oxidation–reduction potentials, on addition of ATP. Experiments have been carried out using reduction by substrates or poising with naturally occurring redox couples, such as NADH–NAD or succinate–fumarate (Gutman et al., 1975), as well as with artificial oxidation–reduction mediators and potentiometry (Ohnishi, 1973). The interpretation of the observed changes in oxidaton state, however, is not without ambiguities in a complex particulate system. The intricacies of redox potentiometry and the use of artificial mediators has been alluded to above (Section II,A,2). Moreover, irrespective of the way in which the oxidation state is adjusted, the observed shifts in oxidation state brought about by ATP may simply be the consequence of reverse electron transfer induced by ATP, so that the component whose oxidation state is seen to undergo changes may not be the primary target of ATP. In the presence of mediators, the system supposedly is sufficiently "redox-buffered" ("potential-clamped") that reverse electron transfer should not interfere. However, for this to hold, the redox mediators must react rapidly with all components of the system. The situation is well illustrated by the example of *Nitrobacter* (Ingledew and Chappel, 1975), where ATP was found to produce a shift of $+90$ mV in the midpoint potential of a c type cytochrome when the redox couple nitrite–nitrate was present; this change was abolished by the presence of the mediators (10 μM) phenazine methosulfate and diaminodurol. On the other hand, the midpoint potential of an a_1 type cytochrome underwent a shift of -70 mV in the presence of the two mediators, but not in the presence of the nitrite–nitrate couple. Similarly, it was observed that such potential shifts depended on the mediator concentration (Wikström and Lambowitz, 1974). Support for the approach of measuring ATP-dependent midpoint potential shifts is seen in experiments that showed analogous behavior of the respective redox components in the presence and absence of mediators, with the oxidation state adjusted by substrate couples in the latter instance (Dutton and Wilson, 1974). For details of all these considerations, the reader is referred to recent discussions particularly concerned with this subject (Dutton and Wilson, 1974; Wikström, 1973). It appears that the conclusions drawn from observations of ATP-induced apparent midpoint potential shifts of components of the electron transfer system should be considered with reservation. We must, however, remain alert to their potential significance. The present status of the Fe–S centers of the electron transfer system is that with redox potentiometry, ATP-dependent changes in midpoint potentials were only reported for centers 1a and 2 of NADH dehydrogenase (Ohnishi, 1973). The former becomes 80 mV more negative—a shift from -380 to -460 mV—while the latter turns 110 mV more positive—a change from -20 to $+90$ mV. Thus, NADH dehydrogenase would encompass a really remarkable span of oxidation reduction potentials. However, it has also been pointed out that these apparent midpoint potential shifts are in the direction opposite of that expected from the probable location of these Fe–S centers on the membrane (Dutton and Wilson, 1974) if they were involved in site I phosphorylation. Indeed, experiments have been reported (Gutman et al., 1975) that indicate ATP-induced changes in oxidation state opposite to those found by redox potentiometry with mediators. ATP was found to cause uncoupler-sensitive reduction of center 1 in beef heart submitochondrial particles poised with NADH–NAD and blocked with cyanide and piericidin (Gutman et al., 1975). Also, piericidin-blocked submitochondrial particles were cycled in the presence of NADH and excess oxygen, so that a final state was reached with all Fe–S centers of NADH dehydrogenase reoxidized except center 2, which remained partly reduced. On addition of ATP, there was a substantial reoxidation of this center, which extended to several minutes (Gutman et al., 1972). On the other hand, Albracht and Slater (1971) observed in submitochondrial particles from beef heart under various conditions reoxidation by 5 mM ATP of Fe–S center 1 and 2 of NADH dehydrogenase, the HiPIP component of succinate dehydrogenase, and of the Fe–S center of Complex III.

Note Added in Proof: An article has appeared by Ohnishi (1976) in which a number of the experimental data referred to above are collected, expanded and discussed in the light of different viewpoints. Although some experimental uncertainties have been clarified, ambiguities remain and no unique interpretation can be given to the observations made to date.

Although, obviously, there are apparent or real contradictions in these reports, we must consider that the number of experimental results available today is limited and that there were substantial differences in the experimental conditions used by the various investigators, so that a common denominator may yet be found as work progresses. However, the essential feature, which stands out beyond discrepancies and controversies seems to be the large energy gap between Fe–S centers 1 and 2 within the NADH dehydrogenase molecule that spans the midpoint potential range between the NADH–NAD couple and the cytochrome b–ubiquinone region. Schemes, incorporating centers 1 and 2 in a transmembrane e^- transfer loop according to the chemiosmotic theory have been proposed (Mitchell, 1972; Garland et al., 1972).

As mentioned above, the second major approach toward unraveling the possible role played by the Fe–S centers of NADH dehydrogenase in phosphorylation at site I was that of studying in yeast development and breakdown of this site in relation to that of the components of NADH dehydrogenase. The outcome of this work did not specifically implicate any one component in site I (Grossman et al., 1974; Cobley et al., 1975). The answer was simply that a complete NADH dehydrogenase of the type consistently found in mammalian mitochondria and containing at least the more prominent Fe–S centers, namely 1–3, is necessary for phosphorylation at site I to occur. This does not imply that this is the only condition that must be met. Although the result was anticipated by the statement just made, the path to this result in the course of the years was not a smooth one. However, since this subject is discussed to some extent in Volume I of this treatise (Volume I, Chapter 1), some of the difficulties will only be mentioned briefly. Without doubt, the most serious shortcoming in the earlier work was the lack of sensitivity and resolution in EPR spectroscopy, at that time exclusively done at $>77°K$. A second area of confusion derived from the fact that nonheme iron, as determined chemically, was not always clearly distinguished from specific iron–sulfur components, which we are concerned with when we consider the NADH dehydrogenase system. A third area of controversy and difficulty in interpretation resulted from the growth conditions applied, namely whether the yeast was grown in batch (Ohnishi, 1972) or in continuous culture (Garland, 1970). Since it was found that *Candida utilis* elaborates a mammalian type of NADH dehydrogenase, concomitant with phosphorylation at site I, when passing from the logarithmic to the stationary growth phase (Grossman et al., 1974), it is clear that growth in batch culture, which implies continuously changing conditions, is bound to introduce a number of additional variables that are hard to control.

An interesting by-product of these studies was the observation (Grossman et al., 1974; Cobley et al., 1975) that at some stages of growth, when

3. MITOCHONDRIAL ELECTRON TRANSFER SYSTEM

the mammalian type of NADH dehydrogenase was being synthesized, center 3 (or 3 plus 4) as measured from its resonance at $g = 1.86$ was closer to its finally reached maximal level than centers 1 and 2. This implies that all three or four centers are not produced exactly in synchrony, but that there is a transition period when more molecules have centers 3 (or 3 plus 4) than 1 plus 2. It would be interesting to know the composition and properties of such intermediate forms.

IV. OUTLOOK

The insight that has been gained in the past few years concerning the number, relationships, constitution, and function of the various Fe–S centers of the mitochondrial electron transfer system stimulates our curiosity as to analogies in related biological systems. As shown in Fig. 1 and as is also obvious from the many references to the work on the Fe–S centers of yeast mitochondria and submitochondrial particles, there is very little if any difference between the patterns and behavior on oxidation–reduction of Fe–S centers of mitochondria from different species. It may be well to recall here that additional Fe–S centers occur in mitochondria of specialized tissues, such as the Fe–S protein of adrenal cortex mitochondria (Volume I, Chapter 8), which is involved in reactions of steroid metabolism. A new development in this area is the finding of an analogous protein in mitochondria of the renal cortex of chicken (Pedersen *et al.*, 1976) that is involved in the 25-hydroxyvitamin D_3—1α-hydroxylase system.

With respect to plants and photosynthetic bacteria, if the photosynthetic electron transfer systems of chloroplasts or bacteria have anywhere near the number of Fe–S components as the mitochondrial one, much work and excitement still lies ahead of us. A beginning has been made with the discovery of the bound ferredoxins (Malkin and Bearden, 1971; Evans *et al.*, 1972, 1975), which seem, for the photosynthetic electron transfer apparatus, the counterparts of the mitochondrial membrane-bound Fe–S centers. Also, it has become clear that the ferredoxin type of Fe–S center of high potential, represented in the Rieske Fe–S center in mitochondria, is a widespread component of electron transfer systems (Malkin and Aparicio, 1975). The same seems to be true for the high potential type (HiPIP) (Ruzicka and Beinert, 1974; Dus *et al.*, 1967; DerVartanian *et al.*, 1974) of component as it occurs in the succinate dehydrogenase system (Beinert *et al.*, 1974, 1975).

If we consider the variety of types of Fe–S centers that nature utilizes, namely those with [2Fe–2S] and [4Fe–4S] clusters, those with multiple clusters of either kind, the Fe–S flavin combinations, or those of the

HiPIP type, and if we consider that there is the implicit potential for a two-electron change between the extreme oxidation states, depending on protein constraint, we come to realize that the Fe–S structures are not only the most numerous but also the most diversified electron carriers of the mitochondrial electron transfer system that we know to date. They span a range of midpoint redox potentials from about -400 to $+300$ mV. From the work at very low temperatures and extreme powers, there are inklings that we are not yet at the end of the road, although this work poses great difficulties as to the actual significance and meaning of the observations made under such conditions.

The quantity of labile sulfur in submitochondrial particles from beef heart is 7.1–7.3 nmoles per milligram of protein (Hanstein and Hatefi, 1970). We have to realize, however, that this may be only a minimal value, as some labile sulfur in the particles may not have responded to the analysis. A rough estimate of the quantity of Fe–S groups associated with the now known components of the electron transfer system should then be possible, if we assume stoichiometry of the respective Fe–S centers with cytochrome c_1 (Rieske protein) and with acid-extractable or nonextractable flavin, for the dehydrogenases. There is considerable uncertainty, however, concerning the acid-extractable flavin. First, it had been known for some time (Singer and Cremona, 1964) that this is composed of FAD and FMN, with FAD accounting for two-thirds. It may be recalled that FMN is the prosthetic flavin of NADH dehydrogenase. The source of FAD must then be ETF dehydrogenase and flavoproteins that are largely but not entirely solubilized during the preparation of particles, such as the flavoproteins of fatty and oxidation and lipoate dehydrogenase. If we assume 0.26 nmoles of acid-extractable flavin per milligram of protein (Singer and Cremona, 1964) and consider one-third of this to be FMN, we would have 0.09 nmoles of NADH dehydrogenase flavin. From a comparison of the intensities of the EPR signals of center 3 of NADH dehydrogenase and of the Fe–S center of ETF dehydrogenase in beef heart submitochondrial particles, we assume that the concentration of ETF dehydrogenase flavin is about one-half of this, or 0.04 nmoles/mg. For acid-nonextractable flavin and cytochrome c_1, we use the values 0.20 (T. P. Singer, personal communication) and 0.46 nmoles/mg (Orme-Johnson et al., 1974b), respectively. If we take, according to our present knowledge, the number of Fe–S groups for ETF dehydrogenase as 4 per flavin, that for succinate dehydrogenase as 8 per flavin, and that of the Rieske protein as 2 per cytochrome c_1, we arrive at a sum of $0.16 + 1.60 + 0.92 = 2.7$ Fe–S groups per milligram for these three categories. We then would have $7.2 - 2.7 = 4.5$ Fe–S groups per milligram left for NADH dehydrogenase and unknown or other centers. One of these is the Fe–S protein of the outer mitochondrial membrane, which is present in beef heart submitochondrial particles at ~0.3 nmoles per milligram of protein (Albracht and Heidrich, 1975; S. P. J. Albracht, personal communication). Since this is likely to be a [2Fe–2S] center, we have to reserve 0.6 nmoles of Fe–S groups per milligram for this. This leaves 3.9 Fe–S groups per milligram for NADH dehydrogenase and unknown centers. The number of Fe–S groups associated with Complex I is given as 16–18 per flavin (Hatefi *et al.*, 1974), whereas in the soluble dehydrogenase of Baugh and King (1972) as many as 28 Fe–S groups per FMN are found. This would bring the quantity of Fe–S groups in NADH dehydrogenase to anywhere from 0.09×16 to 0.09×28 or 1.44 to 2.5 nmoles per milligram of protein. Thus, we come close to the value of total labile sulfur, but there is room left for unknowns, particularly if they should occur at the concentration level of FMN or FAD in particles. We also must consider that there may be a number of undetectable Fe–S centers, as they are present in other complex Fe–S proteins, such as bacterial nitrogenase (Volume I, Chapters 2 and 3; and this volume, Chapter 2). Although obviously much uncertainty remains, calculations of the kind just presented may give us some feeling for the approximate limits within which we are constrained here.

The balance that was just presented does not directly apply to mitochondria, since some Fe–S centers are solubilized when membrane fragments are prepared, e.g., the soluble HiPIP species (Ruzicka and Beinert, 1974).

These last considerations project, of course, new avenues of research for the future. The past few years have increasingly brought to light signs of interaction between paramagnetic components. It is almost certain that, as we become thoroughly familiar with the EPR spectra of the basic components of the electron transfer system, we will become increasingly alerted to those minor shoulders and permutations in the spectra, which may be indicative of interactions. We may learn to enhance such minor features by proper manipulations of the biological conditions and the capabilities of EPR spectroscopy. An example is the detection of the interaction of ubisemiquinone with another yet unidentified paramagnetic species (Ruzicka *et al.*, 1975). The recognition and definition of such interactions may be of no minor importance for considerations of the path and mechanism of electron transfer. The potential of this direction of exploration for deepening our understanding of details of the system is probably not yet generally appreciated.

Another avenue of research promises to become of great value for the area considered in this chapter. In the past, biochemists working with Fe–S structures have generally not considered the Fe–S group as an autonomous prosthetic group. It was rather felt that it is part of the protein and tinkering with this delicate structure can only result in the release of iron and sulfide. However, progress in the chemistry of model compounds for Fe–S proteins (this volume, Chapter 7) showed that this defeatist attitude is not justified, and ways have indeed been found to remove and exchange Fe–S clusters intact (this volume, Chapter 7). This constitutes enormous progress and raises hopes that we may be able to disentangle the multiple clusters of complex Fe–S proteins and determine their exact composition. Obviously, application of existing methods to multiple clusters, possibly even containing EPR silent Fe–S centers, is likely to experience rough going initially.

The detection of the Fe–S protein of the outer mitochondrial membrane and the possibility afforded therewith, namely of a simple test of particles for contamination with outer-membrane fragments, directs our attention to the topology of Fe–S centers in mitochondria. Promising advances have been made by use of lipophilic chelating agents for iron (Harmon and Crane, 1974). The application to Fe–S centers of this approach, which, of course, is not *a priori* specific for Fe–S components, is still in its beginnings. However, by this approach, by advances in preparation and separation methods, or by assessment of differential interactions of Fe–S centers in membrane vesicles with added oxidizing or reducing agents, paramagnetic ions, or spin labels, more detailed information on the location of the Fe–S centers is likely to come forth.

As in other fields of biochemistry, contributions to the field of Fe–S proteins can be expected by use of microbial mutants in exploring genetic and functional relationships. An example, concerned with the relationship of the membrane-bound HiPIP type of Fe–S center and the succinate dehydrogenase system was discussed above (De Kok *et al.*, 1975).

The discussion of future trends has thus far mainly been concerned with structure. Information on kinetic behavior is required for our understanding of functional details. Since low-temperature EPR spectroscopy has remained the only effective and practical method for measuring oxidation–reduction of Fe–S centers, kinetics monitored by low-temperature EPR must be studied. This implies the use of rapid freeze-quenching techniques (Smith *et al.*, 1973; Ballou and Palmer, 1974; Beinert *et al.*, 1976). Therefore, the prediction of increased activity in this field is probably justified.

ACKNOWLEDGMENTS

I am indebted to Drs. S. P. J. Albracht, D. Bäckström, and N. R. and W. H. Orme-Johnson for supplying information in advance of publication, to Drs. Y.

Hatefi, F. J. Ruzicka, and T. P. Singer for analytical data and clarifying discussions, to Dr. T. Ohnishi for permitting me to quote extensively from her publications, and to Drs. N. R. Orme-Johnson and F. J. Ruzicka for reading parts of the manuscript.

REFERENCES

Ackrell, B. A. C., Kearney, E. B., and Mayr, M. (1974). *J. Biol. Chem.* **249**, 2021.
Albracht, S. P. J. (1974). *Biochim. Biophys. Acta* **347**, 183.
Albracht, S. P. J., and Dooijewaard, G. (1975). *In* "Electron Transfer Chains and Oxidative Phosphorylation" (E. Quagliarello *et al.*, eds.), p. 49. North-Holland Publ., Amsterdam.
Albracht, S. P. J., and Heidrich, H.-G. (1975). *Biochim. Biophys. Acta* **376**, 231.
Albracht, S. P. J., and Slater, E. C. (1971). *Biochim. Biophys. Acta* **245**, 508.
Bäckström, D., Hoffström, I., Gustafsson, I., and Ehrenberg, A. (1973). *Biochem. Biophys. Res. Commun.* **53**, 596.
Baginsky, M. L., and Hatefi, Y. (1971). *Biochemistry* **10**, 2517.
Ballou, D. P., and Palmer, G. A. (1974). *Anal. Chem.* **46**, 1248.
Baugh, R. F., and King, T. E. (1972). *Biochem. Biophys. Res. Commun.* **49**, 1165.
Beinert, H. (1963). *In* "The Enzymes" (P. D. Boyer, H. Lardy, and K. Myrbäck, eds.), 2nd ed., Vol. 7, p. 467. Academic Press, New York.
Beinert, H., and Ruzicka, F. J. (1975). *In* "Electron Transfer Chains and Oxidative Phosphorylation" (E. Quagliarello *et al.*, eds.), p. 37. North-Holland Publ., Amsterdam.
Beinert, H., and Sands, R. H. (1960). *Biochem. Biophys. Res. Commun.* **3**, 41.
Beinert, H., Palmer, G., Cremona, T., and Singer, T. P. (1965). *J. Biol. Chem.* **240**, 475.
Beinert, H., Ackrell, B. A. C., Kearney, E. B., and Singer, T. P. (1974). *Biochem. Biophys. Res. Commun.* **58**, 564.
Beinert, H., Ackrell, B. A. C., Kearney, E. B., and Singer, T. P. (1975). *Eur. J. Biochem.* **54**, 185.
Beinert, H., Hansen, R. E., and Hartzell, C. R. (1976). *Biochim. Biophys. Acta* **423**, 339.
Carter, C. W., Jr., Kraut, J., Freer, S. T., Alden, R. A., Sieker, L. C., Adman, E., and Jensen, L. H. (1972). *Proc. Natl. Acad. Sci. U.S.A.* **69**, 3526.
Chance, B., Wilson, D. F., Dutton, P. L., and Erecínska, M. (1970). *Proc. Natl. Acad. Sci. U.S.A.* **66**, 1175.
Cobley, J. G., Grossman, S., Singer, T. P., and Beinert, H. (1975). *J. Biol. Chem.* **250**, 211.
Coles, C. J., Tisdale, H. D., Kenney, W. C., and Singer, T. P. (1972). *Physiol. Chem. Phys.* **4**, 301.
Crane, F. L., and Beinert, H. (1956). *J. Biol. Chem.* **218**, 717.
Davis, K. A., and Hatefi, Y. (1971). *Biochemistry* **10**, 2509.
De Kok, J., Muller, J. L. M., and Slater, E. C. (1975). *Biochim. Biophys. Acta* **387**, 441.
DerVartanian, D., Baugh, R. F., and King, T. E. (1973). *Biochem. Biophys. Res. Commun.* **50**, 629.
DerVartanian, D. V., Morgan, T. V., and Brantner, R. V. (1974). *Biochim. Biophys. Acta* **347**, 497.
Dus, K., De Klerk, H., Sletten, K., and Bartsch, R. G. (1967). *Biochim. Biophys. Acta* **140**, 291.

Dutton, P. L., and Wilson, D. F. (1974). *Biochim. Biophys. Acta* **346**, 165.
Erbes, D. F., Burris, R. H., and Orme-Johnson, W. H. (1975). *Proc. Natl. Acad. Sci. U.S.A.* **72**, 4795.
Evans, M. C. W., Telfer, A., and Lord, A. V. (1972). *Biochim. Biophys. Acta* **267**, 530.
Evans, M. C. W., Siliva, C. K., Bolton, J. R., and Cammack, R. (1975). *Nature (London)* **256**, 668.
Flatmark, T., Ruzicka, F. J., and Beinert, H. (1976). *FEBS Lett.* **63**, 51.
Freer, S. T., Alden, R. A., Carter, C. W., Jr., and Kraut, J. (1975). *J. Biol. Chem.* **250**, 46.
Fry, K. T., and San Pietro, A. (1962). *Biochem. Biophys. Res. Commun.* **9**, 218.
Garland, P. B. (1970). *Biochem. J.* **118**, 329.
Garland, P. B., Clegg, R. A., Downie, J. A., Gray, T. A., Lawford, H. G., and Skyrme, J. (1972). *Mitochondria Biomembr., Fed. Eur. Biochem. Soc., Meet., 8th, 1972* Vol. 28, BBA Libr., p. 105.
Grimmelickhuijzen, C. J. P., and Slater, E. C. (1972). *Biochim. Biophys. Acta* **256**, 24.
Grossman, S., Cobley, J. G., Singer, T. P., and Beinert, H. (1974). *J. Biol. Chem.* **249**, 3819.
Gutman, M., Singer, T. P., and Beinert, H. (1972). *Biochemistry* **11**, 556.
Gutman, M., Beinert, H., and Singer, T. P. (1975). In "Electron Transfer Chains and Oxidative Phosphorylation" (E. Quagliarello, ed.), p. 55. North-Holland Publ., Amsterdam.
Hanstein, W. G., and Hatefi, Y. (1970). *Arch. Biochem. Biophys.* **138**, 87.
Harmon, H. J., and Crane, F. L. (1974). *Biochim. Biophys. Res. Commun.* **59**, 326.
Hatefi, Y., and Hanstein, W. G. (1973). *Biochemistry* **12**, 3515.
Hatefi, Y., Hanstein, W. G., Davis, K. A., and You, K. S. (1974). *Ann. N.Y. Acad. Sci.* **227**, 504.
Hatefi, Y., and Bearden, A. J. (1976). *Biochem. Biophys. Res. Commun.* **69**, 1032.
Ingledew, W. J., and Chappel, J. B. (1975). *Fed. Proc., Fed. Am. Soc. Exp. Biol.* **34**, 488.
Ingledew, W. J., and Ohnishi, T. (1975). *FEBS Lett.* **54**, 167.
Kearney, E. B. (1957). *J. Biol. Chem.* **229**, 363.
Kearney, E. B., Ackrell, B. A. C., Mayr, M., and Singer, T. P. (1974). *J. Biol. Chem.* **249**, 2016.
Keilin, D. (1966). "The History of Cell Respiration and Cytochrome." Cambridge, Univ. Press, London and New York.
King, T. E. (1964). *Biochem. Biophys. Res. Commun.* **16**, 511.
King, T. E., and Howard, R. L. (1967). In "Methods in Enzymology" (R. W. Estabrook and M. E. Pullman, eds.), Vol. 10, p. 275. Academic Press, New York.
Lee, I. Y., and Slater, E. C. (1974). *Biochim. Biophys. Acta* **347**, 14.
Leigh, J. S., Jr., and Chance, B. (1974). *Fed. Proc., Fed. Am. Soc. Exp. Biol.* **33**, 1289.
Leigh, J. S., Jr., and Erecínska, M. (1975). *Biochim. Biophys. Acta* **387**, 95.
Lindsay, J. G., and Wilson, D. F. (1972). *Biochemistry* **11**, 4613.
Löw, H., and Vallin, I. (1963). *Biochim. Biophys. Acta* **69**, 361.
Lusty, C. J., Machinist, J. M., and Singer, T. P. (1965). *J. Biol. Chem.* **240**, 1804.
Mackey, L. N., Kuwana, T., and Hartzell, C. R. (1973). *FEBS Lett.* **36**, 326.
Mahler, H. R., and Elowe, D. (1953). *J. Am. Chem. Soc.* **75**, 5769.
Mahler, H. R., and Elowe, D. (1954). *J. Biol. Chem.* **210**, 165.
Malkin, R., and Aparicio, P. J. (1975). *Biochem. Biophys. Res. Commun.* **63**, 1157.

Malkin, R., and Bearden, A. J. (1971). *Proc. Natl. Acad. Sci. U.S.A.* **68,** 16.
Mathews, R., Charlton, S., Sands, R. H., and Palmer, G. (1974). *J. Biol. Chem.* **249,** 4326.
Mitchell, P. (1972). *Mitochondria Biomembr., Fed. Eur. Biochem. Soc., Meet. 8th, 1972* Vol. 28, BBA Libr., p. 353.
Ohnishi, T. (1972). *FEBS Lett.* **24,** 305.
Ohnishi, T. (1973). *Biochim. Biophys. Acta* **301,** 105.
Ohnishi, T. (1975). *Biochim. Biophys, Acta* **387,** 475.
Ohnishi, T. (1976). *Eur. J. Biochem.* **64,** 91.
Ohnishi, T., Asakura, T., Wohlrab, H., Yonetani, T., and Chance, B. (1970). *J. Biol. Chem.* **245,** 901.
Ohnishi, T., Asakura, T., Yonetani, T., and Chance, B. (1971). *J. Biol. Chem.* **246,** 5960.
Ohnishi, T., Wilson, D. F., Asakura, T., and Chance, B. (1972). *Biochem. Biophys. Res. Commun.* **46,** 1631.
Ohnishi, T., Winter, D. B., Lim, J., and King, T. E. (1973). *Biochem. Biophys. Res. Commun.* **53,** 231.
Ohnishi, T., Leigh, J. S., Ragan, C. I., and Racker, E. (1974a). *Biochem. Biophys. Res. Commun.* **56,** 775.
Ohnishi, T., Winter, D., Lim, J., and King, T. E. (1974b). *Biochem. Biophys. Res. Commun.* **61,** 1017.
Ohnishi, T., Leigh, J. S., Winter, D. B., Lim, J., and King, T. E. (1974c). *Biochem. Biophys. Res. Commun.* **61,** 1026.
Ohnishi, T., Ingledew, W. J., and Shiraishi, S. (1975). *Biochem. Biophys. Res. Commun.* **63,** 894
Ohnishi, T., Ingledew, W. J., and Shiraishi, S. (1976). *Biochem. J.* **153,** 39.
Orme-Johnson, N. R., Orme-Johnson, W. H., Hansen, R. E., Beinert, H., and Hatefi, Y. (1971a). *Biochem. Biophys. Res. Commun.* **44,** 446.
Orme-Johnson, N. R., Hansen, R. E., and Beinert, H. (1971b). *Biochem. Biophys. Res. Commun.* **45,** 871.
Orme-Johnson, N. R., Hansen, R. E., and Beinert, H. (1974a). *J. Biol. Chem.* **249,** 1922.
Orme-Johnson, N. R., Hansen, R. E., and Beinert, H. (1974b). *J. Biol. Chem.* **249,** 1928.
Papa, S., Lorusso, M., and Guerrieri, F. (1975). *Biochim. Biophys. Acta* **387,** 425.
Pederson, J. I., Ghazarian, J. G., Orme-Johnson, N. R., and De Luca, H. F. (1976). *J. Biol. Chem.* **251,** 3933.
Prince, R. C., and Dalton, P. L. (1976). *FEBS Lett.* **65,** 117.
Prince, R. C., Lindsay, J. G., and Dutton, P. L. (1975). *FEBS Lett.* **51,** 108.
Rieske, J. S. (1967). *In* "Methods of Enzymology" (R. W. Estabrook and M. E. Pullman, eds.), Vol. 10, p. 357. Academic Press, New York.
Rieske, J. S., Hansen, R. E., and Zaugg, W. S. (1964). *J. Biol. Chem.* **239,** 3017.
Ruzicka, F. J., and Beinert, H. (1974). *Biochem. Biophys. Res. Commun.* **58,** 556.
Ruzicka, F. J., and Beinert, H. (1975a). *Biochem. Biophys. Res. Commun.* **66,** 622.
Ruzicka, F. J., and Beinert, H. (1975b). *Fed. Am. Soc. Exp. Biol. Proc.,* **34,** 579.
Ruzicka, F. J., Beinert, H., Schepler, K. L., Dunham, W. R., and Sands, R. H. (1975). *Proc. Natl. Acad. Sci. U.S.A.* **72,** 2886.
Sands, R. H., and Beinert, H. (1960). *Biochem. Biophys. Res. Commun.* **3,** 47.
Singer, T. P. (1966). *Compr. Biochem.* **14,** 127.
Singer, T. P., and Cremona, T. (1964). *In* "Oxygen in the Animal Organism," (F. Dickens and E. Niel, eds.), p. 179. Pergamon, Oxford.

Singer, T. P., Kearney, E. B., Ackrell, B. A. C., and Beinert, H. (1975). *Proc. Fed. Eur. Biochem. Soc., Meet., 10th, 1975, Symposium 8.* p. 173.
Slater, E. C., Berden, J. A., and Wegdam, H. J. (1970). *Nature (London)* **226**, 1248.
Smith, B. E., Lowe, D. J., and Bray, R. C. (1973). *Biochem. J.* **135**, 331.
Wikström, M. K. F. (1973). *Biochim. Biophys. Acta* **301**, 155.
Wikström, M. K. F., and Lambowitz, A. M. (1974). *FEBS Let.* **40**, 149.
Yu, C. A., Yu, L., and King, T. E. (1974). *J. Biol. Chem.* **249**, 4905.
Zeylemaker, W. P., DerVartanian, D. V., and Veeger, C. (1965). *Biochim. Biophys. Acta* **99**, 183.

CHAPTER 4

Biosynthesis of Iron–Sulfur Proteins

JAMES W. BRODRICK and JESSE C. RABINOWITZ

    I. Introduction... 101
   II. Bacterial Protein Synthesis...................................... 103
  III. Regulation of Synthesis of Ferredoxin......................... 106
  IV. Conversion of Apoferredoxin to Holo- or Native Ferredoxin...... 107
   V. Synthesis of Specific Proteins *in Vitro*......................... 110
  VI. Immunological Studies with Clostridial Ferredoxin.............. 111
 VII. Enzymatic Assays for Ferredoxin................................ 112
VIII. Synthesis of Clostridial Ferredoxin or Apoferredoxin *in Vitro*..... 113
       References.. 118

In contrast to the wealth of information that has been accumulated in recent years concerning the physical and chemical properties of iron–sulfur proteins, the reactions involved in the biosynthesis of these proteins, the formation of their iron–sulfur clusters, and the regulation of these processes have received limited attention. This review will deal primarily with results obtained with only two types of iron–sulfur proteins—clostridial ferredoxins and plant ferredoxins. This limitation is dictated by the lack of information about the biosynthesis of other types of iron–sulfur proteins. It is hoped that this assembly of current information concerning the biosynthesis of iron–sulfur proteins will stimulate further efforts in this area.

I. INTRODUCTION

Ferredoxins are rather atypical proteins with respect to their size. Clostridial ferredoxin, a monomeric protein with a molecular weight of about 6000 daltons, is one of the smallest known proteins with any type of active or catalytic site. Many proteins, especially those with defined enzymatic activities, are oligomeric complexes of a small integral number of subunits. Two, four, six, or more separate polypeptide chains, which may be identical, similar (as in the two α and β chains of hemoglobin), or quite different in molecular weight and primary structure, are held together in the oligomeric form by noncovalent interactions and may be dissociated under conditions that leave the individual polypeptides intact.

The large number of known multisubunit proteins is illustrated by the tabulation of Darnall and Klotz (1975) of over 500 well-characterized multisubunit proteins for which reliable molecular weight data for both subunit and complex have been accumulated. No protein is listed with a molecular weight lower than 11,466 (insulin, composed of two subunits of molecular weight 5733) and only four proteins (including insulin) are listed with subunit molecular weights below 10,000. Examples of monomeric proteins with molecular weights below 10,000 daltons are also relatively rare. Milk lipase, horse kidney metallothionein, bovine pancreas trypsin inhibitor, and adrenocorticotropin, are examples found in this compilation. Cytochrome c, a widely distributed monomeric protein, has a molecular weight of 14,000 daltons. Ferredoxin, with a molecular weight of 6263 daltons and a polypeptide molecular weight of only 5560 daltons, may therefore be considered an unusually small protein.

The small size of ferredoxin, especially the clostridial type, makes it necessary to consider all possible types of biosynthetic pathways that have been described to date for synthesis of proteins and polypeptides. For instance, small peptide antibiotics are synthesized by nonribosomal enzyme complexes of high molecular weight (Lipmann, 1971; Kurahashi, 1974). The mechanism of this process is not yet completely understood, but it is clear that the sequence of the peptide is determined by the enzymatic specificity of the protein complex in which synthesis occurs and not by any template such as the messenger RNA in ribosomal protein synthesis. The largest of this group, gramicidin, is a linear pentadecapeptide. Thus, one must consider the possibility that the clostridial ferredoxin polypeptide, not quite four times as large, is synthesized by a similar mechanism.

The major pathway of protein biosynthesis in all cells is the ribosomal pathway, which is reviewed in some detail below. An additional step in normal protein biosynthesis, posttranslational modification, has recently been recognized. The types of modification that have been described include the following: (a) modification of amino acid residues, as in the conversion of proline to hydroxproline in collagen, and methylation of lysine, which occurs in many types of proteins; (b) covalent addition of nonprotein components, such as phosphate and carbohydrate moieties; (c) covalent addition of cofactors that participate in catalysis, such as pyridoxal phosphate and covalently bound porphyrin rings; and (d) posttranslational cleavage of a precursor polypeptide, such as in zymogen activation. Recent results concerning the synthesis of insulin and other polypeptide hormones indicate that many are derived by specific proteolytic cleavage from precursors of higher molecular weight (Tager and Steiner, 1974). This type of mechanism should also be considered in the study of the biosynthesis of any protein of low molecular weight, such as ferredoxin.

II. BACTERIAL PROTEIN SYNTHESIS

For the benefit of readers unfamiliar with recent work on protein synthesis, a brief and generalized description of current concepts concerning mechanism and regulation of protein biosynthesis, based mainly on the process as it occurs in bacterial systems, is presented in this section. For more extensive information on the process in bacteria, as well as the variations recognized to occur in eukaryotic sources, recent reviews (Lengyel and Söll, 1969; Lucas-Lenard and Lipmann, 1971; Haselkorn and Rothman-Denes, 1973) are recommended. No attempt at completeness will be made here; the fact that the book "Ribosomes" (Nomura et al., 1974) concerning just one of the numerous macromolecules involved in protein synthesis contains over 900 pages illustrates the complexity of this subject.

Proteins are synthesized in a process in which over 130 known macromolecules participate. The sequence of amino acids in any polypeptide is determined by the sequence of triplets of nucleotides in the messenger RNA that is characteristic of each protein. The "stage" upon which protein synthesis takes place is the ribosome—a macromolecular nucleoprotein containing about two-thirds RNA and one-third protein by weight. Ribosomes are composed of two subunits of unequal size, which are identified by their sedimentation coefficients. In bacteria, the small and large subunits are designated 30 S and 50 S, respectively. The 30 S subunit contains one large RNA molecule and twenty-one different proteins, while the 50 S subunit contains one large and one small RNA, with no fewer than thirty-three proteins. Eukaryotic ribosomes are larger and even more complex.

The biosynthesis of a polypeptide may be divided into three distinct phases: (1) initiation, a process through which the assembly of all the components necessary for peptide bond formation is achieved; (2) elongation, the cyclical process of addition of amino acids, one at a time, to the growing polypeptide chain; and (3) termination, a poorly understood process during which the completed polypeptide is released and the ribosome becomes available for subsequent initiation of a new polypeptide.

During elongation, each amino acid is added to the growing polypeptide by a transfer reaction from an aminoacyl derivative of a specific RNA of about 25,000 daltons called a transfer RNA. Amino acids are "activated" by becoming attached to specific transfer RNA's by highly selective enzymes, one for each amino acid, known as aminoacyl transfer RNA synthetases, in a process driven by the free energy of hydrolysis of ATP. The selection of the correct amino acid at a given position in the growing polypeptide is dictated by a sequence of three nucleotides in each transfer RNA that is complementary to the triplet nucleotide sequence for that amino acid in the messenger RNA. These processes require the participation of three well-characterized soluble proteins, as well as that of the ribosome.

Details of the mechanism of the initiation process are still somewhat controversial, but the general features of the process are known. At least three protein factors, the ribosomal subunits, the messenger RNA, and a unique transfer RNA that is aminoacylated with methionine, followed by enzymatic N-formylation of the methionine, combine to form a specific initiation complex that is able to participate in subsequent elongation cycles.

Thus, all bacterial proteins contain an *N*-formylmethionine residue at the N-terminus at the time of synthesis. In most instances, this residue is removed by enzymatic deformylation followed by cleavage of the methionine by an aminopeptidase. However, some bacterial proteins have been found to contain *N*-formylmethionine at their N-termini. An example of this is rubredoxin, an iron-containing electron transfer protein from *Clostridium pasteurianum* (McCarthy and Lovenberg, 1970).

In vivo, the biosynthetic process occurs on structures known as polyribosomes, composed of messenger RNAs containing several ribosomes at different stages of polypeptide synthesis. Polyribosomes have been isolated from several bacterial species and many eukaryotic sources and have been consistently shown to be highly active substrates for polypeptide synthesis *in vitro*. The importance of polyribosomes is further emphasized by the observation that many bacterial messenger RNA's are polycistronic; that is, a single messenger RNA molecule may code for the synthesis of several proteins. Thus, the size of a protein such as ferredoxin may not be indicative of the size of its messenger RNA or of the polyribosomes engaged in its synthesis, since the information necessary for its synthesis may reside on a polycistronic messenger RNA.

4. BIOSYNTHESIS OF IRON–SULFUR PROTEINS

It is at the level of messenger RNA synthesis that metabolites are known to affect the rate of specific protein synthesis in bacteria. Thus, Jacob and Monod many years ago devised a model by which lactose stimulates synthesis of the polycistronic messenger RNA coding for three specific proteins required for its metabolism in *Escherichia coli*, a process known as induction. In other instances the presence of a metabolite has been shown to be at least indirectly responsible for inhibition of synthesis of the messenger RNA coding for enzymes required for its synthesis, a process known as repression. Both modes of control have been studied in great detail with several systems in *E. coli* and other bacteria. Indeed, an entire book (Beckwith and Zipser, 1970) was devoted to results of studies on the control of lactose utilization in *E. coli*.

In view of the foregoing generalized discussion of protein biosynthesis and its control, several considerations of particular interest arise with respect to the detailed mechanism of biosynthesis of the iron–sulfur proteins. These considerations are clarified to some extent later in this review, but most of the questions that arise cannot be answered at this time because of the lack of available information. First, because of their small size, one might question whether the iron–sulfur proteins, especially clostridial ferredoxins, are indeed synthesized by the normal ribosomal system utilized for synthesis of larger proteins. It is conceivable that a polypeptide of only fifty-five amino acid residues could be assembled by an enzymatic process similar to that involved in the synthesis of peptide antibiotics or by a similar but as yet unrecognized mechanism.

Even if iron–sulfur proteins are synthesized by the ribosomal system, several questions arise about the details of the process:

a. Is a special type of processing involved in forming this relatively small polypeptide, such as a proteolytic cleavage of a larger precursor polypeptide?

b. Since the messenger RNA for ferredoxin would be extremely small (minimum about 175 nucleotides), is its synthesis directed by a polycistronic messenger RNA that also codes for other proteins?

c. Is the primary polypeptide formed in prokaryotes one that is initiated with N-formylmethionine?

d. Is the iron–sulfur cluster introduced in an enzymatic or nonenzymatic process?

e. At what point in the biosynthesis of the polypeptide is the iron–sulfur cluster introduced?

f. What factors are involved in the regulation of the biosynthesis of the apoprotein?

III. REGULATION OF SYNTHESIS OF FERREDOXIN

Two systems have been described in which factors affecting the induction of ferredoxin synthesis have been recognized. In the first instance, concentration of iron salts in the growth medium has been found to exert a differential control of synthesis of both bacterial ferredoxin and flavodoxin, a flavoprotein than can at least partially replace ferredoxin in most of the reactions in which it participates. This response to iron is an example of control at the level of protein biosynthesis, since ferredoxin and flavodoxin from *C. pasteurianum* have been shown to be quite distinct in amino acid sequence (Tanaka et al., 1966; Fox et al., 1972). Knight and Hardy (1966) observed that less than 5% of the amount of ferredoxin in control cells was present in cultures of *C. pasteurianum* grown in "iron-poor" medium containing 0.44 mg/ml of iron, conditions under which the cultures yielded maximal amounts of flavodoxin. A similar situation was observed in *Peptostreptococcus elsdenii* by Mayhew and Massey (1969), who also demonstrated that no flavodoxin was present in these bacteria when grown on "iron-rich" medium. Flavodoxin has also been isolated from *Desulfovibrio gigas* and *D. vulgaris* (Dubordieu and Le Gall, 1970), and from *Rhodospirillum rubrum* (Cusanovich and Edmondson, 1971) grown on "iron-poor" media. Thus, in these several organisms, iron salts can both induce the synthesis of ferredoxin and at the same time repress the synthesis of flavodoxin, a protein that can replace ferredoxin functionally, but does not contain iron.

In the second example, synthesis of ferredoxin is ultimately regulated by illumination with visible light. Numerous examples of light induction of chloroplast proliferation have been reported in plants and algae. The stimulation of visible light of synthesis of individual proteins in several systems has been recognized. Of interest in this review are the reports of Armstrong et al. (1971) and Haslett et al. (1973) in which the effect of light on the biosynthesis of ferredoxin was examined in the unicellular blue-green alga *Chlamydomonas reinhardi* and in etiolated beans (*Phaseolus vulgaris*), respectively. Armstrong et al. (1971) demonstrated that cultures of *C. reinhardi* that had been synchronized by periodic exposure to light and dark exhibited periodic synthesis of various chloroplast components, including ferredoxin. The increase in ferredoxin content during the synthetic cycle was limited only by inhibition of cytoplasmic protein synthesis (by addition of cycloheximide) but not by chloramphenicol, which inhibits chloroplast protein synthesis, indicating that in this organism ferredoxin is quite likely being synthesized by cytoplasmic ribosomes. Similarly, Haslett et al. (1973) showed that induction by light of synthesis of *P. vulgaris* ferredoxin was not inhibited by chloramphenicol or by lincomycin, another known inhibitor of chloroplast protein synthesis.

A related effect of light on synthesis of ferredoxin has been described in the photosynthetic bacterium *R. rubrum*. This organism is notable among photosynthetic bacteria in that it can be grown either anaerobically under photosynthetic conditions or aerobically without light. Interestingly, *R. rubrum* was found by Shanmugam et al. (1972) to contain two ferredoxins differing in iron–sulfur content, amino acid composition, and molecular weight. Both were produced when the organism was grown photosynthetically, but only one (Fd II) could be isolated from cells grown heterotropically in the dark. These results, together with those of Cusanovich and Edmondson (1971), indicate that synthesis of one of the ferredoxins from *R. rubrum* is controlled both by light and iron concentration in the growth medium.

IV. CONVERSION OF APOFERREDOXIN TO HOLO- OR NATIVE FERREDOXIN

Although apoferredoxin, the iron- and sulfide-free protein, may be converted to native ferredoxin in nonenzymatic reactions by reaction with a compound of unknown chemical structure prepared by mixing a ferrous or ferric salt in aqueous mercaptoethanol with sodium sulfide (Hong and Rabinowitz, 1967), it is not known whether native ferredoxin is formed *in vivo* in an analogous reaction. Formation of ferredoxin from apoferredoxin and the iron–sulfur model compounds (Que et al., 1975; Job and Bruice, 1975) has not yet been described, but one might anticipate that it will be possible to demonstrate this reaction *in vitro*. Although the formation of ferredoxin *in vitro* from the reduced apoprotein and the active iron–sulfur donor occurs spontaneously, the structure of the protein at the time that the iron and sulfur are introduced *in vivo* is not known, and reduced apoferredoxin may not be the normal intermediate in the biosynthesis of ferredoxin. N-Formylmethionylapoferredoxin and methionylapoferredoxin are very likely precursors of intermediates in the biological synthesis of the native ferredoxin. It is conceivable that enzymatic reactions may be involved in the introduction of the iron–sulfur clusters into the apoprotein.

It is of interest to consider the sources of the reactants necessary for formation of native ferredoxin *in vivo*, and indeed whether the reagents necessary for reconstitution *in vitro* occur *in vivo*. Iron occurs as a ubiquitous constituent of all living cells, and sulfide has been shown to be present in all plants, bacteria, and fungi that have been studied. Its formation results from the action of sulfite reductase, an enzyme in the normal pathway for reduction of assimilated sulfate (Siegel, 1975). However, it is unlikely that FeS, Fe_2S_3, or other iron salts are utilized directly for ferredoxin formation *in vivo*, since these compounds are extremely insoluble (Neilands, 1974). The mixture used for the reconstitution of ferredoxin *in vitro* contains a sulfhydryl reducing agent that is probably involved in forming a water-soluble complex with the iron and sulfide. It is possible that a protein or a smaller sulfhydryl-containing metabolite such as glutathione functions in the reaction *in vivo* as a component of the soluble iron–sulfur intermediate that serves as the donor of the inorganic components of the newly formed native ferredoxin.

The iron and inorganic sulfide in bacterial ferredoxin can be removed by addition of a mercurial followed by gel filtration (Malkin and Rabinowitz, 1966) or by precipitation with trichloroacetic acid either in the presence or absence of a sulfhydryl reducing agent (Hong and Rabinowitz, 1967), forming reduced or oxidized apoferredoxin, respectively. These proteins are water-soluble, contain little or no iron or inorganic sulfide, and have no activity in the phosphoroclastic or other enzymatic assays for ferredoxin. The reports of Rabinowitz and co-workers (Hong and Rabinowitz, 1967, 1970c; Malkin and Rabinowitz, 1966) have demonstrated the formation of enzymatically active ferredoxin, chemically indistinguishable from native ferredoxin, by addition of iron and sulfide in stoichiometric amounts, in the presence of a sulfhydryl reductant, to unfolded, reduced apoferredoxin prepared by reduction of oxidized apoferredoxin with β-mercaptoethanol in urea or by isolation of apoferredoxin in the presence of β-mercaptoethanol. Similar methods of reconstitution have been applied to plant (Fee and Palmer, 1971) and animal (Orme-Johnson *et al.*, 1968) iron–sulfur proteins. Hong and Rabinowitz (1970c) found no evidence for forms containing intermediate levels of iron and sulfide once the reconstituted ferredoxin was purified of reactants, indicating that the fully active protein containing eight iron and eight sulfide atoms per molecule is the only stable product of reconstitution.

Regardless of the detailed mechanism of formation of active ferredoxin *in vivo*, it is clear that the amino acid sequence of the polypeptide is important to the stability of ferredoxin. This has been evident for some time, since ferredoxins of differing amino acid sequences from various clostridial species display widely different stabilities, at least under aerobic conditions. It has also been demonstrated directly by measurement of the stabilities of derivatives of *Clostridium acidi-urici* apoferredoxin containing amino acid substitutions, additions, and deletions.

Hong and Rabinowitz (1970a) first showed that oxidized *C. acidi-urici* apoferredoxin is an ideal substrate for modification at the amino terminus, since this polypeptide contains no other α-amino groups (Rall et al., 1969). They prepared acetyl-, *N-tert*-butyloxycarbonyl- (BOC), and acetamidoapoferredoxins, which were then reconstituted and purified of reactants by DEAE-cellulose chromatography and crystallization. All three derivatives, although nearly fully biologically active when freshly prepared, showed markedly reduced stabilities with respect to both loss of A_{390} and biological activity in the phosphoroclastic enzymatic assay system (Section VII). This result indicates that a free N-terminal α-amino group is important for stability of the parent molecule. In addition, Hong and Rabinowitz (1970a) prepared aminoacylferredoxins by reacting apoferredoxin with each of several BOC amino acids, removing the blocking group, and reconstituting the aminoacylferredoxin. Although the half-life of these derivatives varied with the amino acid used, all were markedly less stable than reconstituted, unmodified ferredoxin. Since the derivatives tested included methionyl ferredoxin, these results, taken with those on blocking of the N-terminal amino group, indicate that enzymatic cleavage of formate and methionine from the *N*-formylmethionyl polypeptide, a normal intermediate in the biosynthesis of a bacterial protein, is a prerequisite for subsequent formation of stable ferredoxin *in vivo*.

Lode et al. (1974, 1976) have extended these studies by use of Edman degradation and subsequent replacement of the N-terminal or penultimate amino acid residue by various other amino acids. They found that the nature of the N-terminal amino acid markedly affected the stability of *C. acidi-urici* ferredoxin derivatives, although all were fully biologically active. In addition, they showed that a large penultimate residue, such as Phe^2, Trp^2, or Leu^2, was required to form a stable derivative from the modified apoferredoxin, since [Gly^2]apoferredoxin could not be reconstituted, while the previously mentioned derivatives were relatively stable. These results are consistent with the elegant X-ray crystallographic studies of Adman et al. (1973), which indicated that the penultimate residue (usually an aromatic amino acid in clostridial ferredoxin) is in juxtaposition to one of the iron–sulfur clusters in crystalline *Micrococcus aerogenes* ferredoxin and is probably important, therefore, in maintenance of a stable cluster structure.

Two groups have reported on possible enzyme-mediated reconstitution of ferredoxin. Jeng and Mortenson (1968) observed that the exchange of [^{35}S]sulfide into *C. pasteurianum* ferredoxin is dependent on the presence of *C. pasteurianum* crude extract and on ferredoxin. They hypothesized that this exchange reaction, although small in extent, is involved in the enzymatic reconstitution of ferredoxin. Finazzi-Agro et al. (1971) and Tomati et al. (1974) have obtained preliminary evidence for the net reconstitution of parsley ferredoxin from apoferredoxin, $FeCl_3$, thiosulfate, and dithiothreitol in the presence of rhodanese (thiosulfate:cyanide sulfurtransferase). These authors suggested that rhodanese was "activating" sulfur in some way, since rhodanese and thiosulfate could replace the Na_2S normally added to the reconstitution system. However, it has been demonstrated that rhodanese can catalyze the formation of S^{2-} from thiosulfate and reduced lipoate (Villarejo and Westley, 1963). 2,3-Dimercaptopropanol, but not cysteine or reduced glutathione, can replace the dihydrolipoate. Since lipoic acid, dithiothreitol, and 2,3-dimercaptopropanol share the common property of formation of a stable ring structure in the oxidized state, it is likely that the above authors were in fact observing the formation of S^{2-} from thiosulfate by rhodanese in the presence of dithiothreitol, in turn providing substrate for chemical reconstitution of parsley ferredoxin. Thus, the evidence on possible enzymatic formation of ferredoxin from apoferredoxin *in vivo* remains sketchy.

V. SYNTHESIS OF SPECIFIC PROTEINS *IN VITRO*

Since the first description (Schweet et al., 1958) of synthesis of hemoglobin by lysates of rabbit reticulocytes, many investigators have demonstrated synthesis *in vitro* of specific proteins. The criteria used to characterize the proteins synthesized fall into three main categories—enzymatic, immunological, and chemical.

The appearance of a new enzyme activity after incubation of a mixture of components, in which *in vitro* protein synthesis occurred, was first reported for bacteriophage T4 lysozyme by Salser et al. (1967). Since that time, Zubay and Chambers (1969) and Schweiger and Gold (1969) have developed active DNA-dependent *in vitro* systems using *E. coli* components for the synthesis of β-galactosidase and several enzymes specifically coded for by phage T4 DNA, respectively. Many other recent reports have expanded upon the original observations made by these workers.

The rationale behind demonstrating specific protein synthesis *in vitro* by enzymatic assay is indeed a simple one. To a protein synthesizing system lacking the enzymatic activity to be studied is added the genetic information in the form of DNA or messenger RNA. After incubation, a measurable amount of enzymatic activity is produced. As long as it can be demonstrated that appearance of the enzymatic activity is dependent on protein synthesis, synthesis *in vitro* of the enzyme in question is established. Demonstration of the chemical, physical, or enzymatic properties characteristic of the enzyme would then further confirm the synthesis *in vitro* of a specific protein. Enzymatic criteria are perhaps the most rigorous demonstration of specific protein synthesis *in vitro*.

Many reports on the use of immune precipitation to measure specific protein synthesis *in vitro* have appeared in recent years. Examples are the tyrosine aminotransferase system from rat hepatoma cells (Beck et al., 1972) and the synthesis of ovalbumin by chick oviduct polyribosomes (Rhoads et al., 1973). Application of immunological techniques to *in vitro* protein synthesis systems is limited by the extent to which denatured or fragmented proteins or polypeptides that contain some incorrect amino acids due to misreading (as observed at high Mg^{2+} concentrations) are reactive. As an example, Boime and Leder (1972) have shown that encephalomyocarditis virus RNA directed the synthesis of a family of discrete polypeptides bearing the same N-terminus but with many distinct C-termini in an *in vitro* system derived from Ehrlich ascites tumor cells. None of the products was as large as the polypeptide produced *in vivo* from the same RNA molecule. In order to be truly specific, an immunological assay for specific protein synthesis *in vitro* would have to be able to distinguish between such fragments and the parent protein.

Physical or chemical characterizations of specific proteins synthesized *in vitro* have recently been based largely on gel electrophoretic techniques, although analysis of tryptic peptides by chromatography is still being used occasionally. Recent improvements in the resolving power of SDS (sodium dodecyl sulfate)–polyacrylamide gel electrophoresis (Laemmli, 1970) and the use of a slab of acrylamide gel for electrophoresis (Studier, 1972, 1973) have led to the increasing acceptance of coincidence of a labeled band on an autoradiograph with a stained band of the protein in question as the sole criterion for demonstrating synthesis *in vitro* of a specific protein. O'Farrell and Gold (1973) have recently studied the regulation of synthesis *in vitro* of many bacteriophage T4 proteins by slab–SDS gel electrophoresis in combination with comparison to *in vivo* protein patterns produced with various T4 phage strains carrying nonsense mutations.

VI. IMMUNOLOGICAL STUDIES WITH CLOSTRIDIAL FERREDOXIN

Antisera to clostridial ferredoxin have been described by Nitz et al. (1969) and by Hong and Rabinowitz (1970b). The antisera of Nitz et al. (1969) were probably directed toward denatured or degraded *C. pasteurianum* ferredoxin, since they reacted more strongly with alkylated or trichloroacetic acid-precipitated ferredoxin than with the active molecule. Antisera to performic acid oxidized *C. pasteurianum* ferredoxin have also been prepared (Nitz et al., 1969). This antiserum was found to react weakly to native *C. pasteurianum* ferredoxin. Later studies (Kelly and Levy, 1971) indicated that the positions Ala^1–Ser^7 and Ala^{51}–Glu^{55} account for most, if not all, of the antigenic activity of the oxidized polypeptide, and that the synthetic peptide corresponding to Pro^{19}–Val^{31} in *C. pasteurianum* ferredoxin has no haptenic activity.

Hong and Rabinowitz (1970b) prepared antisera to *C. acidi-urici* and *C. pasteurianum* ferredoxins, both of which reacted much more strongly with native ferredoxin than with apoferredoxin. Neither ferredoxin would cross-react with antiserum prepared against the other. The antisera could be used in quantitative microcomplement fixation to detect as little as 30 ng of ferredoxin. Both ferredoxins were considered to be weakly antigenic, since optimal dilutions of antisera for microcomplement fixation were only 1:800 to 1:1000.

Although it is possible to prepare antisera that are relatively specific for native ferredoxin and could thus be used for quantitative measurement by microcomplement fixation of ferredoxin synthesized *in vitro*, the lack of absolute specificity renders this method less useful for assay of ferredoxin synthesized *in vitro*, since it cannot be determined with certainty which form of the molecule one is observing. An additional problem is that the level of sensitivity obtainable with microcomplement fixation is well below that of the radiochemical methods. Perhaps the more sensitive radioimmunoassay methodology (Skelley et al., 1973) can be applied in this area.

VII. ENZYMATIC ASSAYS FOR FERREDOXIN

All known ferredoxins may be reversibly reduced and oxidized by partially purified enzyme systems. In many cases, the reactions can be utilized for the quantitative determination of ferredoxin based on its catalytic activity in the reaction, resulting in the formation of a specific, easily measurable product. Since the ferredoxin functions in a manner analogous to that of a coenzyme, such systems can serve as the basis of a very sensitive assay for active ferredoxin.

Three enzymatic assays for ferredoxin, varying in both sensitivity and specificity, have been described. Tagawa and Arnon (1962) demonstrated that both clostridial and spinach ferredoxins stimulate NADP reduction by ferredoxin-depleted spinach chloroplasts illuminated with visible light. They also devised a more purified system, consisting of ferredoxin–NADP reductase from spinach chloroplasts, *Desulfovibrio vulgaris* hydrogenase, and a small amount of benzylviologen, which was strongly dependent upon either *C. pasteurianum* or spinach ferredoxin for production of NADPH with H_2 as hydrogen donor. The quantitative dependence of this assay system on ferredoxin was not mentioned in this report. This system forms the basis (Brodrick, 1974) of a quantitative assay in which approximately 10 ng of *C. acidi-urici* ferredoxin can be detected. The usefulness of this assay is limited, however, by the necessity of performing the measurements in anaerobic cuvettes under H_2 and by the difficulty in preparing the protein reagents necessary for the assay.

In a modification of the standard assay for plant-type ferredoxin with ferredoxin-depleted chloroplasts, Haslett *et al.* (1973) measured ferredoxin-dependent formation of O_2 by illumination of chloroplasts with visible light, using an O_2 electrode. This assay system detects as little as 1 μg of spinach ferredoxin. Clostridial ferredoxin would probably function in this system, but its activity is not reported.

The most commonly used enzymatic assay for clostridial ferredoxin is the clastic assay system (Rabinowitz, 1972) in which pyruvate is converted to acetyl-CoA and acetyl phosphate in a reaction catalyzed in the presence of ferredoxin by a crude extract of *C. pasteurianum* depleted of ferredoxin by treatment with DEAE-cellulose. The formation of acetyl-CoA and acetyl phosphate (measured at 540 nm as the ferric hydroxamate) is dependent on ferredoxin, and as little as 0.1 μg may be detected. Flavodoxin can replace clostridial ferredoxin in this system, as can dyes of low redox potential such as methylviologen; but plant-type ferredoxin is inactive.

The major advantage of using an enzymatic assay for detection of ferredoxin synthesized *in vitro* is that enzymatically active ferridoxin would be the only form detected. One could thus study the formation of the iron–sulfur clusters as well as the formation of the polypeptide chain. However, none of these enzymatic assay systems is sufficiently sensitive to be useful in experiments of protein synthesis *in vitro*. In addition, the demonstrable lack of absolute specificity of these assays for ferredoxin may introduce some doubt as to what protein is responsible for the enzymatic activity observed after incubation of *in vitro* protein synthesis mixtures.

VIII. SYNTHESIS OF CLOSTRIDIAL FERREDOXIN OR APOFERREDOXIN *IN VITRO*

The clostridial ferredoxin polypeptide apoferredoxin has unique properties that are advantageous for demonstration of its synthesis *in vitro* by chemical means. First, few proteins of such low molecular weight occur in significant amounts in these bacteria. Ferredoxin accounts for approximately 1% of the total soluble protein in *C. acidi-urici*. Second, apoferredoxin is a highly acidic polypeptide that adheres very strongly to ion exchange resins such as DEAE-cellulose or DEAE-Sephadex. A third property of importance is that the clostridial ferredoxins known to date do not contain all twenty of the amino acids commonly found in proteins. Ferredoxin from *C. pasteurianum* contains no leucine, methionine, histidine, arginine, or tryptophan (Lovenberg et al., 1963) (Table I). All of these properties have been used in our demonstration of synthesis of the apoferredoxin polypeptide *in vitro* by polyribosomes prepared from *C. pasteurianum* (Brodrick, 1974), thus demonstrating that clostridial ferredoxin, like the vast majority of cellular polypeptides, is synthesized by the ribosomal system.

Polyribosomes that were highly active in overall incorporation of radioactive amino acids into protein were prepared by a procedure involving (a) rapid harvest of a log phase culture grown under specific condition that had been shown to yield highly active lysates containing many polyribosomes; (b) quick lysis of the culture with egg white lysozyme and detergents; and (c) purification of polyribosomes from the crude lysate by centrifugation through a discontinuous sucrose gradient. The preparation obtained directed the synthesis of many specific proteins in the entire molecular weight range found in *C. pasteurianum* crude extracts, as judged by discontinuous SDS–polyacrylamide slab gel electrophoresis with an acrylamide gradient followed by autoradiography of radioactive protein bands produced by incorporation of amino acids *in vitro*. The amount of incorporation observed was greatly reduced if only the amino acids Leu, His, Arg, and Trp were omitted from the assay mixture. Furthermore, amino acid incorporation

TABLE I
AMINO ACID COMPOSITION OF *C. pasteurianum* FERREDOXIN[a]

Aliphatic	Aliphatic hydroxy	Aromatic	Basic	Acidic	Amide	S
Leu 0		Trp 0	Arg 0			Met 0
			His 0			
Ala 8	Ser 5	Phe 1	Lys 1	Asp 5	Asn 3	Cys 8
Val 6	Thr 1	Tyr 1		Glu 2	Gln 2	
Ile 5						
Gly 4						
Pro 3						

[a] From Lovenberg et al. (1963) and Tanaka et al. (1966).

by this polyribosome preparation was stimulated by a ribosome salt-wash fraction and messenger RNA from *C. pasteurianum* (Stallcup and Rabinowitz, 1973). The polyribosomes were capable of initiation of new polypeptides, since formyl-Met derived from formylmethionyl-tRNA^{fMet} was incorporated intact into protein. Many specific protein bands were observed after SDS-gel electrophoresis and autoradiography of assay mixtures containing formyl-[^{35}S]Met-tRNA^{fMet} as the only source of radioactive label.

Chemical identification of apoferredoxin synthesized *in vitro* by *C. pasteurianum* polyribosomes was based on several of the unique properties of ferredoxin mentioned previously. Assay mixtures for protein synthesis *in vitro* were incubated with a complete amino acid mixture containing [^{14}C]Ala, [^{14}C]Ile, and [^{14}C]Val, three amino acids that occur in high mole percentage in *C. pasteurianum* ferredoxin (Table I). The proteins in the mixture were then reduced and alkylated with iodoacetamide in 8 M urea. In some experiments, the mixture was digested overnight with trypsin prior to reduction and alkylation. This treatment results in the formation of a 52 amino acid peptide, since *C. pasteurianum* ferredoxin contains only a single lysine residue at position 3 and no arginine (Fig. 1). The alkylated mixture was then applied to a small column of DEAE-Sephadex A-25 equilibrated with 0.1 M NH$_4$OAc. After washing with intermediate concentrations of NH$_4$OAc, apoferredoxin was specifically eluted with 0.3 M NH$_4$OAc. After concentration by lyophilization, the fractions containing alkylated apoferredoxin were subjected to electrophoresis on a highly cross-linked, discontinuous SDS-polyacrylamide slab gel containing 8 M urea, modified from a technique previously described for separation of low molecular weight proteins (Swank and Munkres, 1971). After drying of the gel slab and autoradiography, apoferredoxin synthesized *in vitro* was identified by comparison of positions of bands of autoradiographic density with the position of stained carrier apoferredoxin. When such comparison was made, it was found that a double band of autoradiographic density was aligned with the stained apoferredoxin band, perhaps indicating apoferredoxin and its formylmethionyl derivative. However, in experiments where the protein mixture was subjected to trypsin digestion prior to chromatography on DEAE-Sephadex, a single autoradiographic density band corresponding in both shape and size to carrier apoferredoxin (after trypsin treatment) was observed, and the bands of density at the position of unmodified apoferredoxin were absent.

Synthesis of apoferredoxin by *C. pasteurianum* polyribosomes was confirmed in two ways. First, amino acid incorporation experiments similar to those described above were performed in which Leu, Arg, His,

```
(Amino terminal)  Ala¹ -     Ile    -
                  Tyr  -     Phe³⁰ -
                  Lys  -     Val   -
                  ILE  -     ILE   -
                  ALA  -     ALA^Asp -
                  ASP  -     ASP   -
                  Ser  -     Thr   -
                  CYS  -     CYS   -
                  Val  -     Ile   -
                  Ser¹⁰-     Asp   -
                  CYS  -     CYS⁴⁰ -
                  GLY  -     GLY   -
                  Ala  -     Asn   -
                  CYS  -     CYS   -
                  ALA  -     ALA   -
                  Ser  -     Asn   -
                  Glu  -     Val   -
                  CYS  -     CYS   -
                  PRO  -     PRO   -
                  VAL²⁰-     VAL   -
                  Asn  -     Gly⁵⁰ -
                  ALA  -     ALA   -
                  Ile  -     Pro   -
                  Ser  -     Val   -
                  GLN  -     GLN   -
                  Asp  -     Glu   -    (Carboxyl terminal)
                  Ser  -
```

Fig. 1. Amino acid sequence of *C. pasteurianum* ferredoxin. Capital letters are used to emphasize the symmetry of sequence of *C. pasteurianum* ferredoxin relative to the two halves of the molecule (i.e., these are identical). From Tanaka et al. (1966).

and Trp, amino acids not present in *C. pasteurianum* ferredoxin, were omitted. In this instance, autoradiographic density observed after electrophoresis was not significantly reduced in the area of the stained apoferredoxin band, whereas the intensity at other electrophoretic mobilities was markedly reduced when compared to control mixtures containing all twenty amino acids. Additional confirmation was achieved by comparison of the autoradiographic density patterns obtained after purification and electrophoresis of trypsin-digested amino acid incorporation assay mixtures in which the only source of radioactive label was either isoleucine or leucine. Since *C. pasteurianum* ferredoxin contains no leucine (Table I) no autoradiographic density would be expected at the position of trypsin-digested apoferredoxin when radioactive leucine is used. Comparison of peak areas after tracing the autoradiographs with a microdensitometer revealed that the density with leucine label was only 10% of that with isoleucine, indicating that roughly 90% of the protein corresponding to the position of apoferredoxin was, in fact, apoferredoxin.

Two other reports claiming demonstration of synthesis *in vitro* of *C. pasteurianum* ferredoxin have appeared previously. In the first,

4. BIOSYNTHESIS OF IRON–SULFUR PROTEINS

Trakatellis and Schwartz (1969), using a crude extract from *C. pasteurianum* containing several grams of ribosomes, observed a small amount of incorporation of radioactive amino acids into protein after incubation at 37°C. The sole criterion used by these authors for demonstration of synthesis of ferredoxin in these crude extracts was purification of the product by chromatography on DEAE-cellulose and reprecipitation to constant specific activity. In the second report, Nepokroeff and Aronson (1970) prepared ribosomes and high-speed supernate by ultracentrifugation of crude extracts of *C. pasteurianum* cells that had been lysed by incubation with egg white lysozyme. Transfer RNA aminoacylated with radioactive amino acids served as the source of label in protein synthesis incubation mixtures containing crude ribosomes and high-speed supernate fraction from *C. pasteurianum*. Partial purification of the protein was achieved by acetone fractionation followed by performic acid oxidation and digestion of the product with trypsin. Examination of radioactive tryptic peptides was then used as the criterion for demonstration of apoferredoxin synthesis *in vitro*. In neither report was any attempt at quantitation made. However, it appears that in both cases, the overall incorporation of amino acids into protein was very low, since large amounts of ribosomes were necessary for significant incorporation of radioactivity.

In the polyribosome dependent system form *C. pasteurianum* (Brodrick, 1974), approximately 1 nmole of [^{14}C]valine was incorporated into protein by 0.25 mg polyribosomes under optimal conditions, when messenger RNA and crude initiation factors from *C. pasteurianum* (Stallcup and Rabinowitz, 1973) were included. This level of incorporation compares favorably with that observed by others in translating bacteriophage messenger RNA with ribosome dependent systems from *E. coli*. When Ala, Ile, and Val were used as sources of radioactive label in a complete mixture of amino acids for synthesis of *C. pasteurianum* apoferredoxin, approximately 3% of the total TCA precipitable counts were recovered after DEAE-Sephadex chromatography, of which of the order of 5–10% had the electrophoretic mobility of carrier apoferredoxin. This amount of synthesis agrees well with the amount of ferredoxin in *C. pasteurianum*, which we estimate to be about 0.2–0.3% of the total protein.

The *in vitro* protein-synthesizing system described here is a more precisely defined one than has been previously available. The results obtained with it contribute to the most convincing demonstration to date of *in vitro* synthesis of apoferredoxin. However, the results obtained so far answer almost none of the questions raised earlier in this discussion (Section II) concerning the details and the unique aspects involved in the synthesis of iron–sulfur proteins.

REFERENCES

Adman, E. T., Sieker, L. C., and Jensen, L. H. (1973). *J. Biol. Chem.* **248**, 3987.
Armstrong, J. J., Surzycki, S. J., Moll, B., and Levine, R. P. (1971). *Biochemistry* **10**, 692.
Beck, J.-P., Beck, G., Wong, K. Y., and Tomkins, G. M. (1972). *Proc. Natl. Acad. Sci. U.S.A.* **69**, 3615.
Beckwith, J. R., and Zipser, D., eds. (1970). "The Lactose Operon." Cold Spring Harbor Lab., Cold Spring Harbor, New York.
Boime, I., and Leder, P. (1972). *Arch. Biochem. Biophys.* **153**, 706.
Brodrick, J. W. (1974). Ph.D. Dissertation, University of California, Berkeley.
Cusanovich, M. A., and Edmondson, D. E. (1971). *Biochem. Biophys. Res. Commun.* **45**, 327.
Darnall, D. W., and Klotz, I. M. (1975). *Arch. Biochem. Biophys.* **166**, 651.
Dubourdieu, M., and Le Gall, J. (1970). *Biochem. Biophys. Res. Commun.* **38**, 965.
Fee, J. A., and Palmer, G. (1971). *Biochim. Biophys. Acta* **245**, 175.
Finazzi-Agro, A., Cannella, C., Graziani, M. T., and Cavallini, D. (1971). *FEBS Lett.* **16**, 172.
Fox, J. L., Smith, S. S., and Brown, J. R. (1972). *Z. Naturforsch., Teil B* **27**, 1096.
Haselkorn, R., and Rothman-Denes, L. B. (1973). *Annu. Rev. Biochem.* **42**, 397.
Haslett, B. G., Cammack, R., and Whatley, F. R. (1973). *Biochem. J.* **136**, 697.
Hong, J.-S., and Rabinowitz, J. C. (1967). *Biochem. Biophys. Res. Commun.* **29**, 246.
Hong, J.-S., and Rabinowitz, J. C. (1970a). *J. Biol. Chem.* **245**, 4988.
Hong, J.-S., and Rabinowitz, J. C. (1970b). *J. Biol. Chem.* **245**, 4995.
Hong, J.-S., and Rabinowitz, J. C. (1970c). *J. Biol. Chem.* **245**, 6574.
Jeng, D., and Mortenson, L. E. (1968). *Biochem. Biophys. Res. Commun.* **32**, 984.
Job, R. C., and Bruice, T. C. (1975). *Proc. Natl. Acad. Sci. U.S.A.* **72**, 2478.
Kelly, B., and Levy, J. G. (1971). *Biochemistry* **10**, 1763.
Knight, E., Jr., and Hardy, R. W. F. (1966). *J. Biol. Chem.* **241**, 2752.
Kurahashi, K. (1974). *Annu. Rev. Biochem.* **43**, 445.
Laemmli, U. K. (1970). *Nature (London)* **227**, 680.
Lengyel, P., and Söll, D. (1969). *Bacteriol. Rev.* **33**, 264.
Lipmann, F. (1971). *Science* **173**, 875.
Lode, E. T., Murray, C. L., Sweeney, W. V., and Rabinowitz, J. C. (1974). *Proc. Natl. Acad. Sci. U.S.A.* **71**, 1361.
Lode, E. T., Murray, C. L. Rabinowitz, J. C. (1976). *J. Biol. Chem.* **251**, 1675.
Lovenberg, W., Buchanan, B. B., and Rabinowitz, J. C. (1963). *J. Biol. Chem.* **238**, 3899.
Lucas-Lenard, J., and Lipmann, F. (1971). *Annu. Rev. Biochem.* **40**, 409.
McCarthy, K. F., and Lovenberg, W. (1970). *Biochem. Biophys. Res. Commun.* **40**, 1053.
Malkin, R., and Rabinowitz, J. C. (1966). *Biochem. Biophys. Res. Commun.* **23**, 822.
Mayhew, S. G., and Massey, V. (1969). *J. Biol. Chem.* **244**, 794.
Neilands, J. B. (1974). *In* "Microbial Iron Metabolism" (J. B. Neilands, ed.), p. 4. Academic Press, New York.
Nepokroeff, C., and Aronson, A. I. (1970). *Biochemistry* **9**, 2074.
Nitz, R. M., Mitchell, B., Gerwing, J., and Christensen, J. (1969). *J. Immunol.* **103**, 319.
Nomura, M., Tissieres, A. and Lengyel, P., eds. (1974). "Ribosomes." Cold Spring Harbor Lab., Cold Spring Harbor, New York.

O'Farrell, P. Z., and Gold, L. M. (1973). *J. Biol. Chem.* **248,** 5512.
Orme-Johnson, W. H., Hansen, R. E., Beinert, H., Tsibris, J. C. M., Bartholomaus, R. C., and Gunsalus, I. C. (1968). *Proc. Natl. Acad. Sci. U.S.A.* **60,** 368.
Que, L., Jr., Holm, R. H., and Mortensen, L. E. (1975). *J. Am. Chem. Soc.* **97,** 463.
Rabinowitz, J. C. (1972). *In* "Methods in Enzymology" (A. San Pietro, ed.), Vol. 24, Part B, p. 431. Academic Press, New York.
Rall, S. C., Bolinger, R. E., and Cole, R. D. (1969). *Biochemistry* **8,** 2486.
Rhoads, R. E., McKnight, G. S., and Schimke, R. T. (1973). *J. Biol. Chem.* **248,** 2031.
Salser, W., Gesteland, R. F., and Bolle, A. (1967). *Nature (London)* **215,** 588.
Schweet, R., Lamfrom, H., and Allen, E. (1958). *Proc. Natl. Acad. Sci. U.S.A.* **44,** 1029.
Schweiger, M., and Gold, L. M. (1969). *Cold Spring Harbor Symp. Quant. Biol.* **34,** 763.
Shanmugam, K. T., Buchanan, B. B., and Arnon, D. I. (1972). *Biochim. Biophys. Acta* **256,** 477.
Siegel, L. M. (1975). *In* "Metabolism of Sulfur Compounds" (D. Greenberg, ed.), p. 217. Academic Press, New York.
Skelley, D. S., Brown, L. P., and Besch, P. K. (1973). *Clin. Chem.* **19,** 146.
Stallcup, M. R., and Rabinowitz, J. C. (1973). *J. Biol. Chem.* **248,** 3209.
Studier, F. W. (1972). *Science* **176,** 367.
Studier, F. W. (1973). *J. Mol. Biol.* **79,** 237.
Swank, R. T., and Munkres, K. D. (1971). *Anal. Biochem.* **39,** 462.
Tagawa, K., and Arnon, D. I. (1962). *Nature (London)* **195,** 537.
Tager, H. S., and Steiner, D. F. (1974). *Annu. Rev. Biochem.* **43,** 509.
Tanaka, M., Nakashima, T., Benson, A., Mower, H., and Yasunobu, K. T. (1966). *Biochemistry* **5,** 1666.
Tomati, U., Matarese, R., and Federici, G. (1974). *Phytochemistry* **13,** 1703.
Trakatellis, A. C., and Schwartz, G. (1969). *Proc. Natl. Acad. Sci. U.S.A.* **63,** 436.
Villarejo, M., and Westley, J. (1963). *J. Biol. Chem.* **238,** 4016.
Zubay, G., and Chambers, D. A. (1969). *Cold Spring Harbor Symp. Quant. Biol.* **34,** 753.

CHAPTER 5

Role of Iron-Sulfur Proteins in Formate Metabolism

RUDOLF K. THAUER, GEORG FUCHS, and KURT JUNGERMANN

I. Introduction.. 121
II. Role of Formate in Metabolism................................ 122
    A. Formate as Intermediate or Substrate in Energy Metabolism... 122
    B. Formate as Intermediate in Anabolism....................... 128
    C. Formate as a By-Product of Metabolism..................... 128
III. Ferredoxin and Formate Metabolism............................ 129
    A. Ferredoxin-Dependent Formate Dehydrogenases.............. 129
    B. Ferredoxin-Dependent Monocarboxylic Acid Cycle in *Clostridium kluyveri*.. 137
    C. Ferredoxin as an Indirect Electron Donor in CO_2 Reduction to Formate... 138
    D. Ferredoxin-Dependent Activation of Pyruvate:Formate Lyase.. 141
IV. Formate Dehydrogenases and Formate Metabolism............... 143
    A. Iron–Sulfur Protein Formate Dehydrogenases 143
    B. Formate Dehydrogenases Not Known to Be Iron–Sulfur Proteins 149
    C. Comparison of Formate Dehydrogenases..................... 149
V. Concluding Remarks... 150
    References.. 151

I. INTRODUCTION

In 1964, Brill *et al.* first demonstrated that in cell-free extracts of "*Methanobacillus omelianskii*" and of *Clostridium acidi-urici*, the oxidation of formate to CO_2 is dependent on ferredoxin. In 1971, the formate dehydrogenase from *Pseudomonas oxalaticus* was found to be an iron–sulfur flavoprotein (Höpner and Trautwein, 1971). Since then numerous reports on a direct or indirect involvement of iron–sulfur proteins in formate metabolism have been published. We shall focus attention upon recent work in this selected area. This chapter (1) reviews the enzymes and reactions involved in formate metabolism, (2) discusses the role of ferredoxin in formate metabolism, (3) describes the properties and function of iron–sulfur protein formate dehydrogenases, and (4) discusses the properties of formate dehydrogenases, for which it is not known whether they contain nonheme iron or acid-labile sulfur.

This chapter does not describe the properties of aldehyde oxidase from mammalian liver, which is known to be an iron–sulfur flavoprotein and which catalyzes the oxidation of formaldehyde to formate. The enzyme is not specific for formaldehyde; due to the high K_m for formaldehyde of 3.8×10^{-1} M (Palmer, 1962), the oxidation of formaldehyde via the oxidase is not considered to be of physiological importance (for a review, see Massey, 1973).

II. ROLE OF FORMATE IN METABOLISM

In the following three subsections, the role of formate in metabolism is reviewed briefly without considering the involvement of iron–sulfur proteins. This is to introduce the reader into this special subject; a recent review is not available.

The enzymes known to catalyze the formation and catabolism of formate are summarized in Table I. Of the enzymes listed the formate dehydrogenase(s) from aerobic bacteria and the aldehyde oxidase(s) from mammalian liver have definitively been shown to be iron–sulfur proteins. The activities of formate dehydrogenase from several anaerobic bacteria are also known to be dependent on ferredoxin. The formate dehydrogenases from anaerobic bacteria, yeasts, and plants (see Section IV,B); the glutathione-independent formaldehyde dehydrogenase from *Pseudomonas* AM1 (Johnson and Quayle, 1964); and the formyl-FH$_4$ dehydrogenase from liver (Kutzbach and Stokstad, 1971) have not been characterized with respect to their iron and sulfur content. It is therefore not known whether or not they are iron–sulfur proteins. The transferases, hydrolases, lyases, and ligases listed in Table I probably do not contain nonheme iron and acid-labile sulfur, since all iron–sulfur proteins known to date are involved in oxidoreduction reactions.

A. Formate as Intermediate or Substrate in Energy Metabolism

Many aerobic bacteria and several yeasts are capable of growth upon methane, methanol, or formate. In the energy metabolism of these organisms, the single carbon compounds are oxidized to CO_2. Methanol, formaldehyde, formylglutathione (when the formaldehyde dehydrogenase is glutathione dependent), and formate are intermediates. They are formed from the growth substrates via a methane monooxygenase, a methanol oxidase or dehydrogenase, formaldehyde dehydrogenase, and S-formylglutathione hydrolase, respectively. The oxidation of formate to CO_2 is mediated by formate dehydrogenase (for reviews, see Ribbons et al., 1970; Quayle, 1972).

One carbon compounds can also be metabolized by anaerobic bacteria. The methanogenic bacteria grow on methanol or formate or CO_2 plus H_2 as energy source. Methanol is disproportionated to CO_2 and methane; formaldehyde(?), formyl tetrahydrofolate, formate, and a methylcorrinoid are assumed to be intermediates. When formate is the substrate, 3 moles of CO_2 and 1 mole of methane are formed from 4 moles of formate. Formate appears to be first oxidized to CO_2, which is then in part reduced to methane. When CO_2 plus H_2 are the substrates only methane is generated. The initial step in this process is most likely the reduction of CO_2 to formate. Free formate, however, has been questioned as an intermediate in methane synthesis (for reviews, see Stadtman, 1967; Pine, 1971; McBride and Wolfe, 1971; Wolfe, 1971; Bryant et al., 1971).

Clostridium thermoaceticum and several other anaerobic bacteria growing on carbohydrates or purines utilize CO_2 as an electron acceptor in energy metabolism and reduce it to acetate. Formate and formyl tetrahydrofolate, which are formed from CO_2 via formate dehydrogenase and formyl tetrahydrofolate synthetase have been shown to be intermediates (for review, see Ljungdahl and Wood, 1969).

Enterobacteriaceae and many other bacteria anaerobically ferment carbohydrates to pyruvate, which is then cleaved to acetyl-CoA or acetyl phosphate and formate. Pyruvate cleavage is catalyzed by an enzyme named pyruvate formate lyase (Chase and Rabinowitz, 1968). The enzyme of *Escherichia coli* (Knappe et al., 1974), *Clostridium butyricum* (Thauer et al., 1972), *Streptococcus faecalis* (Lindmark et al., 1969), and *Rhodospirillum rubrum* (Jungermann and Schön, 1974) catalyzes the CoA-dependent cleavage of pyruvate to acetyl-CoA and formate, while the pyruvate formate lyase of *Micrococcus lactilyticus* (McCormick et al., 1962a,b) has been reported to mediate the phosphate-dependent breakdown of pyruvate to acetyl phosphate and formate. The formate formed from pyruvate is either oxidized to CO_2 and H_2 via formate hydrogen lyase [formate dehydrogenase plus electron carrier(s) plus hydrogenase (see Douglas et al., 1974)] or is excreted into the medium (Piéchaud et al., 1967; Pichinoty, 1969).

TABLE I
ENZYMES CATALYZING THE FORMATION AND UTILIZATION OF FORMATE

Enzyme	Reaction presumably catalyzed *in vivo*[a]	Organisms	References
Oxidoreductases			
Formate dehydrogenases	$HCOO^- + X_{ox} \rightarrow CO_2 + X_{red}^{2-} + H^+$ $X = NAD$; ferredoxin; flavodoxin; F_{420}; cytochromes; quinones(?)	Animals, plants, yeasts, aerobic bacteria, and anaerobic bacteria except for those listed below	See Table VI
	$CO_2 + X_{red}^{2-} + H^+ \rightarrow HCOO^- + X_{ox}$ $X =$ ferredoxin, NADP, NAD(?) $F_{420}(?)$, flavodoxin(?)	*Clostridium pasteurianum, C. thermoaceticum, C. formicoaceticum*, methanogenic bacteria(?)	
Formaldehyde dehydrogenase EC 1.2.1.1 (specific for formaldehyde)	H_2CO + glutathione + NAD $\rightarrow$ formylglutathione + NADH	Aerobic bacteria, yeasts, animals	Strittmatter and Ball, 1955; Harrington and Kallio, 1960; Johnson and Quayle, 1964; Rose and Racker, 1962, 1966; Sahm and Wagner, 1973; Kato et al., 1972, 1974; Uotila and Koivusalo, 1974a; Schütte et al., 1975; Fujii and Tonomura, 1972
(not specific for formaldehyde)	$H_2CO + H_2O + X_{ox} \rightarrow HCOO^- + X_{red} + H^+$ $X =$ dichlorophenolindophenol; the physiological electron acceptor is not known	*Pseudomonas* AM 1	Johnson and Quayle, 1964

Enzyme	Reaction	Organism	Reference
Aldehyde dehydrogenases (not specific for formaldehyde) EC 1.2.1.3/4/5	$H_2CO + NAD(P) + H_2O \rightarrow HCOO^- + NAD(P)H + H^+$ (probably not of physiological importance)	Animals, bacteria	Jakoby, 1963
Aldehyde oxidase (not specific for formaldehyde) EC 1.2.3.1	$H_2CO + H_2O + O_2 \rightarrow HCOO^- + H_2O_2 + H^+$ (probably not of physiological importance)	Animals	Massey, 1973
Catalase EC 1.11.1.6	$HCOO^- + H_2O_2 + H^+ \rightarrow CO_2 + 2 H_2O$	Aerobic organisms	Oro and Rappoport, 1959; Schulman and Richert, 1959; Vaisey et al., 1961; Nicholls and Schonbaum, 1963; Leek et al., 1972; Halliwell, 1974
Formyl-FH$_4$ dehydrogenase EC 1.5.1.6	Formyl $FH_4 + NADP + H_2O \rightarrow CO_2 + FH_4 + NADPH$ (in the absence of NADP, formate is formed)	Animals (liver)	Kutzbach and Stokstad, 1968, 1971
Nonenzymatic	Glyoxylate$^- + H_2O_2 \rightarrow HCOO^- + CO_2 + H_2O$ (involved in photorespiration)	Plants and algae	Zelitch, 1972; Elstner and Heupel, 1973
Transferases			
CoA transferases	Formyl CoA + succinate $\rightarrow$ formate + succinyl-CoA	Pseudomonas oxalaticus	Quayle et al., 1961; Quayle, 1963
	Formyl CoA + acetate $\rightarrow$ formate + acetyl-CoA (nonphysiological reaction)	Clostridium kluyveri	Sly and Stadtman, 1963a
Formate kinase EC 2.7.2.6	Formyl phosphate + ADP $\rightarrow$ formate + ATP (function is not known)	Clostridium cylindrosporum	Sly and Stadtman, 1963b

(continued)

TABLE I (*Continued*)

Enzyme	Reaction presumably catalyzed *in vivo*[a]	Organisms	References
Hydrolases			
S-Formylglutathione hydrolase	S-Formylglutathione + $H_2O \rightarrow$ formate + glutathione	Aerobic bacteria, yeasts, animals	Kato *et al.*, 1974; Uotila and Koivusalo, 1974b; Schütte *et al.*, 1975
Kynurenine formamidase EC 3.5.1.9	Formylkynurenine + $H_2O \rightarrow$ formate + kynurenine	Animals, yeasts, *Xanthomonas pruni*	Mehler and Knox, 1950; Jakoby, 1954; Bailey and Wagner, 1974
Peptide deformylase	N-Formylmethionylaminoacyl-tRNA + $H_2O \rightarrow$ methionylaminoacyl-tRNA + formate	Prokaryotes	Adams, 1968; Livingston and Leder, 1969; Takeda and Webster, 1968; Lucas-Lenard and Lipmann, 1971
N-Formylmethionine deformylase EC 3.5.1.31	N-Formylmethionine + $H_2O \rightarrow$ formate + methionine	*Euglena gracilis*	Aronson and Lugay, 1966
N-Formylaspartate deformylase EC 3.5.1.8	N-Formylaspartate + $H_2O \rightarrow$ formate + aspartate	*Pseudomonas* sp. (catabolism of histidine and imidazolyl acetate)	Ohmura and Hayaishi, 1957
GTP-8-formylhydrolase (GTP cyclohydrolase) EC 3.5.4.16	GTP + 2 $H_2O \rightarrow$ formate + 7,8-dihydroneopterin triphosphate	Plants, yeasts, bacteria	Burg and Brown, 1968; Wolf and Brown, 1969; Elstner and Suhadolnik, 1971, 1975; Cone *et al.*, 1974; Plowman *et al.*, 1974

Enzyme	Reaction	Organism	References
Lyases			
Pyruvate formate lyases	Pyruvate + CoA → acetyl-CoA + formate	Many (not all) anaerobic bacteria	Chase and Rabinowitz, 1968; Knappe et al., 1969, 1974; Lindmark et al., 1969; Nakayama et al., 1971; Thauer et al., 1972; Jungermann and Schön, 1974
	Pyruvate + P_i → acetyl phosphate + formate	Micrococcus lactilyticus	McCormick et al., 1962a,b
Oxalate decarboxylase EC 4.1.1.2	Oxalate^{2-} + H$^+$ → formate$^-$ + CO_2	Fungi, yeasts	Shimazono and Hayaishi, 1957; Emiliani and Bekes, 1964; Lillehoj and Smith, 1965
Ligases			
Formyl tetrahydrofolate synthetase EC 6.3.4.3	Formate + FH_4 + ATP → formyl-FH_4 + ADP + P_i	Animals, plants, yeasts, bacteria, except for those listed below	For reviews, see Huennekens, 1968; Rader and Huennekens, 1973
	Formyl-FH_4 + ADP + P_i → formate + FH_4 + ATP	Clostridium cylindrosporum, Methanosarcina barkeri(?), Micrococcus aerogenes(?)	Blaylock and Stadtman, 1966; for reviews, see Barker, 1961; Stadtman, 1967; Whiteley, 1967

[a] For functions, see Sections II,A–C.

Pseudomonas oxalaticus can metabolize oxalate, which is oxidized to 2 moles of CO_2 with molecular oxygen as terminal electron acceptor. Oxalyl-CoA, formyl-CoA, and formate are intermediates. Formate is generated from formyl-CoA via CoA transfer to succinate and is oxidized to CO_2 via formate dehydrogenase (for reviews, see Kornberg and Elsden, 1961, Kornberg, 1966).

Several *Pseudomonas* species can grow on histidine or β-imidazolyl-4(5) acetate as energy source (Hayaishi et al., 1957; Tabor and Mehler, 1954). N-Formyl glutamate and N-formyl aspartate have reported to be intermediates in the energy metabolism of these organisms. Both formamide compounds are hydrolyzed to formate and the respective amino acid. Deformylation is mediated by N-formyl aspartate deformylase (Ohmura and Hayaishi, 1957).

B. Formate as Intermediate in Anabolism

In the anabolism of clostridia formate rather than serine (the usual precursor) has been shown to be the major precursor of one carbon unit positions such as the positions 2 and 8 of the purines, the S-methyl group of methionine, and the 5-methyl group of thymidine (Decker et al., 1967; Jungermann et al., 1968, 1970a; Thauer et al., 1972). The formate required for the synthesis of the one carbon unit positions is formed either from CO_2 (Jungermann et al., 1970b) or from pyruvate via pyruvate cleavage to acetyl-CoA and formate (Thauer et al., 1970a, 1972). Recently, a paper appeared suggesting that CO_2 reduction to formate might even be of importance in CO_2 fixation in plants (Kent, 1972a,b). However, the evidence is indirect and needs to be confirmed.

C. Formate as a By-Product of Metabolism

In the anabolism of most cells, formate is formed in small amounts during the synthesis of D-*erythro*-dihydroneopterin triphosphate from GTP. This reaction, which is catalyzed by GTP-8-formylhydrolase (GTP-cyclohydrolase), is the initial step in the biosynthesis of riboflavin, the pteridines, a pteridine cofactor for phenylalanine hydroxylation, the azapteridine ring, and the benzimidazole ring (Burg and Brown, 1968; Wolf and Brown, 1969; Elstner and Suhadolnik, 1971, 1975; Cone et al., 1974; Plowman et al., 1974).

In animals and yeasts and in the bacterium *Xanthomonas pruni*, formate is generated as a by-product of the synthesis of the pyridine nucleotides NAD and NADP. These coenzymes are formed from tryptophan via formylkynurenine, which is hydrolyzed to kynurenine and formate. This reaction is mediated by kynurenine formamidase (for reviews, see Saxton et al., 1968; Feigelson and Brady, 1974).

In prokaryotes, formate is released in minor amounts during protein synthesis, which is initiated by formylmethionine. The formyl group is hydrolytically removed via a peptide deformylase during chain elongation or after chain termination. Thus, in prokaryotes, 1 mole of formate is formed per mole of monomeric protein synthesized (for review, see Lucas-Lenard and Lipmann, 1971).

In aerobic organisms, the demethylation of methylamines and of methyl ethers is catalyzed by monooxygenases. The methyl group is oxidized to formaldehyde, which either reacts with tetrahydrofolate to form methylene tetrahydrofolate (Rader and Huennekens, 1973) or is oxidized to formate. In plants and algae, formate also appears to be formed from glyoxylate. During photorespiration, glyoxylate is assumed to be nonenzymatically oxidized with H_2O_2 to formate and CO_2 (Zelitch, 1972; Elstner and Heupel, 1973). The importance of this nonenzymatic reaction in photorespiration is, however, still unclear.

Formate formed as a by-product is either oxidized to CO_2 or activated to formyl tetrahydrofolate. The latter is then used for the synthesis of one-carbon-unit positions such as the positions 2 and 8 of the purines, the S-methyl group of methionine, and the 5-methyl group of thymidine (for review, see Rader and Huennekens, 1973). The oxidation of formate to CO_2 is mediated by a specific formate dehydrogenase (for literature, see Table VI) or nonspecifically by catalase (for literature, see Table I). The synthesis of formyl tetrahydrofolate is catalyzed by formyltetrahydrofolate synthetase. The formation of formate as a by-product of metabolism in all organisms may be the reason, why formate dehydrogenases and formyltetrahydrofolate synthetases are found almost ubiquitously.

III. FERREDOXIN AND FORMATE METABOLISM

The general properties and functions of ferredoxins have been reviewed in Volumes I and II of this treatise. The role of ferredoxin in formate metabolism will be discussed in the following section.

A. Ferredoxin-Dependent Formate Dehydrogenases

1. REDUCED FERREDOXIN:CO_2 OXIDOREDUCTASE OF *Clostridium pasteurianum*

Clostridium pasteurianum is a strictly anaerobic bacterium that can grow on glucose as sole carbon and energy source. Fermentation products are acetate, butyrate, CO_2, and hydrogen. Either molecular nitrogen or ammonium ions can be utilized as nitrogen source. In the metabolism of this organism ferredoxin-mediated oxidation–reduction reactions are of central importance (for review, see Mortenson and Nakos, 1973). In the energy metabolism, ferredoxin accepts electrons from NADH (generated in the glyceraldehyde dehydrogenase reaction) and pyruvate via NADH: ferredoxin oxidoreductase (Jungermann et al., 1971a, 1973) and pyruvate: ferredoxin oxidoreductase (Mortenson, 1963), respectively. Approximately 90% of the reduced ferredoxin is reoxidized in the catabolic hydrogenase reaction. Approximately 10% is utilized for biosynthetic purposes (Jungermann et al., 1974), one of which is the synthesis of one-carbon units from CO_2 via ferredoxin-dependent CO_2 reduction to formate (Fig. 1).

a. FUNCTION. Cell-free extracts of *C. pasteurianum* were first shown by Mortenson (1966) to mediate the oxidation of formate to CO_2 with oxidized ferredoxin as electron acceptor. Jungermann et al. (1970b) then demonstrated that the extracts also catalyze the reverse reaction, the reduction of CO_2 to formate with reduced ferredoxin as electron donor. Later, evidence was obtained indicating that, *in vivo*, only the reduction of CO_2 to formate of the two reactions is of physiological importance (Thauer et al., 1974). The formate generated from CO_2 is required for the synthesis of one-carbon-unit positions, which are not formed from serine or glycine (the usual precursors) in this organism (Thauer et al., 1972). The enzyme catalyzing CO_2 reduction was named CO_2 reductase

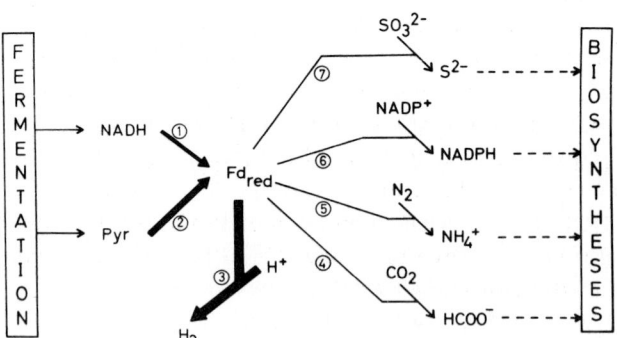

Fig. 1. Short scheme of the ferredoxin-dependent metabolism of *C. pasteurianum* growing on glucose as sole carbon and energy source. (1) NADH:ferredoxin oxidoreductase; (2) pyruvate:ferredoxin oxidoreductase; (3) ferredoxin hydrogenase; (4) reduced ferredoxin:CO_2 oxidoreductase; (5) nitrogenase; (6) reduced ferredoxin:NADP oxidoreductase; (7) sulfite reductase. (For literature, see Jungermann et al., 1974; Laishley et al., 1971.)

(reduced ferredoxin:CO_2 oxidoreductase) (Jungermann et al., 1970b) in order to distinguish it from the then known formate dehydrogenases, which catalyzed the oxidation of formate rather than the reduction of CO_2.

b. PROPERTIES. The CO_2 reductase has partially been purified (seventyfold). It still contains over 90% impurities at this stage. The enzyme has a molecular weight greater than 250,000 daltons (estimated on Sephadex G-200) (R. K. Thauer, unpublished). It catalyzes the following three reactions (Thauer et al., 1975a):

1. The reduction of CO_2 to formate with either reduced clostridial ferredoxin [$E_0' = -380$ mV; number (n) of electrons carried = 2], reduced plant ferredoxin ($E_0' = -430$ mV; $n = 1$), or reduced methylviologen ($E_0' = -440$ mV; $n = 1$).
2. The oxidation of formate to CO_2 with either ferredoxin, methylviologen, benzylviologen ($E_0' = -360$ mV; $n = 1$), methylene blue ($E_0' = 11$ mV; $n = 2$), or ferricyanide ($E_0' = 360$ mV; $n = 1$).
3. An isotopic exchange between CO_2 and formate in the absence of ferredoxin.

The active species of CO_2 i.e., CO_2 or HCO_3^- (H_2CO_3) utilized by the enzyme has been shown to be CO_2 rather than HCO_3^- (Thauer et al., 1975b). The kinetic properties of the enzyme are summarized in Table II. They have been obtained from initial velocity studies of the three reactions (for example, see Fig. 2).

c. REVERSIBLE INHIBITORS. CO_2 reductase from *C. pasteurianum* is reversibly inhibited by pseudohalides (azide, cyanate, thiocyanate), halides (fluoride, chloride, bromide, iodide), and oxoacids of nitrogen and chlorine (nitrite, nitrate, chlorate) (Table III) (Thauer et al., 1975a). Azide was found to be the most potent inhibitor. The rate of CO_2 reduction to formate, of isotopic exchange, and of formate oxidation are equally affected by the inhibitors. The dissociation constants (K_I) of the enzyme inhibitor complex increase in the group of pseudohalides in the order azide < cyanate < thiocyanate, and in the group of halides from fluoride to iodide. Nitrate is less inhibitory than nitrite. The same relative order is observed for the stability of complexes of these anions with transition metals of weak "A class" character (Pearson, 1963; Ahrland et al., 1958). Of this group of metals, Mn, Fe, Co, Zn, and Mo must be considered biologically important (Schwarz, 1974). The inhibition data thus suggest that one of these metals is involved in the binding of the ligands to CO_2 reductase.

TABLE II

Kinetic Properties of Reduced Ferredoxin: CO_2 Oxidoreductase from *C. pasteurianum*[a]

Reaction	$S_{0.5V}$ (mM)	V_{max}^{app} (munits/mg protein)	pH optimum
CO_2 reduction			
CO_2[b]	0.3	—	—
Fd (*C. pasteurianum*)	0.003	47	6.5–7.5
Fd (spinach)	0.004	23	—
Methylviologen	—	50	—
Formate oxidation			
Formate	2	—	—
Fd (*C. pasteurianum*)	0.005	45	—
Methylviologen	<0.01	—	—
Benzylviologen	<0.01	200	6.5–7
Isotopic exchange			
CO_2[b]	0.5	—	—
Formate	0.5	48	6 (V_{max}^{app})

[a] Determined in imidazole acetate buffer, pH 6.8, 100 mM at 35°C, using a ferredoxin- and nucleotide-free extract of *C. pasteurianum* (Thauer et al., 1975a,b); 1 unit = 1 μmole/minute.

[b] Note that CO_2 rather than HCO_3^- (H_2CO_3) has been shown to be the active species of "CO_2" utilized by the enzyme (Thauer et al., 1975b).

d. METALS. CO_2 reductase of *C. pasteurianum* is most probably a molybdoenzyme. Both the synthesis of the active enzyme and the synthesis of formate in growing cultures have been shown to be dependent on molybdenum (Thauer et al., 1973, 1974). Tungsten did not substitute for molybdenum, nor did it antagonize the effect of molybdenum even at tungsten concentrations one thousand times higher than those of molybdenum. This is in contrast to the formate dehydrogenase activity in *E. coli* (Pinsent, 1954; Enoch and Lester, 1972), which is decreased by the addition of tungsten, and to the formate dehydrogenase activities of *C. thermoaceticum* (Ljungdahl and Andreesen, 1975) and *Clostridium formicoaceticum* (Andreesen et al., 1974), which are increased by the addition of tungsten. A positive effect of selenium on the synthesis of the active CO_2 reductase was not observed. Concentrations of selenite of higher than 10^{-6} M even inhibited the growth of *C. pasteurianum*. This was surprising, as the synthesis of formate dehydrogenase of *E. coli* (Pinsent, 1954; Shum and Murphy, 1972), *C. thermoaceticum* (Andreesen and Ljungdahl, 1973), and *C. formicoaceticum* (Andreesen et al., 1974)

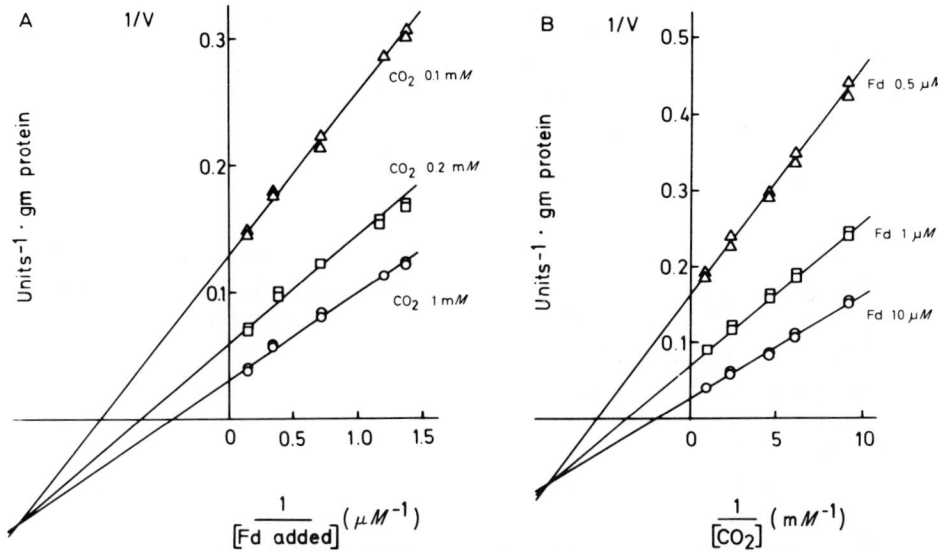

FIG. 2. Reduced ferredoxin:CO_2 oxidoreductase from *C. pasteurianum:* Double reciprocal plots of the rate of CO_2 reduction to formate versus either the ferredoxin concentration (A) or the CO_2 concentration (B). Assay conditions: 100 mM imidazole acetate buffer, pH 6.8 at 35°C; ferredoxin from *C. pasteurianum* (Fd_{ox}/Fd_{red}; $E_0' = -380$ mV) was kept reduced with molecular hydrogen (H^+/H_2; $E_0' = -420$ mV) via hydrogenase. Therefore, only approximately 50% of the ferredoxin added was in the reduced state.

TABLE III
INHIBITORS OF REDUCED FERREDOXIN: CO_2 OXIDOREDUCTASE FROM *C. pasteurianum*[a]

Inhibitor	K_I	Type of inhibition with respect to		
		Formate	CO_2	Ferredoxin
N_3^-	0.004	Competitive	Competitive	Noncompetitive
OCN^-	0.3	Competitive	Competitive	—
NO_2^-	0.4	Noncompetitive	—	—
SCN^-	1	Competitive	Competitive	—
ClO_3^-	3	Noncompetitive	—	—
F^-	5	Competitive	—	—
NO_3^-	6	Noncompetitive	—	—
Cl^-	> 5	—	—	—
Br^-	> 5	—	—	—
I^-	> 10	—	—	—

[a] From Thauer *et al.* (1975a).

have been shown to be dependent on low concentrations of selenium in the growth medium. A contamination with selenium of the reagents used to prepare the synthetic growth medium of *C. pasteurianum* was ruled out by showing that the synthesis of formate dehydrogenase of *E. coli* growing on the medium of *C. pasteurianum* was dependent on the addition of selenite (10^{-10}–10^{-9} M) (Thauer et al., 1974).

e. IRREVERSIBLE INHIBITORS. From the spectrochemical series, which orders ligands according to the stability of the respective transition metal complexes, a strong inhibitory effect of cyanide was to be expected (Vallee and Wacker, 1970). However, a reversible inhibition was not observed. Instead, the enzyme seems to be irreversibly inactivated in the presence of low concentrations of cyanide (10^{-5} M) (Thauer et al., 1973). The CO_2-reducing activity, the exchange activity, and the formate-oxidizing activity are equally affected (Thauer et al., 1975a). The inactivation rate is markedly affected by the pH, increasing with increasing pH. A plot of the apparent inactivation rate constants versus the pH is sigmoidal and resembles the titration curve of a weak acid with a pK of 9–10. As cyanide has a pK_a of 9.3 (Perrin, 1969) the data suggest that the cyanide anion rather than the undissociated acid is the inactivating agent (Thauer et al., 1975a).

f. PROTECTION FROM INACTIVATION. CO_2 reductase from *C. pasteurianum* is a very labile enzyme which is rapidly inactivated by heat (55°C), by trace amounts of molecular oxygen and, as has been described above, by very low concentrations of cyanide. Azide and cyanate (the other anions were not tested) were found to protect the enzyme from inactivation. In the presence of high concentrations (10 mM) of these monovalent anions, the rate of inactivation by heat (55°C), molecular oxygen, and cyanide decreases by a factor of more than 100 (Thauer et al., 1975a). Based on this observation, a purification procedure is being worked out using buffers containing 10 mM azide in all steps.

2. FORMATE DEHYDROGENASE OF *Clostridium acidi-urici*

Clostridium acidi-urici is an obligate anaerobe, which can grow on hypoxanthine, xanthine, or urate as an energy source. Xanthine is fermented to 4 moles of ammonia, 3 moles of CO_2, and 1 mole of acetate; in addition, small amounts of formate usually accumulate. With the more oxidized purine uric acid, the yield of CO_2 is increased to 3.5 moles and the yield of acetate is decreased to 0.7 moles, whereas with the more reduced purine hypoxanthine, the yield of CO_2 is decreased to 2 moles and the yield of acetate is increased to 1.25 moles (Barker, 1961). In addition, small amounts of butyrate are formed during hypoxanthine fermentation (Schulman et al., 1972).

a. FUNCTION. Both uric acid- and hypoxanthine-grown cells contain a formate dehydrogenase. During growth on uric acid, the function of the enzyme is most likely the oxidation of formate to CO_2 (Barker, 1961). During growth on hypoxanthine, 9% of the acetate formed is totally synthesized from CO_2 (Schulman et al., 1972). The reduction of CO_2 to formate is believed to be the initial step in this process. Whether formate oxidation during fermentation of uric acid and CO_2 reduction to formate during fermentation of hypoxanthine are mediated by the same enzyme is not known.

b. PROPERTIES. Cell-free extracts of uric acid grown cells of *C. acidi-urici* have been shown to catalyze a ferredoxin-dependent reduction of NAD with formate (Brill et al., 1964; Kearny and Sagers, 1972) and the reduction of CO_2 to formate in the presence of both reduced ferredoxin and NADH (Thauer, 1973). Neither reduced ferredoxin nor NADH were effective when tested alone. The requirement of both NADH and reduced ferredoxin in CO_2 reduction is not understood.

Formate dehydrogenase from *C. acidi-urici* has been partially purified (approximately thirtyfold) from uric acid-grown cells by Kearny and Sagers (1972). The molecular weight of the protein is at least 200,000 daltons. The partially purified enzyme has been reported to mediate the reduction of benzylviologen with formate and an isotopic exchange between CO_2 and formate in the absence of either NAD or ferredoxin. A reduction of CO_2 to formate using substrate amounts of reduced benzylviologen was not observed. The partially purified formate dehydrogenase did not mediate the reduction of NAD with formate either in the absence or presence of ferredoxin. It was not reported whether ferredoxin was reduced in these experiments. The partially purified enzyme is inactivated by irradiation with light and inhibited by low concentrations of cyanide. The effect of azide and other pseudohalides on the enzyme activities has not been determined.

Clostridium cylindrosporum is an obligate anaerobe that can grow on purines as energy source. This organism forms the same products as does *C. acidi-urici*, except that the yield of formate is higher (0.5–1.0 mole per mole decomposed) and 0.14–0.45 mole of glycine accumulate (Barker, 1961). Bradshaw and Reeder (1964) reported that in cell-free extracts of

C. cylindrosporum, the oxidation of formate can be coupled via ferredoxin to the reduction of uric acid. This finding indicates that the formate dehydrogenase from this organism is a ferredoxin-dependent enzyme. A detailed study of this formate dehydrogenase is not available.

3. FORMATE DEHYDROGENASE ACTIVITY IN CELL-FREE EXTRACTS OF *"Methanobacillus omelianskii"* AND OF THE S-ORGANISM ISOLATED FROM *"M. omelianskii"*

"Methanobacillus omelianskii" grows on ethanol plus CO_2 with the formation of acetate and methane. It has been shown to be a symbiontic mixture of two strictly anaerobic organisms named S-organism and strain MOH. The S-organism ferments ethanol to acetate and H_2, and strain MOH reduces CO_2 with H_2 to methane (Bryant *et al.*, 1967).

Brill *et al.* (1964) reported that in cell-free extracts of *"M. omelianskii,"* formate oxidation is a ferredoxin-dependent reaction. Formate oxidation to CO_2 in extracts of the S-organism was recently shown to be dependent on NAD rather than on ferredoxin (Reddy *et al.*, 1972). Strain MOH does not appear to contain a formate dehydrogenase. This has been concluded from the finding that strain MOH does not use formate as electron donor for growth and methane formation (Bryant *et al.*, 1968). It must be considered, however, that the inability to grow on formate might be a reflection of the lack of a specific carrier for formate in the membrane rather than of a formate dehydrogenase in the cells. Carbonic acids have been shown to penetrate the bacterial cell membrane only in the undissociated form (Harold and Levin, 1974; Riebeling *et al.*, 1975). Formic acid has a pK of 3.7. At the pH of the growth medium of near 7, less than 0.1% of the formic acid are in the undissociated form. It therefore cannot be excluded that the ferredoxin-dependent formate dehydrogenase described by Brill *et al.* (1964) is a component of strain MOH rather than of the S-organism. No definite explanation for the apparent discrepancy between the work of Reddy *et al.* (1972) and Brill *et al.* (1964) can be given.

Tzeng *et al.* (1975) have recently described a formate dehydrogenase from *Methanobacillus ruminantium.* This enzyme is dependent on factor 420 (Cheeseman *et al.*, 1972) rather than on ferredoxin as electron acceptor. In *Methanobacterium vannielii,* the formation of formate dehydrogenase is stimulated by selenium; tungsten has been reported to promote the growth of this organism (Stadtman, 1974; T. C. Stadtman, personal communication). This suggests that formate dehydrogenases in methane bacteria might be similar to those found in *C. thermoaceticum* and *C. formicoaceticum* (see Section III,C). Other data on formate dehydrogenases from methanogenic bacteria have not been published.

4. FORMATE DEHYDROGENASE ACTIVITY IN CELL-FREE EXTRACTS OF
 Bacillus polymixa, Klebsiella pneumoniae, AND *Sarcina ventriculi*

Nitrogen fixation with formate as electron donor has been demonstrated in cell-free extracts of *B. polymixa* (Fisher and Wilson, 1970; Yoch, 1973) and *K. pneumoniae* (Yoch, 1974). Both organisms appear to contain a ferredoxin- (or flavodoxin-) dependent formate dehydrogenase. A partially purified formate dehydrogenase from *S. ventriculi* was shown to mediate the reduction of ferredoxin by formate (Stephenson and Dawes, 1971). In crude extracts formate oxidation was not ferredoxin dependent. The enzyme is susceptible to a variety of inhibitors such as cyanide, nitrite, and nitrate (Bauchop and Dawes, 1959, 1968). The physiological function of these formate dehydrogenases has not been elucidated.

B. Ferredoxin-Dependent Monocarboxylic Acid Cycle in *Clostridium kluyveri*

Clostridium kluyveri grows on ethanol, acetate, and CO_2 or crotonate and CO_2 as the sole carbon and energy source (Thauer *et al.*, 1968). Formate derived from CO_2 has been shown to be the predominant precursor of tetrahydrofolate-activated one-carbon units in this organism (Decker *et al.*, 1967; Jungermann *et al.*, 1968, 1970a). The reduction of CO_2 to formate proceeds via a mechanism different from that observed in *C. pasteurianum*. The reaction is not catalyzed by a CO_2 reductase (a formate dehydrogenase activity could not be demonstrated in cell-free extracts of *C. kluyveri*), but by two enzymes, a ferredoxin-dependent pyruvate synthase and a pyruvate:formate lyase (Thauer *et al.*, 1970a, 1972). Pyruvate synthase catalyzes the reductive synthesis of pyruvate from acetyl-CoA and CO_2 with reduced ferredoxin as electron donor (Andrew and Morris, 1965); the active species of "CO_2" utilized by pyruvate synthase has been shown to be CO_2 rather than HCO_3^- or H_2CO_3 (Thauer *et al.*, 1975c). Pyruvate formate lyase catalyzes the cleavage of pyruvate to acetyl-CoA and formate (Fig. 3). CO_2 is thus converted to formate via a cyclic mechanism. The cycle has been designated the reductive monocarboxylic acid cycle. The evidence for the operation of the cycle consists of (1) demonstration in cell-free extracts of the two component enzymes, pyruvate synthase and pyruvate formate lyase; (2) demonstration *in vitro* of a production of formate from CO_2 and reduced ferredoxin in the presence of acetyl CoA and its inhibition by glyoxylate, which specifically inactivates pyruvate synthase (Thauer *et al.*, 1970b); and (3) the finding that *in vivo* pyruvate formate lyase

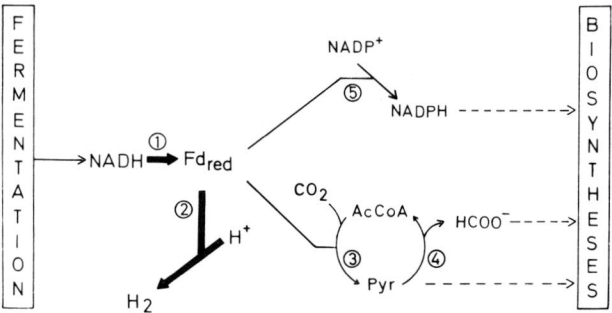

Fig. 3. Short scheme of the ferredoxin-dependent metabolism of *C. kluyveri* growing on ethanol, acetate, and CO_2 as sole carbon and energy source. (1) NADH: ferredoxin oxidoreductase; (2) ferredoxin hydrogenase; (3) pyruvate synthase; (4) pyruvate formate lyase; (5) reduced ferredoxin:NADP oxidoreductase.

is "repressed" under conditions under which CO_2 is not incorporated into formate, i.e., in cells grown in the presence of formate.

The cyclic mechanism of CO_2 reduction to formate uses the power of reduced ferredoxin (Fig. 3), as does the noncyclic mechanism found in *C. pasteurianum*. Reduced ferredoxin in *C. kluyveri* is generated solely via ferredoxin reduction with either NADH or NADPH (Jungermann *et al.*, 1969, 1971b; Thauer *et al.*, 1969, 1971; Gottschalk and Chowdhury, 1969). Which of the two pyridine nucleotides is the physiological electron donor is not definitively known. However, NADH rather than NADPH has been shown to be the reductant of ferredoxin in *C. pasteurianum* and *C. butyricum*, suggesting that this might also be the case in *C. kluyveri*. This is substantiated by the finding that the kinetic and regulatory properties of NADH:ferredoxin oxidoreductase of *C. kluyveri* and of the two saccharolytic clostridia are almost identical (Jungermann *et al.*, 1973).

C. Ferredoxin as an Indirect Electron Donor in CO_2 Reduction to Formate

1. CO_2 Reduction to Formate in *C. thermoaceticum*

Clostridium thermoaceticum is a strictly anaerobic bacterium that grows on glucose or fructose as energy source. Three moles of acetate are produced per hexose in the fermentation. One of the three is formed by a total synthesis of acetate from CO_2 (Ljungdahl and Wood, 1969; Schulman *et al.*, 1972). The initial step in acetate formation from CO_2 presumably is its reduction to formate by a soluble, NADP-specific formate dehydrogenase (Li et al., 1966; Andreesen and Ljungdahl, 1971; Thauer, 1972). The formate is reduced to acetate via tetrahydrofolate derivatives and a methylcorrinoid (Ljungdahl and Wood, 1969; Andreesen et al., 1973). The methylcorrinoid is probably converted to acetate by a process involving transcarboxylation from pyruvate (Schulman et al., 1973).

a. FERREDOXIN AS INDIRECT ELECTRON DONOR. In the electron-donating reactions of the fermentation, the hexose is oxidized to two acetate and two CO_2 via an NAD-specific glyceraldehyde–phosphate dehydrogenase and via a ferredoxin-specific pyruvate dehydrogenase (Thauer, 1972). In the electron-accepting part of the fermentation, NADPH rather than NADH or reduced ferredoxin is required for CO_2 reduction to formate. It must therefore be assumed that NADPH is either regenerated via NADP reduction with reduced ferredoxin or via transhydrogenation between NADH and NADP. As cell-free extracts of *C. thermoaceticum* contain an active reduced ferredoxin:NADP oxidoreductase rather than a transhydrogenase, reduced ferredoxin must be considered as the physiological electron donor for NADP reduction (Thauer, 1972) (Fig. 4).

Why reduced ferredoxin is not the direct electron donor in CO_2 reduction to formate as it is in *C. pasteurianum* is not known. A specific requirement for the reduced pyridine nucleotide as a hydride donor in the reaction mechanism of the enzyme is not very likely, as the formate dehydrogenase from *C. thermoaceticum* has been shown to catalyze a rapid reduction of CO_2 to formate with reduced methylviologen, which is a one-electron-transferring redox dye (Andreesen and Ljungdahl, 1971; Thauer, 1972).

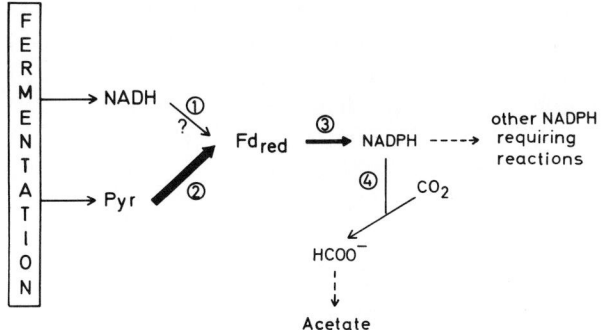

Fig. 4. Short scheme of the ferredoxin-dependent metabolism of *C. thermoaceticum* growing on either glucose or fructose. (1) NADH:ferredoxin oxidoreductase; (2) pyruvate:ferredoxin oxidoreductase; (3) reduced ferredoxin:NADP oxidoreductase; (4) formate dehydrogenase.

The presence of ferredoxin in *C. thermoaceticum* has been questioned. Poston and Stadtman (1967) demonstrated an atypical ferredoxin, whereas Li *et al.* (1966) were unable to detect ferredoxin. Instead, they obtained evidence for a flavodoxin type of protein. Recently, however, Le Gall *et al.* (unpublished, cited in Andreesen *et al.*, 1973) found two ferredoxin-type proteins in *C. thermoaceticum*. The controversial findings can easily be explained assuming that flavodoxin rather than ferredoxin is synthesized by the cells when growing on medium with low concentrations of iron and that ferrodoxin is formed when the cells are grown on media with a high iron concentration (Knight *et al.*, 1966). The medium used by Li *et al.* (1966) contained only trace amounts of iron (no iron was added); the medium used by Le Gall *et al.* contained 39 mg $Fe(NH_4)_2(SO_4)_2 \cdot 6H_2O$ per liter (Andreesen *et al.*, 1973).

b. PROPERTIES OF FORMATE DEHYDROGENASE FROM *C. thermoaceticum*. The enzyme has recently been purified (Andreesen and Ljungdahl, 1974). It differs from the ferredoxin specific CO_2 reductase from *C. pasteurianum* in several respects. The synthesis of the formate dehydrogenase from *C. thermoaceticum* is dependent on selenium plus molybdenum or tungsten in the growth medium. Tungsten not only replaces but is even better than molybdenum (Andreesen *et al.*, 1973). The partially purified enzyme contains acid-labile selenide (Andreesen and Ljungdahl, 1973) and tungsten when grown in the presence of tungsten (Ljungdahl and Andreesen, 1975). In contrast, the synthesis of the enzyme from *C. pasteurianum* is not dependent on selenium in the growth medium, and tungsten does not substitute for molybdenum in stimulating the synthesis of the active CO_2 reductase (Thauer *et al.*, 1973, 1974). The formate dehydrogenase from *C. thermoaceticum* catalyzes an isotopic exchange between CO_2 and formate only in the presence of the second substrate NADP (Li *et al.*, 1966), whereas the exchange reaction catalyzed by CO_2 reductase from *C. pasteurianum* is not dependent on the second substrate ferredoxin (Thauer *et al.*, 1975a,b). This indicates that the catalytic mechanism of the two enzymes may be essentially different. The kinetic properties of the enzyme are summarized in Table IV.

2. CO_2 REDUCTION TO FORMATE IN *C. formicoaceticum*

Like *C. thermoaceticum*, the recently described *C. formicoaceticum* belongs to the group of bacteria that can use CO_2 as electron acceptor and reduce it to acetate (Andreesen *et al.*, 1970, 1974; Schulman *et al.*, 1972). When growing on pyruvate ferredoxin is the ultimate electron donor in all reactions involved in CO_2 reduction to acetate; pyruvate

TABLE IV
KINETIC PROPERTIES OF FORMATE DEHYDROGENASE FROM *C. thermoaceticum*

Reaction	$S_{0.5V}$ (mM)	pH optimum
CO_2 reduction[a]		
HCO_3^- added	11	—
NADPH	0.02	7
Formate oxidation[b]		
Formate	0.2	—
NADP	0.1	7–9
Methylviologen	2	—

[a] From Thauer (1972).
[b] From Andreesen and Ljungdahl (1974).

oxidation to acetyl-CoA has been shown to be mediated by a pyruvate: ferredoxin oxidoreductase rather than by pyruvate:formate lyase (Linke, 1969; Andreesen et al., 1974). The first reaction in the sequence leading from CO_2 to acetate is most likely the reduction of CO_2 to formate via a soluble formate dehydrogenase. Cell-free extracts of *C. formicoaceticum* have been reported to mediate the reversible reduction of CO_2 to formate with reduced methylviologen as electron donor (Andreesen et al., 1974). The nature of the physiological electron donor is still uncertain. There is no indication so far that ferredoxin is directly involved. The slow oxidation of formate in the presence of pyridine nucleotides (only a small fraction of the activity observed with methylviologen as electron acceptor) was not stimulated by ferredoxin from either *C. formicoaceticum* or *C. pasteurianum*. Andreesen et al. (1974) therefore suggested that an unknown factor may be separated from the formate dehydrogenase during extract preparation.

Addition of tungsten or molybdenum together with selenite to the growth medium promotes the growth of *C. formicoaceticum* and specifically increases the formate dehydrogenase activity (Andreesen et al., 1974). As with *C. thermoaceticum*, tungsten is better than molybdenum. These findings indicate that the formate dehydrogenases of the two clostridia are very similar.

D. Ferredoxin-Dependent Activation of Pyruvate:Formate Lyase

Pyruvate:formate lyase catalyzes the formation of acetyl-CoA and formate from pyruvate and CoA. In *E. coli* (Knappe et al., 1974), *S.*

faecalis (Lindmark et al., 1969), and *R. rubrum* (Jungermann and Schön, 1974), the enzyme serves mainly to mediate acetyl-CoA generation for ATP synthesis in catabolism, while the clostridial lyase (Thauer et al., 1972) functions mainly to furnish formate for C_1-unit formation in anabolism.

The pyruvate:formate lyase from *E. coli* (Knappe et al., 1969, 1972) and *C. butyricum* (Thauer et al., 1972) have been reported to exist in interconvertible active and inactive forms. The activation process requires *S*-adenosylmethionine (SAM) and a reducing system with reduced ferredoxin or reduced flavodoxin as the physiological reductants. The activation is mediated by an activating enzyme, which has partially been purified from *E. coli* (Knappe et al., 1969; Vetter and Knappe, 1971). The pyruvate formate lyase is reversibly inactivated when either SAM or reduced ferredoxin are omitted from the system (Fig. 5) (Wood and Jungermann, 1972). Inactivation occurs also when *S*-adenosylmethionine is antagonized by *S*-adenosylhomocysteine or when ferredoxin is in the oxidized rather than in the reduced state. This behavior indicates that both an oxidative and a nonoxidative chemical process may participate in the interconversion of pyruvate:formate lyase. The physiological function and the mechanism of interconversion are not understood (for a discussion, see Wood and Jungermann, 1972; Knappe et al., 1974). The role of ferredoxin in the activation of pyruvate formate lyase by *S*-adenosylmethionine thus remains unclear.

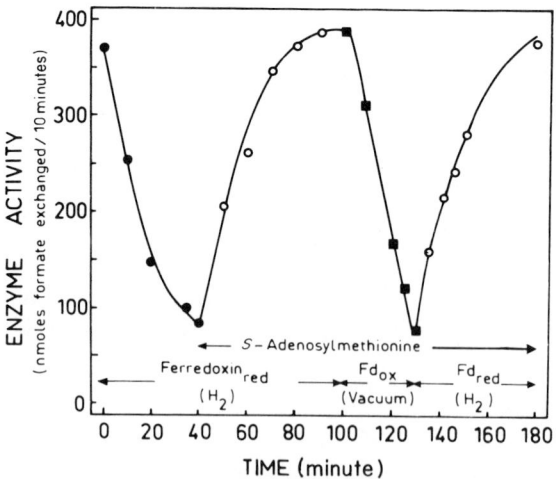

Fig. 5. Reversible inactivation of the pyruvate formate lyase in cell-free extracts of *C. butyricum*. The enzyme is reversibly inactivated in the absence of *S*-adenosylmethionine (SAM) or reduced ferredoxin. (From Wood and Jungermann, 1972.)

IV. FORMATE DEHYDROGENASES AND FORMATE METABOLISM

A. Iron–Sulfur Protein Formate Dehydrogenases

A large variety of aerobic bacteria are known that can grow solely at the expense of single carbon compounds such as methane, methanol, or formate. Examples are *Pseudomonas methanica, Methanomonas methanooxidans, Methylococcus capsulata, Pseudomonas* AM1, and *Pseudomonas oxalaticus* (for reviews, see Ribbons *et al.*, 1970; Quayle, 1972). In the energy metabolism of these organisms, the C1 compounds are oxidized to CO_2. As far as can be judged, they contain a soluble iron–sulfur protein formate dehydrogenase that is specific for NAD as electron acceptor (Kaneda and Roxburgh, 1959; Johnson and Quayle, 1964; Johnson *et al.*, 1964; Höpner and Trautwein, 1972).

1. Function

The activity of the NAD-specific formate dehydrogenase has been shown to be of the order required to account for the rate of C_1 compound oxidation observed *in vivo* (Johnson and Quayle, 1964). Thus, it was assumed that the enzyme catalyzes the terminal step in C_1 compound oxidation to CO_2. Höpner and Trautwein (1972) pointed out, however, that besides the soluble formate dehydrogenase, the bacteria contain an insoluble formate oxidase. They suggested that the physiological role of the NAD-dependent soluble enzyme is to provide NADH for biosynthesis, while the particles are considered to contain the respiratory chain.

2. Molecular Properties

The soluble formate dehydrogenase from *P. oxalaticus* has recently been purified 185-fold from formate–pyruvate-grown bacteria by conventional methods at pH 5.6 under anaerobic conditions (Höpner and Trautwein, 1972; Höpner *et al.*, 1975). It was found to be an iron–sulfur flavoprotein with a molecular weight of approximately 320,000 daltons. SDS electrophoresis shows two chains with molecular weights 100,000 and 59,000 daltons. The enzyme contains 2 moles FMN, 17–20 moles of nonheme iron, and approximately the same amount of acid-labile sulfide per molecule (Höpner *et al.*, 1975). Even at pH 5.6, the relative optimum of stability, the enzyme loses iron and sulfide, indicating that during purification some of the iron and sulfur may have been lost. X-ray fluorescence analysis of the purified enzyme indicated that no metal other than iron was present in an amount greater than 0.1 mole/mole. The enzyme is greenish brown with an absorption spectrum similar to other iron–sulfur flavoproteins. The spectrum shows a broad decrease of extinction between 320 and 600 nm with shoulders at 360 and 420 nm (Fig. 6).

Cell-free extracts of *P. oxalaticus* contain small amounts of a second NAD-specific formate dehydrogenase (FDH II), which has a molecular weight of approximately 175,000 daltons and contains 1 FMN, 8–10 iron atoms, and about the same amount of acid-labile sulfur per mole. It has been suggested that the low molecular weight form (FDH II) is formed by dissociation of the high molecular weight form (FDH I) (320,000 daltons) (Höpner et al., 1975).

3. Kinetic Properties

The purified formate dehydrogenase I catalyzes the following reactions (Höpner and Trautwein, 1972; Höpner et al., 1975): (1) the oxidation of formate with either NAD, methylviologen, benzylviologen, methylene blue, dichlorophenolindophenol, ferricyanide, or molecular oxygen as electron acceptor, (2) the reduction of CO_2 to formate with NADH as the electron donor; and (3) the reduction of methylviologen, benzylviologen, methylene blue, dichlorophenolindophenol, ferricyanide, and molecular oxygen by NADH. CO_2 rather than HCO_3^- (or H_2CO_3) has been shown to be the active species of CO_2 utilized by the enzyme (Ruschig et al., 1975). The kinetic properties of the enzyme are summarized in Table V. The formate dehydrogenase from *P. oxalaticus* has an $S_{0.5V}$ for

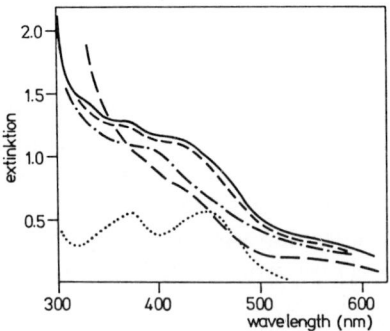

Fig. 6. Spectra of formate dehydrogenase (FDH) from *P. oxalaticus*. (1) (———) 4.85 mg/ml FDH in 0.05 M histidine hydrochloride buffer, pH 5.6; (2) (— — —) same as 1, reduced by 0.16 mM dithionite; (3) (– – –) same as 2, reoxidized by O_2; (4) (· — · —) same as 1, 10 minutes after addition of formate (8 mM); (5) (· · · ·) supernate of 1 after acidification by perchloric acid, 5× enlarged. Conditions: anaerobic, 0°–3°C. (From Höpner and Trautwein, 1972.)

TABLE V
Kinetic Properties of Formate Dehydrogenase from *P. oxalaticus*

Reaction	$S_{0.5V}$ (mM)	V_{max}^{app} [a] (units/mg protein)	pH optimum
Formate oxidation			
Formate	0.08[b]	—	—
NAD	0.6[b]	42	7.5[b]
DCPIP[d]	—	20	—
O_2	—	2	—
CO_2 reduction			
CO_2	40[c]	4	
NADH	—	—	
Diaphorase reaction			
NADH	—	—	
DCPIP	—	20	

[a] Determined with purified formate dehydrogenase of *P. oxalaticus* at 20°C ((Höpner *et al.*, 1975); 1 unit = 1 μmole/minute.
[b] From Quayle (1966).
[c] From Ruschig *et al.* (1975). The active species of "CO_2" utilized by the enzyme has been shown to be CO_2 rather than HCO_3^-.
[d] DCPIP = dichlorophenolindophenol.

formate of 0.08 mM and can be used for the microestimation of formate (Johnson *et al.*, 1964; Quayle, 1966; Höpner and Knappe, 1970).

4. Inhibitors

The formate dehydrogenase is inhibited by azide, cyanide, fluoride, and hypophosphite (Quayle, 1966). Inhibition by azide is competitive with respect to formate (Höpner and Trautwein, 1971), indicating that a metal is probably involved in the binding of formate to the enzyme.

5. Inactivation

The enzyme is inactivated by light irradiation and by formate in the absence of NAD. In both cases, the inactivation concerns only the reaction that involves formate and not the diaphorase reaction (Höpner and Trautwein, 1972). When the enzyme is dialyzed against histidine buffer, pH 5.6, containing 1.2 M ammonium sulfate, 5 mM sulfide, 0.5 mM EDTA, and some charcoal, an inactive deflavo enzyme is formed that can be fully reactivated by the addition of saturating concentrations of FMN. FAD binds to the deflavo enzyme without restoring its activity. The deflavo enzyme does not mediate the oxidation of formate with NAD. It does, however, catalyze the oxidation of formate and NADH with

TABLE VI
Properties of Formate Dehydrogenases from Different Sources

Source	Coenzyme	Metals, flavins	Molecular weight	Inhibitors	References
Anaerobic bacteria					
Clostridium pasteurianum[a]	Fd	Mo	>250,000	N_3^-, OCN^-, NO_2^-, SCN^-, CN^-[b], NO_3^-, F^-, PO_2^-	Mortenson, 1966; Jungermann et al., 1971a; Thauer et al., 1973, 1974, 1975a,b
Clostridium thermoaceticum	NADP	Mo/W, Se	~290,000	CN^-,[b] SO_3^{2-}, PO_2^-	Li et al., 1966; Thauer, 1972; Andreesen and Ljungdahl, 1973, 1974; Andreesen et al., 1973; Ljungdahl and Andreesen, 1975
Clostridium formicoaceticum	Not known NAD(?)	Mo/W, Se	—	?	Andreesen et al., 1974
Clostridium acidi-urici[c]	Fd(?)	?	>200,000	CN^-	Kearny and Sagers, 1972; Thauer, 1973
Clostridium cylindrosporum	Fd	?	—	?	Bradshaw and Reeder, 1964
Methanobacillus omelianskii	Fd(?)	?	—	?	Brill et al., 1964
S-organism	NAD	?	—	?	Reddy et al., 1972
Methanobacter ruminantium	F_{420}	?	—	?	Tzeng et al., 1975
Methanococcus vannielii	—	W, Se(?)	—	—	Stadtman, 1974; T. C. Stadtman, personal communication
Desulfovibrio vulgaris	Cyt_{553}	—	—	CN^-	Yagi, 1969; Riederer-Henderson and Peck, 1970
Klebsiella pneumoniae	Fd	?	—	?	Yoch, 1974

Organism	Electron acceptor	Cofactors	MW	Inhibitors	References
Escherichia coli	Cytochrome (b type)	Mo, Se, Fe(?), flavin(?)	>230,000	CN^-, N_3^-, NO_3^-, NO_2^-, H_2CO, PO_2^-	Pinsent, 1954; Bovarnick, 1955; Itagaki et al., 1962; Linnane and Wrigley, 1963; Fukuyama and Ordal, 1965; Azoulay and Marty, 1970; Lester and DeMoss, 1971; Enoch and Lester, 1972; 1974a,b; Ruiz-Herrera et al., 1972; Shum and Murphy, 1972; Douglas et al., 1974
Proteus rettgeri	?	?	—	?	Kröger et al., 1974
Vibrio succinogenes	Cyt b	Mo	—	N_3^-	Kröger, 1975
Bacillus polymixa	Fd(?)	?	—	?	Fisher and Wilson, 1970; Yoch, 1973
Rhodopseudomonas palustris (grown phototrophically)	NAD	?, Flavin	—	N_3^-, CN^-	Quadri and Hoare, 1968; Yoch and Lindstrom, 1969; Tkacheva, 1972
Aerobic bacteria					
Pseudomonas oxalaticus[a,b]	NAD	Fe/S FMN	320,000 and 175,000	N_3^-, CN^-, F^-, PO_2^-	Johnson et al., 1964; Quayle, 1966; Höpner and Trautwein, 1971, 1972; Höpner et al., 1973, 1975; Höpner and Knappe, 1974; Ruschig et al., 1975
Pseudomonas AM1	NAD	?	—	N_3^-, CN^-	Johnson and Quayle, 1964
Nitrobacter agilis	Cytochrome(?)	?	—	CN^-, N_3^-	Malavolta et al., 1962; O'Kelley and Nason, 1970
Yeasts					
Candida boidinii	NAD	?	74,000	CN^-, N_3^-, OCN^-, NO_3^-, SCN^-, NO_2^-	Sahm and Wagner, 1973; Schütte et al., 1975
Candida sp.	NAD	?	—	—	Fujii and Tonomura, 1972;
Saccharomyces sp.					Kato et al., 1972, 1974

(*continued*)

TABLE VI (Continued)

Source	Coenzyme	Metals, flavins	Molecular weight	Inhibitors	References
Plants					
Beta vulgaris	NAD	?	—	N_3^-, CN^-	Mazelis, 1960; Leek et al., 1972; Halliwell, 1974
Spinacia oleraceae					
Phaseolus aureus	NAD	?	92,000	N_3^-, CN^-, NO_3^-	Davison, 1951; Nason and Little, 1955; Peacock and Boulter, 1970
Pisum sp.	NAD	?	—	N_3^-, CN^-	Mathews and Vennesland, 1950; Adler and Sreenivasaya, 1937; Krakow et al., 1963
Crucifera	NAD	?	—	N_3^-, CN^-	Mazelis, 1960
Animals		Formate oxidation is catalyzed by catalase			Mathews and Vennesland, 1950; Oro and Rappoport, 1959; Schulman and Richert, 1959; for review on catalase, see Nicholls and Schonbaum 1963

[a] The active species of "CO_2" utilized or released by the enzyme has been shown to be CO_2 rather than HCO_3^- or H_2CO_3.
[b] Inactivation by cyanide.
[c] Inactivation by irradiation with light.

ferricyanide with unchanged specific activities. This finding is surprising; the interaction of the deflavo enzyme, at least with NADH, disagrees with the common experience (Höpner et al., 1975).

B. Formate Dehydrogenases Not Known to Be Iron–Sulfur Proteins

The properties of formate dehydrogenases from different organisms are summarized in Table VI. Only two of the many enzymes have been purified to homogeneity, i.e., the enzyme from P. oxalaticus (Höpner et al., 1975) and from Candida boidinii (Schütte et al., 1975). The formate dehydrogenase from P. oxalaticus has been shown to be an iron–sulfur flavoprotein. The formate dehydrogenase from C. boidinii is probably also a metalloprotein (see Section IV,C). The metal involved, however, has not yet been determined. All the other formate dehydrogenases listed in Table VI have only been partially purified or studied as to their properties in cell-free extracts. It is therefore not possible to decide whether they contain nonheme iron and acid-labile sulfur.

C. Comparison of Formate Dehydrogenases

A comparison of formate dehydrogenases from different sources (Table VI) shows many differences, e.g., in coenzyme specificity (NAD, NADP, ferredoxin, cytochrome, F_{420}), in metal and flavin composition (molybdenum, tungsten, selenium), and in molecular weight ($\geq$250,000 daltons in prokaryotes versus <100,000 daltons in eukaryotes). However, all formate dehydrogenases have one property in common; they are inhibited by azide and other monovalent anions known to be effective ligands to transition metals (Schwarzenbach, 1961; Nelson, 1972; Beck and Fehlhammer, 1972), suggesting that they are all metalloenzymes. In the formate dehydrogenase from the aerobic P. oxalaticus, no metal other than iron was detected in stoichiometric amounts (Höpner and Trautwein, 1972). It must therefore be assumed that in this enzyme, iron is the site of binding of the monovalent anions. The formate dehydrogenases from the anaerobic bacteria contain molybdenum or tungsten. Indirect evidence is available indicating that they also may contain iron; the formate dehydrogenase from C. thermoaceticum has been reported to have a spectrum similar to that of nitrate reductase from E. coli (Andreesen and Ljungdahl, 1974), which is known to be an iron–sulfur molybdoprotein (Forget, 1974), and the synthesis of formate dehydrogenase in growing E. coli has been reported to be dependent on iron (Fukuyama and Ordal, 1965). The metal attacked by azide or by one of the other ligands in these enzymes is not known.

The formate dehydrogenases from anaerobic bacteria have been shown to be inactivated by cyanide (Thauer et al., 1974; J. R. Andreesen, unpublished). The inactivation appears to be irreversible. From the pH dependence of the rate of inactivation, it was concluded that the cyanolysis of a disulfide is the most probable mechanism of inactivation (Thauer et al., 1975a). The formate dehydrogenase from P. oxalaticus is also affected by cyanide (Quayle, 1966). Whether this iron–sulfur flavoprotein is reversibly inhibited, reversibly inactivated, or irreversibly inactivated by the nucleophile is not known. It is interesting to note that xanthine oxidases (dehydrogenases) and aldehyde oxidases, which are known to be iron–sulfur molybdoflavoproteins, are irreversibly inactivated by cyanide with the concomitant formation of thiocyanate (Massey and Edmondson, 1970; Branzoli and Massey, 1974).

The active species of "CO_2" utilized or released by formate dehydrogenase has been determined for the enzyme from C. pasteurianum (Thauer et al., 1975b) and P. oxalaticus (Ruschig et al., 1975). CO_2 rather than HCO_3^- (H_2CO_3) was found to be the substrate or product of the two enzymes. Whether this is also the case for the other formate dehydrogenases is not known. This point will have to be examined at least for those formate dehydrogenases that have different coenzyme specificity, as it is known that closely related reactions can differ in the active species of "CO_2" utilized. Thus, HCO_3^- has been reported to be the substrate of phosphoenolpyruvate carboxylase (Maruyama et al., 1966; Cooper and Wood, 1971), while CO_2 appears to be the substrate of phosphoenolpyruvate carboxykinase and phosphoenolpyruvate carboxytransphosphorylase (Cooper et al., 1968).

V. CONCLUDING REMARKS

Iron–sulfur proteins have been shown to be involved in the interconversion of formate and CO_2 either as coenzyme (ferredoxin) or as enzyme (iron–sulfur protein formate dehydrogenases). The involvement appears not to be obligatory. The coenzyme used by formate dehydrogenases varies from organism to organism (see Table VI), and at least the formate dehydrogenases from eukaryotes are probably not iron–sulfur proteins (Schütte et al., 1975). Thus, a mechanistic or functional necessity for a participation of iron–sulfur proteins in formate metabolism appears not to exist.

ACKNOWLEDGMENT

This work was supported by a grant from the Deutsche Forschungsgemeinschaft, Bonn-Bad Godesberg, and by the Fonds der Chemischen Industrie, GFR.

REFERENCES

Adams, J. M. (1968). *J. Mol. Biol.* **33**, 571.
Adler, E., and Sreenivasaya, M. (1937). *Hoppe-Seyler's Z. Physiol. Chem.* **249**, 24.
Ahrland, S., Chatt, J., and Davies, N. R. (1958). *Q. Rev., Chem. Soc.* **12**, 265.
Andreesen, J. R., and Ljungdahl, L. G. (1971). *Bacteriol. Proc.* p. 166.
Andreesen, J. R., and Ljungdahl, L. G. (1973). *J. Bacteriol.* **116**, 867.
Andreesen, J. R., and Ljungdahl, L. G. (1974). *J. Bacteriol.* **120**, 6.
Andreesen, J. R., Gottschalk, G., and Schlegel, H. G. (1970). *Arch. Mikrobiol.* **72**, 154.
Andreesen, J. R., Schaupp, A., Neurauter, C., Brown, A., and Ljungdahl, L. G. (1973). *J. Bacteriol.* **114**, 743.
Andreesen, J. R., El Ghazzawi, E., and Gottschalk, G. (1974). *Arch. Microbiol.* **96**, 103.
Andrew, I. G., and Morris, J. G. (1965). *Biochim. Biophys. Acta* **97**, 176.
Aronson, J. N., and Lugay, J. C. (1969). *Biochem. Biophys. Res. Commun.* **34**, 311.
Azoulay, E., and Marty, B. (1970). *Eur. J. Biochem.* **13**, 168.
Bailey, C. B., and Wagner, C. (1974). *J. Biol. Chem.* **249**, 4439.
Barker, H. A. (1961). *In* "The Bacteria" (I. C. Gunsalus and R. Y. Stanier, eds.), Vol. 2, p. 151. Academic Press, New York.
Bauchop, T., and Dawes, E. A. (1959). *Biochim. Biophys. Acta* **36**, 294.
Bauchop, T., and Dawes, E. A. (1968). *J. Gen. Microbiol.* **52**, 195.
Beck, W., and Fehlhammer, W. P. (1972). *Inorg. Chem., Ser. One* **2**, 253.
Blaylock, B. A., and Stadtman, T. C. (1966). *Arch. Biochem. Biophys.* **116**, 138.
Bovarnick, M. (1955). *In* "Methods in Enzymology" (S. P. Colowick and N. O. Kaplan, eds.), Vol. 1, p. 539. Academic Press, New York.
Bradshaw, W. H., and Reeder, D. J. (1964). *Bacteriol. Proc.* p. 110.
Branzoli, U., and Massey, V. (1974). *J. Biol. Chem.* **249**, 4346.
Brill, W. J., Wolin, E. A., and Wolfe, R. S. (1964). *Science* **144**, 297.
Bryant, M. P., Wolin, E. A., Wolin, M. J., and Wolfe, R. S. (1967). *Arch. Mikrobiol.* **59**, 20.
Bryant, M. P., McBride, B. C., and Wolfe, R. S. (1968). *J. Bacteriol.* **95**, 1118.
Bryant, M. P., Tzeng, S. F., Robinson, I. M., and Joyner, A. E. (1971). *Adv. Chem. Ser.* **105**, 23.
Burg, A. W., and Brown, G. M. (1968). *J. Biol. Chem.* **243**, 2349.
Chase, T., and Rabinowitz, J. C. (1968). *J. Bacteriol.* **96**, 1065.
Cheeseman, P., Toms-Wood, A., and Wolfe, R. S. (1972). *J. Bacteriol.* **112**, 527.
Cone, J. E., Plowman, J., and Guroff, G. (1974). *J. Biol. Chem.* **249**, 5551.
Cooper, T. G., and Wood, H. G. (1971). *J. Biol. Chem.* **246**, 5488.
Cooper, T. G., Tchen, T. T., Wood, H. G., and Benedict, C. R. (1968). *J. Biol. Chem.* **243**, 3857.
Davison, D. C. (1951). *Biochem. J.* **49**, 520.
Decker, K., Jungermann, K., Thauer, R. K., and Hunt, S. V. (1967). *Biochim. Biophys. Acta* **141**, 202.
Douglas, M. W., Ward, F. B., and Cole, J. A. (1974). *J. Gen. Microbiol.* **80**, 557.
Elstner, E. F., and Heupel, A. (1973). *Biochim. Biophys. Acta* **325**, 182.
Elstner, E. F., and Suhadolnik, R. J. (1971). *J. Biol. Chem.* **246**, 6973.
Elstner, E. F., and Suhadolnik, R. J. (1975). *In* "Methods in Enzymology" (S. P. Colowick and N. O. Kaplan, eds.), Vol. 43, p. 515. Academic Press, New York.
Emiliani, E., and Bekes, P. (1964). *Arch. Biochem. Biophys.* **105**, 488.

Enoch, H. G., and Lester, R. L. (1972). *J. Bacteriol.* **110**, 1032.
Enoch, H. G., and Lester, R. L. (1974a). *Fed. Proc., Fed. Am. Soc. Exp. Biol.* **33**, 1577.
Enoch, H. G., and Lester, R. L. (1974b). *Biochem. Biophys. Res. Commun.* **61**, 1234.
Feigelson, P., and Brady, F. D. (1974). *In* "Molecular Mechanisms of Oxygen Activation" (O. Hayaishi, ed.), p. 87. Academic Press, New York.
Fisher, R. J., and Wilson, P. W. (1970). *Biochem. J.* **117**, 1023.
Forget, P. (1974). *Eur. J. Biochem.* **42**, 325.
Fujii, T., and Tonomura, K. (1972). *Agric. Biol. Chem.* **36**, 2297.
Fukuyama, T., and Ordal, E. J. (1965). *J. Bacteriol.* **90**, 673.
Gottschalk, G., and Chowdhury, A. (1969). *FEBS Lett.* **2**, 342.
Halliwell, B. (1974). *Biochem. J.* **138**, 77.
Harold, F. M., and Levin, E. (1974). *J. Bacteriol.* **117**, 1141.
Harrington, A. A., and Kallio, R. E. (1960). *Can. J. Microbiol.* **6**, 1.
Hayaishi, O., Tabor, H., and Hayaishi, T. (1957). *J. Biol. Chem.* **227**, 161.
Höpner, T., and Knappe, J. (1970). *In* "Methoden der enzymatischen Analyse" (H. U. Bergmeyer, ed.), 2nd ed., Vol. 2, p. 1509. Verlag Chemie, Weinheim.
Höpner, T., and Trautwein, A. (1971). *Abstr. Commun., 7th Meet. Eur. Biochem. Soc.* p. 240.
Höpner, T., and Trautwein, A. (1972). *Z. Naturforsch., Teil B* **27**, 1075.
Höpner, T., Ruschig, U., and Müller, U. (1973). *Hoppe-Seyler's Z. Physiol. Chem.* **354**, 216.
Höpner, T., Müller, U., Ruschig, U., and Willnow, P. (1976). *In* "Flavins and Flavoproteins" (T. P. Singer, ed.), Vol. IV (in press).
Huennekens, F. M, (1968). *In* "Biological Oxidations" (T. P. Singer, ed.), p. 439. Wiley (Interscience), New York.
Itagaki, E., Fujita, T., and Sato, R. (1962). *J. Biochem. (Tokyo)* **52**, 131.
Jakoby, W. B. (1954). *J. Biol. Chem.* **207**, 657.
Jakoby, W. B. (1963). *In* "The Enzymes" (P. D. Boyer, H. Lardy, and K. Myrbäck, eds.), 2nd ed., Vol. 7, p. 203. Academic Press, New York.
Johnson, P. A., and Quayle, J. R. (1964). *Biochem. J.* **93**, 281.
Johnson, P. A., Jones-Mortimer, M. C., and Quayle, J. R. (1964). *Biochim. Biophys. Acta* **89**, 351.
Jungermann, K., and Schön, G. (1974). *Arch. Microbiol.* **99**, 109.
Jungermann, K., Thauer, R. K., and Decker, K. (1968). *Eur. J. Biochem.* **3**, 351.
Jungermann, K., Thauer, R. K., Rupprecht, E., Ohrloff, C., and Decker, K. (1969). *FEBS Lett.* **3**, 144.
Jungermann, K., Schmidt, W., Kirchniawy, H., Rupprecht, E., and Thauer, R. K. (1970a). *Eur. J. Biochem.* **16**, 424.
Jungermann, K., Kirchniawy, H., and Thauer, R. K. (1970b). *Biochem. Biophys. Res. Commun.* **41**, 682.
Jungermann, K., Leimenstoll, G., Rupprecht, E., and Thauer, R. K. (1971a). *Arch. Mikrobiol.* **80**, 370.
Jungermann, K., Rupprecht, E., Ohrloff, C., Thauer, R. K., and Decker, K. (1971b). *J. Biol. Chem.* **246**, 960.
Jungermann, K., Thauer, R. K., Leimenstoll, G., and Decker, K. (1973). *Biochim. Biophys. Acta* **305**, 268.
Jungermann, K., Kirchniawy, H., Katz, N., and Thauer, R. K. (1974). *FEBS Lett.* **43**, 203.
Kaneda, T., and Roxburgh, J. M. (1959). *Can. J. Microbiol.* **5**, 187.

Kato, N., Tamaoki, T., Tani, Y., and Ogata, K. (1972). *Agric. Biol. Chem.* **36**, 2411.
Kato, N., Tani, Y., and Ogata, K. (1974). *Agric. Biol. Chem.* **38**, 675.
Kearny, J. J., and Sagers, R. D. (1972). *J. Bacteriol.* **109**, 152.
Kent, S. S. (1972a). *J. Biol. Chem.* **247**, 7288.
Kent, S. S. (1972b). *J. Biol. Chem.* **247**, 7293.
Knappe, J., Schacht, J., Möckel, W., Höpner, T., Vetter, H., and Edenharder, R. (1969). *Eur. J. Biochem.* **11**, 316.
Knappe J., Blaschkowski, H. P., and Edenharder, R. (1972). In "Metabolic Interconversion of Enzymes" (O. Wieland, E. Helmreich, and H. Holzer, eds.), p. 319. Springer-Verlag, Berlin and New York.
Knappe, J., Blaschkowski, H. P., Gröbner, P., and Schmitt, T. (1974). *Eur. J. Biochem.* **50**, 253.
Knight, E., D'Eustachio, A. J. D., and Hardy, R. W. F. (1966). *Biochim. Biophys. Acta* **113**, 626.
Kornberg, H. L. (1961). *Essays Biochem.* **2**, 1.
Kornberg, H. L., and Elsden, S. R. (1961). *Adv. Enzymol.* **23**, 401.
Krakow, G., Ludowieg, J., Mather, J. H., Normore, W. M., Tosi, L., Udaka, S., and Vennesland, B. (1963). *Biochemistry* **2**, 1009.
Kröger, A. (1975). In "Electron Transfer Chains and Oxidative Phosphorylation," (E. Quagliariello, ed.), pp. 265–270. North-Holland, Amsterdam.
Kröger, A., Schimkat, M., and Niedermaier, S. (1974). *Biochim. Biophys. Acta* **347**, 273.
Kutzbach, C., and Stokstad, E. L. R. (1968). *Biochem. Biophys. Res. Commun.* **30**, 111.
Kutzbach, C., and Stokstad, E. L. R. (1971). In "Methods in Enzymology" (D. B. McCormick and L. D. Wright, eds.), Vol. 18-B, p. 793. Academic Press, New York.
Laishley, E. J., Lin, P. M., and Peck, H. D. (1971). *Can. J. Microbiol.* **17**, 889.
Leek, A. E., Halliwell, B., and Butt, V. S. (1972). *Biochim. Biophys. Acta* **286**, 299.
Lester, R. L., and DeMoss, J. A. (1971). *J. Bacteriol.* **105**, 1006.
Li, L. F., Ljungdahl, L. G., and Wood, H. G. (1966). *J. Bacteriol.* **92**, 405.
Lillehoj, E. B., and Smith, F. G. (1965). *Arch. Biochem. Biophys.* **109**, 216.
Lindmark, D. G., Paolella, P., and Wood, N. P. (1969). *J. Biol. Chem.* **244**, 3605.
Linke, H. A. B. (1969). *Arch. Mikrobiol.* **64**, 203.
Linnane, A. W., and Wrigley, C. W. (1963). *Biochim. Biophys. Acta* **77**, 408.
Livingston, D. M., and Leder, P. (1969). *Biochemistry* **8**, 435.
Ljungdahl, L. G., and Andreesen, J. R. (1975). *FEBS Lett.* **54**, 279.
Ljungdahl, L. G., and Wood, H. G. (1969). *Annu. Rev. Microbiol.* **23**, 515.
Lucas-Lenard, J., and Lipmann, F. (1971). *Annu. Rev. Biochem.* **40**, 409.
McBride, B. C., and Wolfe, R. S. (1971). *Adv. Chem. Ser.* **105**, 11.
McCormick, N. G., Ordal, E. J., and Whiteley, H. R. (1962a). *J. Bacteriol.* **83**, 887.
McCormick, N. G., Ordal, E. J., and Whiteley, H. R. (1962b), *J. Bacteriol.* **83**, 899.
Malavolta, E., Delwiche, C. C., and Burge, W. D. (1962). *Biochim. Biophys. Acta* **57**, 347.
Maruyama, H., Easterday, R. L., Chang, H. C., and Lane, M. D. (1966). *J. Biol. Chem.* **241**, 2405.
Massey, V. (1973). In "Iron-Sulfur Proteins" (W. Lovenberg, ed.), Vol. 1, p. 301. Academic Press, New York.
Massey, V., and Edmondson, D. (1970). *J. Biol. Chem.* **245**, 6595.

Mathews, M. B., and Vennesland, B. (1950). *J. Biol. Chem.* **186,** 667.
Mazelis, M. (1960). *Plant Physiol.* **35,** 386.
Mehler, A. H., and Knox, W. E. (1950). *J. Biol. Chem.* **187,** 431.
Mortenson, L. E. (1963). *Annu. Rev. Microbiol.* **17,** 115.
Mortenson, L. E. (1966). *Biochim. Biophys. Acta* **127,** 18.
Mortenson, L. E., and Nakos, G. (1973). *In* "Iron-Sulfur Proteins" (W. Lovenberg, ed.), Vol. 1, p. 37. Academic Press, New York.
Nakayama, H., Midwinter, G. G., and Krampitz, L. O. (1971). *Arch. Biochem. Biophys.* **143,** 526.
Nason, A., and Little, H. N. (1955). *In* "Methods in Enzymology" (S. P. Colowick and N. O Kaplan, eds.), Vol. 1, p. 536. Academic Press, New York.
Nelson, S. M. (1972). *Inorg. Chem., Ser. One* **5,** 175.
Nicholls, P., and Schonbaum, G. R. (1963). *In* "The Enzymes" (P. D. Boyer, H. Lardy, and K. Myrbäck, eds.), 2nd ed., Vol. 8, 147. Academic Press, New York.
Ohmura, E., and Hayaishi, O. (1957). *J. Biol. Chem.* **227,** 181.
O'Kelley, J. C., and Nason, A. (1970). *Biochim. Biophys. Acta* **205,** 426.
Oro, J., and Rappoport, D. A. (1959). *J. Biol. Chem.* **234,** 1661.
Palmer, G. (1962). *Biochim. Biophys. Acta* **56,** 444.
Peacock, D., and Boulter, D. (1970). *Biochem. J.* **120,** 763.
Pearson, R. G., (1963). *J. Am. Chem. Soc.* **85,** 3533.
Perrin, D. D. (1969). "Dissociation Constants of Inorganic Acids and Bases in Aqueous Solution." IUPAC, Butterworth, London.
Pichinoty, F. (1969). *Ann. Inst. Pasteur, Paris* **117,** 3.
Piéchaud, M., Puig, J., Pichinoty, F., Azoulay, E., and LeMinor, L. (1967). *Ann. Inst. Pasteur, Paris* **112,** 24.
Pine, M. J. (1971). *Adv. Chem. Ser.* **105,** 1.
Pinsent, J. (1954). *Biochem. J.* **57,** 10.
Plowman, J., Cone, J. E., and Guroff, G. (1974). *J. Biol. Chem.* **249,** 5559.
Poston, J. M., and Stadtman, E. R. (1967). *Biochem. Biophys. Res. Commun.* **26,** 550.
Quadri, S. M. H., and Hoare, D. S. (1968). *J. Bacteriol.* **95,** 2344.
Quayle, J. R. (1963). *Biochem. J.* **87,** 368.
Quayle, J. R. (1966). *In* "Methods in Enzymology" (W. A. Wood, ed.), Vol. 9, p. 360. Academic Press, New York.
Quayle, J. R. (1972). *Adv. Microb. Physiol.* **7,** 119.
Quayle, J. R., Keech, D. B., and Taylor, G. A. (1961). *Biochem. J.* **78,** 225.
Rader, J. I., and Huennekens, F. M. (1973). *In* "The Enzymes" (P. D. Boyer, ed.), 3rd ed., Vol. 9, Part B, p. 197. Academic Press, New York.
Reddy, C. A., Bryant, M. P., and Wolin, M. J. (1972). *J. Bacteriol.* **110,** 126.
Ribbons, D. W., Harrison, J. E., and Wadzinski, A. M. (1970). *Annu. Rev. Microbiol.* **24,** 135.
Riebeling, V., Thauer, R. K., and Jungermann, K. (1975). *Eur. J. Biochem.* **55,** 445.
Riederer-Henderson, M. A., and Peck, H. D. (1970). *Bacteriol. Proc.* p. 70.
Rose, Z. B., and Racker, E. (1962). *J. Biol. Chem.* **237,** 3279.
Rose, Z. B., and Racker, E. (1966). *In* "Methods in Enzymology" (W. A. Wood, ed.), Vol. 9, p. 357. Academic Press, New York.
Ruiz-Herrera, J., Alvarez, A., and Figueroa, I. (1972). *Biochim. Biophys. Acta* **289,** 254.

Ruschig, U., Müller, U., Willnow, P., and Höpner, T. (1976). Submitted for publication.
Sahm, H., and Wagner, F. (1973). *Arch. Mikrobiol.* **90**, 263.
Saxton, R. E., Rocha, V., Rosser, R. J., Andreoli, A. J., Shimoyama, M., Kosaka, A., Chandler, J. L. R., and Gholson, R. K. (1968). *Biochim. Biophys. Acta* **156**, 77.
Schulman, M., Parker, D. J., Ljungdahl, L. G., and Wood, H. G. (1972). *J. Bacteriol.* **109**, 633.
Schulman, M., Ghambeer, R. K., Ljungdahl, L. G., and Wood, H. G. (1973). *J. Biol. Chem.* **248**, 6255.
Schulman, M., and Richert, D. A. (1959). *J. Biol. Chem.* **234**, 1781.
Schütte, H., Flossdorf, J., Sahm, H., and Kula, M. (1975). *Eur. J. Biochem.* **62**, 151.
Schwarz, K. (1974). *Fed. Proc., Fed. Am. Soc. Exp. Biol.* **33**, 1748.
Schwarzenbach, G. (1961). *Adv. Inorg. Chem. Radiochem.* **3**, 257.
Shimazono, H., and Hayaishi, O. (1957). *J. Biol. Chem.* **227**, 151.
Shum, A. C., and Murphy, J. C. (1972). *J. Bacteriol.* **110**, 447.
Sly, W. S., and Stadtman, E. R. (1963a). *J. Biol. Chem.* **238**, 2632.
Sly, W. S., and Stadtman, E. R. (1963b). *J. Biol. Chem.* **238**, 2639.
Stadtman, T. C. (1967). *Annu. Rev. Microbiol.* **21**, 121.
Stadtman, T. C. (1974). *Science* **183**, 915.
Stephenson, M. P., and Dawes, E. A. (1971). *J. Gen. Microbiol.* **69**, 331.
Strittmatter, P., and Ball, E. G. (1955). *J. Biol. Chem.* **213**, 445.
Tabor, H., and Mehler, A. H. (1954). *J. Biol. Chem.* **210**, 559.
Takeda, M., and Webster, R. E. (1968). *Proc. Natl. Acad. Sci. U.S.A.* **60**, 1487.
Thauer, R. K. (1972). *FEBS Lett.* **27**, 111.
Thauer, R. K. (1973). *J. Bacteriol.* **114**, 443.
Thauer, R. K., Jungermann, K., Henninger, H., Wenning, J., and Decker, K. (1968). *Eur. J. Biochem.* **4**, 173.
Thauer, R. K., Jungermann, K., Rupprecht, E., and Decker, K. (1969). *FEBS Lett.* **4**, 108.
Thauer, R. K., Rupprecht, E., and Jungermann, K. (1970a). *FEBS Lett.* **8**, 304.
Thauer, R. K., Rupprecht, E., and Jungermann, K. (1970b). *FEBS Lett.* **9**, 271.
Thauer, R. K., Rupprecht, E., Ohrloff, C., Jungermann, K., and Decker, K. (1971). *J. Biol. Chem.* **246**, 954.
Thauer, R. K., Kirchniawy, F. H., and Jungermann, K. A. (1972). *Eur. J. Biochem.* **27**, 282.
Thauer, R. K., Fuchs, G., Schnitker, U., and Jungermann, K. (1973). *FEBS Lett.* **38**, 45.
Thauer, R. K., Fuchs, G., and Jungermann, K. (1974). *J. Bacteriol.* **118**, 758.
Thauer, R. K., Fuchs, G., and Käufer, B. (1975a). *Hoppe Seyler's Z. Physiol. Chem.* **356**, 653.
Thauer, R. K., Käufer, B., and Fuchs, G. (1975b). *Eur. J. Biochem.* **55**, 111.
Thauer, R. K., Käufer, B., and Scherer, P. (1975c). *Arch. Microbiol.* **104**, 237.
Tkacheva, Z. G. (1972). *Biol. Nauki (Moscow)* **15**, 102.
Tzeng, S. F., Bryant, M. P., and Wolfe, R. S. (1975). *J. Bacteriol.* **121**, 192.
Uotila, L., and Koivusalo, M. (1974a). *J. Biol. Chem.* **249**, 7653.
Uotila, L., and Koivusalo, M. (1974b). *J. Biol. Chem.* **249**, 7664.
Vaisey, E. B., Cheldelin, V. H., and Newburgh, R. W. (1961). *Arch. Biochem. Biophys.* **95**, 63.
Vallee, B. L., and Wacker, W. E. C. (1970). *In* "The Proteins" (H. Neurath, ed.), 2nd ed., Vol. 5, p. 5. Academic Press, New York.

Vetter, J., and Knappe, J. (1971). *Hoppe-Seyler's Z. Physiol. Chem.* **352**, 443.
Whiteley, H. R. (1967). *Arch. Mikrobiol.* **59**, 315.
Wolf, W. A., and Brown, G. M. (1969). *Biochim. Biophys. Acta* **192**, 468.
Wolfe, R. S. (1971). *Adv. Microb. Physiol.* **6**, 107.
Wood, N. P., and Jungermann, K. (1972). *FEBS Lett.* **27**, 49.
Yagi, T. (1969). *J. Biochem. (Tokyo)* **66**, 473.
Yoch, D. C. (1973). *J. Bacteriol.* **116**, 384.
Yoch, D. C. (1974). *J. Gen. Microbiol.* **83**, 153.
Yoch, D. C., and Lindstrom, E. S. (1969). *Arch. Mikrobiol.* **67**, 182.
Zelitch, I. (1972). *Arch. Biochem. Biophys.* **150**, 698.

CHAPTER 6

X-Ray Analysis of High-Potential Iron–Sulfur Proteins and Ferredoxins†

CHARLES W. CARTER, JR.

I. Introduction	158
A. Review of the Known Iron–Sulfur Protein Structures	158
B. Summary of Current Results	159
II. $Fe_4S_4^*$ Active Centers and the Three-State Hypothesis	160
A. Elucidation of the $Fe_4S_4^*$ Structure in HiPIP	160
B. $Fe_4S_4^*$ Clusters in *P. aerogenes* Ferredoxin	162
C. $Fe_4S_4^*$ Clusters in Other Proteins	163
D. The Three-State Hypothesis	163
E. Analogous Synthetic $Fe_4S_4^*$ Clusters	164
III. Comparison of the Protein-Bound and Synthetic $Fe_4S_4^*$ Clusters	165
A. Protein Structure Refinement	165
B. Iron Oxidation Levels in the Clusters	166
C. Mean Interatomic Distances and Angles in Protein-Bound and Analogue $Fe_4S_4^*SR_4$ Clusters	166
D. Are All Paired-Spin State $Fe_4S_4^*$ Clusters the Same Size?	170
E. How Symmetrical Are the Protein-Bound $Fe_4S_4^*$ Clusters?	171
IV. Characteristic Features of the Cluster Binding Cavities	172
A. Polypeptide Chain Folding	172
B. Nonpolar Amino Acid Side Chains	177
C. Aromatic Side Chains in Juxtaposition with the $Fe_4S_4^*$ Clusters	182
D. $Fe_4S_4^*$ Coordination Geometry: NH---S Hydrogen Bonds	184
V. Analysis of Reduced versus Oxidized HiPIP	187
A. Preparation and Stabilization of HP_{ox} and HP_{red} Crystals	188
B. Structural Changes in the $Fe_4S_4^*$ Cluster	189
C. Changes in the Cluster Binding Cavity	193
D. Is There a Conformational Change?	195

†*Abbreviations used:* S*, inorganic sulfur; S_γ, cysteinyl sulfur; HP_{ox}, oxidized *Chromatium* HiPIP; HP_{red}, reduced *Chromatium* HiPIP; Fd_{ox}, oxidized *Peptococcus aerogenes* ferredoxin; Fd_{red}, reduced *P. aerogenes* ferredoxin; DMSO, dimethyl sulfoxide.

VI. Conclusions and Review.................................... 196
  A. Thermodynamic and Mechanistic Considerations.............. 197
  B. Is There Cooperativity in Ferredoxin?...................... 199
  C. Do HiPIP and Ferredoxin Have a Common Ancestor?......... 200
References.. 202

I. INTRODUCTION

A. Review of Known Iron–Sulfur Protein Structures

Of many possible and interesting candidates for crystal structure analysis among the iron–sulfur proteins, only three have been studied successfully by this method. They include *Clostridium pasteurianum* rubredoxin (Watenpaugh et al., 1971, 1973). *Chromatium vinosum* high-potential iron protein (HiPIP) (Carter et al., 1971, 1974a,b), and *Peptococcus aerogenes* ferredoxin (Adman et al., 1973, 1975, 1976). The active sites of these proteins contain, respectively, one iron and no inorganic sulfur atoms, four iron and four inorganic sulfur atoms, and eight iron and eight inorganic sulfur atoms.

Selection of these three proteins has been dictated almost entirely by practicality; they have been crystallized in forms suitable for X-ray studies. It is perhaps not the most logical selection. For example, there is no representative of that class of ferredoxins that contains two iron and two inorganic sulfur atoms. Magnetic and spectroscopic properties of two–iron ferredoxins have been studied exhaustively (Tsibris and Woody, 1970; and Vols. I and II of this treatise). The important task of rationalizing these measurements with active-site geometries of these proteins remains unfinished because they are still unknown. A two-iron ferredoxin has recently been crystallized (C. W. Carter, unpublished) and X-ray analysis is now in progress (J. Kraut, personal communication).

Nevertheless, the three known crystal structures represent a fortunate sampling of the iron–sulfur proteins in at least two respects. First, HiPIP and ferredoxin display markedly different thermodynamic properties and presumably have different biological functions. The reduced forms of HiPIP and ferredoxin oxidize with midpoint potentials of $+0.35$ V and -0.40 V, respectively. Yet, there is a rather startling similarity between the $Fe_4S_4^*$ cluster geometry in these two proteins (Carter et al., 1972; Adman et al., 1973). Thus, we have been stimulated by a desire to understand how oxidoreduction properties of the $Fe_4S_4^*$ clusters depend on features of surrounding polypeptide structure in the two proteins. Second, similar stereochemical relationships are observed in the cluster binding cavities of both proteins (Carter et al., 1974b; Adman et al., 1975;

Carter, 1976). Such coincidences afford presumptive evidence that these relationships have functional significance. In this way, the two structures together reveal to us considerably more of their secrets than might have been expected to emerge from analysis of either one alone.

B. Summary of Current Results

I shall consider almost exclusively the X-ray analyses of *Chromatium* HiPIP and *P. Aerogenes* ferredoxin limiting discussion to those results that pertain to polypeptide–cluster interactions and how they affect the thermodynamic properties and possible electron transfer mechanisms of the $Fe_4S_4^*$ clusters in these proteins. The reader is referred to a previous review (Jensen, 1974a) for a thorough discussion of crystal structure determination and refinement of both structures and for an excellent account of results published prior to 1974. The most important recent contributions have been the presentation and description of high resolution atomic models for both proteins (Carter *et al.*, 1974a,b; Freer *et al.*, 1975; Adman *et al.*, 1975, 1976). New information concerning three important areas of iron–sulfur protein research has come directly from these descriptions:

1. They reveal the geometry of $Fe_4S_4^*$ clusters in the two proteins with surprising precision.
2. For each protein, they also provide an essentially complete picture of coordination stereochemistry, including both polypeptide main chain elements and amino acid side chains that impinge on the cluster from the surrounding protein structure.
3. It is possible to observe changes in the $Fe_4S_4^*$ cluster geometry when oxidized HiPIP is reduced chemically by addition of one electron (Carter *et al.*, 1974b).

These three types of observations suggest several ways in which the polypeptide might modify properties of the bound $Fe_4S_4^*$ clusters in the two proteins. Hence, at least to a first approximation, we can rationalize the dramatic functional differences between them. Comparison of the two HiPIP oxidation states offers the additional possibility of understanding how dynamic aspects of oxidoreduction in iron–sulfur proteins depend on their structures.

Elucidation of structure–function relationships in these two small iron–sulfur proteins can make important contributions to understanding how iron–sulfur prosthetic groups participate in more complex enzymes. As noted by Orme-Johnson (1973), structural studies are essential to the interpretation of electronic and magnetic spectra used to analyze these integrated systems.

A vigorous crop of new questions has replaced those that could be answered solely on the basis of protein crystal structure data. Fortunately for this reviewer, iron–sulfur protein research has enjoyed an extensive "adaptive radiation" of this structural information. Particularly noteworthy is the set of active-site analogues for one-, two-, and four-iron complexes prepared by Holm and co-workers (Lane et al., 1975; Mayerle et al., 1975; Herskovitz et al., 1972), and the exhaustive analysis to which they have been submitted (Herskovitz et al., 1972; Averill et al., 1973; Holm et al., 1974a,b; Frankel et al., 1974; DePamphilis et al., 1974; Que et al., 1974; Bruice et al., 1975; Job and Bruice, 1975). Holm has reviewed many of these studies (Holm, 1974, and this volume).

II. $Fe_4S_4{}^*$ ACTIVE CENTERS AND THE THREE-STATE HYPOTHESIS

A. Elucidation of the $Fe_4S_4{}^*$ Structure in HiPIP

A single, tight cluster of electron-dense iron atoms was apparent in HiPIP even at the earliest stages of X-ray analysis (Kraut et al., 1968; Strahs and Kraut, 1968). These studies demonstrated that anomalous scattering of X-rays by the iron atoms could be detected with sufficient accuracy to reveal structural features even though it represents only a small percentage (about 10%) of the total scattering. Anomalous scattering arises from the interaction of X-rays with inner-shell electrons having transition energies close to that of the incident X-rays. For CuK_α radiation, the effect is pronounced for iron atoms and negligible for carbon, nitrogen, oxygen, and sulfur. As a result, it is possible to calculate by difference methods Patterson (interatomic vector) and electron density maps showing just those electrons that participate in anomalous scattering (Strahs and Kraut, 1968). The electron density will be centered on the atomic positions of the anomalous scatterers, in this case the four iron atoms.

The ability to visualize just these four atoms became quite convincing when the HiPIP anomalous scattering, or Bijvoet difference density map was recalculated at higher resolution (Fig. 1) (Carter et al., 1971). Interatomic Fe–S* ($\sim$2.3 Å) and Fe–Fe ($\sim$2.8 Å) distances in the cluster are somewhat longer than typical interatomic distances in proteins. As a consequence, the four iron atoms are individually resolved from one another even at the lowest significant difference density level (Fig. 1a). These maps provide an unequivocal qualitative model for the locations

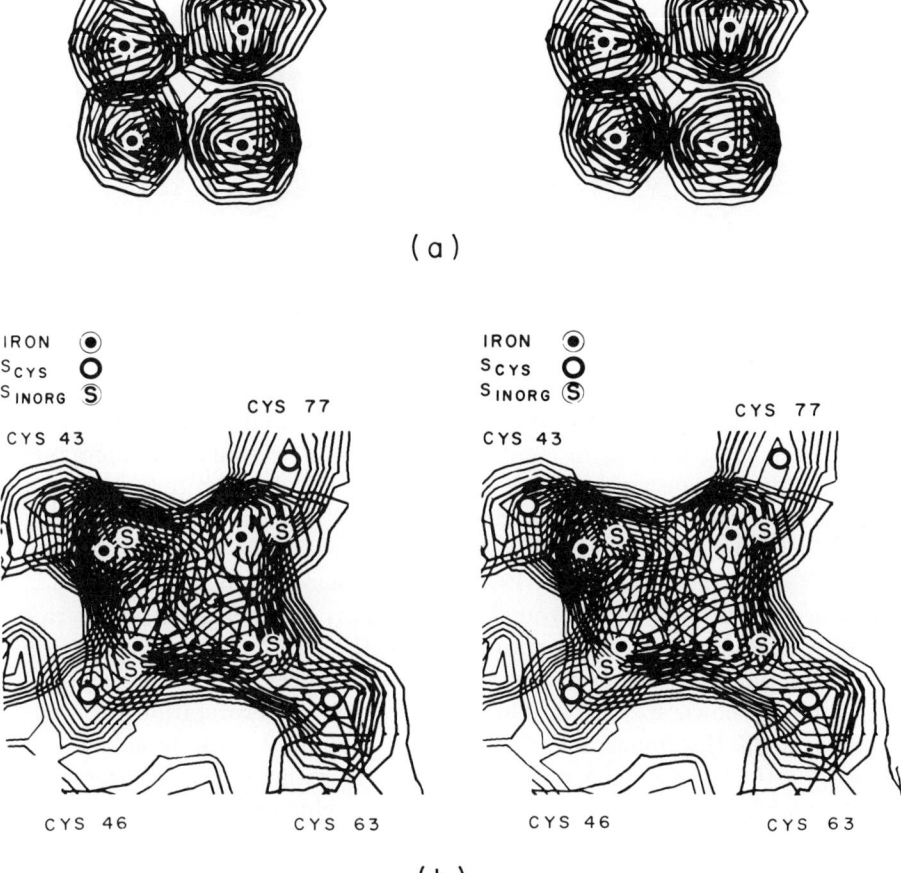

Fig. 1. Geometry of the $Fe_4S_4^*$ cluster in oxidized HiPIP as revealed in density maps. (a) Bijvoet difference density maps showing just the four iron positions. (b) Fo or electron density maps showing all twelve cluster atoms. These and subsequent pairs are meant to be viewed with a stereoscopic viewer. (From Carter et al., 1971.)

of all twelve cluster atoms (4 Fe, 4 S*, 4 S_γ), including the covalent linkage to the protein via the four cysteinyl residues. However, the impression of atomic resolution has since proven to be quantitatively misleading, in that the iron positions apparent in these maps are about 0.09 Å further from the center of the cluster than are the current refined positions, which we believe to be more realistic (Carter et al., 1972). This point is worth noting because similarly large estimates of Fe–Fe distances in both ferredoxin clusters were made in the initial assignment of atomic positions (Adman et al., 1973).

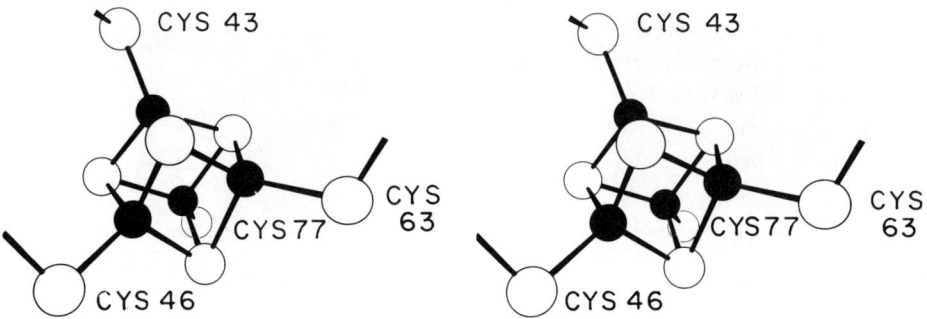

Fig. 2. The $Fe_4S_4{}^*$ cluster in HiPIP. Ball-and-stick representation of the *Chromatium* HiPIP $Fe_4S_4{}^*$ cluster is based on electron density maps in Fig. 1. Black spheres represent iron atoms. (From Carter *et al.*, 1971.)

The cluster geometry revealed by these maps is that of two concentric, interpenetrating tetrahedra consisting of the four iron and four inorganic sulfur atoms, respectively. Each iron atom is also bonded to a cysteinyl sulfur atom from the polypeptide chain and is therefore surrounded by three inorganic and one organic sulfur ligands.

Until recently (Yang *et al.*, 1975), no rigorous attempt had been made to describe bonding in such an $Fe_4S_4{}^*$ cluster. Lacking such a description, it has become customary to represent the cluster with the four iron and four inorganic sulfur atoms at alternate corners of a cube. One such drawing is shown in Fig. 2. This "cubane"-like interpretation may not reflect all of the important electronic interactions in the actual cluster itself, particularly since it does not suggest the rather close iron–iron contacts that characterize all six of the known clusters (Table I, Section III,C).

B. $Fe_4S_4{}^*$ Clusters in *P. aerogenes* Ferredoxin

Initial X-ray work on *P. aerogenes* ferredoxin (Sieker and Jensen, 1965) contradicted the prevailing view that the eight iron atoms of this protein were arranged in a linear array (Blomstrom *et al.*, 1964). However, the presence of two distinct active sites in this protein was not even suspected until they were actually observed in high-resolution electron density maps (Sieker *et al.*, 1972). This important discovery—that eight iron ferredoxins actually contain two separate $Fe_4S_4{}^*$ clusters, each apparently indistinguishable from that which had already been described in HiPIP—has had a dramatic impact on subsequent iron–sulfur protein research. A common structural theme can now be seen to underly a previously bewildering array of iron–sulfur proteins with apparently variable iron and sulfur content and often conflicting data regarding their oxidoreduction stoichiometries (Orme-Johnson, 1973). Moreover, by showing that the same prosthetic group can have radically different oxidoreduction properties in different proteins, this discovery, *ipso facto*, implicated the polypeptide moiety as an active, coparticipant with the cluster in iron–sulfur protein function.

C. $Fe_4S_4^*$ Clusters in Other Proteins

Ferredoxins that contain only four iron and four inorganic sulfur atoms have been isolated from at least two bacterial sources, *Desulfovibrio gigas* (Laishley et al., 1969) and *Bacillus polymyxa* (Yoch and Valentine, 1972). Although there has been no structural study of either protein, their biological, spectroscopic, and magnetic properties justify the conclusion that these proteins contain one $Fe_4S_4^*$ cluster similar to those found in *P. aerogenes* ferredoxin. Of course, it should be noted that until X-ray structures are reported for these proteins, no quantitative conclusions can be made concerning their active-site geometries. Reference to these proteins is important in the present context because their existence implies that the unusually low potentials of clostridial-type ferredoxins are unrelated to the fact that they contain two $Fe_4S_4^*$ clusters.

D. The Three-State Hypothesis

To explain the large (0.75 V) difference between the midpoint electrode potentials of HiPIP ($E_0' = +0.35$ V) and ferredoxin ($E_0' = -0.40$ V) in light of the fact that both proteins contain the same prosthetic group, Carter et al. (1972) proposed the three-state hypothesis (Fig. 3). This paradigm has served to underline the fact that, to a first approximation, the polypeptide chain in each protein need not exert an unreasonable thermodynamic influence on the oxidoreduction properties of the cluster, either by applying severe stereochemical constraints on the cluster or by coupling oxidoreduction to an energetically imposing conformational change. Rather, since different pairs of oxidation states are accessible in each protein, most of the observed potential difference between the HiPIP and ferredoxin reactions is due to intrinsic potential differences between the three states, C^-, C, and C^+, of the $Fe_4S_4^*$ cluster itself. However, Cammack (1973) has demonstrated that reduction of Hp_{red} to the C^- state can occur under mildly denaturing conditions (80% DMSO) and at a potential at least 0.3 V lower than that observed for the analogous reduction of Fd_{ox}. This result, together with measurements on model compounds (reviewed by Holm, Chapter 7) indicate that half-cell potentials are indeed modulated by subtle influences inside the proteins. Possible

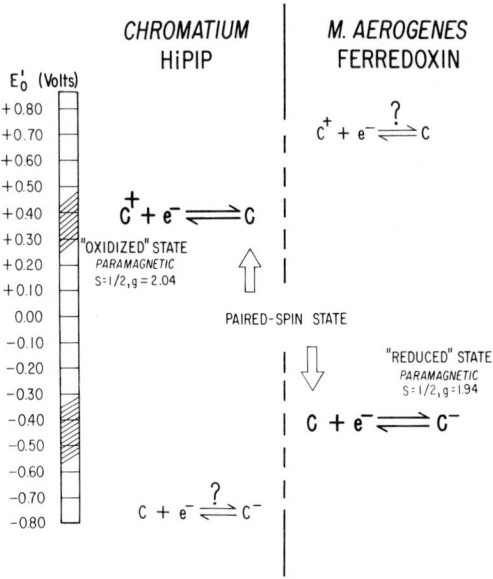

Fig. 3. The three-state hypothesis for relative oxidation levels of the $Fe_4S_4^*$ cluster in iron–sulfur proteins (Carter et al., 1972). The paired–spin state, C, occurs in both reduced HiPIP and oxidized ferredoxin. Half-cell reactions marked with a question mark have not been observed in the indicated proteins except under non-physiological conditions (Cammack, 1973).

"fine tuning" mechanisms employed by the proteins have been discussed by Carter et al. (1972, 1974b) and others (Adman et al., 1975).

E. Analogous Synthetic $Fe_4S_4^*$ Clusters

Shortly after the structure of the two $Fe_4S_4^*$ clusters in ferredoxin were reported (Sieker et al., 1972), Herskovitz et al. (1972) described the synthesis, structure, and properties of an inorganic $Fe_4S_4^*$ cluster whose structure was almost indistinguishable from those found inside the two proteins. Magnetic and spectroscopic properties of this compound showed it to be a surprisingly close analogue of the clusters in reduced HiPIP and oxidized ferredoxin (Herskovitz et al., 1972). Similar compounds have been synthesized subsequently with a variety of organic thiolate ligands. These studies (reviewed by Holm, Chapter 7) have greatly enhanced our understanding of the protein-bound clusters. Most importantly, they firmly establish the relative oxidation levels assumed in the three-state hypothesis by demonstrating that spectroscopic properties of both Hp_{red} and Fd_{ox} can be very closely duplicated in a single, synthetic analogue.

III. COMPARISON OF THE PROTEIN-BOUND AND SYNTHETIC $Fe_4S_4^*$ CLUSTERS

Meaningful comparisons between different protein-bound and synthetic $Fe_4S_4^*$ cluster geometries have been made possible by two crucial, but quite distinct developments.

1. Crystallographic refinement of the protein structures has provided reasonably precise atomic coordinates for the protein-bound clusters (Freer et al., 1975; Adman et al., 1976). These results indicate the limits within which legitimate comparison can be made.

2. Absolute oxidation levels of the protein-bound clusters have been unambiguously established by synthesis and spectroscopic study of inorganic analogues having known stoichiometries (Herskovitz et al., 1972). Both developments deserve a brief elaboration.

A. Protein Structure Refinement

Crystallographic refinement techniques have only recently been applied to proteins (Jensen, 1974b). Nevertheless, they are routinely employed in crystal structure analysis of low molecular weight compounds. Were it not for the considerable computing costs required, they could be used routinely in protein structure determinations. In principle, these methods provide for systematic adjustment of atomic positions so as to arrive at a structure that best agrees with observed X-ray data. Crystallographic refinement produces more reliable atomic positions as well as estimates for their uncertainty. In practice, both objectives have been realized to some extent as refinement methods have been applied to proteins. However, refinement of any protein must still be considered to be at an intermediate stage. Higher resolution intensity data are potentially accessible to measurement, and incorporating these data into the refinement should materially reduce the uncertainty in atomic positions (Jensen, 1974b). Moreover the indices of refinement—the so called crystallographic "R factors"[†]—for the proteins compare rather poorly with those customarily attained for small molecule crystal structures [about 20% for HiPIP and ferredoxin, compared with 3.6% for the earliest reported cluster analogue (Herskovitz et al., 1972)]. Nevertheless, $R \cong 20\%$ for the refined protein structures should also be compared to $R \cong 45\%$ for the unrefined protein structures.

Estimates of uncertainty in cluster geometries have been obtained for both proteins by averaging chemically equivalent bond lengths and

[†] The crystallographic R factor, $R = \Sigma ||F_0| - |F_0||/\Sigma |F_c|$, where $|F|$ and $|F_c|$ are the absolute values of the observed and calculated structure factor amplitudes.

angles. There is rather good agreement between these estimates and those which can be obtained from refinement procedures (Freer et al., 1975). As a result, several significant conclusions regarding cluster geometries are well supported by X-ray crystal data for the proteins.

B. Iron Oxidation Levels in the Clusters

The true oxidation levels of the Hp_{red} and Fd_{ox} $Fe_4S_4^*$ clusters were somewhat surprising when they were first established. The inorganic analogue having the closest optical and magnetic analogies to these protein-bound clusters crystallized as a dianion, $Fe_4S_4(SR)_4^{2-}$. Making the reasonable assumption that S* atoms could be represented by S^{2-}, Herskovitz et al. (1972) inferred that the four iron atoms included, formally, two ferrous and two ferric iron atoms. In retrospect, the discovery of mixed-valence states for iron should not have been surprising. Indeed, mixed-valence properties in the protein-bound clusters follow from the fact that they undergo one electron oxidoreduction reactions. It is worth noting that this "trapped valence state" description is purely formal, inasmuch as Holm et al. (1974b) have shown that intracluster electron exchange in the inorganic dianion occurs with a frequency greater than 10^{16} sec^{-1}. Finally, the reader should bear in mind that this compound is an analogue for the Hp_{red} and Fd_{ox} states, which are both "paired-spin" states according to the three-state hypothesis. Crystal structures for analogue C$^-$ and C$^+$ clusters have not been reported.

C. Mean Interatomic Distances and Angles in Protein-Bound and Analogue $Fe_4S_4^*(SR)_4$ Clusters

The current protein crystal structure data presented in Table I convincingly demonstrate the identity of the HiPIP and ferredoxin clusters with one another and with their synthetic analogues. Variation in mean values of each structural parameter from compound to compound do not exceed twice the rms deviations observed for that parameter in the protein-bound clusters, whose structures are known with less precision. This comparison implies that all six different $Fe_4S_4^*(SR)_4$ cores are geometrically very nearly congruent.

Several authors have noted that the $Fe_4S_4^*$ cores deviate substantially from the geometry of a simple cube having iron and sulfur atoms at alternate corners (Carter et al., 1972; Averill et al., 1973; Adman et al., 1973). Specifically, the four iron atoms are much closer to each other ($\overline{Fe-Fe} \cong 2.75$ Å) than are the four inorganic sulfur atoms ($\overline{S^*-S^*} \cong 3.50$ Å). This unusual geometry can be rationalized by viewing it as a

TABLE I
Mean Interatomic Distances and Angles in Protein-Bound and Analogue $Fe_4S_4^*SR_4$ Clusters

Bond	HP_{ox}[a]	HP_{red}[a]	$Fd_{ox}I$[b]	$Fd_{ox}II$[b]	$Fe_4S_4^*SCH_2\text{–}C_6H_5$[c]	$Fe_4S_4^*S\text{–}C_6H_5$[d]
Fe–Fe						
Mean	2.72	2.81	2.73	2.67	2.75	2.74
Range	2.68–2.78	2.74–2.87	2.66–2.83	2.65–2.77	2.776(2)–2.732(4)	2.730(2)–2.739(4)
rms deviation	0.04	0.04	0.06	0.07	—	—
Fe–S*						
Mean	2.26	2.32	2.23	2.22	2.28	2.29
Range	2.10–2.39	2.18–2.45	1.98–2.40	1.97–2.44	2.239(4)–2.310(8)	2.267(4)–2.296(8)
rms deviation	0.08	0.09	0.12	0.14	—	—
Fe–S$_\gamma$						
Mean	2.20	2.22	2.22	2.25	2.251	2.263
Range	2.17–2.22	2.19–2.26	1.98–2.43	1.98–2.45	—	—
rms deviation	0.02	0.03	0.16	0.17	—	—
S*····S*						
Mean	3.55	3.65	3.52	3.49	3.61	3.61
Range	3.49–3.64	3.50–3.75	3.41–3.60	3.37–3.68	—	—
rms deviation	0.06	0.11	0.07	0.12	—	—
S$_\gamma$····S$_\gamma$						
Mean	6.32	6.40	6.31	6.29	6.42	6.39
Range	6.06–6.66	6.02–6.65	5.60–6.78	5.86–6.54	5.95–6.71	6.30–6.53
rms deviation	0.20	0.22	0.37	0.23	0.30	0.10
Fe–S*/Fe–Fe	0.83	0.81	0.82	0.83	0.83	0.84

(continued)

TABLE I (continued)

Bond	HP_{ox}^a	HP_{red}^a	$Fd_{ox}I^b$	$Fd_{ox}II^b$	$Fe_4S_4 \cdot SCH_2-C_6H_6^c$	$Fe_4S_4 \cdot S-C_6H_6^d$
Fe–S*–Fe						
Mean	74	76	75	75	73.8	73.5
Range	72–76	72.80	—	—	—	—
rms deviation	1.3	2.4	2	2	—	—
S*–Fe–S*						
Mean	104	104	103	104	104.1	104.3
Range	101–109	99–107	—	—	—	—
rms deviation	2.4	2.6	—	—	—	—
S*–Fe–S$_\gamma$						
Mean	115	116	115	115	115.1	114.4
Range	107–120	106–126	—	—	—	—
rms deviation	4.9	5.3	3	3	—	—

[a] Carter et al., 1972.
[b] L. H. Jensen, private communication.
[c] Averill et al., 1973.
[d] Que et al., 1974.

compromise between conflicting architectural demands arising from the interatomic angles one might expect to observe about the iron and inorganic sulfur atoms, respectively (Carter et al., 1972). Thus, on the one hand, X–S–X bond angles between 90° and 105° are normally observed for sulfides (Pauling, 1960); and one might expect to find similar Fe–S*–Fe angles in the Fe_4S_4* clusters. On the other hand, each iron atom is surrounded by four sulfur atoms, and one might expect this environment to favor S*–Fe–S* angles close to the tetrahedral angles 109° 28'. These two angular requirements cannot both be satisfied simultaneously in the same geometry. Rather, two hypothetical limiting structures can be envisioned, as illustrated in Fig. 4A and 4B. The two limiting structures can be characterized by the ratio $\overline{Fe-S^*}/\overline{Fe-Fe}$. Limiting values of this ratio are 0.71 (Fig. 4A) and 0.92 (Fig. 4B). Observed values for this ratio are all about 0.83 (Table I, line 16), indicating that all six Fe_4S_4* cores assume very nearly the same intermediate geometry. It should also be noted that Fe–S*–Fe angles are all about 74° in this configuration, a value in excellent agreement with that expected on the basis of recent valence bond calculations for atoms forming spd hybrid orbitals (Pauling, 1975).

Perhaps the most striking feature of the cluster geometry is the existence of six rather close Fe–Fe contacts in the Fe_4S_4* core. These distances ($\overline{Fe-Fe} \cong 2.75$ Å) strongly suggest that iron–iron interactions must

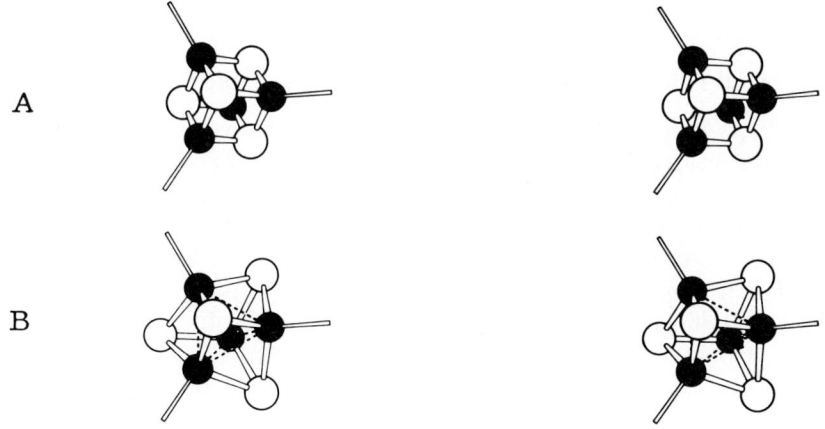

Fig. 4. Two hypothetical limiting structures for Fe_4S_4*SR_4 geometry. (A) A perfect cube of iron (filled circles) and sulfur (open circles). In such a compound, all S*–Fe–S* angles and all Fe–S*–Fe angles would be equal to 90°. (B) A similar structure constructed from four, perfectly tetrahedral iron atoms. Such a compound would have Fe–S*–Fe angles equal to 66°. The tendency for iron to assume tetrahedral bond angles introduces close iron–iron contacts in B (dashed lines).

assume a prominent place in any electronic description of the cluster. However, it is probably an oversimplification to assume, as was stated previously (Carter et al., 1971), that these contacts result from iron–iron bonding. The argument advanced in the preceding paragraph suggests an alternative possibility. The tendency of S*–Fe–S* angles toward the tetrahedral value may help to impose close iron–iron contacts, thereby introducing a repulsive component along the directions indicated by dashed lines between iron atoms in Fig. 4B. Experimental evidence (Carter et al., 1974b) favors the latter interpretation, and theoretical calculations consistant with it have recently been described (Yang et al., 1975). It is quite likely, therefore, that those iron–iron interactions most affected by biophysically important electron transfer reactions are actually antibonding ones. Further discussion of this point is deferred to Section VI below.

More detailed conclusions about cluster geometry are difficult to justify entirely on the basis of the protein structure data. Greater precision of the atomic positions can be anticipated to result from further refinement using higher-resolution X-ray data. Improved protein models will, one hopes, help to illuminate several unresolved questions suggested by data in Table I regarding subtle differences of size and shape between the six Fe_4S_4* clusters represented therein. However, in view of the importance of such differences to the stereochemistry of the HiPIP oxidoreduction cycle (Carter et al., 1972, 1974b) (Section V), it is worthwhile to pose two of these questions here.

D. Are All Paired-Spin State Fe_4S_4* Clusters the Same Size?

There is a statistically significant difference between the mean Fe–Fe distances in HP_{ox} ($\overline{Fe-Fe}$ = 2.72 Å) and HP_{red} ($\overline{Fe-Fe}$ = 2.81). This qualitative conclusion is quite secure and does not depend solely on positional parameters reflected in Table I or on further protein structure refinement (Carter et al., 1974b) (Section V). Clusters represented in columns 2–6 in Table I all have the same paired-spin oxidation state. Yet, their mean Fe–Fe distances vary over a greater range than that spanned by the HP_{ox} and HP_{red} distances (HP_{red} has $\overline{Fe-Fe}$ = 2.81 Å; $Fd_{ox}II$ has $\overline{Fe-Fe}$ = 2.67 Å). Furthermore, and in contrast to a previous observation based on ferredoxin coordinates that were somewhat less well refined (Carter et al., 1972), the two ferredoxin Fe_4S_4* clusters do not seem to differ significantly in size from the HP_{ox} cluster (Jensen, 1974a). Indeed, assuming that all Fe–Fe distances in a given cluster ought to be identical, then the standard deviation of that distance is the rms deviation/$\sqrt{5}$ = σ_m. Choosing σ_m based on the rms deviation for $Fd_{ox}II$

we find that Fe–Fe values for HP_{red} differ from those of $Fd_{ox}I$ and $Fd_{ox}II$ by 2.6 σ_m and 4.5 σ_m respectively and those for $Fd_{ox}I$ and $Fd_{ox}II$ differ from each other by 1.9 σ_m. Resolution of these intriguing and apparently meaningful differences is a compelling reason for extending X-ray analysis of the proteins to higher resolution. As we shall see below (Section IV,D), cysteinyl ligands to the HP_{ox}, HP_{red}, $Fd_{ox}I$ and $Fd_{ox}II$ clusters differ in the extent to which their $S\gamma$ atoms accept N–H---S hydrogen bonds from main chain amide groups (Carter et al., 1974b; Adman et al., 1975). One possibility is that these differences influence the extent of iron–iron bonding, and hence the overall size of the respective $Fe_4S_4^*$ cores. This possibility is strengthened by the fact that there is a statistically significant difference between Fe–Fe values for two synthetic clusters differing only in those ligands analogous to cysteinyl $S\gamma$ atoms (Table I, columns 5 and 6).

E. How Symmetrical Are the Protein-Bound $Fe_4S_4^*$ Clusters?

One might expect that for a compound such as $Fe_4S_4^*SR_4$, in which each iron is chemically equivalent, the four iron atoms would be found at the corners of a regular tetrahedron and hence would display a tetrahedral point group symmetry T (23) or Td (43m). In particular, all six Fe–Fe and all twelve Fe–S* bonds would be the same. Structures of the two synthetic clusters are sufficiently precise to justify the conclusion that neither has strict Td point group symmetry (Averill et al., 1973; Que et al., 1974). Deviations from Td symmetry are indicated in Table I by two sets of $\overline{\text{Fe–Fe}}$ distances and two sets of $\overline{\text{Fe–S}^*}$ distances. These deviations flatten the $Fe_4S_4^*$ "cube" along a direction roughly parallel to four Fe–S* bonds, the resulting geometry becoming more nearly tetragonal (point group = D_{2d}, 42m). Ibers (Averill et al., 1973) has pointed out the difficulty of assessing the true symmetry of a generalized M_4L_4 core even in precise, small molecule crystal structures. These difficulties notwithstanding, it should be noted here that data for each paired-spin, protein-bound cluster (HP_{red}, $Fd_{ox}I$, and $Fd_{ox}II$) are consistent with the conclusion that they all have elements of tetragonal symmetry. Carter et al. (1972, 1974b) have noted that an orientation for the HP_{red} cluster can be shown that displays two sets of Fe–S* distances—one, a set of eight, which averages 2.35 Å; the other, an approximately parallel set of four, which average 2.25 Å. Examination of the two Fd_{ox} clusters (Carter, 1976) indicates that they both have a similar, unique orientation. That is, all three paired-spin state clusters are apparently flattened by a relative shortening of four roughly parallel Fe–S* bonds and are perhaps of even lower symmetry. Moreover, the ($HP_{ox} - HP_{red}$) difference Fourier map shows strong elements of 222 symmetry in the region of the $Fe_4S_4^*$ cluster, strengthening the suggestion that the HP_{red} cluster has an appreciable tetragonal distortion from tetrahedral symmetry. As yet, no synthetic analogue for the HP_{ox} cluster has been crystallized, so it is impossible to specify the symmetry properties of a free C^+ state (actual net charge probably -1) cluster. However, the HP_{ox} cluster appears to be more nearly of tetrahedral symmetry than is the reduced cluster (Carter et al., 1974b) (Section V,B).

These considerations suggest that polypeptide–cluster interactions actually preserve some symmetry elements of model compound geometry in both proteins, and hence underscore the current interest in the precise details which characterize the protein cluster binding cavities.

IV. CHARACTERISTIC FEATURES OF THE CLUSTER BINDING CAVITIES

Both the HiPIP and the ferredoxin $Fe_4S_4^*$ clusters are immersed in cavities bounded principally by nonpolar amino acid side chains. Their environments would therefore be characterized by low dielectric constants, and hence are apparently similar to the solvents in which spectra and midpoint reduction potentials of model $Fe_4S_4^*SR_4$ clusters have been measured. So we cannot ascribe the distinctive, biological properties of the protein-bound clusters to bulk or solvent effects. Rather, their uniqueness apparently results from the fact that backbone amide groups penetrate the oil drop at specific points with geometries appropriate for hydrogen bond formation to specific cysteinyl and inorganic sulfur atoms (Carter et al., 1974b; Adman et al., 1975). Discussion of these characteristic features of the cluster binding cavities will follow after a brief description of polypeptide chain folding in each protein.

A. Polypeptide Chain Folding

The $Fe_4S_4^*$ cluster is a large chromophore that is covalently bound to the protein by cysteinyl sulfur atoms, which must occupy the corners of a nearly regular tetrahedron. One might ask whether or not any unusual polypeptide chain conformations are required in order to create an internal cavity that is complementary to such an imposing moiety. If so, then of course one would also want to know how these constraints influence the thermodynamic and or kinetic aspects of cluster function.

Chain tracings of the two proteins, drawn to the same scale for comparison, are shown in Fig. 5. *Chromatium* HiPIP has nearly three times

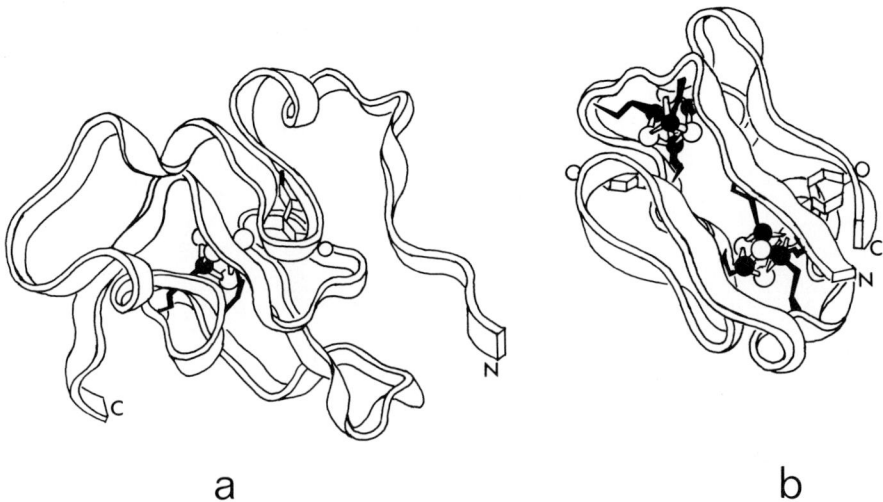

FIG. 5. Chain tracings of *Chromatium* HiPIP (a) and *P. aerogenes* ferredoxin (b). Both molecules are drawn to the same scale and with one of the ferredoxin $Fe_4S_4^*$ clusters (cluster II) and the HiPIP cluster in the same relative orientation with respect to the two tyrosine residues associated with them.

as many amino acid residues (85) with which to form a cavity for its single cluster as has ferredoxin to form one of its two cavities (54/2 = 27). Not surprisingly, then, chain conformations in the two proteins show little obvious similarity (however, see Section VI,C).

1. HiPIP

A careful inspection for conformational strain in oxidized HiPIP (Carter *et al.*, 1974a; Freer *et al.*, 1975) revealed no evidence for strained or unusual conformations in this protein. On the contrary, the polypeptide chain conformation may be described, almost in its entirety, as a sequence (Fig. 6) of α helical or extended (β) conformations, and hairpin turns (Venkatachalam, 1968) that are surprisingly close in their detailed geometry to those predicted to be of lowest energy. More than 75% of the main-chain hydrogen bonding sites are bonded either within secondary structures (47%) or to water (30%). Thus, nearly 80% of the folded structure has a hydrogen-bonded pattern similar to that expected for a solvated extended chain into which secondary structures have been introduced. The HiPIP main-chain conformation is that of a typical globular protein in most respects, a possible exception being the unusually high (17) number of hairpin turns. It is invariably observed, and therefore apparently essential to Fe–Sγ complex formation, that sequences con-

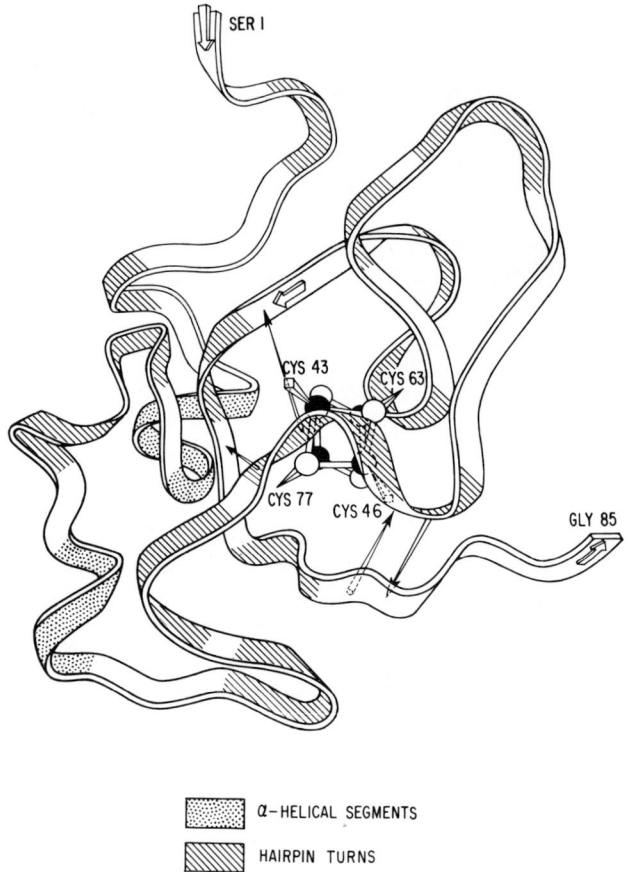

Fig. 6. Secondary structures in *Chromatium* HiPIP. Unmarked segments assume an extented configuration. Arrows denote tertiary main chain-to-main chain hydrogen bonds within the cluster binding cavity.

taining cysteinyl ligands assume hairpin-turn conformations (Adman *et al.*, 1975).

Hydrogen bonded secondary structures help to frame the binding cavity in HiPIP (Fig. 7). The cavity is constructed from the carboxyl-terminal segment containing residues 43–80 and from a short segment, residues 17–20 of the amino terminal portion. Two sides of the cavity are provided by antiparallel β sheet structures that contain cysteine residues 63 and 77. A third side is constructed from two linked hairpin turns and contains the remaining pair of cysteine residues, 43 and 46.

Residues 1–42 (segment on the left in Fig. 6) fold up upon the C terminal, cluster-binding segment, firmly "trapping" its conformation.

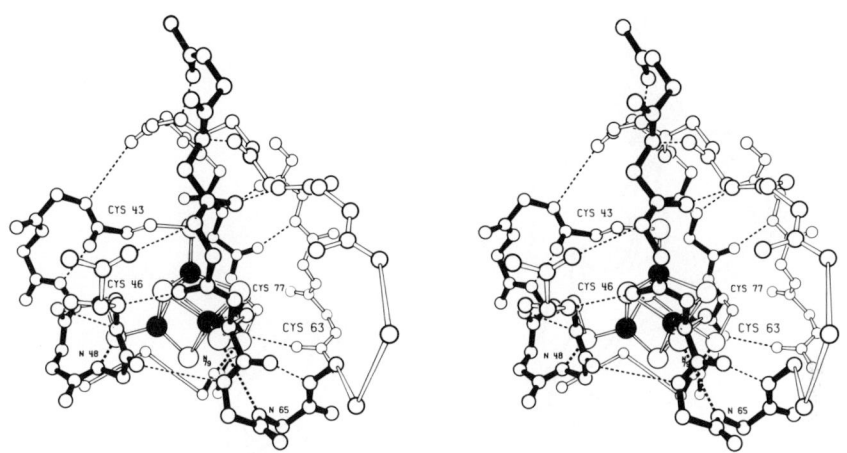

Fig. 7. Hydrogen-bonded secondary structures form the framework of the HiPIP cluster binding cavity. Chain segments containing the four cysteine residues 43, 46, 63, and 77, and the subsequent hairpin turns are shown by *heavy solid bonds*. Intrapeptide hydrogen bonds are indicated by *thin broken lines;* peptide-to-cluster hydrogen bonds (Section IV,D) are indicated by *heavy broken lines*. (From Carter *et al.,* 1974b.)

In view of this arrangement, it would obviously be difficult to alter the cluster chemically without first unfolding a major part of the molecule. So it is understandable that HiPIP is considerably more stable than *P. aerogenes* ferredoxin with respect to acid catalyzed decomposition of the cluster (Dus *et al.,* 1967; Maskiewicz *et al.,* 1975).

The protective, N-terminal sheath also contains Tyr 19, an invariant residue in the sequences of four HiPIPs (Tedro *et al.,* 1976, and S. Tedro personal communication). Close contacts between its aromatic ring and $Fe_4S_4^*$ cluster atoms have focussed attention on this residue as a possible component in the electron transfer pathway (Sections IV,C and V,C).

2. FERREDOXIN

The ferredoxin chain is so short that in order to create two complete cavities most chain segments must contribute to the formation of both cavities. Dual chain segment roles in ferredoxin arise in a rather curious and economical way. The N- and C-terminal segments (residues 1–25 and 26–54) each separately assume almost exactly the same main chain conformation (Fig. 8). Each half chain constitutes only a partial cluster binding cavity, affording three cysteinyl ligands to one of the two clusters and a single ligand to the other. The intact molecule, therefore, displays

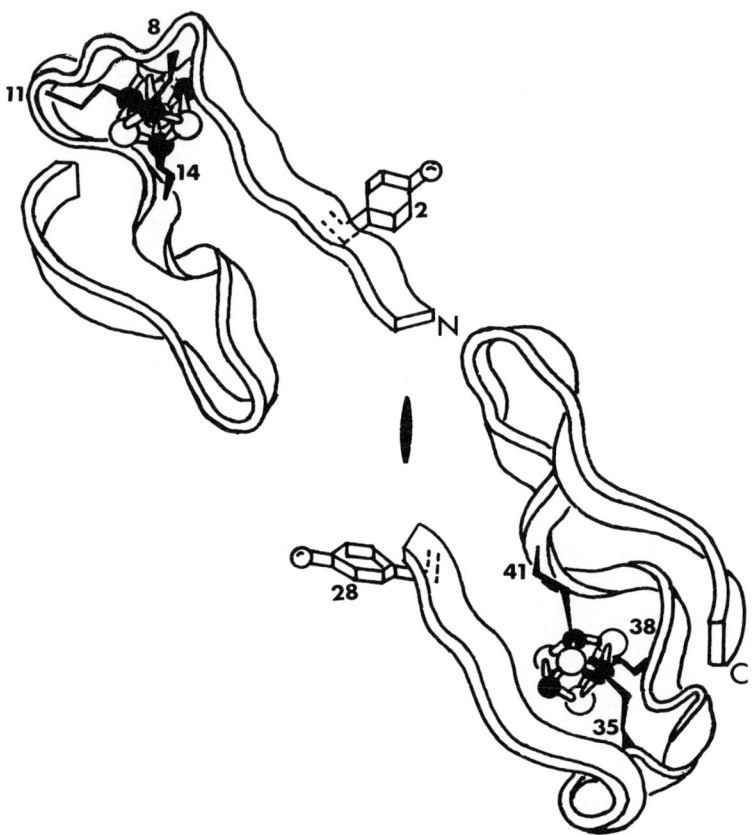

Fig. 8. Exploded view of *P. aerogenes* ferredoxin chain, showing elements of approximate twofold rotation symmetry relating the cluster binding cavities I and II.

an approximate intramolecular dyad axis (Adman *et al.*, 1973). This unexpected result is made possible by extensive repetition of amino acid sequences in the first and second halves of the molecule (Fig. 9).

A distance of about 12 Å separates the two $Fe_4S_4^*$ clusters, and the molecule is elongated along the line joining them. This distance is spanned by six short segments. Four of the six occur in pairs of antiparallel β structure, which are themselves associated in antiparallel fashion along one side of the molecule. Each pair contains two cysteines (bound to the same cluster) at one end and a semi-invariant tyrosine residue close to the other end. The remaining two strands occur as short (antiparallel) stretches of irregular helix. These six strands form a cylinder within which the two $Fe_4S_4^*$ clusters are suspended. Local, twofold rotation axes of the

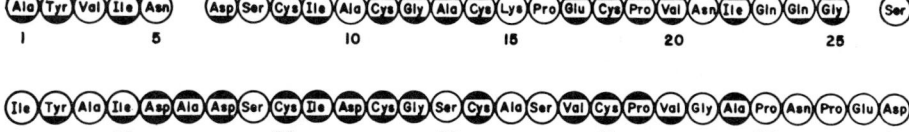

Fig. 9. Repetition of amino acid sequences in the first and second halves of *P. aerogenes* ferredoxin. Invariant (shaded top and bottom) and semi-invariant (shaded bottom) residues are based on five clostridial ferredoxin sequences (from Adman *et al.*, 1973).

clusters are roughly parallel to each other and to the cylinder axis, but they are not colinear (Fig. 5).

No strained or distorted conformations have been reported to occur in ferredoxin. Two residues, Ala 10 and Asp 37, assume the unusual conformation characteristic of a left-handed 3.0_{10} helix (Adman *et al.*, 1975). This conformation is unusual, not because it is strained, but because it occurs in a confined region of conformation space that is separated by a moderately high potential energy barrier from more common configurations, such as the right-hand helix and extended chain. On this basis, one would expect folding equilibria involving a left-handed helical configuration to be much slower than those involving the rest of the chain. Similar conformations are found in HiPIP, although not near the cluster-binding segment, and in most other proteins (Carter *et al.*, 1974a; C. W. Carter, Jr., unpublished observations). They invariably connect more extensive pieces of secondary structure. These observations suggest that the special significance, if any, of left-handed helical conformations in ferredoxin is related to the mechanism of folding, rather than to the mechanism of action in this protein.

Neither protein is refined sufficiently to permit identification of such distortions as grossly nonplanar peptide bonds or badly bent bond angles. However, it is difficult to imagine how such features could occur without also introducing unfavorable conformational angles. So, on the basis of the available data, one must conclude that strained stereochemistry is minimal in the two proteins.

B. Nonpolar Amino Acid Side Chains

Hydrophobic side chains virtually surround the $Fe_4S_4^*$ clusters in both proteins. The HiPIP cluster is in contact with an elaborate constellation of aromatic and other nonpolar amino acid side chains (Fig. 10). Intramolecular side-chain packing places one or more hydrophobic groups near each inorganic sulfur atom. Side chains of Leu 17, Tyr 19, Phe 48, Met 49,

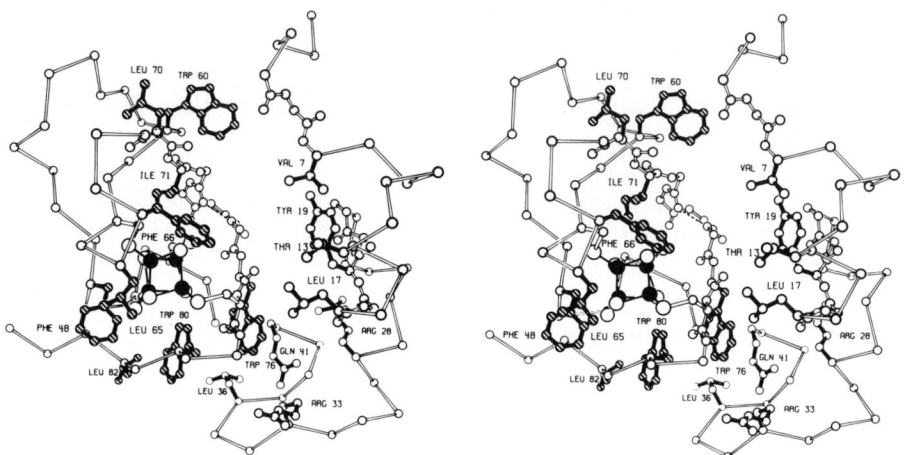

Fig. 10. Nonpolar amino acids close to the HiPIP $Fe_4S_4^*$ cluster. The molecule has been rotated from the orientation of Fig. 6 by 180° about a vertical axis passing through the α-carbon of Cys 43. The segment consisting of residues 1–41 is on the right and has been separated from the rest of the molecule by first translating, then rotating it to the right by 10° about a vertical axis passing through the α-carbon of Gln 41. Nonpolar side chains are emphasized by *heavy black bonds*. Side chains contributed by residues 42–85 are distinguished by *striped atoms*. The entire peptide unit is included wherever hydrogen bonds are formed across the interface. (From Carter *et al.*, 1974a.)

Leu 65, Phe 66, Ile 71, Trp 76, and Trp 80 all approach to within 5 Å of cluster atoms. Specific interactions of each S* atom are not completely similar for all four positions, but occur in pairs. The two S* atoms in the forefront of Fig. 10 are entirely surrounded by nonpolar side chains—Leu 17, Tyr 19, Phe 48, Leu 65, Phe 66, and Ile 71—whereas each of the remaining two S* atoms can accept an N–H---S hydrogen bond from a backbone amide group as well as being covered by the side chains of Met 49, Trp 76, and Trp 80.

These side-chain interactions effectively prevent access of solvent molecules to the cluster in the HiPIP crystal structure. Recent titration studies indicate that a similar statement can be made about the protein in solution (Maskiewicz *et al.*, 1975; Antanaitis and Moss, 1975).

A comparison of the amino acid sequences of high-potential iron proteins from *Thiocapsa pfennigii* (Tedro *et al.*, 1974) and *Rhodopseudomonas gelatinosa* (Tedro *et al.*, 1976) with that of *Chromatium* (Dus *et al.*, 1973) tends to underscore the importance of these features of the cluster-binding cavity (Table II). The three proteins have limited homology—only 16 of the 85 residues in *Chromatium* HiPIP are invariant

TABLE II
Important Amino Acids in Cluster-Binding Segments of Three Bacterial HiPIP's[a]

Chromatium vinosum	Thiocapsa pfennigii	Rhodopseudomonas gelatinosa
Leu 17[b]	**Leu**	**Leu**
Tyr 19[b]	**Tyr**	**Tyr**
Cys 43	**Cys**	**Cys**
Cys 46	**Cys**	**Cys**
Phe 48	Phe	Leu
Met 49	Ile	Phe
Trp 60	Trp	
Cys 63	**Cys**	**Cys**
Leu 65	**Leu**	**Leu**
Phe 66	Tyr	Phe
Gly 68	**Gly**	**Gly**
Ile 71	Val	Val
Gly 75	**Gly**	**Gly**
Trp 76	**Trp**	**Trp**
Cys 77	**Cys**	**Cys**
Trp 80	**Trp**	**Trp**

[a] Amino acids in **boldface** are invariant in the three sequences. (From Carter et al., 1974b.)

[b] These amino acids do not occur in the cluster-binding segment, but they lie close to the cluster.

in all three species. However, the spacing of the four cysteines within the amino acid sequences is nearly identical. The *T. pfennigii* and *R. gelatinosa* HiPIP sequences have deletions of four and two residues, respectively, in the segment between Cys 46 and Cys 63. These deletions do not seem likely to disrupt the hydrogen-bonded framework shown in Fig. 7, because the corresponding segment in *Chromatium* HiPIP is a surface loop containing residues 55–61. Moreover, the cluster-binding cavity contains eleven of the sixteen invariant amino acids (Leu 17, Tyr 19, Cys 43, Cys 46, Cys 63, Leu 65, Gly 68, Gly 75, Trp 76, Cys 77, and Trp 80); and four additional hydrophobic residues are replaced by conservative substitutions (Phe 48, Met 49, Phe 66, and Ile 71).

In contrast to the obvious importance of the amino acids that form the cluster-binding portion of the molecule, there is no indication from the amino acid sequences that surface features such as charged side chains are conserved. In fact, the isoeletric points of the *Chromatium* (pI′ = 3.68) and *R. gelatinosa* (pI′ = 9.50) proteins (Dus et al., 1967) suggest that at physiological pH values the two proteins would be both strongly and oppositely charged. The most crucial features of the HiPIP molecule, as indicated by their tendency to be preserved in several related amino acid sequences, appear to be concerned with the shape and nonpolar character of the cluster-binding cavity.

Parenthetically, it should be noted that some surface feature should be involved in specific interaction with physiological oxidases and reductases. In this regard, recent sequence information (Tedro *et al.*, 1976) suggests that absolute invariance of Asp 22 in four HiPIP sequences may implicate the region near this side chain in such physiological interactions. Asp 22 is, of course, quite near to Tyr 19, which is thought to participate in the oxidation–reduction mechanism.

Nonpolar side chains surrounding the ferredoxin clusters differ somewhat from those found in contact with the HiPIP cluster. Specifically, the ferredoxin cavity contains chiefly aliphatic side chains (Adman *et al.*, 1973), whereas the HiPIP environment includes many aromatic side chains. Aliphatic side-chain packing arrangements at either end of the (prolate) ellipsoidal molecule and between the two $Fe_4S_4^*$ clusters are shown in Figs. 11 and 12, respectively. These views do not suggest any obvious pathway for electron exchange between clusters in the same molecule and suggest that the clusters are protected, albeit tenuously, from contact with solvent water molecules. As was also the case for HiPIP, amino acid side chains in these positions are highly resistant to evolutionary substitution (Table III). Two particularly noteworthy exceptions

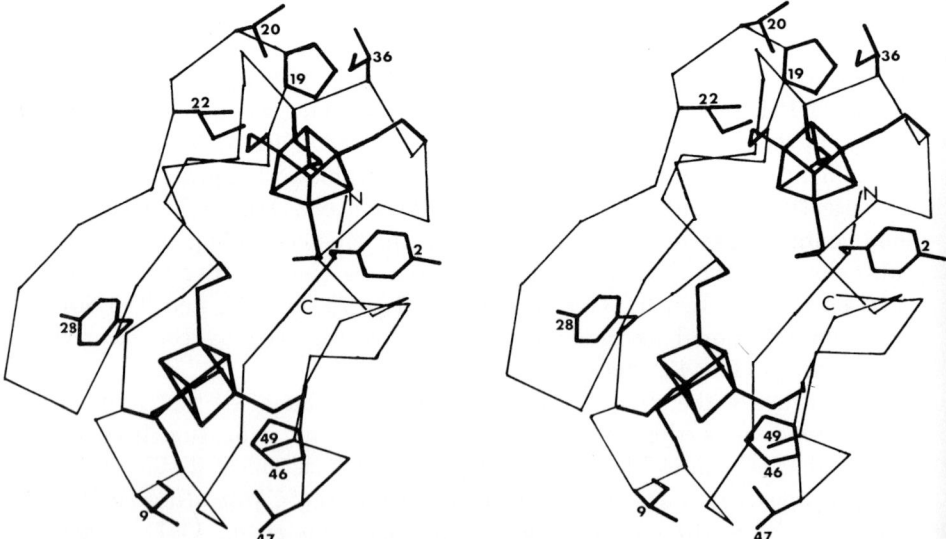

Fig. 11. Aliphatic side-chain packing at either end of the ferredoxin molecule. (Coordinates courtesy of Professor L. H. Jensen.)

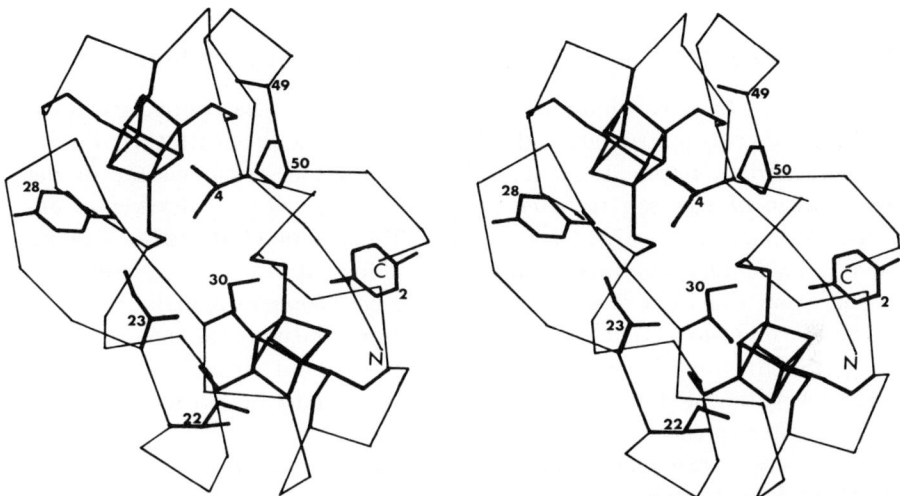

Fig. 12. Aliphatic side chain packing in between Fe_4S_4 clusters in ferredoxin. Note the strong element of approximate twofold rotation symmetry in this and the previous figure. (Coordinates courtesy of Professor L. H. Jensen.)

TABLE III
Aliphatic Amino Acid Side Chains in Contact with Ferredoxin Clusters[a]

Cluster I	Cluster II
Fig. 11	
ILE 9	Ile 36
Pro 46	**Pro 19**
VAL 47	**Val 20**
ALA 49	Ile 22
Fig. 12	
Ile 4	ILE 30
Ala 49	Ile 22
PRO 50	ILE 23 (Glu[b])

[a] Adapted from Adman et al., 1973. Amino acids in **boldface** are invariant in seven bacterial ferredoxin sequences (Tanaka et al., 1974); amino acids in capital letters only are highly conserved.

[b] There is an unresolved discrepancy concerning this residue which appears as ile in the X-ray structure (Adman et al., 1976) and as Glu in the chemical sequence (Tsunoda et al., 1968).

to the aliphatic character of ferredoxin amino acid side chains, i.e., Tyr 2 and 28 (Figs. 11 and 12), will be discussed in the following section.

C. Aromatic Side Chains in Juxtaposition with the $Fe_4S_4{}^*$ Clusters

There has been considerable interest in the fact that both ferredoxin and HiPIP contain aromatic side chains in close contact with their $Fe_4S_4{}^*$ clusters and in implications that these residues may participate in the mechanism of electron transfer to and from the cluster (Packer et al., 1972; Adman et al., 1973; Carter et al., 1974b; Lode et al., 1974; Tanaka et al., 1974; DePamphilis et al., 1974). This interest derives from the observation that Tyr 2 and Tyr 28 in P. aerogenes ferredoxin, and Tyr 19 in Chromatium HiPIP enjoy a particularly intimate association with an $Fe_2S_2{}^*$ face of the cluster in the protein crystal structures (Fig. 13) together with the fact that these residues tend to be conserved in the amino acid sequences of homologous proteins.

Geometric arrangements in the two proteins are similar but not identical. In both proteins, the tyrosine side chain and cluster are in distant parts of the amino acid sequence, and the association between them, therefore, depends only on tertiary interactions (Section IV,A). The principal interaction of the tyrosine ring with the cluster occurs via an inorganic

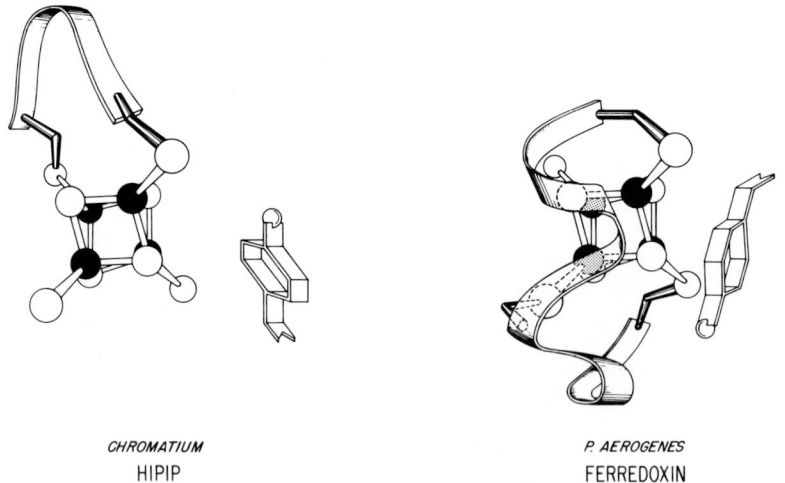

CHROMATIUM
HIPIP

P. AEROGENES
FERREDOXIN

Fig. 13. The relative orientations of Tyr 19 in Chromatium HiPIP and Tyr 28 in P. aerogenes ferredoxin with the respective $Fe_4S^*{}_4$ clusters. The ribbon represents that part of the polypeptide chain that contains closely spaced cysteine residues Cys–X–X–Cys in HiPIP and Cys–X–X–Cys–X–X–Cys in ferredoxin. (From Carter et al., 1974b.)

sulfur atom in both cases. One of the two chemically equivalent C_δ ring carbon atoms is closest to the cluster, and in each protein these atoms lie about 3.5 Å from one of the inorganic sulfur atoms. However, three differences should be noted between the arrangements in HiPIP and in ferredoxin.

1. In HiPIP, the ring is offset from the cluster so that while the $C_\delta 1$ and $C_\varepsilon 1$ atoms lie about 3.5–4.0 Å from the sulfur atom S^*3, the $C_\delta 2$ and $C_\varepsilon 2$ atoms are 5.6–6.0 Å away. In contrast, both tyrosine rings in ferredoxin are much more nearly centered on the corresponding inorganic sulfur atom. In fact the $C_\delta 1$, C_δ, $C_\varepsilon 1$, and $C_\varepsilon 2$ atoms of Tyr 28, which appears in Fig. 13, are all between 3.4 and 3.8 Å from the inorganic sulfur atom. Moreover, the ring plane in ferredoxin is more nearly parallel to the plane of the $Fe_2S_2^*$ face. A curious consequence is that $C_\delta 1$ and $C_\delta 2$ of the ring make equally close contacts with three cluster atoms, that is with $S_\gamma 8,35$ and $S_\gamma 14,41$ as well as with $S^*123,678$ (Table IV).

2. The aromatic side chains are oriented differently in the two proteins with respect to the axis defined by the β-carbon and hydroxyl oxygen atoms. In HiPIP, this axis is approximately parallel with a line connecting two iron atoms; in ferredoxin, this axis is rotated by approximately 90°, so that it is nearly parallel to a line connecting two inorganic sulfur atoms.

3. In HiPIP, the Tyr 19 hydroxyl group appears to accept a hydrogen bond from the backbone NH of Asn 72 (Fig. 16) and to donate a hydrogen bond to a bound water molecule, WAT 140. No similar hydrogen-bonding arrangement can be found in ferredoxin, the homologous contact being in each case with C_α of an invariant glycine (Gly 12, Gly 39). Moreover, the aromatic residue in question is an invariant tyrosine in all three HiPIP sequences (Table II), but it is frequently phenylalanine or histidine in ferredoxin sequences (Tanaka et al., 1974).

TABLE IV
CONTACT DISTANCES BETWEEN AROMATIC SIDE CHAINS AND $Fe_4S_4^*$ CLUSTER ATOMS

HP_{ox}		Fd_{ox} I		Fd_{ox} II	
Tyr 19	(Å)	Tyr 28	(Å)	Tyr 2	(Å)
$C_\delta 1-S^*3$	3.7	$C_\delta 1-S^*123$	3.7	$C_\delta 1-S^*678$	3.5
		$C_\delta 2-S^*123$	3.6	$C_\delta 2-S^*678$	4.2
$C_\delta 1-S_\gamma 77$	5.0	$C_\delta 1-S_\gamma 8$	4.1	$C_\delta 1-S_\gamma 35$	4.0
		$C_\delta 2-S_\gamma 14$	3.8	$C_\delta 2-S_\gamma 41$	3.8

The full significance of these differences is obscure at present. Carter et al. (1974b) (Section V,C) have proposed an electron transport mechanism for HiPIP utilizing the hydrogen-bonded interactions of the tyrosine hydroxyl group. The absence of similar interactions in ferredoxin rules out such a mechanism for reduction of this protein. Two laboratories have recently presented separate pieces of evidence questioning the participation of the conserved aromatic amino acids in electron transport at all. First, Lode et al. (1974) have replaced Tyr 2 by leucine in *Clostridium acidi-urici* ferredoxin using partial digestion and resynthesis methods. They found that this replacement did not impair biological activity; that is, the synthetic protein could still accept two electrons per molecule in an enzymatic reaction. In the second experiment, Tanaka et al. (1974) sequenced a new and presumably active clostridial ferredoxin (*Clostridium M-E*) having an arginine residue in position 30 (28). Although these experiments suggest that aromatic side chains are dispensible for electron transport, both have an alternative explanation, which is that intramolecular electron transfer is fast enough that only one "active site" is required for the transfer of two electrons, one to each cluster. The second experiment admits of yet another interpretation, perhaps more interesting from a structural viewpoint, than either of the previous ones. Arginine is, of course, a conservative amino acid replacement for histidine and is accessible to it via a single nucleotide base change. Histidine has been observed previously in position 2, but not in position 30 of related ferredoxins (Tanaka et al., 1974). The rather striking novelty of this replacement may provide an important clue as to the evolutionary importance of aromatic side chains in positions 2 and 30 of ferredoxin without actually ruling them out as a site for electron transfer. Recall the distinctly symmetric interactions of tyrosine C_δ atoms with cysteinyl sulfur atoms in *P. aerogenes* ferredoxin. Obvious geometric and electronic similarities exist between the arginine guanido group and the C_β, C_γ, $C_\delta 1$, and $C_\delta 2$, atoms of an aromatic side chain. Thus, the guanido group is actually surprisingly well suited in two respects to occupy the binding pocket reserved for tyrosine in *P. aerogenes* ferredoxin; first, it can present a planar, four center, delocalized π-electon system to the inorganic sulfur atom in analogy with the aromatic ring; and second, judging from homologous distances given in Table IV, one might also expect it to donate N–H---S hydrogen bonds to one or both S_γ atoms.

D. $Fe_4S_4{}^*$ Coordination Geometry: NH---S Hydrogen Bonds

Discovery of NH---S hydrogen bonds from the backbone chain to cysteinyl and inorganic sulfur atoms in HiPIP has provided the most important key to understanding how the biological $Fe_4S_4^*$ clusters differ from their synthetic analogues. Carter et al. (1974b) observed that amide groups of residues in the $n + 2$ position following Cys 46, Cys 63, and Cys 77 were oriented by reasonably standard hairpin turn geometry so as to donate hydrogen bonds to the S_γ atoms immediately preceeding them (Fig. 7); that N to S distances for these and other NH---S contacts appeared to shorten appreciably upon reduction of oxidized HiPIP, implicating these bonds in stabilization of the increased negative charge; and, finally, noted that additional NH---S hydrogen bonds observed in ferredoxin but not in HiPIP could account for the ability of the former protein to undergo further reduction to the C⁻ (formal charge —3) state. Subsequent analysis has clarified the number and precise stereochemistry of NH---S* and NH---S_γ hydrogen bonds, not only in ferredoxin but in rubredoxin as well (Adman et al., 1975), providing additional support for these conclusions.

1. NH---S Hydrogen Bonds

Identification of hydrogen bonds from crystal structure data is always presumptive, invoking two empirical criteria: (1) deviation of the donor –H--- acceptor system from colinearity, which in the ideal case is appreciable but small (Donohue, 1968); and (2) shortening of the H--- acceptor distance to below the sum of their van der Waals radii. Donohue (1969) observed N---S distances in the range 3.25–3.55 Å in small compounds. Owing to the intermediate stages of protein refinement, the errors associated with protein coordinates are approximately 0.1–0.2 Å. Bearing this range in mind, observed geometries for all possible NH---S_γ, and most NH---S* interactions are close to expected values (Table V).

2. NH—S Contacts and Polypeptide Secondary Structure

In view of the obvious possibility that these NH---S contacts may influence the thermodynamic and kinetic properties of the $Fe_4S_4^*$ clusters, it is worthwhile to note exactly how they are related to secondary structures in HiPIP and ferredoxin. It appears that in this respect the two proteins form very similar NH---S_γ bonds to the cysteinyl sulfur atoms, but have a rather different set of NH---S* bonds to the inorganic sulfur atoms.

HiPIP and ferredoxin have in common the manner in which NH---S_γ hydrogen bonds result from slight modifications of hairpin turn geometry. Three of the four cysteine residues linked to each cluster—Cys 46, 63, and 77 in HiPIP; Cys 8, 11, and 45 in Fd_{ox} I; and Cys, 35, 38, and 18 in Fd_{ox} II—are followed by such configurations, in which the S_γ atom accepts a hydrogen bond from the amide nitrogen of residue $n + 2$. One such S_γ

TABLE V
NH---S Hydrogen Bonds in HiPIP and Ferredoxin

	Bond length (Å)			Bond length (Å)		
	HP_{ox}	$HP_{red}{}^a$		$Fd_{ox}\ I^b$		$Fd_{ox}\ II^b$
			NH---S_γ hydrogen bonds			
N48-S_γ46	3.7	3.6	N10-S_γ8	3.6	N37-S_γ35	3.3
N81-S_γ46	3.7	3.7	N28-S_γ8	3.4	N2-S_γ35	3.5
N65-S_γ63	3.5	3.1	N13-S_γ11	3.6	N40-S_γ38	3.4
N79-S_γ77	3.5	3.3	N47-S_γ45	3.7	N20-S_γ18	>4.0
			N49-S_γ45	3.6	N22-S_γ18	3.9
			NH---S* hydrogen bonds^d			
N49-S*2^c	4.0	3.6	N9-S*124	3.7	N36-S*567	3.7
N77-S*1	3.7	3.4	N11-S*124	>4.0	N38-S*678	3.3
			N12-S*123	3.1	N39-S*678	3.5
			N14-S*234	3.9	N41-S*578	3.1

^{a} Carter et al., 1974b.
^{b} Adman et al., 1975.
^{c} C. W. Carter, Jr., unpublished.
^{d} Numbers refer to the three iron atoms to which S* is bonded in their order of appearance in the amino acid sequences, e.g., S*124 is bonded to Fe8, Fe11, Fe45.

atom (Cys 46) in HiPIP, and two each for the ferredoxin clusters (Cys 8 and 45 for Fd_{ox} I, and Cys 35 and 18 for Fd_{ox} II) accept an additional hydrogen bond from another amide group elsewhere in the chain. One S_γ atom in each cluster (S_γ43 in HiPIP, S_γ14 and 41 in ferredoxin) cannot accept an NH---S hydrogen bond for want of an appropriate NH donor.

There are rather pronounced differences in the location of NH---S* hydrogen bonds in the two proteins. These differences are evidently related to the relative spacing of cysteine residues liganded to the Fe_4S_4* clusters. The eight cysteine residues in ferredoxin occur in two groups of four, having the spacing Cys 8,35 XX Cys 11,38 XX Cys 14,41 XXX Cys 18,45. As was pointed out in Section IV,A,2, the first triplet of each group is bonded to one cluster, and the last is joined to the neighboring cluster. HiPIP contains two cysteine residues in a similar arrangement Cys 43 XX Cys 46; but the remaining two, Cys 63 and Cys 77, are well separated from this pair and from each other. These relationships are illustrated schematically in Fig. 14.

Apparent similarity between Cys XX Cys sequences in the two proteins is misleading with regard to detailed coordination geometries. Adman et al. (1975) point out that cysteinyl sulfur atoms linked by such a sequence need not be coordinated to an Fe_4S_4* cluster in the same

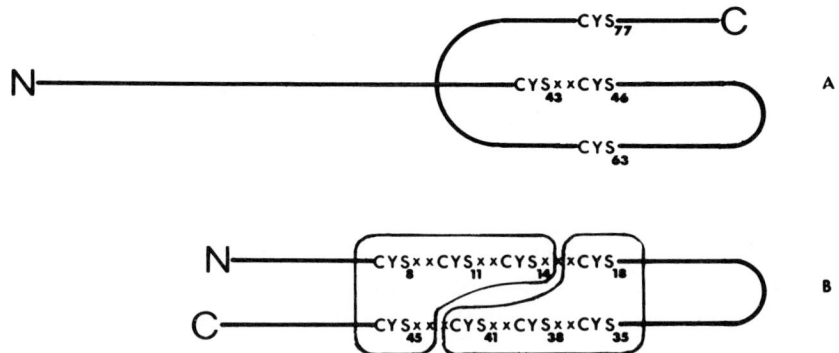

Fig. 14. Relative locations of cysteine residues in HiPIP and ferredoxin. (A) Cysteine residues in the HiPIP sequence; (B) (from Adman et al., 1973) cysteine residues in the ferredoxin sequences. Note the approximate twofold symmetry axis.

way. Rather, there are at least two different possibilities, each characterized by a different sequence of main-chain conformational angles. One such conformational sequence is found in the ferredoxin cavities; another occurs in the HiPIP cavity. The essential difference between NH---S* contacts in these two arrangements is that all three backbone amide groups in the HiPIP Cys 43 XX Cys 46 sequence participate in intrachain hydrogen bonding (Fig. 7) and are not directed toward the cluster, whereas all twelve amide groups in the two ferredoxin Cys XX Cys XX Cys sequences are directed toward either $S\gamma$ or S* sulfur atoms in the Fe_4S_4* clusters. Indeed, all of the NH---S* contacts in ferredoxin originate in the Cys XX Cys XX Cys sequences. In contrast, the two NH---S* contacts in HiPIP involve amide nitrogens (N 49, N 77) within the antiparallel β sheets that enclose the Fe_4S_4* cluster on opposite sides (Fig. 7).

It is possible only to guess at what implications these apparently significant differences might have. An attractive hypothesis, borne out for HiPIP in the comparison of its two oxidation states, is that NH---S* contacts help to fix the directions in which Fe_4S_4* cluster atoms move during electron transfer. Further discussion of this point will be deferred to Section V,C.

V. ANALYSIS OF REDUCED VERSUS OXIDIZED HiPIP

Reduced iron–sulfur proteins are difficult to stabilize for long enough time periods to permit extensive X-ray intensity measurements. For this reason, all published structure models are of proteins in their oxidized states (Rb_{ox}, Watenpaugh et al., 1973; HP_{ox}, Carter et al., 1974a; Fd_{ox},

Adman et al., 1976). HiPIP has the highest midpoint potential of the three proteins, and it has proven to be easiest to maintain crystals of this protein in the reduced state. Moreover, HP_{ox} and HP_{red} crystals are isomorphous, thereby permitting comparison by difference Fourier methods. A difference Fourier map between oxidized and reduced *Chromatium* HiPIP shows that the HP_{red} $Fe_4S_4{}^*$ cluster shrinks slightly upon oxidation. Changes in cluster geometry indicated in this map agree quite well with results of independent least squares refinement of cluster atom parameters in both oxidation states and hence almost certainly represent real differences between HP_{red} and HP_{ox} (Carter et al., 1974b).

A. Preparation and Stabilization of HP_{ox} and HP_{red} Crystals

At issue in this study (Carter et al., 1974b) is not so much the interpretation of X-ray data, per se, but rather the demonstration of protein oxidation states during the actual X-ray intensity measurements. The following points are to be made in this regard:

1. Although HiPIP is isolated and crystallized in the fully reduced state, these initially reduced crystals become extensively oxidized by air within hours after conventional mounting inside a sealed glass capillary. Oxidation was detected by redissolving the irradiated crystal and measuring its optical absorption spectrum. This result made it necessary to use individual pretreatment and mounting procedures for both HP_{ox} and HP_{red} crystals to insure a valid comparison between the two.

2. HP_{ox} crystals were prepared by oxidizing reduced crystals with ferricyanide prior to mounting in the capillary. HP_{red} crystals were stabilized by a saturating concentration of dithiothreitol and then mounted in a bath of this reducing mother liquor.

3. In order to verify that data crystals had retained their intended oxidation states, they were removed from the capillary, washed, and redissolved for spectrophotometric analysis. Three kinds of data crystals could be distinguished on the basis of their previous preparation and their optical spectra after irradiation: fully, chemically oxidized (HP_{ox}); incompletely air oxidized (HiPIP); and fully reduced (HP_{red}) crystals.

4. The three different kinds of crystals gave consistently different X-ray patterns, as evidenced by difference electron density maps and by relative overall temperature factor corrections.† The two crystal types

† Relative overall temperature factor correction constants C appear in the expression $\exp[-C^2 (\sin^2 \theta/\lambda^2)]$ and are employed routinely in scaling together intensity data sets from two isomorphous protein crystals. They compensate for such resolution-dependent factors as differences in static disorder and thermal motion between crystals.

6. IRON–SULFUR PROTEIN CRYSTAL STRUCTURES

designated as oxidized both had similar overall temperature factor constants relative to HP_{red} crystals, and a difference map using coefficients (HP_{ox}–HiPIP) was essentially featureless. On the other hand, substantial and virtually identical features appeared in difference maps using either of the coefficients (HP_{ox}–HP_{red}) or (HiPIP–HP_{red}). Comparison of the three difference Fourier maps also gave reassuring evidence that overall systematic errors, for example, in the relative scale and temperature factor constants, were negligible in comparison to features that could be ascribed to structural differences. The consistency of these results strongly supports the conclusion that HP_{ox} and HP_{red} X-ray data were actually measured from crystals of oxidized and reduced *Chromatium* HiPIP, respectively.

B. Structural Changes in the $Fe_4S_4^*$ Cluster

1. AGREEMENT BETWEEN THE DIFFERENCE FOURIER AND RESULTS OF STRUCTURE REFINEMENT

The $Fe_4S_4^*$ cluster and its four covalently linked cysteine S_γ atoms are shown in Fig. 15. This figure is simply a close-up view of the cluster from exactly the same viewpoint as it is seen in Fig. 7 (right-hand view).

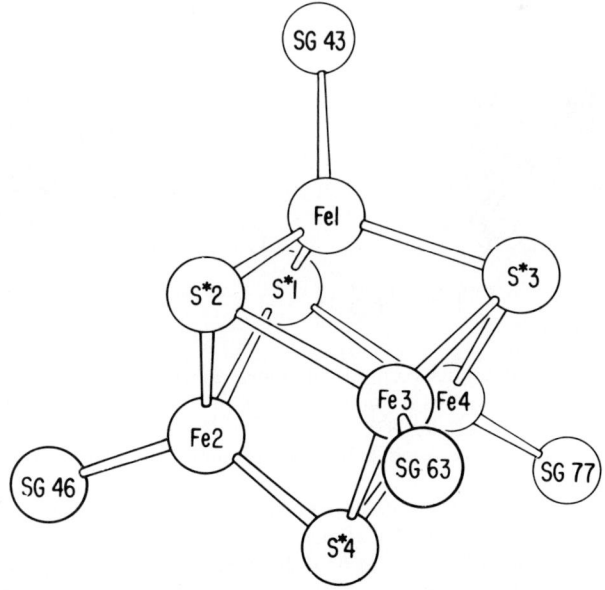

Fig. 15. Close-up view of iron–sulfur cluster in HiPIP as it appears in Fig. 7 (right) showing individual atom identifiers.

Its purpose is to display the individual atom identification symbols and numbers to be used in subsequent discussion and to aid in correlating the orientation of the cluster as it is shown in the $HP_{ox} - HP_{red}$ difference Fourier (Fig. 16) with its orientation in the other figures.

The most striking features of the oxidized minus reduced difference Fourier are two columns of positive density on the left and right sides of the cluster and perpendicular to the sections illustrated. Two less prominent columns of negative density lie in the same orientation at the top and bottom sides of the cluster.

An immediate conclusion to be drawn from the excess of positive density within the boundaries of the cluster is that in the oxidized state the cluster is somewhat smaller than it is in the reduced state. Moreover, from the shape of the columns of positive density, it is evident that the greatest shortening takes place between the top and bottom sides of the cluster as it appears in Fig. 16. That is, the faces bound on the top by Cys 46 and Cys 63 and on the bottom by Cys 43 and Cys 77 move closer together upon oxidation. Stated in another way, oxidation of the cluster causes shortening of the bond Fe1–S*2 and the three other Fe–S*

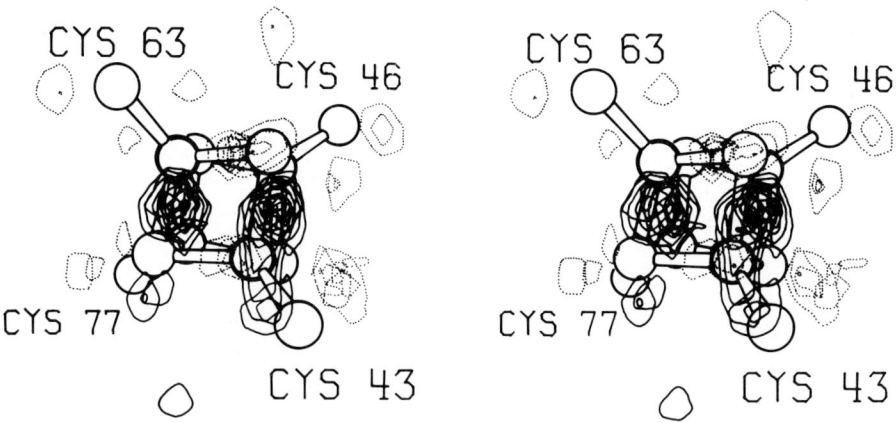

Fig. 16. The oxidized minus reduced difference Fourier for *Chromatium* HiPIP. The map is shown superimposed on a drawing of the reduced Fe$_4$S$_4$* cluster as determined by refinement of the HP$_{red}$ crystal structure. The view is along a skew vector running approximately through the centers of the top and bottom faces. Contour lines enclosing positive (solid lines) and negative (broken lines) difference density were drawn at ±0.225, ±0.300, ±0.375, ±0.450, and ±0.525 electrons per Å^3 on an absolute scale. Sulfur (S$_\gamma$) atoms of the four cysteine ligands are labeled with their residue numbers in order to help distinguish the four iron atoms. Further, to aid in correlating this view of the cluster with those shown in other figures, observe that S*2 is in the upper right corner and S*3 is in the lower left corner of the Cys 43–Cys 63 face.

bonds approximately parallel to it. In addition, a lesser contraction occurs in a direction perpendicular to the plane of the page, or along the bond Fe1–S*1 and the three other Fe–S* bonds approximately parallel to it. The weaker columns of negative density seen in the Cys 46–Cys 63 and the Cys 43–Cys 77 bound faces suggest that simultaneously the left and right faces in Fig. 16 move apart slightly upon oxidation, i.e. slight expansion occurs in the direction of Fe1–S*3. Results from refinement of the twelve cluster atoms in HP_{ox} and HP_{red} by the method of least squares (Freer *et al.*, 1975) support this interpretation of the difference Fourier and provide rough quantitative estimates of the effects observed. In Table VI are listed individual Fe–S* bond distances for both the oxidized and reduced clusters. They are grouped into three sets of four roughly parallel Fe–S* bonds mentioned above, and the mean bond distances in each direction are also given. The rms deviation within each set is about 0.07 Å, consistent with an estimate of the current precision of these data based upon their behavior during the last few cycles of refinement.

There is very good agreement between the quantitative estimates obtained by structure refinements and the qualitative interpretation of the HP_{ox}–HP_{red} difference Fourier given above. The most significant struc-

TABLE VI
INTERATOMIC DISTANCES IN IRON–SULFUR CLUSTERS OF OXIDIZED AND REDUCED HiPIP[a]

Iron–sulfur cluster	HP_{ox}	HP_{red}
Fe1–S*2	2.32	2.45
Fe2–S*1	2.24	2.41
Fe3–S*3	2.16	2.40
Fe4–S*4	2.17	2.27
Mean	2.22	2.38
Fe1–S*1	2.24	2.35
Fe2–S*2	2.17	2.28
Fe3–S*4	2.37	2.37
Fe4–S*3	2.22	2.33
Mean	2.25	2.33
Fe1–S*3	2.34	2.34
Fe2–S*4	2.27	2.28
Fe3–S*2	2.25	2.18
Fe4–S*1	2.25	2.20
Mean	2.28	2.25

[a] From Carter *et al.* 1974b.

tural change that accompanies HP_{red} oxidation and that can be observed by X-ray crystallography is a net contraction of the $Fe_4S_4{}^*$ cluster principally involving movement toward each other of two opposite $Fe_2S_2{}^*$ faces.

2. Apparent Symmetry in the $Fe_4S_4{}^*$ Cluster Rearrangement

There is no evidence in these results to indicate that oxidation affects any of the cluster atoms uniquely. On the contrary, rearrangement of the $Fe_4S_4{}^*$ cluster atoms occurs with approximate twofold symmetry along what might be expected to be twofold axes in the cluster itself. Twofold symmetric elements are obvious in the HP_{ox}–HP_{red} difference map (Fig. 16). They imply that all four atoms in HP_{ox} are affected equally by the loss of an electron. The reader may recall that the two ferrous and two ferric iron atoms in synthetic $Fe_4S_4{}^*$ analogues of HP_{red} exchange by electron delocalization reactions that are faster than 10^{16} sec^{-1} (Holm et al., 1974b). In all likelihood, the protein-bound clusters in both HP_{red} and HP_{ox} also have a similarly delocalized electronic structure.

There have been repeated observations of two slightly inequivalent magnetic resonance spectra arising from the paramagnetic "hole" in HP_{ox} (Phillips et al., 1970; Anderson et al., 1975; Antanaitis and Moss, 1975). There is no ready explanation from the HP_{red} or HP_{ox} structures for these observations; they probably arise from an as yet unrecognized asymmetry in the cluster binding cavity.

It is notable, in view of the evidence cited above regarding electron delocalization in $Fe_4S_4{}^*$ analogues, that the difference Fourier is not of higher apparent symmetry. Each of the three pairs of opposite $Fe_2S_2{}^*$ faces is associated with a different pattern of difference density (Fig. 16). This result is qualitatively different from what one would expect, for example, if all six iron–iron distances shortened by the same amount in HP_{ox} relative to their distances in HP_{red}. Rearrangement of the $Fe_4S_4{}^*$ core must therefore be *nonuniform*.

This qualitative observation is of considerable significance because it does not depend on the precision of atomic coordinates. It is worthwhile, therefore, to pursue an interesting implication: the nonuniform net contraction of the cluster in HP_{ox} must either increase or decrease its apparent symmetry. If we compare the mean Fe–S* distances obtained from least squares refinement for HP_{red} (Table VI), one set of four is clearly shorter than the other two—2.25, 2.23, 2.38 Å. For HP_{ox}, the corresponding values are much closer to each other—2.28, 2.25, 2.22 Å, given in the same order. These values show clearly that it is the oxidized, or C$^+$, state of the cluster that has higher symmetry.

An important corollary is that the HiPIP cluster binding cavity must limit the possible orientations of the HP_{red} cluster, which is of lower symmetry. This conclusion follows from the approximate symmetry of the HP_{ox}–HP_{red} difference Fourier; a mixture of HP_{red} molecules whose clusters were oriented differently would produce a difference map with similar density at all six faces, i.e., a map with higher than apparent 222 symmetry. Of course, it is possible that nonbiological crystal packing forces are responsible for the apparent "freezing out" of a single cluster orientation. However, examination of changes within the cavity itself suggests that this is not the case by providing a convincing basis for rationalizing a unique orientation for the reduced cluster.

C. Changes in the Cluster Binding Cavity

Features of the (HP_{ox}–HP_{red}) difference Fourier outside the $Fe_4S_4^*$ cluster region are difficult to interpret because they are scarcely above the background noise level of the map. However, when refined atomic positions of selected atoms in the cluster binding cavity are compared in the two oxidation states, there appear to be small readjustments in the positions of certain groups, notably nonpolar side chains and amide groups making NH---S contacts. Individual movements are only suggestive. Nevertheless, taken as a whole they constitute a convincing set of indicators that the cavity is delicately poised to accommodate the cluster rearrangement. These proposed structural changes will be described with reference to the HiPIP "active" site as represented in Fig. 17.

1. Changes Affecting NH---S Contacts

Changes in NH---S distances indicate that these interactions contribute substantially to the distribution of increased negative charge in the reduced (formal charge probably −2) as opposed to the oxidized (formal charge probably −1) cluster (Table V). All such contacts shorten upon reduction, the mean change in length being 0.23 Å. NH---S_γ contacts, as a class, shorten less ($\bar{\Delta} = 0.18$ Å); NH---S* contacts shorten more ($\bar{\Delta} = 0.35$ Å). That this effect is not simply due to the expansion of the cluster is indicated by the fact that the N atoms in each case move directly toward the respective sulfur atoms.

As was noted in Section IV,D, a single NH---S* contact originates in each of the two β sheets that enclose the cluster from opposite sides. In HP_{ox}, these two NH---S* contact distances lie outside the specified limits for NH---S hydrogen bonds. They shorten to within hydrogen bond limits in HP_{red} as the cluster expands against the two β sheets, principally by the outward movement of opposite $Fe_2S_2^*$ faces. Hence, they might possibly represent molecular dipole–charge interactions in

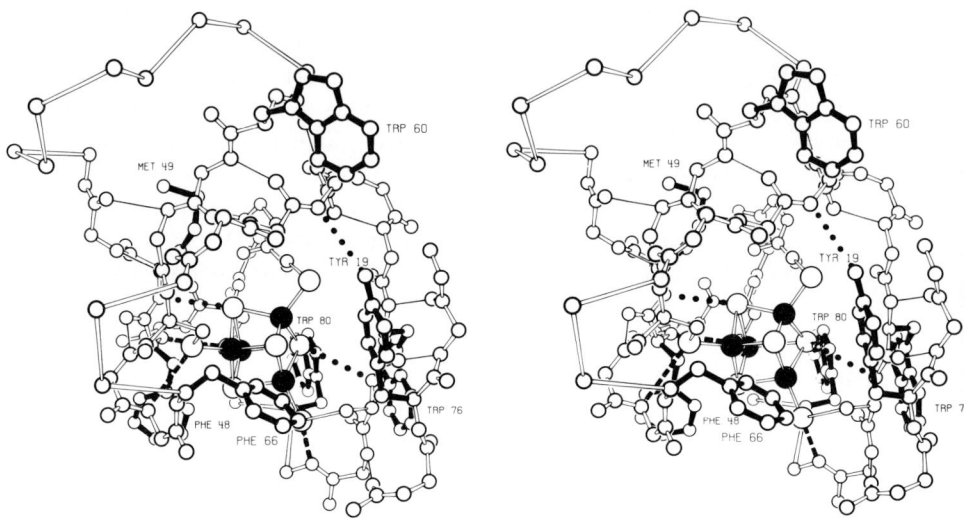

Fig. 17. Proposed changes in the HiPIP active site during oxidation–reduction. NH---S_γ hydrogen bonds (heavy broken lines) shorten by ~0.2 Å; NH---S* and NH---OH_{19} contacts (heavy dotted lines) apparently form new hydrogen bonds in HP_{red} by shortening by ~0.4 Å.

HP_{ox}, and become hydrogen bonds in HP_{red}. The important point is that these NH---S* contacts can form stronger hydrogen bonds if and only if the cluster expands in the preferred direction. In this respect, they bear a remarkable resemblance to similar hydrogen bond donors that stabilize the tetrahedral intermediate formed during serine protease catalysis by forming hydrogen bonds to the negatively charged oxyanion (Robertus *et al.*, 1972; Rühlmann *et al.*, 1973). In any case, the presence of "free" amide groups within the cluster binding cavities must introduce into the prevailing nonpolar environment of the negatively charged cluster discrete directions of unusually high molecular polarizability. It will be of considerable interest to determine whether these sharp variations in polarizability also guide the stereochemistry of ferredoxin reduction in the same manner as they seem to do in HiPIP.

It is interesting, and perhaps more than coincidental, that the four Fe–S* bonds that experience the greatest change in bond length during the oxidation–reduction cycle (Table VI) are approximately perpendicular to the intramolecular interface (Fig. 10). That is, the greatest reduction in size of the cluster also occurs in the direction that would best effect closer side chain packing in this interface. However, current protein coordinates are probably too imprecise to permit identification of changes that would support this idea, and no consistant pattern has emerged from a preliminary search for such changes.

2. CHANGES AFFECTING TYROSINE 19

Expansion of the cluster has an unusual and quite unexpected influence on the Tyr 19 side chain. In view of the fact that $C_\delta 1$ of residue 19 is in van der Waals contact with the HP_{ox} cluster ($C_\delta 1 - S^*3 = 3.7$ Å), one might expect that the larger, HP_{red} cluster would push this side chain further away, particularly as this movement would require only a slight rotation about the C_β–C_γ bond of this residue. Contrary to this expectation, it appears to move closer to the HP_{red} cluster ($C_\delta 1 - S^*3 = 3.4$ Å) by rotation about the C_α–C_β bond. A concomitant and equally unexpected change occurs in the hydrogen bonding network described in Section IV,C involving the tyrosine hydroxyl group, the amide group of residue 72 and the bound water WAT 140. Movement of the ring toward the cluster brings the tyrosine hydroxyl group ~ 0.3 Å closer (3.0 Å in HP_{ox}, 2.7 Å in HP_{red}) to a hydrogen bond donor, NH 72, and ~ 0.2 Å further away from a presumed hydrogen bond acceptor, WAT 140.

Carter et al. (1974b) noted that an electrostatic charge–dipole interaction between the negatively charged cluster and the polarizable tyrosine ring could account for both observations. Such an interaction might force extra electron density onto the hydroxyl group, thereby making it a better hydrogen bond acceptor, and result in a weakly attractive interaction between the aromatic ring and the cluster. Oxidation of the cluster would weaken this interaction by lowering its net negative charge. They speculated further that proximity of the tyrosine hydroxyl group to a hydrogen bond donor might stabilize a phenoxy radical transition state by facilitating transfer of the hydroxyl proton to WAT 140 forming H_3O^+. A one electron oxidation of the tyrosine—either by the HP_{ox} cluster at $C_\delta 1$ or by some group on the physiological oxidase acting at $C_\varepsilon 2$ or OH, both of which are exposed to solvent—would produce a phenoxy radical. Thus, the tyrosine could mediate electron transport both to and from the $Fe_4S_4{}^*$ cluster. The absence of a similar hydrogen bonding network precludes such a mechanism for ferredoxin.

D. Is There a Conformational Change?

There is no evidence in the HP_{ox}–HP_{red} difference Fourier map that any significant conformational change accompanies cluster rearrangement. Lacking such evidence, we must beg an important question, for it is not likely that the polypeptide remains unaffected by the expansion and contraction of the cluster. Either small changes (of the order of 0.1–

0.2 Å) do occur elsewhere in the protein, i.e., cluster rearrangement actually induces a facile isomerization of the polypeptide, or else the protein is unable to make such an isomerization and therefore is itself at a higher energy in one oxidation state than it is in the other.

It is obviously of fundamental importance to our understanding of iron–sulfur proteins and, indeed, of protein structure in general to know in what proportions each statement is true in HiPIP. In the former case, the polypeptide would exert a minimal influence on thermodynamic properties of the cluster, such as its midpoint reduction potential. However, even a subtle conformational change might influence the rate of reaction substantially, by promoting the necessary Frank–Condon rearrangement. Specifically, one can easily envision the surrounding protein stabilizing an intermediate transition state between the reduced and oxidized structures by favoring certain vibrational modes of the cluster, especially since, as noted in Section V,C, the greatest change in Fe–S* distances occurs in a direction approximately perpendicular to the "intramolecular interface" (Fig. 10). On the other hand, if the polypeptide is rigid, as in the latter case, then differential strain of the protein in HP_{ox} and HP_{red} would necessarily influence the midpoint potential of the $Fe_4S_4^*$ cluster by contributing to the overall free energy of the reaction.

VI. CONCLUSIONS AND REVIEW

Many observations have emerged from X-ray crystal structure analysis of HiPIP and ferredoxin. A unifying theme for these observations is provided by the three-state hypothesis (Section II,D), for the compelling fascination of these two proteins resides in the fact that they have evolved ways of making the same prosthetic group do two different things. The question persists, how do they do it? There are obviously connections between the crystal structure data and answers to this question. In this concluding section, I want first to sketch some plausible connections that we seem to have found.

It is important to recognize that two separate problems are involved in understanding why the protein-bound clusters behave as they do. First, one would like to rationalize the absolute differences between reduction potentials for the same reaction observed in the two proteins and in analogue clusters. Why, in other words, does reduction of Fd_{ox} occur at a midpoint potential, $E_0' = -0.4$ V, and not at $E_0' \cong -1.08$, as it does in the analogue $Fe_4S_4^*(SCH_2C_6H_5)_4$ (DePamphilis et al., 1974)? Why does oxidation of HiPIP occur at $E_0' = 0.35$ V and not at $E_0' \cong \pm 0.1$ V, where the corresponding reaction occurs in analogue clusters? Second, one would like to identify the protein features that normally prevent oxidation of the paired spin state in ferredoxin and reduction of the paired spin state in HiPIP. These two transitions are indicated by question marks in Fig. 3.

It is easy enough to confuse these two questions, when in fact they are quite distinct. The former question concerns only thermodynamic influences and can be answered by enumerating the factors that contribute to the overall free-energy change of oxidoreduction. Since these factors depend only on initial and final states, it should be possible to identify them in the structures of HP_{ox} and HP_{red}. The latter question implies that kinetic barriers may prevent the unobserved transitions. Such barriers may involve mechanistic considerations and hence be more difficult to understand on the basis of current X-ray structure data.

A. Thermodynamic and Mechanistic Considerations

Two aspects of the cluster binding cavities in HiPIP and ferredoxin might be expected to alter the free energy of electron transfer to and from the protein-bound clusters:

1. A tendency to form stronger NH---S hydrogen bonds in the presence of an added electron would be expected to lower the enthalpy of the more reduced species and hence raise the midpoint potential above that observed for the free $Fe_4S_4^*$ cluster.

2. An increase or decrease in cluster symmetry during electron transfer may introduce an entropy change if the protein-bound cluster can assume only one of several possible orientations.

Quantitative estimates for both contributions can be made using available data for the model compounds. There is rather good agreement between these estimates and the observed midpoint potentials in HiPIP and ferredoxin. Moreover, they suggest that NH---S hydrogen bonding predominates in determining the midpoint potentials of the protein-bound clusters.

The contribution of a single NH---S hydrogen bond may be estimated by noting that midpoint potentials of analogue clusters with thiolate ligands R = ethyl and R = N-acetyl-L-cysteine-N-methylamide (Bobrick et al., 1974) differ by about 0.3 V when measured in the same aprotic solvent, DMSO (DePamphilis et al., 1974). The latter compound is actually capable of donating four amide NH groups to the four S* atoms in a manner analogous to the NH groups of Cys 11 and Cys 14 in $Fd_{ox}I$, Cys 38 and Cys 41 in $Fd_{ox}II$, and Cys 77 in HP_{red}. Assuming that hydrogen bonding accounts for the entire potential difference between the two model compounds, each hydrogen bond would contribute about

0.075 V or 1.73 kcal mole^{-1}. Ferredoxin clusters can form nine CuCyr NH----S hydrogen bonds, which would raise the potential for reduction by about 0.68 V to -0.41 V from -1.08 V, the value cited above for an analogue cluster that cannot donate NH groups. HiPIP, on the other hand, has only six possible NH donors. Assuming that oxidation of the paired-spin state occurs at about 1.0 V higher than reduction (Holm, 1974), a similar calculation predicts a potential $E_0' = +0.37$ V for HP$_{ox}$/HP$_{red}$. For comparison, the observed potentials for ferredoxin and HiPIP are -0.40 V and $+0.35$ V, respectively. To be sure, the citations of model compound data used in these illustrative calculations have been chosen judiciously so as to obtain the best possible agreement with the fewest assumptions. Nevertheless, the results show convincingly that differences in NH----S hydrogen bonding can indeed regulate the midpoint potentials of the two proteins.

The nonuniform cluster rearrangement observed on oxidation of HP$_{red}$ implies that entropic contributions may also affect the relative potentials of HiPIP, ferredoxin, and the model compounds. Carter et al. (1974b) have offered the following rationalization of the nonuniform expansion–contraction in HiPIP. In the oxidized state of the HiPIP cluster (actual net charge probably -1), a single unpaired electron occupies a degenerate antibonding orbital, and Jahn–Teller splitting is slight. Therefore, the cluster has nearly full tetrahedral T (32) symmetry. Upon reduction the additional electron pairs with the first, Jahn–Teller splitting becomes appreciable, and the cluster symmetry decreases to tetragonal D$_{2d}$ (42m) or lower. Because the orbitals in question are antibonding, the net size of the cluster increases. Recent molecular orbital calculations for an analogue cluster (Yang et al., 1975) are consistent with this interpretation, predicting appreciable ground-state electronic degeneracy and hence a Jahn–Teller distortion in the paired-spin state, which also occurs in HP$_{red}$. Moreover, recent spectroscopic measurements (Tang et al., 1975) support the contention that the protein cluster-binding cavities impose a unique, or at least a preferred, orientation for the remaining approximate symmetry elements of the Fe$_4$S$_4$* cluster. It follows that the C$^+ \to$ C transition in a free cluster would involve a positive entropy change that would be prevented in the HP$_{ox}$/HP$_{red}$ transition. Such an effect could have a subtle but appreciable influence on the relative midpoint potentials of HiPIP and the analogue. To illustrate, assume for simplicity that the HP$_{ox}$ cluster has full tetrahedral symmetry, that the HP$_{red}$ cluster has D$_{2d}$ symmetry, and hence a unique $\bar{4}$ axis, and that in HP$_{red}$ the direction of this axis is restricted to one of three possible orientations. In this case, the entropic contribution at 25°C would be $-T \Delta S = 298R \ln 3 = 650$ cal mole$^{-1} = -0.028$ V for HP$_{ox}$/HP$_{red}$. Having made so many assumptions, it is undoubtedly fortuitous, but this correction actually improves the agreement between observed and calculated potentials for HiPIP ($E'_{0\,calc} = (0.37-0.03) = 0.34$ V, $E'_{0\,obs} = +0.35$ V). Similar considerations might, of course, apply to ferredoxin. However, there is no evidence concerning the symmetry properties of Fd_{red} clusters.

What prevents the protein-bound clusters from assuming three oxidation states? As was noted above, the answer is probably that oxidation and reduction of the paired-spin state proceed by different mechanisms involving different Frank–Condon rearrangements of the cluster. We have come to expect enzymes to provide a low-energy route through the transition states of the reaction they catalyze (Wolfenden, 1976), and it would come as no surprise if HiPIP and ferredoxin had evolved mechanisms for stabilizing different transition states. X-Ray structure analysis of the two HiPIP oxidation states reveals three ways in which the polypeptide might participate in the electron transfer mechanism:

1. The apparently strategic location of NH---S^* hydrogen bonds may provide electrostatic guidance for the cluster rearrangement. In this regard, there must be a delicate balance between kinetic and thermodynamic effects, owing to the fact that hydrogen bonds exhibit both directional and enthalpic characteristics (Section V,C,1).

2. Relative thermal movement of distant, cysteine-containing chain segments may transmit preferred vibrational modes to the four iron atoms (Section V,D).

3. Aromatic side chains may participate in the actual electron transfer (Section V,C,2).

In all three respects, significant differences between the two proteins have been noted in previous sections (Sections IV,C and IV,D). Comparison of the Fd_{ox} and Fd_{red} crystal structures might elucidate other as yet unrecognized differences. Nevertheless, it is evident that the current, static pictures of the two proteins provide a much more satisfactory account of their equilibrium properties than of their reaction mechanisms. Bennett has discussed relevant kinetic studies and mechanistic inferences elsewhere in this volume (Chapter 9).

Two topics raised by X-ray structure analysis of HiPIP and ferredoxin have been omitted from previous discussion. They are presented briefly below as examples of the rather intriguing questions that remain unexplored.

B. Is There Cooperativity in Ferredoxin?

Ferredoxin has a remarkable intramolecular twofold symmetry axis. Rossmann and Argos (1976) have shown that the first and second halves

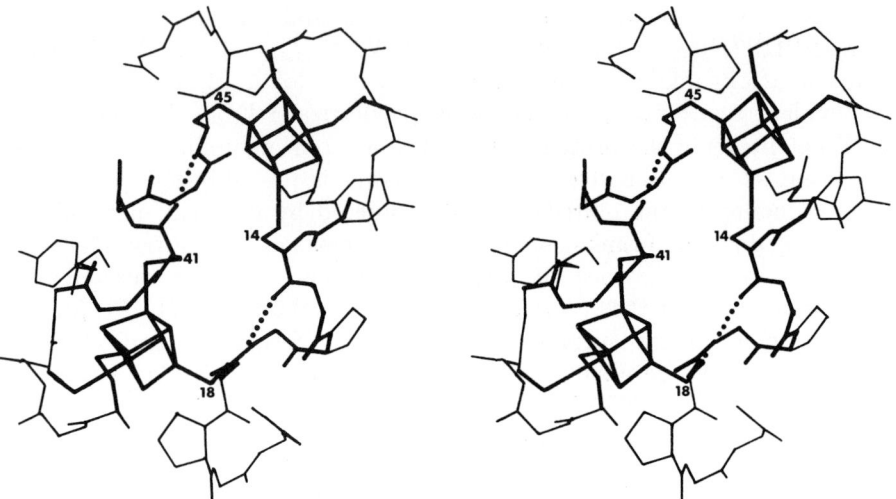

Fig. 18. Intramolecular symmetry and the mechanical linkages between $Fe_4S_4^*$ cluster in *P. aerogenes* ferredoxin. Heavy bonds indicate regions of possible communication between clusters.

of the sequence are related by rotation of almost exactly 180° and that such a rotation based only on main chain atoms will superimpose the omitted $Fe_4S_4^*$ cluster atoms even more exactly than the main-chain atoms. This nearly exact intramolecular symmetry raises the thorny question of cooperativity. Specifically, the Cys–S–Fe linkages of Cys 14 and Cys 45 are mechanically coupled to those of Cys 41 and Cys 18 (Fig. 18). The C_β atoms of residues 14 and 41 are only about 4.0 Å apart, and the chain segments 14–18 and 41–45 are both roughly helical. It is difficult to visualize how a change in the geometry of one cluster could leave the neighboring cluster unaffected, and it is tempting to speculate that these mechanical linkages act to preserve intramolecular symmetry during oxidoreduction. Cooperativity would be especially valuable to ferredoxin, whose biological functions involve two-electron transfers. Previous speculations about cooperativity in bacterial ferredoxins have been noted by Orme-Johnson (1973).

C. Do HiPIP and Ferredoxin Have a Common Ancestor?

Finally, we must consider possible evolutionary relationships between HiPIP and ferredoxin. No detectable amino acid sequence homologies exist between the two proteins (Tedro *et al.*, 1976). However, a prevailing opinion is that homologous tertiary structures are a more sensitive test

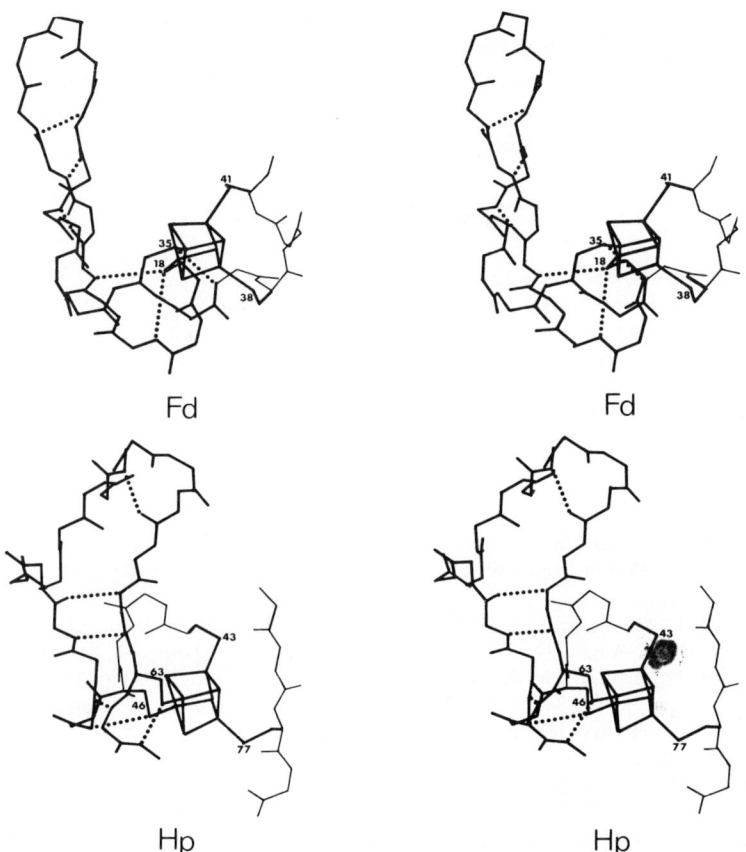

Fig. 19. A common ancestor for HiPIP and ferredoxin? For the preparation of this illustration a least squares fit was obtained for the Fe$_4$S$_4$* clusters using equivalence of iron atoms attached to Cys 46, 63, 77, and 43 of HiPIP to those attached to Cys 18, 35, 38, and 41 of ferredoxin, respectively.

of evolutionary relatedness (Rossmann and Argos, 1976). Alignment of homologous regions of the two molecules is illustrated in Fig. 19. Chain segments included in this alignment—residues 46–63 in HiPIP and 18–35 in ferredoxin—represent only a small portion of either cluster-binding cavity. However, they are exactly the same length, they assume very similar twisted β-sheet conformations, and they provide two of the four cysteinyl ligands to the cluster. The following consideration suggests that this alignment may be more than fortuitous. The two Fe$_4$S$_4$* clusters have the same orientation with respect to this common chain segment, as specified both by the unique direction defined by four short Fe---S* bonds (Section III,E) and by the location of the three cysteine residues involved in NH---S$_\gamma$ hydrogen bonding (Section IV,D,2) (Carter, 1976). There are many more facets of these two questions than can be explored further here. Indeed, they are rather pointed reminders that the HiPIP and ferredoxin crystal structures continue to hold tantalizing secrets about their function and evolution.

ACKNOWLEDGMENTS

It is a pleasure to thank Lyle H. Jensen for atomic coordinates of *P. aerogenes* ferredoxin, Lyle H. Jensen and Michael G. Rossmann for manuscripts prior to publication, Joseph Kraut for many valuable discussions, and Max Perutz for comments on the manuscript. Special thanks are due also to Jan Hermans, John McQueen, Joel Sussman, and the University of North Carolina Molecular Graphics Laboratory for invaluable assistance in preparing illustrations.

REFERENCES

Adman, E. T., Sieker, L. C., and Jensen, L. H. (1973). *J. Biol. Chem.* **248**, 3987.
Adman, E. T., Watenpaugh, K. D., and Jensen, L. H. (1975). *Proc. Natl. Acad. Sci. U.S.A.* **72**, 4854.
Adman, E. T., Sieker, L. C., and Jensen, L. H. (1976). *J. Biol. Chem.* **25**, 3801.
Anderson, R. E., Anger, G., Petersson, L., Ehrenberg, A., Cammack, R., Hall, D. O., Mullinger, R., and Rao, K. K. (1975). *Biochim. Biophys. Acta* **376**, 63.
Antanaitis, B. C., and Moss, T. H. (1975). *Biochim. Biophys. Acta* **405**, 262.
Averill, B. A., Herskovitz, T., Holm, R. H., and Ibers, J. A. (1973). *J. Am. Chem. Soc.* **95**, 3523.
Blomstrom, D. C., Knight, E., Jr., Phillips, W. D., and Weiher, J. F. (1964). *Proc. Natl. Acad. Sci. U.S.A.* **51**, 1085.
Bobrik, M. A., Que, L., Jr., and Holm, R. H. (1974). *J. Am. Chem. Soc.* **96**, 285.
Bruice, T. C., Maskiewicz, R., and Job, R. (1975). *Proc. Natl. Acad. Sci. U.S.A.* **72**, 231.
Cammack, R. (1973). *Biochem. Biophys. Res. Commun.* **54**, 548.
Carter, C. W., Jr. (1976). In preparation.
Carter, C. W., Jr., Freer, S. T., Xuong, Ng.H., Alden, R. A., and Kraut, J. (1971). *Cold Spring Harbor Symp. Quant. Biol.* **36**, 381.
Carter, C. W., Jr., Kraut, J., Freer, S. T., Alden, R. A., Sieker, L. C., Adman, E. T., and Jensen, L. H. (1972). *Proc. Natl. Acad. Sci. U.S.A.* **69**, 3526.
Carter, C. W., Jr., Kraut, J., Freer, S. T., Xuong, Ng. H., Alden, R. A., and Bartsch, R. G. (1974a). *J. Biol. Chem.* **249**, 4212.
Carter, C. W., Jr., Kraut, J., Freer, S. T., and Alden, R. A. (1974b). *J. Biol. Chem.* **249**, 6339.
DePamphilis, B. V., Averill, B. A., Herskovitz, T., Que, L., Jr., and Holm, R. H. (1974). *J. Am. Chem. Soc.* **96**, 4159.
Donohue, J. (1968). In "Structural Chemistry and Molecular Biology" (A. Rich and N. Davidson, eds.), p. 443. Freeman, San Francisco, California.
Donohue, J. (1969). *J. Mol. Biol.* **45**, 231.
Dus, K., De Klerk, H., Sletten, K., and Bartsch, R. G. (1967). *Biochim. Biophys. Acta* **140**, 291.

Dus, K., Tedro, S., and Bartsch, R. G. (1973). *J. Biol. Chem.* **248,** 7318.
Frankel, R. B., Herskovitz, T., Averill, B. A., Holm, R. H., Krusic, P. J., and Phillips, W. D. (1974). *Biochem. Biophys. Res. Commun.* **58,** 974.
Freer, S. T., Alden, R. A., Carter, C. W., Jr., and Kraut, J. (1975). *J. Biol. Chem.* **250,** 46.
Herskovitz, T., Averill, B. A., Holm, R. H., Ibers, J. A., Phillips, W. D., and Weiher, J. F. (1972). *Proc. Natl. Acad. Sci. U.S.A.* **69,** 2437.
Holm, R. H. (1974). *Endeavour* **34,** 38.
Holm, R. H., Phillips, W. D., Averill, B. A., Mayerle, J. J., and Herskovitz, T. (1974a). *J. Am. Chem. Soc.* **96,** 2109.
Holm, R. H., Averill, B. A., Herskovitz, T., Frankel, R. B., Gray, H. B., Siiman, O., and Grunthaner, F. J. (1974b). *J. Am. Chem. Soc.* **96,** 2644.
Jensen, L. H. (1974a). *Annu. Rev. Biochem.* **43,** 461.
Jensen, L. H. (1947b). *Annu. Rev. Biophys. Bioeng.* **3,** 81.
Job, R. C., and Bruice, T. C. (1975). *Proc. Natl. Acad. Sci. U.S.A.* **72,** 2478.
Kraut, J., Strahs, G., and Freer, S. T. (1968). *In* "Structural Chemistry and Molecular Biology" (A. Rich and N. Davidson, eds.), p. 55. Freeman, San Francisco, California.
Laishley, E. J., Travis, J., and Peck, H. D., Jr. (1969). *J. Bacteriol.* **98,** 302.
Lane, R. W., Ibers, J. A., Frankel, R. B., and Holm, R. H. (1975). *Proc. Natl. Acad. Sci. U.S.A.* **72,** 2868–2872.
Lode, E. T., Murray, C. L., Sweeney, W. V., and Rabinowitz, J. C. (1974). *Proc. Natl. Acad. Sci. U.S.A.* **71,** 1361.
Lovenberg, W., ed. (1973). "Iron–Sulfur Proteins," Vols. 1 and 2. Academic Press, New York.
Maskiewicz, R., Bruice, T. C., and Bartsch, R. G. (1975). *Biochem Biophys Res. Commun.* **65,** 407.
Mayerle, J. J., Denmark, S. E., DePamphilis, B. V., Ibers, J. A., and Holm, R. H. (1975). *J. Am. Chem. Soc.* **97,** 1032.
Orme-Johnson, W. H. (1973). *Annu. Rev. Biochem.* **42,** 159.
Packer, E. L., Sternlicht, H., and Rabinowitz, J. C. (1972). *Proc. Natl. Acad. Sci. U.S.A.* **69,** 3278.
Pauling, L. (1960). "The Nature of the Chemical Bond," 3rd ed., p 112. Cornell University Press, Ithaca, New York.
Pauling, L. (1975). *Proc. Natl. Acad. Sci. U.S.A.* **72,** 4200.
Phillips, W. D., Poe, M., McDonald, C. C., and Bartsch, R. G. (1970). *Proc. Natl. Acad. Sci. U.S.A.* **67,** 682.
Que, L., Jr., Bobrik, M. A., Ibers, J. A., and Holm, R. H. (1974). *J. Am. Chem. Soc.* **96,** 4168.
Robertus, J. D., Kraut, J., Alden, R. A., and Birktoft, J. J. (1972). *Biochemistry* **11,** 4293.
Rossmann, M. G., and Argos, P. (1976). *J. Mol. Biol.* **105,** 75–95.
Rühlmann, A., Kukla, D., Schwager, P., Bartels, K., and Huber, R. (1973). *J. Mol. Biol.* **77,** 417.
Sieker, L. C., and Jensen, L. H. (1965). *Biochem. Biophys. Res. Commun.* **20,** 33.
Sieker, L. C., Adman, E. T., and Jensen, L. H. (1972). *Nature (London)* **235,** 40.
Strahs, G., and Kraut, J. (1968). *J. Mol. Biol.* **35,** 503.
Tanaka, M., Haniu, M., Yasunobu, K. T., Jones, J. B., and Stadtman, T. C. (1974). *Biochemistry* **13,** 5284.
Tang, S-P. W., Spiro, T. G., Antanaitis, B. C., Moss, T. H., Holm, R. H., Herskovitz, T., and Mortensen, L. E. (1975). *Biochem. Biophys. Res. Commun.* **62,** 1.

Tedro, S., Meyer, T. E., and Kamen, M. D. (1974). *J. Biol. Chem.* **249,** 1182.
Tedro, S., Meyer, T. M., and Kamen, M. D. (1976). *J. Biol. Chem.* **251,** 129.
Tsibris, J. C. M., and Woody, R. W. (1970). *Coord. Chem. Rev.* **5,** 417.
Tsunoda, J. N., Yasunobu, K. T., and Whitely, H. R. (1968). *J. Biol. Chem.* **243,** 6262.
Venkatachalam, C. M. (1968). *Biopolymers* **6,** 1425.
Watenpaugh, K. D., Sieker, L. C., Herriott, J. R., and Jensen, L. H. (1971). *Cold Spring Harbor Symp. Quant. Biol.* **36,** 359.
Watenpaugh, K. D., Sieker, L. C., Herriott, J. R., and Jensen, L. H. (1973). *Acta Crystallogr., Sect. B* **29,** 943.
Wolfenden, R. V. (1976). *Annu. Rev. Biophys. Bioeng.* **5,** 271.
Yang, C. Y., Johnson, K. H., Holm, R. H., and Norman, J. G., Jr. (1975). *J. Am. Chem. Soc.* **97,** 6596.
Yoch, D. C., and Valentine, R. C. (1972). *J. Bacteriol.* **110.** 1211.

CHAPTER 7

Synthetic Analogues of the Active Sites of Iron–Sulfur Proteins*

R. H. HOLM and JAMES A. IBERS

I. Introduction	206
A. Characterized Active Sites in Iron–Sulfur Proteins	207
B. Solution Studies of Iron(II,III) Complexes with Thiols and Sulfide	211
C. Structures of Coordination Units in Nonanalogue Iron–Sulfur Complexes	214
II. Synthesis and Structures of 1-Fe, 2-Fe, and 4-Fe Active-Site Analogues	217
A. Synthesis	218
B. Analogue Structures and Comparisons with Proteins	220
III. Physical Properties of Analogues	228
A. Oxidation–Reduction and Analogue–Protein Oxidation Level Equivalencies	228
B. Electronic Features	233
IV. Chemical Reactivity of Analogues	256
A. Thiolate Ligand Substitution	257
B. Active-Site Core Extrusion Reactions	260
C. Protonation and Acid Solvolysis	264
V. Iron–Sulfur Units as Redox Centers	265
A. Structural Aspects	265
B. Redox Potential Dependence on Ligand Structure	268
VI. Perspectives and Conclusions	271
References	272

* Abbreviations used in this chapter: BM, Bohr magneton; Cys-S, coordinated cysteinate; DMF, N,N-dimethylformamide; DMSO, dimethyl sulfoxide; edt, $(SCH_2CH_2S)^{2-}$; Fd, ferredoxin; HP, high-potential iron protein; mnt, $S_2C_2(CN)_2$; Rd, rubredoxin; S*, sulfide; salen, bis(salicylaldehyde)ethylenediimine dianion; S_2-o-xyl, o-xylyl-α,α'-dithiolate dianion; tfd, $S_2C_2(CF_3)_2$; X, amino acid residue; z, generalized charge on ionic species.

I. INTRODUCTION

The last decade has witnessed remarkable advances in the extent of stereochemical and electronic definition of the coordination units present in metalloproteins, metalloenzymes, and other metal-containing molecules of biological origin. Those units known or postulated to be involved in electron transfer and/or catalytic transformations of substrates are hereafter termed "active sites." Although detailed and incisive physicochemical studies on substances of enhanced chemical purity and high biological activity have contributed substantially, the dominant factor responsible for this accelerated progress continues to be the expanding availability of protein X-ray crystallographic results at increased atomic resolution. While these results are not yet at the stage of conveying molecular parameters with the precision normally afforded by X-ray studies of nonbiological metal compounds and complexes, those obtained at a resolution of about 2.5 Å or higher generally do reveal the primary coordination number, identity of coordinated ligands, and overall stereochemistry of ligation within the coordination unit. In addition, determination of tertiary structure together with the primary structure, usually obtained initially from amino acid sequencing, allows an assessment of the active site environment. On the basis of crystallographic data and allied physicochemical information, biological coordination units can be classified into three structural types: (i) mononuclear, involving ligation by protein side chains (and possibly water or hydroxide) (for a tabulation of metal-binding groups in proteins, see Vallee and Wacker, 1970); (ii) mononuclear, involving ligation by a porphyrin or corrin ring with axial coordination by protein side chains (heme proteins and enzymes). (iii) polynuclear, with metals bridged by sulfide, oxide, or hydroxide and remaining positions occupied by protein side chains. Among the examples of these three types the following may be cited: (i) bovine carboxypeptidase A (Lipscomb, 1973; Quicho and Lipscomb, 1971), thermolysin (B. W. Mathews et al., 1974), horse liver alcohol dehydrogenase (Eklund et al., 1974); (ii) hemoglobin and myoglobin (Antonini and Brunori, 1971), cytochromes b (Mathews et al., 1972) and c (Dickerson et al., 1971; Takano et al., 1973; Salemme et al., 1973a,b); (iii) ferritin (Crichton, 1973; Harrison et al., 1974).

In addition to facilitating interpretations of electronic properties and, where appropriate, catalytic function, the availability of compositional and stereochemical information for biological coordination units permits in some cases an approach to the study of these features that is complementary to investigations of the biological molecules themselves. When reduced to practice, this approach necessitates the synthesis of relatively low molecular weight complexes, which, ideally, are obtainable in crystalline form and approach or duplicate the biological unit in terms of composition, ligand types, structure, and oxidation level(s). Such models, or synthetic analogues, of course, cannot simulate the environmental effects of and whatever structural constraints are imposed by the normal protein conformation. Indeed, this may be considered an advantage of synthetic analogues, for, being unencumbered by the protein, they should reflect the intrinsic properties of the coordination unit unmodified by the protein milieu. Appropriate comparisons of properties of analogues and biological molecules can potentially afford a measure of the extent to which the properties of the former have been altered—and possibly the molecular origins of these alterations—upon passing from simple synthetic species to holoproteins and enzymes. Such information is of value in further examination of the entatic state hypothesis (Vallee and Williams, 1968; Williams, 1971). In our laboratories, the synthetic analogue approach has been applied to a study of the active sites of iron–sulfur proteins, with emphasis on full structural, electronic, and reactivity characterization of the analogues together with the consistent theme of utilizing comparative properties as a means of detecting the influence of protein structure and environment. Many of the chemical and biological aspects of these proteins have been described in Volumes I and II (Lovenberg, 1973) and are updated and expanded by the contents of this volume and by more recent reviews (Hall et al., 1973a,b, 1974; Llinás, 1973; Mason and Zubieta, 1973; Orme-Johnson, 1973). A brief review of our studies of synthetic analogues has been given elsewhere (Holm, 1975). A more detailed account is presented in this chapter.

A. Characterized Active Sites in Iron–Sulfur Proteins

Comprehensive tabulations of iron–sulfur proteins and enzymes and certain of their physical properties have been assembled by Hall et al. (1973a,b, 1974) and Orme-Johnson (1973). The presently well-defined proteins of molecular weight about 6000–20,000 contain one, two, four, or eight iron atoms and an equivalent amount of sulfide per molecule, except for the 1-Fe (rubredoxin†) proteins, which lack sulfide. As shown by amino acid analyses and sequences (Yasunobu and Tanaka, 1973), all proteins contain a number of cysteinyl residues that equals or exceeds the number of iron atoms present. The biological function of many of these proteins has not been clarified in satisfactory detail. However, the available evidence suggests that all proteins have at least two redox states coupled by one-electron transfer reactions per active site, usually at rela-

† Designation of proteins as rubredoxins (Rd), ferredoxins (Fd), and "high-potential" (HP) proteins follows the currently recommended nomenclature of iron–sulfur proteins [*Biochemistry* **12**, 3582 (1973)].

tively low potentials (E_0' — 0.06 to —0.4 V *in vitro*), and function as electron carriers rather than as catalytic binding centers for chemical transformations (see also Section VI).

1. ACTIVE-SITE STRUCTURES

The structures of the active sites of 1-Fe, 4-Fe, and 8-Fe proteins have been unequivocally established by X-ray diffraction methods. These structures will be described briefly at this point; more detailed examination is deferred to Section II,B (see also Chapter 6). All rubredoxins (molecular weight ~6,000) are of bacterial origin. The structure 1, markedly distorted from idealized T_d symmetry, has been determined for the [Fe(S-Cys)$_4$] site in Rd$_{ox}$ from *Clostridium pasteurianum* (Watenpaugh *et al.*, 1973; Jensen, 1973, 1974). Proteins of the 4-Fe type have been isolated from bacterial sources† (molecular weight ~6000–10,000 and

contain the [Fe$_4$S$_4$*(S-Cys)$_4$] site. The structure of HP from the photosynthetic bacterium *Chromatium vinosum* has been determined in two oxidation levels, HP$_{red}$ and HP$_{ox}$ (Carter *et al.*, 1972, 1974a,b; Freer *et al.*, 1975). The active center occurs as the compact tetranuclear unit **2** somewhat distorted from cubic symmetry. The 8-Fe proteins (molecular weight ~6000–14,000) have been obtained exclusively from bacteria. The active sites in *Peptococcus aerogenes* Fd$_{ox}$ are organized into two apparently identical units **2** separated by about 12 Å and dimensionally very similar to the site in HP$_{red}$ (Carter *et al.*, 1972; Adman *et al.*, 1973). There is as yet no confirmatory evidence from X-ray crystallography on the active site structures of 2-Fe proteins which have been isolated from

† The first isolation of a 4-Fe protein (apparently containing structure **2**) from plant chloroplast (spinach) membranes has recently been reported (Malkin *et al.*, 1974).

algae, higher plants, bacteria, and mammalian organs. However, an extensive body of physicochemical data is fully consistent with the minimal composition [$Fe_2S_2^*$(S-Cys)$_4$] and the structure **3** containing two tetrahedrally coordinated iron atoms bridged by sulfide (Dunham *et al.*, 1971b; Palmer, 1973; Sands and Dunham, 1975). This structure, which was originally proposed nearly ten years ago on the basis of limited physical evidence (Brintzinger *et al.*, 1966; Gibson *et al.*, 1966), has been rendered even more probable by results from the synthetic analogue approach (Sections II and III). Structure **1** is a type (i) biological coordination unit, whereas structures **2** and **3** are of type (iii).

Several regularities are evident in structures **1–3**. Iron is coordinated only to sulfide and/or thiolate (Cys-S) sulfur and only with approximate tetrahedral stereochemistry. Metal bridging is by sulfide; Cys-S functions exclusively as a terminal (unidentate) ligand. In further discussions of structures **2**, **3**, and other oligomers of general type $M_nA_nL_m$ ($m \geq n$) containing bridging ligands A and terminal donor atoms L, the entire unit is considered a *cluster* and the integral substructure M_nA_n the *core*. Where useful for clarity, core sulfur is designated by S^*, as in preceding formulas.

Based on evaluations of spectroscopic and magnetic properties (Hall *et al.*, 1973a,b, 1974; Lovenberg, 1973; Orme-Johnson, 1973; Palmer, 1973; Tsibris and Woody, 1970), there is no clear evidence of major departures from the aforementioned active-site structural features in other low molecular weight proteins for which X-ray data are lacking. Consequently, structures **1–3** are considered entirely representative of those present in each group of n-Fe protein. In particular, the properties of the recently recognized group of 4-Fe proteins from nonphotosynthetic bacteria, such as *Bacillus polymyxa* (Stombaugh *et al.*, 1973), *Bacillus stearothermophilus* (Mullinger *et al.*, 1975), and *Desulfovibrio desulfuricans* (Zubieta *et al.*, 1973), are consistent with structure **2**. Small active site structural differences may exist among proteins of a given n-Fe type, but it will prove extremely difficult to separate the consequences of these differences in site properties from those resulting from variations in protein primary structure, conformations, and associated environmental effects. The somewhat closer correspondence of electronic properties of plant and algal proteins with one another than with 2-Fe bacterial (Gunsalus and Lipscomb, 1973) and mammalian (adrenal) ferredoxins (Kimura, 1968; Estabrook *et al.*, 1973) is a case in point. Along similar lines, it has been proposed by Carter *et al.* (1974b) that protein structure plays a dominant role in facilitating the HP_{red}/HP_{ox} redox process in which the HP_{ox} active site possesses a higher oxidation level than can normally be attained in 4-Fe and 8-Fe Fd proteins (see Section III,A).

2. Chelate Structures

X-ray determinations of 1-Fe, 4-Fe, and 8-Fe proteins reveal that these molecules are fundamentally metal complexes, albeit rather elaborate ones, with their metal centers or cores bound to a single polypeptide chain by Cys–S–Fe interactions, which in turn fold and cross-link the chain, forming large chelate rings as part of the tertiary structure. Here attention is drawn to certain chelate structural patterns that have emerged from both X-ray and primary structure determinations (Yasunobu and Tanaka, 1973). Spacings of cysteinyl residues in n-Fe proteins from similar organisms are generally highly conserved. Representative arrangements for rubredoxin (**4**, *C. pasteurianum* Rd), high-potential proteins (**5**, *Chromatium* HP), 8-Fe ferredoxins (**6**, *P. aerogenes* Fd), and plant and algal 2-Fe ferredoxins (**7**, spinach Fd) are schematically illustrated. Each contains at least one Cys–X–X–Cys fragment which evidently has a highly flexible $\overbrace{S \cdots S}$ "bite" distance, inasmuch as it forms

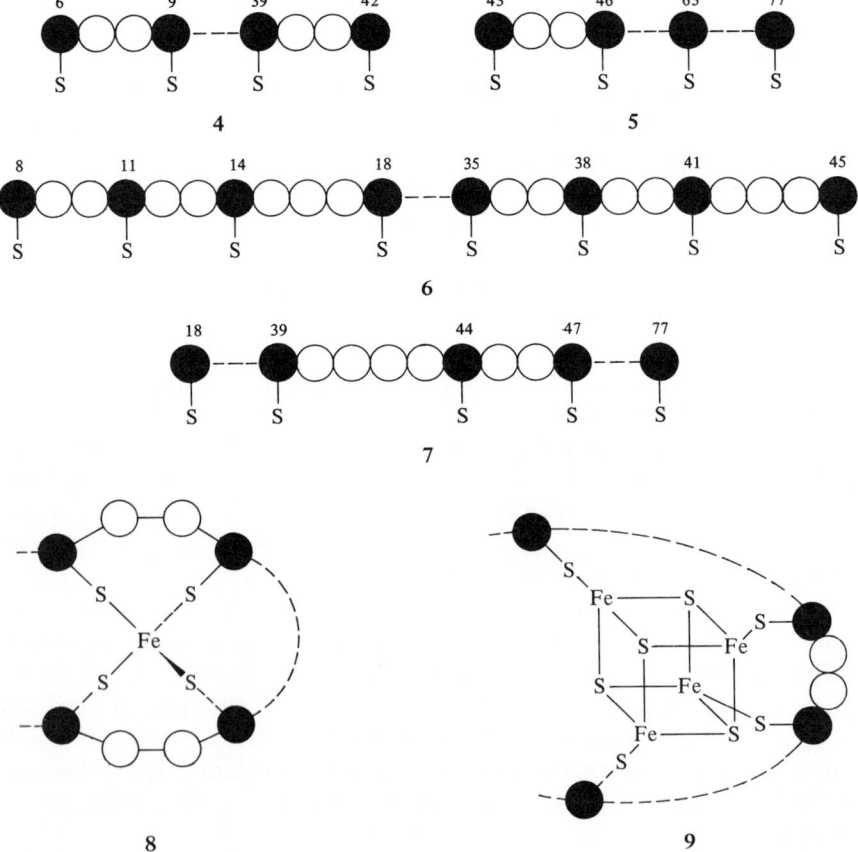

two chelate rings with the single iron atom in Rd (**8**, bite ~ 4 Å), and one and two chelate rings with two and three iron atoms of the $Fe_4S_4^*$ core (**9**, bite ~ 7 Å) of HP and *P. aerogenes* Fd, respectively. Coordination of cysteinyl residues in the latter is such that 8–11–14–45 bind to one core and 18–35–38–41 bind to the other (Adman *et al.*, 1973), resulting in an unexpected "cross-threaded" structure with two Cys–X–X–Cys rings per cluster. The chelate structure in 2-Fe proteins is unknown. In view of the arrangements **8** and **9** and the likelihood, based on analogue structures (Section II,B.), that a face of **2** and the core of **3** have similar dimensions, it is difficult to speculate on the most probable mode of chelation. However, some ambiguity has been removed by the recent finding that the protein from *Equisetum*, which has Val instead of Cys at position 18, exhibits the properties of a normal plant Fd (Kagamiyama *et al.*, 1975). Hence, it is likely that Cys residues at 39, 44, 47, and 77 are involved in chelation to the $Fe_2S_2^*$ core of all plant-type Fds. Sequences of adrenal (5 Cys) and bacterial (6 Cys) Fds (Tanaka *et al.*, 1973, 1974) are not highly homologous nor are they closely related to those of plant-type Fd. Each does contain one Cys–X–X–Cys run, which, by analogy to other proteins, could be involved in chelation. Identity of other coordinated residues cannot be established from present information.

The foregoing brief survey of salient structural aspects of protein active sites makes evident the following two minimum requirements for any synthetic iron–sulfur complex designed as an analogue of these sites: (1) close approach to or duplication of the compositional and stereochemical features of units **1**, **2**, and **3**, with terminal ligation by thiolate groups only; (2) stabilization in the same total oxidation state(s) as in the proteins, with these states interconverted by one-electron redox processes. More refined models would in addition include the features (3) terminal ligation by simple cysteinyl groups or small cysteinyl peptides containing the spacings in **8** and **9**, or, more desirably, (4) ligation by peptides that reproduce the sequence of a cysteinyl-containing fragment of a native protein.

B. Solution Studies of Iron(II,III) Complexes with Thiols and Sulfide

In view of requirement (1), it is appropriate to inquire whether in the extensive studies of iron–thiol complexes that preceded or (prior to 1972) were contemporaneous with definition of protein sites **1–3**, any species demonstrably similar to these coordination units were obtained. The great majority of these studies, some of which have been summarized (McAuliffe and Murray, 1972), have dealt with aqueous solution systems and did not involve product isolation. Ligands investigated include mercaptoacetate (Leussing and Kolthoff, 1953; Leussing and Tischer, 1963; Lappin and McAuley, 1975), 2,3-dimercaptopropan-1-ol (Leussing and Mislan, 1960; Leussing and Jayne, 1962), cysteine (Tanaka et al., 1955; Page, 1955; Leussing et al., 1960; Tomita et al., 1967, 1968a,b), and penicillamine (Bell et al., 1971; Stadtherr and Martin, 1972). These studies have yielded useful information concerning stoichiometries of complexes, polynuclear complex formation, equilibrium constants, and a kinetic scheme for the Fe(III)-catalyzed oxidation of cysteine to cystine (Taylor et al., 1966; Gilmour and McAuley, 1970). However, in no case have they afforded precise definition of the complexes formed in terms of stereochemistry or ligation modes with a polyfunctional agent such as cysteine.

Bis- and triscysteinato complexes of Fe(II) and Fe(III), respectively, have been isolated (Schubert, 1932, 1933; Mathur et al., 1966; Tomita et al., 1968a,b; Murray and Newman, 1975) but elucidation of their structures has not been pursued by X-ray diffraction. Electronic spectra of many of these Fe(III) complexes generated in solution exhibit intense charge transfer absorptions at and below about 500–600 nm, but these spectra are neither stereochemically informative nor similar to those of Rd_{ox} or other proteins. The only structural information available for complexes with such ligands pertains to $Fe(SCH_2CO_2) \cdot H_2O$, which has been shown by X-ray diffraction to be polymeric with distorted octahedral Fe(II) coordination (Jeannin et al., 1972).

Somewhat more promising results have been obtained with solutions containing Fe(III), sulfide, and mono- or difunctional thiols, cysteine derivatives, or glutathione (Yang and Huennekens, 1970; Sugiura and Tanaka, 1972; Sugiura et al., 1972, 1975). At a 1:1 Fe(III)/S^{2-} mole ratio in the presence of excess thiol, spectra similar to distinctive absorptions of 2-Fe Fd_{ox} proteins in the 300–500 nm region were developed. The binuclear structure **10** was suggested as the possible chromophore in the cases of glutathione and 2-mercaptoethanol. These results provided the first definite evidence that absorption spectral properties of any iron–sulfur protein could be closely approached in synthetic complexes derived from simple thiols. Unfortunately, these complexes appear to be of limited stability and none has been isolated or otherwise characterized.

10

A somewhat more interesting approach has been followed in several cases by the synthesis of model cysteinyl peptides (Table I), which re-

TABLE I
Model Peptides of Iron–Sulfur Proteins

No.	Peptide	Related protein[f]	Fe complex
11	Boc-Gly-(Cys-Gly-Gly)$_2$-Cys-Gly-NH$_2$[a]	—	4-Fe[h]
12	Boc-Gly-(Cys-Gly-Gly)$_3$-Cys-Gly-NH$_2$[a]	—	1-Fe,[g] 4-Fe[h]
13	Boc-Cys-Thr-Leu-Cys-Gly-Cys-Pro-Leu-Cys-Gly-OMe[b]	*P. aerogenes* Rd	1-Fe[b,i]
14	H-Ser-Cys-Val-Ser-Cys-OH[c]	*C. pasteurianum* Fd	2-Fe[j]
15	Z-Cys(Bzl)-Gly-Asn-Cys(Bzl)-Ala-OEt[d]	*C. pasteurianum* Fd	—
16	Z-Asn-Val-Cys(Bzl)-Pro-Val-Gly-OEt[d]	*C. pasteurianum* Fd	—
17	H-Cys-Gly-Asn-Cys-Ala-Asn-Val-Cys-Pro-Val-Gly-OEt[e]	*C. pasteurianum* Fd	—[k]

[a] Anglin and Davison, 1975b.
[b] Ali *et al.*, 1973; see also Ali and Weinstein, 1971.
[c] Yajima *et al.*, 1971.
[d] Schöberl *et al.*, 1973a.
[e] Schöberl *et al.*, 1973b.
[f] Yasunobu and Tanaka, 1973.
[g] Anglin and Davison, 1975a.
[h] Que *et al.*, 1974b.
[i] Ali *et al.*, 1972.
[j] Sugiura and Tanaka, 1972.
[k] None reported.

produce the spacings of Cys residues (**11, 12**—compare with **4–7**) in, or are synthetic fragments (**13–17**) of, native proteins. The use of these or similar peptides in model systems is not yet extensive; complexes derived from **11** and **12** are described in Sections III and IV. Reaction of pentapeptide **14** with Fe(III) and sulfide also produces a chromophore similar to that of 2-Fe Fd_{ox} proteins. Peptide **13** was synthesized with the intention of simulating the Rd chromophore **8** but without the long segment linking the two Cys–X–X–Cys portions of the chain. Aerial oxidation of its Fe(II) reaction product afforded a red-brown color, but the spectrum of the species separated by chromatography exhibited only continuous absorption from 600 to 200 nm, in contrast to the featured spectrum of Rd_{ox} in this region. The nature of this species must be considered unestablished.

C. Structures of Coordination Units in Nonanalogue Iron–Sulfur Complexes

The types, structures, and electronic properties of metal complexes of sulfur ligands have been amply reviewed elsewhere (Livingstone, 1965; Abel and Crosse, 1967; Jørgensen, 1968; Mehrotra et al., 1968; Coucouvanis, 1970; Eisenberg, 1970; McAuliffe and Murray, 1972; Lippard, 1973; Vahrenkamp, 1975; Vergamini and Kubas, 1976). In the chemistry of no other element is the coordinative versatility of sulfur ligands as well developed as in the chemistry of iron. To emphasize this point, the varied types of Fe–S coordination units in synthetic complexes established by X-ray diffraction are collected in Table II [for a tabulation of structures of Fe(III) complexes, see Cotton (1972)]. Also included is a partial list of structural information for minerals and related compounds (see also Ward, 1970; Flahaut, 1972). In addition to mononuclear units (**18–20**) of conventional stereochemistry, sulfide, disulfide, and thiolate ligands alone or in combination are capable of generating binuclear (**21–26**), trinuclear (**27**), or tetranuclear (**28–31**) units. Indeed, it is the propensity of these ligands to bridge two or more metal centers that is their dominant structural characteristic with iron and other transition elements. Detailed consideration of the structures in Table II is beyond the scope of this chapter. It is observed here, however, that despite the evident rich structural diversity, there are only a few examples of complexes whose stereochemistries are related to the active sites **1–3**. These include the single case of a discrete tetrahedral complex,† the Fe(II) species Fe[(SP-$Me_2)_2N]_2$ and four complexes that contain the $Fe_4S_4{}^*$ core with the "cubane-type" stereochemistry similar to that in **3**. There are no species

† The structure of the Rd_{red} analogue $[Fe(S_2\text{-}o\text{-}xyl)_2]^{2-}$ (Section II,A) has recently been solved; see footnote, p. 266.

TABLE II
Summary of Structural Types of Coordination Units in Synthetic Iron–Sulfur Complexes

Type[a]	Examples	References
Mononuclear		
Tetrahedral **18**	Fe[(SPMe$_2$)$_2$N]$_2$	Churchill and Wormald, 1971
Square pyramidal **19**	Fe(mnt)$_2$(NO) Fe(tfd)$_2$(AsPh$_3$), [Fe(tfd)$_2$(OPPh$_3$)]$^-$	Rae, 1967 Epstein et al., 1970
Trigonal-octahedral **20**	Fe(S$_2$COR)$_3$, Fe(S$_2$CNR$_2$)$_3$ Fe(S$_2$C-p-tol)$_2$(S$_3$C-p-tol) [Fe(S$_2$CNR$_2$)$_3$]$^+$	Healy and White, 1972[b] Coucouvanis and Lippard, 1969 Martin et al., 1974
Binuclear		
21	Fe$_2$S$_2$(CO)$_6$	Wei and Dahl, 1965ab
22	[Fe(SEt)(CO)$_3$]$_2$ [Fe(SEt)(NO)$_2$]$_2$ [(C$_5$H$_5$)Fe(SPh)(CO)]$_2$ [(C$_5$H$_5$)Fe(SMe)(CO)]$_2^+$ [Fe(SCH$_2$CH$_2$NMe)$_2$(CH$_2$)$_{2,3}$]$_2$	Dahl and Wei, 1963 Thomas et al., 1958 Ferguson et al., 1968 Connelly and Dahl, 1970 Hu and Lippard, 1974
23	[Fe(mnt)$_2$]$_2^{2-}$ [Fe(tfd)$_2$]$_2^-$ [Fe(edt)$_2$]$_2^{2-}$	Hamilton and Bernal, 1967 Schultz and Eisenberg, 1973 Snow and Ibers, 1973

TABLE II (*Continued*)

Type[a]	Examples	References
24 [Fe(S₂CSEt)₂(SEt)]₂ structure	[Fe(S₂CSEt)₂(SEt)]₂	Coucouvanis et al., 1970
25 Fe—(SR)₃—Fe	[Fe₂(SMe)₃(CO)₆]⁺	Schultz and Eisenberg, 1973
26 Fe(S—S)(SR)₂Fe	[(C₅H₅)FeS(SEt)]₂	Terzis and Rivest, 1973

Trinuclear

27 Fe₃S₂ cluster	Fe₃S₂(CO)₉	Wei and Dahl, 1965b

Tetranuclear

28 [MeSFe₂(CO)₆]₂S structure	[MeSFe₂(CO)₆]₂S	Coleman et al., 1967
29 Fe₄S₃ cubane	[Fe₄S₃(NO)₇]⁻ [c]	Johansson and Lipscomb, 1958
30[d] Fe₄S₄ cubane	[(C₅H₅)FeS]₄ (orthorhombic) [(C₅H₅)FeS]₄ (monoclinic) [Fe₄S₄(tfd)₄]²⁻ Fe₄S₄(NO)₄	Schunn et al., 1966 Wei et al., 1966 Bernal et al., 1972 Gall et al., 1974

TABLE II (Continued)

Type[a]	Examples	References
(structure 31: Fe-S cube with RN groups)	$Fe_4S_2(N\text{-}t\text{-}Bu)_2(NO)_4$	Gall et al., 1974
Extended lattices		
Tetrahedral Fe–S$_4$	$KFeS_2$	Boon and MacGillavry, 1942
	FeS (tetragonal)	Kjekshus et al., 1972
	$CuFeS_2$ (chalcopyrite)	Hall and Stewart, 1973
	$CuFe_2S_3$ (cubanite)	Fleet, 1970
Octahedral Fe–S$_6$	FeS (cubic)	de Médicis, 1970
	FeS_2 (pyrite)	Brostigen and Kjekshus,
	FeS_2 (marcasite)	1969, 1970

[a] Only sulfur ligands are shown.
[b] For comparative structures of high- and low-spin tris(chelate) Fe(III) complexes, see also Leipoldt and Coppens (1973) and Hoskins and Pannan (1975).
[c] The core structural representation as a cube missing one vertex is idealized.
[d] For a discussion of structures of compounds containing Fe$_4$S$_6$ units, see Vergamini and Kubas (1976).

that possess the discrete Fe$_2$S$_2$* core of **2**, although this arrangement is approached in KFeS$_2$, whose structure consists of $\cdots$ FeS$_2$FeS$_2$ $\cdots$ chains formed by edge-shared tetrahedra.

The information contained in the preceding two sections, while not of any direct biological relevance, has been presented in order to reveal the exceptional range of complexation tendencies, formal oxidation states, and structural patterns in the organometallic and coordination chemistry of iron with sulfur ligands. In view of the elegant simplicity of active sites **1–3**, it is perhaps surprising that, neither purposely nor serendipitously, had suitable analogues been isolated in the past in the exceptionally far-ranging studies that have characterized iron–sulfur chemistry. Indeed, those complexes that might fulfill requirements (1) and (2) had not been isolated and identified, whereas those that had been fully characterized and contain structural units similar to the proteins (**18** and **30**, Table II) do not satisfy (1) and, as will be seen, (2) as well.

II. SYNTHESIS AND STRUCTURES OF 1-Fe, 2-Fe, AND 4-Fe ACTIVE-SITE ANALOGUES

Prior to the inception of investigations in these laboratories, begun in 1971, the structures of Rd$_{ox}$ and HP$_{red}$ were known (Herriott et al., 1970;

Watenpaugh et al., 1971; Carter et al., 1971) and that of *P. aerogenes* Fd_{ox} was reported shortly thereafter (Sieker et al., 1972). The existence of the Fe_4S_4* core substructure in the latter two proteins and in [(C_5H_5)-FeS]$_4$ (Table II), the synthesis of tetranuclear dithiolenes $[Fe_4S_4(tfd)_4]^z$ by Balch (1969), and the survival of the substructure in redox reactions (Ferguson and Meyer, 1971; Balch, 1969) all were suggestive of considerable stability of the core, albeit not necessarily in the same oxidation levels as in proteins. Further, the ready reconstitution of 8-Fe (Malkin and Rabinowitz, 1966; Hong and Rabinowitz, 1970b) and 2-Fe proteins (Kimura, 1968; Tsibris et al., 1968a,b; Fee and Palmer, 1971) from the apoprotein, an iron salt, and sulfide indicated that the Fe_4S_4* and Fe_2S_2* cores could be formed under mild conditions in solution with simple inorganic reagents. These observations collectively led to the idea that suitably modified "active site" structures might be inherently stable and capable of existence outside the protein and hence be directly synthesized. This has proven to be the case, and analogues of all three active sites have been obtained. The obvious modification in achieving the simplest analogues is replacement of peptide Cys-S binding groups with small organic thiolates. The premise underlying most of the synthetic experiments has been that the usual substitution lability of Fe(II,III), as manifested in, e.g., protein reconstitution, will lead to those soluble reaction products whose formation is thermodynamically controlled.

A. Synthesis

1. 4-Fe Analogues

A simple and versatile route to the species $[Fe_4S_4(SR)_4]^{2-}$, the first synthetic analogues prepared (Herskovitz et al., 1972a), is afforded by the anaerobic reaction of $FeCl_3$ with 3 equiv of a sodium thiolate in methanol solution followed by treatment of the resultant polymeric solid with a methanolic sulfide solution (NaHS + NaOMe). Dark red-brown solutions are formed from which beautifully crystalline salts of the tetranuclear dianions may be isolated upon addition of quaternary cations. The proposed reaction stoichiometry (Averill et al., 1973) is shown in Eq. (1). The dinegative charge of the product implies the formal

$$4\ FeCl_3 + 12\ NaSR \rightarrow (4/n)[Fe(SR)_3]_n$$
$$\downarrow 4\ NaHS + 4\ NaOMe$$
$$(R_4N)_2[Fe_4S_4(SR)_4] \xleftarrow{R_4N^+} Na_2[Fe_4S_4(SR)_4] + RSSR + 6\ NaSR$$

(1)

oxidation states 2 Fe(II) + 2 Fe(III) and indicates that the overall reaction involves oxidation–reduction. This apparently occurs in the second step, and disulfide has been found in the reaction products. Formation of a partially reduced tetramer seemingly results from intramolecular RS$^-$ → Fe(III) electron transfer, a process that may account for the inherent instability of other types of ferric thiolate complexes (Taylor et al., 1966; Leussing and Newman, 1956; Tomita et al., 1968a). This procedure allows synthesis of a large number of tetramers with R = alkyl and aryl (Averill et al., 1973; DePamphilis et al., 1974; Holm et al., 1974a), all of which are isolated in the dianionic form, and should be generally applicable to preparations utilizing monofunctional thiols. In situ generation of tetramers using a similar reaction system has been reported (Schrauzer et al., 1974; Tano and Schrauzer, 1975).

2. 2-Fe Analogue

Inasmuch as variations in the Fe(III)/thiol mole ratio from that in the preceding preparation were found to afford $[Fe_4S_4(SR)_4]^{2-}$ as the only isolable product, a strategy for the synthesis of binuclear tetrahedral complexes was tested employing potentially chelating dithiols whose bite distances render them incapable of spanning the approximately 7 Å distance separating thiolate sulfur atoms within a tetramer (9). The first reaction attempted utilized 1,2-ethanedithiol and yielded $[Fe(edt)_2]_2{}^{2-}$ despite the presence of NaHS in the reaction mixture. Its physical properties (Herskovitz et al., 1975) and structure (23, Table II) show that this complex is not analogous to site 3. However, the S···S bite distance (3.16 Å) and S–Fe–S bite angle (89.6°), together with an earlier observation that in solution the reaction products of $FeCl_3$ and $HS(CH_2)_nSH$ incorporated sulfide only when $n > 3$ (Sugiura et al., 1972), suggested that a more flexible dithiol would allow near-tetrahedral angles around and normal bond distances to Fe(III) and thus decrease the relative stability of 23. o-Xylene-α,α'-dithiol satisfies these requirements, and its use in the anaerobic methanolic reaction mixture of Eq. (2) led to isolation of salts of a red-brown binuclear Fe(III) dianion, $[Fe_2S_2(S_2\text{-}o\text{-xyl})_2]^{2-}$, having the desired composition (Mayerle et al., 1973, 1975).

$$FeCl_3 + \underset{SH}{\overset{SH}{\text{[o-xylene]}}} \xrightarrow[\text{(2) } R_4N^+]{\text{(1) NaHS/NaOMe}} (R_4N)_2[Fe_2S_2(S_2\text{-}o\text{-xyl})_2] \quad (2)$$

3. 1-Fe Analogues

The successful preparation of $[Fe_2S_2(S_2\text{-}o\text{-xyl})_2]^{2-}$ was followed by employment of the same dithiol in an anaerobic sulfide-free reaction system [Eq. (3)] (Lane et al., 1975, 1976). This reaction first yields the

Fe(II) dianion, pale yellow in solution and isolable

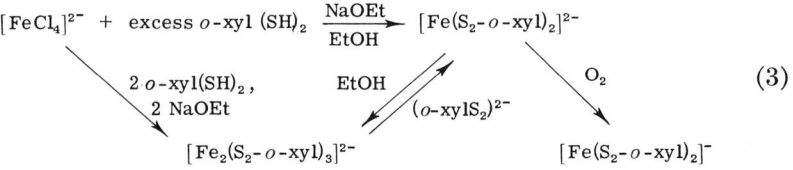

(3)

as the red-brown Na^+/Ph_4As^+ mixed cation salt. Upon treatment with ethanol, this complex is largely converted to a binuclear Fe(II) dianion of unestablished structure. Brief aerial oxidation of a solution of the Fe(II) dianion yields deep red-brown $[Fe(S_2\text{-}o\text{-xyl})_2]^-$, isolable as its Et_4N^+ salt. The Fe(II) complex $[Fe(SPh)_4]^{2-}$ has recently been isolated and a tetrahedral structure established from absorption spectral data (Holah and Coucouvanis, 1975). A model complex containing the cysteinyl coordination in site **1** and the Cys–X–X–Cys spacing in **8** has been generated in DMSO solution by Reaction (4) using the dodecapeptide **12** (Table I). The corresponding Co(II) complex was produced by

$$[Fe(DMSO)_6]^{2+} + Boc\text{-}Gly\text{-}(Cys\text{-}Gly\text{-}Gly)_3\text{-}Cys\text{-}Gly\text{-}NH_2 \rightarrow [Fe(12\text{-peptide})]^{2-}$$
(4)

a similar reaction. Neither species was isolated but the presence of a tetrahedral chromophore in both was established from electronic spectra (Anglin and Davison, 1975a).

These direct preparations involve inexpensive, readily accessible starting materials [excluding reaction (4)] and are simply performed, the only precaution required being exclusion of oxygen. Quaternary cation salts are obtainable in highly crystalline forms that are soluble in polar nonaqueous and mixed aqueous–nonaqueous media. Differently substituted binuclear and tetranuclear complexes are accessible by thiolate substitution reactions (Section IV,A). As is demonstrated in the following section, $[Fe(S_2\text{-}o\text{-xyl})_2]^-$ and $[Fe_4S_4(SR)_4]^{2-}$ approach or contain the structural features of active sites **1** and **2**, respectively, while $[Fe_2S_2(S_2\text{-}o\text{-xyl})_2]^{2-}$ possesses the arrangement previously postulated for site **3**.

B. Analogue Structures and Comparisons with Proteins

Perspective views of the structures of 1-, 2-, and 4-Fe analogues and 1- and 4-Fe protein active site structures are shown in Figs. 1–3; stereoviews of analogues are set out in Fig. 4. In order to facilitate analogue–protein structural comparisons, views from approximately the same perspective are provided in Figs. 1 and 3. Active-site structures and the structure of $Fe[(SPMe_2)_2N]_2$ were calculated from published coordinates

7. ACTIVE-SITE ANALOGUES OF IRON–SULFUR PROTEINS   221

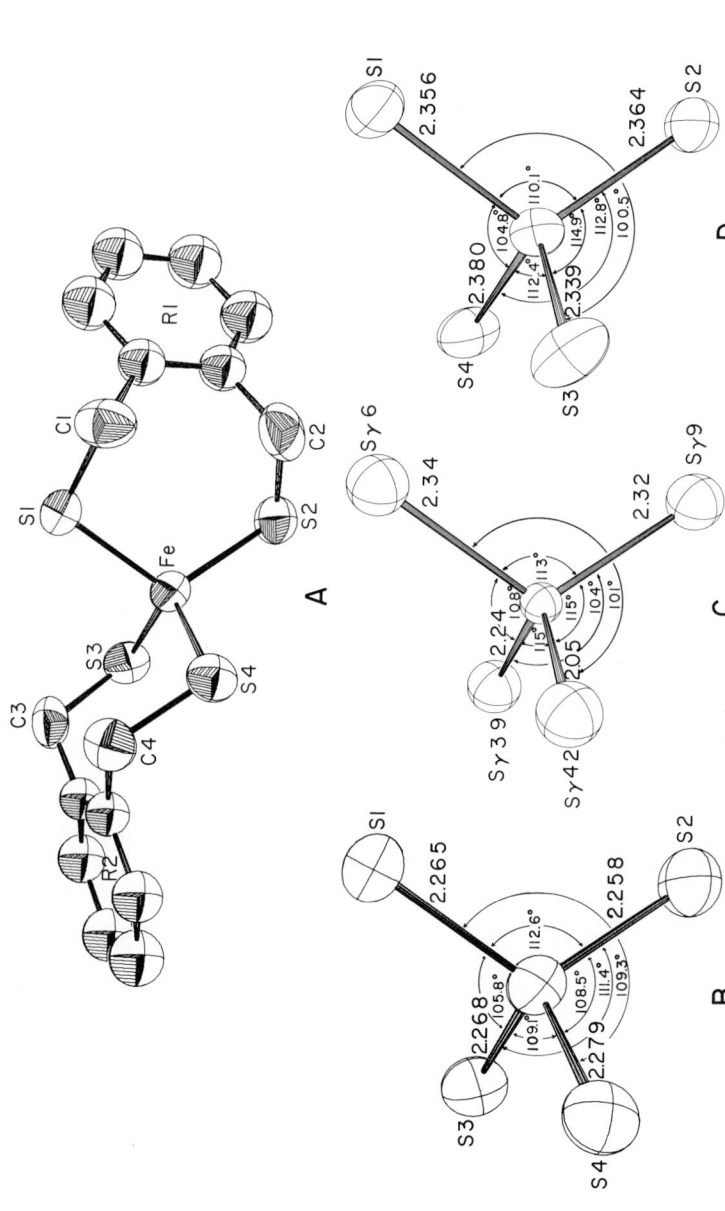

Fig. 1. (A) Overall structure of [Fe(S$_2$-o-xyl)$_2$]$^-$ (hydrogen atoms omitted) illustrating nonplanar chelate rings and chirality of the complex; from Lane et al. (1975). (B) The Fe(III)S$_4$ portion of (A) showing principal distances and angles. (C) The Fe(III)S$_4$ portion of *C. pasteurianum* Rd$_{ox}$ calculated from the coordinates of Watenpaugh et al. (1973) but with slightly different cell constants (K. Watenpaugh, private communication, 1975). (D) The Fe(II)S$_4$ portion of Fe[(SPMe$_2$)$_2$N]$_2$ calculated from the coordinates of Churchill and Wormald (1971). In (A), (B), and (D), the 50% probability ellipsoids are shown, whereas in (C) the sizes of the atoms are arbitrary.

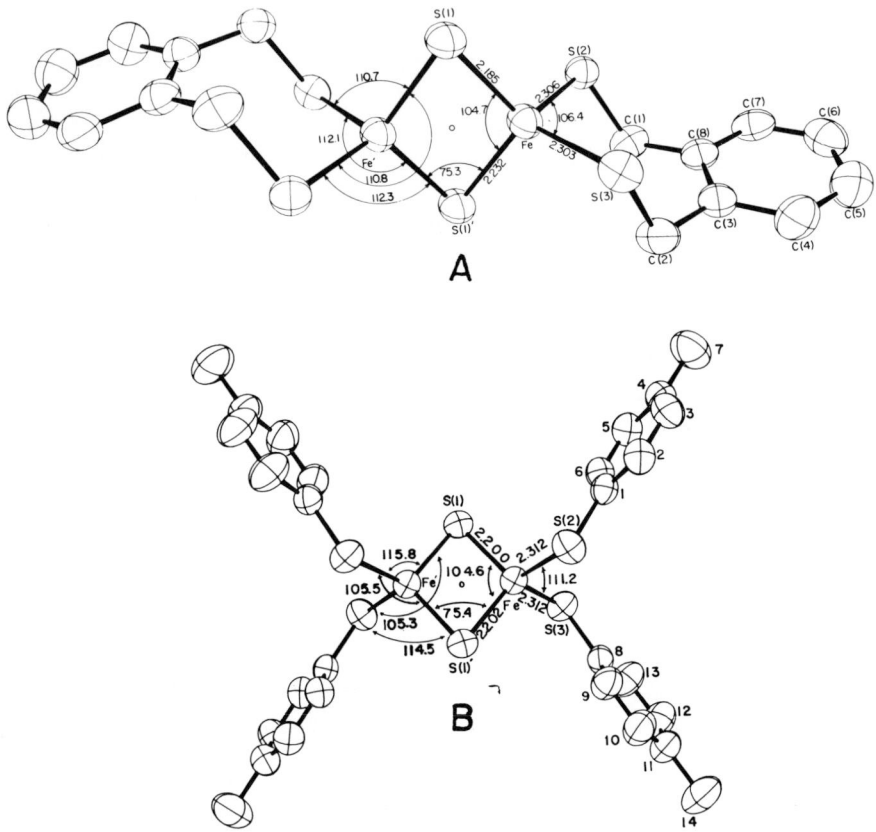

Fig. 2. Structures of the 2-Fe analogues [Fe$_2$S$_2$(S$_2$-o-xyl)$_2$]$^{2-}$ (A) and [Fe$_2$S$_2$(S-p-tolyl)$_4$]$^{2-}$ (B); the 50% probability ellipsoids are shown. Reproduced with permission from Mayerle et al. (1975). J. Am. Chem. Soc. **97**, 1032. Copyright by the American Chemical Society.

or crystallographic data supplied to the authors and were appropriately spatially oriented. Mean distances and angles are collected in Table III. These calculated data for HP$_{red}$ and HP$_{ox}$ differ slightly from values reported in the most recent structural refinement (Freer et al., 1975).

1. 1-Fe Analogue

An overall view of the structure of [Fe(S$_2$-o-xyl)$_2$]$^-$ as determined for its Et$_4$N$^+$ salt is given in Figs. 1A and 4A. The orthorhombic crystals contain two inequivalent anions of nearly identical dimensions in the unit cell; data for the Fe(III)–S$_4$ coordination unit of one anion are given in Fig. 1B. The complex approaches T_d Fe–S$_4$ microsymmetry with small but definite rhombic distortions evident. The average dihedral angle be-

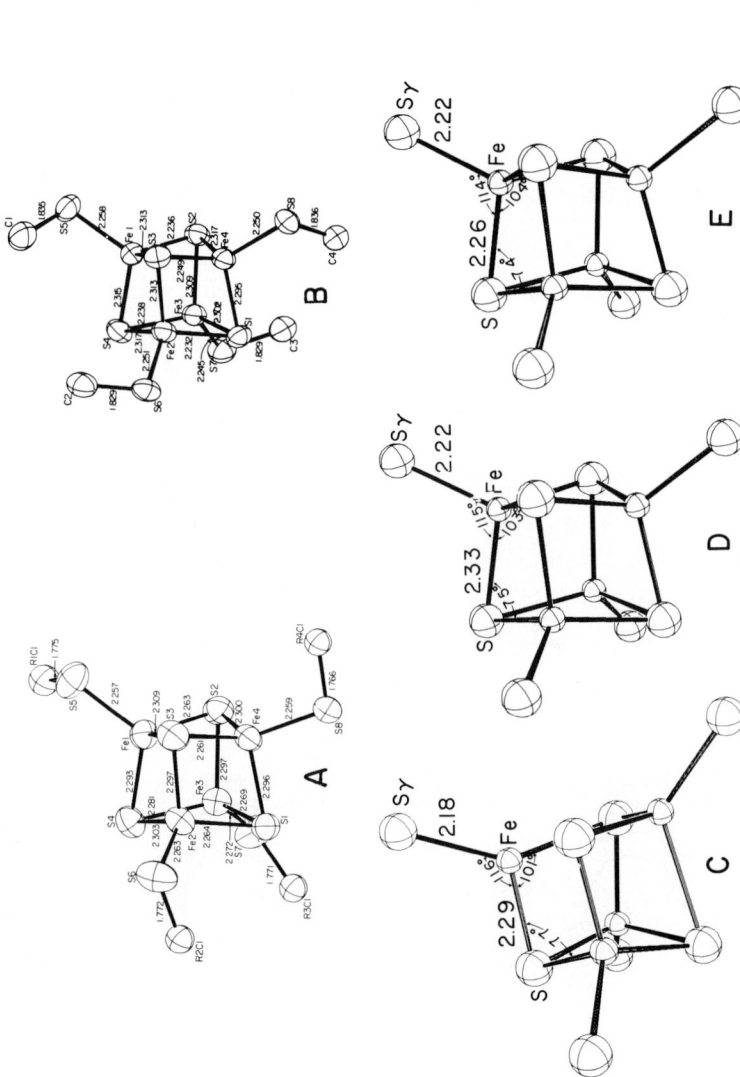

Fig. 3. Views of the Fe$_4$S$_4$*S$_4$ clusters from approximately the same perspective: (A) [Fe$_4$S$_4$(SPh)$_4$]$^{2-}$; (B) [Fe$_4$S$_4$(SCH$_2$-Ph)$_4$]$^{2-}$; (C) *P. aerogenes* Fd$_{ox}$ calculated from the coordinates of Adman *et al.* (1973); (D) *Chromatium* HP$_{red}$, calculated from coordinates supplied by J. Kraut (private communication, 1975); (E) *Chromatium* HP$_{ox}$, calculated from coordinates of Carter *et al.* (1974a). (A) Reproduced by permission from Que *et al.* (1974a). *J. Am. Chem. Soc.* **96**, 4168. (B) Reproduced with permission from Averill *et al.* (1973). *J. Am. Chem. Soc.* **95**, 3523. Copyright by the American Chemical Society.

224

tween Fe–S$_2$ planes is 92.5°, and the intrachelate bite distances and angles average to 3.736 Å and 110.7°, respectively. The chelate rings adopt nonplanar chairlike conformations with obtuse dihedral angles between the mean least-squares planes FeS$_2$–C$_2$S$_2$ and C$_2$S$_2$–C$_2$Ph falling in the ranges 126°–140° and 108°–111°, respectively. In contrast, the coordination unit of *C. pasteurianum* Rd$_{ox}$ as determined by X-ray diffraction is severely distorted from idealized tetrahedral symmetry (Fig. 1C) with much larger angular and distance ranges evident. Indeed, the shortest distance, 2.05 Å, is at least 0.17 Å shorter than any known Fe(III)–SR⁻ distance, and the two longer distances approach or overlap those found in the approximately tetrahedral Fe(II)–S$_4$ unit of Fe[(SPMe$_2$)$_2$N]$_2$ (Fig. 1D). Recently, the structure of lyophilized *P. aerogenes* Rd$_{ox}$ has been investigated by X-ray absorption spectroscopy (Shulman *et al.*, 1975). Analysis of the extended X-ray absorption fine structure (EXAFS) spectrum has led to the conclusion that the Fe–S distances in this protein are closer to being equal than are the distances obtained from diffraction techniques for *C. pasteurianum* Rd$_{ox}$. A similar conclusion has been reached from an EXAFS analysis of the latter protein (Sayers *et al.*, 1976).

The high-spin spherically symmetric Fe(III) ion is devoid of ligand field stabilization effects on stereochemistry and is expected to adopt a strictly regular coordination geometry in the absence of ligand structural constraints and the perturbing influence of crystalline packing forces and other medium effects. While small crystalline and ligand perturbations cannot be fully discounted, [Fe(S$_2$-*o*-xyl)$_2$]⁻ is considered to contain an essentially unconstrained Fe–S$_4$ unit (Lane *et al.*, 1975) whose stereochemistry should be closely approached in Rd$_{ox}$ active sites in the absence of structural constraints imposed by the protein. The essence of the entatic state hypothesis (Vallee and Williams, 1968; Williams, 1971) as applied to metalloproteins is that protein structure at all levels may confer on the active site a local stereochemistry that is less symmetric than that manifested by a structurally unconstrained but otherwise equivalent unit. Inasmuch as [Fe(S$_2$-*o*-xyl)$_2$]⁻ approaches the basic electronic features of Rd$_{ox}$ (Section III,B), and provided that a distorted protein site stereochemistry persists after final structural refinement, a state of entasis has been assigned to Rd$_{ox}$ (Lane *et al.*, 1975). It is our persuasion that demonstration of the presence or absence of an entatic state in metalloproteins is best afforded by the type of analogue–protein struc-

Fig. 4. Stereoviews of 1-, 2-, and 4-Fe analogues: (A) [Fe(S$_2$-*o*-xyl)$_2$]⁻, from Lane *et al.* (1975); (B) [Fe$_2$S$_2$(S$_2$-*o*-xyl)$_2$]²⁻; (C) [Fe$_2$S$_2$(S-*p*-tolyl)$_4$]²⁻; (D) [Fe$_4$S$_4$(SCH$_2$Ph)$_4$]²⁻; (E) [Fe$_4$S$_4$(SPh)$_4$]²⁻. Except for hydrogen atoms, the 50% probability ellipsoids are shown. (C) Reproduced with permission from Mayerle *et al.* (1975). *J. Am. Chem. Soc.* **97**, 1032. (D) Reproduced by permission from Averill *et al.* (1973). *J. Am. Chem. Soc.* **95**, 3523. Copyright by the American Chemical Society.

TABLE III

SELECTED MEAN DISTANCES AND ANGLES OF PROTEIN ACTIVE SITE AND ANALOGUE STRUCTURES[a]

Distance or angle	HP_{ox}[b]	HP_{red}[b]	Fd_{ox}[c]	4-Fe(1)[c,d]	4-Fe(2)[d,e]	2-Fe(1)[f]	2-Fe(2)[f]	Rd_{ox}[g]	1-Fe(1)[h]	1-Fe(2)[i]
Fe–S	2.22(2)	2.22(2)	2.18(6)	2.251(3)	2.263(3)	2.305(2)	2.312(1)	2.24(7)	2.267(3)	2.360(9)
Fe–S*	2.26(2)	2.33(2)	2.29(4)	2.286(10)	2.286(5)	2.21(2)	2.201(1)			
Fe··Fe	2.71(3)	2.82(2)	2.85(4)	2.746(10)	2.736(3)	2.698(1)	2.691(1)			
S*··S*	3.56(3)	3.65(4)	3.53(5)	3.61(1)	3.61(1)	3.498(3)	3.482(3)			
S*–Fe–S*	104(1)	103(1)	101(2)	104.1(1)	104.3(2)	104.73(5)	104.61(4)			
Fe–S*–Fe	73.7(5)	74.6(5)	77(1)	73.81(8)	73.5(1)	75.27(5)	75.39(4)			
S–Fe–S*	114(1)	115(1)	116(2)	114.4(7)	115(3)	111.5(4)	110(3)			

[a] *Abbreviations*: 4-Fe(1), [Fe$_4$S$_4$(SCH$_2$Ph)$_4$]$^{2-}$; 4-Fe(2), [Fe$_4$S$_4$(SPh)$_4$]$^{2-}$; 2-Fe(1), [Fe$_2$S$_2$(S-o-xyl)$_2$]$^{2-}$; 2-Fe(2), [Fe$_2$S$_2$(S-p-tolyl)$_4$]$^{2-}$; 1-Fe(1), [Fe(S$_2$-o-xyl)$_2$]$^-$; 1-Fe(2), Fe[(SPMe$_2$)$_2$N]$_2$. Numbers in parentheses are estimated standard deviations of the mean; Adman et al. (1973) suggest that for Fd$_{ox}$ these numbers should be doubled.
[b] For data sources, see Fig. 3.
[c] Averill et al. (1973).
[d] For these compounds in particular, the statistically significant distortions toward D_{2d} symmetry have been ignored in the calculation of these mean quantities.
[e] Que et al. (1974a).
[f] Mayerle et al. (1975).
[g] For data source see Fig. 1.
[h] Lane et al. (1975).
[i] Churchill and Wormald (1971).

tural comparison evident in Fig. 1, where in the future such comparisons may include the results of diffraction studies on crystalline materials and of EXAFS analyses of lyophilized and frozen solution samples. In this context [Fe(S$_2$-o-xyl)$_2$]$^-$ and other synthetic analogues serve as structurally and electronically symmetrized versions of active sites.

2. 2-Fe Analogues

Because of the absence of X-ray structural information, the active sites of 2-Fe proteins have been particularly attractive objects for clarification by the synthetic analogue approach (Mayerle et al., 1973, 1975). Inspection of Figs. 2A and 4B immediately reveals that as its Et$_4$N$^+$ salt, [Fe$_2$S$_2$(S$_2$-o-xyl)$_2$]$^{2-}$ has the overall structure 2 postulated for 2-Fe Fd sites. The anion is a centrosymmetric dimer of C$_i$ symmetry, which, however, can be regarded as effectively C$_{2h}$ because of the marginal difference between Fe–S(2) and Fe–S(3) bond lengths; it is degraded from D$_{2h}$ symmetry by the markedly unequal Fe–S* bond lengths. The Fe$_2$S$_2$* core is planar with an Fe · · · Fe distance indicative of a stabilizing metal–metal interaction. Inasmuch as the dihedral angle between FeS$_2$ and FeS$_2$* coordination planes is 89.95(5)°, reduction of Fe(III) site symmetry below tetrahedral results from the approximately 0.1 Å difference between Fe–S and Fe–S* bond lengths and the angular variations evident in Fig. 2A. Chelate ring bite distance and angle are 3.690 Å and 106.4°, respectively, and the rings adopt a chairlike conformation. Hence, the dimer can be formally derived from [Fe(S$_2$-o-xyl)$_2$]$^-$ by insertion of the FeS$_2^-$ unit between the chelate rings. On the basis of comparative physical properties (Section III,B), the structure of [Fe$_2$S$_2$(S$_2$-o-xyl)$_2$]$^{2-}$ is concluded to be a suitable minimal representation of the sites of 2-Fe Fd$_{ox}$ proteins (Mayerle et al., 1973, 1975).

[Fe$_2$S$_2$(S-p-tol)$_4$]$^{2-}$ and related arylthiolate dimers may be derived from [Fe$_2$S$_2$(S$_2$-o-xyl)$_2$]$^{2-}$ by ligand substitution reactions (Mayerle et al., 1975) (Section IV,A). The structure of the former (Figs. 2B and 4C) as its Et$_4$N$^+$ salt is quite similar to that of the latter, being a centrosymmetric dimer of essentially perfect C$_{2h}$ symmetry. The presence of aryl- instead of alkylthiolate ligands renders it a somewhat less realistic analogue. However, a significant point that emerges from a structural comparison of the two dimers is that the Fe$_2$S$_2$* core can be transferred from one thiolate ligand environment to another with no significant change in dimensions.

3. 4-Fe Analogues

The structures of (Et$_4$N)$_2$[Fe$_4$S$_4$(SCH$_2$Ph)$_4$] and (Me$_4$N)$_2$[Fe$_4$S$_4$-(SPh)$_4$] have been reported. The stereochemistries of the tetranuclear anions (Figs. 3A,B and 4D,E) are essentially identical and manifest D_{2d} cluster geometry, which is not crystallographically imposed. Distortions from idealized T_d symmetry are quite pronounced. Under D_{2d} symmetry, distances and angles divide into the following sets: Fe–Fe (2 + 4), Fe–S* (4 + 8), and Fe–Fe–Fe, S*–Fe–S*, Fe–S*–Fe (all 4 + 8). Except for the last two types of angles, the intrinsic differences in structural parameters for this symmetry are resolvable. The data in Table III are mean quantities obtained by weighted averaging of these sets. A detailed analysis of cubane-type core geometries is given elsewhere (Averill et al., 1973; Fritchie, 1975). As with the dimers, the tetramer Fe_4S_4* cores are essentially unchanged structurally when ligated by alkyl- and arylthiolates. Another interesting structural feature is seen upon comparing dimer Fe_2S_2* cores and tetramer Fe_2S_2* faces. These fragments exhibit a high degree of dimensional similarity, with corresponding bonded distances and angles differing by no more than 0.11 Å and 2.4°, respectively. The dimer core may be considered a building block of the tetramer; spontaneous dimer → tetramer conversion has been observed under certain conditions (Section IV,B).

The results in Fig. 3 and Table III demonstrate the marked extent of structural similarity between the Fe_4S_4*S_4 clusters of the two analogue dianions and the sites of Fd_{ox} and HP_{red}. Given the somewhat larger uncertainties in the protein determinations, these clusters are virtually congruent with one another. Consequently, the protein cluster symmetry is no higher than D_{2d}, and results quoted at a penultimate stage of refinement for HP_{ox} and HP_{red} (Carter et al., 1974b) suggest that a somewhat lower symmetry may prevail, presumably induced by protein structural constraints. However, it may be safely concluded that $[Fe_4S_4(SR)_4]^{2-}$ complexes are meaningful structural analogues of the HP_{red} and Fd_{ox} sites, all of which have been shown to have the same total oxidation level (Section III). The departure of these structures from T_d symmetry and the small cluster volume contraction observed upon oxidizing HP_{red} to HP_{ox} (Carter et al., 1974b) can be rationalized, at least in part, by a recent theoretical electronic structural model of the tetranuclear clusters (Section III,B,7).

III. PHYSICAL PROPERTIES OF ANALOGUES

A. Oxidation–Reduction and Analogue–Protein Oxidation Level Equivalencies

At present, the single most important biophysical property of the relatively low molecular weight iron–sulfur proteins appears to be the ability to sustain reversible one-electron transfer between two active site oxidation levels, which for a wide variety of proteins have been detected under *in vitro* and, less frequently, under *in vivo* conditions. These proteins are probably the most widely dispersed and numerous common group of electron-transfer substances in biology. Examples of their occurrence in electron-transfer chains are found in the redox coupling of *Pseudomonas oleovorans* Rd (Lode and Coon, 1973), adrenodoxin (Estabrook et al., 1973), and putidaredoxin (Gunsalus and Lipscomb, 1973; Gunsalus et al., 1974) to their respective monooxygenase enzyme systems, and the role of clostridial Fd as an endogenous electron donor to (as well as possible function of Fe-S units as electron transfer centers within) the nitrogenase enzyme complex (Yoch and Valentine, 1972; Burris and Orme-Johnson, 1974). Consequently, redox activity is an obligatory property for any synthetic species purported to be an active site analogue.

As shown by the data in Table IV, 1-Fe, 2-Fe, and 4-Fe synthetic analogues all possess the requisite redox propensities. The electron-transfer series established by polarographic and related electrochemical measurements in nonaqueous media (DePamphilis et al., 1974; Mayerle et al., 1973, 1975; Lane et al., 1975, 1976) are summarized in Eqs. (5)–(7). Cur-

$$[Fe(SR)_4]^{2-} \rightleftharpoons [Fe(SR)_4]^{-}$$
$$Rd_{red} \rightleftharpoons Rd_{ox} \qquad (5)$$
$$\text{(a)} \qquad \text{(b)}$$

$$[Fe_2S_2(SR)_4]^{4-} \rightleftharpoons [Fe_2S_2(SR)_4]^{3-} \rightleftharpoons [Fe_2S_2(SR)_4]^{2-}$$
$$[Fd_{s\text{-red}}] \qquad Fd_{red} \rightleftharpoons Fd_{ox} \qquad (6)$$
$$2\ Fe(II) \qquad Fe(II) + Fe(III) \qquad 2\ Fe(III)$$
$$\text{(a)} \qquad \text{(b)} \qquad \text{(c)}$$

$$[Fe_4S_4(SR)_4]^{4-} \rightleftharpoons [Fe_4S_4(SR)_4]^{3-} \rightleftharpoons [Fe_4S_4(SR)_4]^{2-} \rightleftharpoons [Fe_4S_4(SR)_4]^{-}$$
$$Fd_{red} \rightleftharpoons Fd_{ox} \leftrightarrow [Fd_{s\text{-ox}}]$$
$$[HP_{s\text{-red}}] \leftrightarrow HP_{red} \rightleftharpoons HP_{ox}$$
$$4\ Fe(II) \quad 3\ Fe(II) + Fe(III) \quad 2\ Fe(II) + 2\ Fe(III) \quad Fe(II) + 3\ Fe(III)$$
$$\text{(a)} \qquad \text{(b)} \qquad \text{(c)}$$

$$(7)$$

rent–voltage curves for redox processes in the tetramer series are shown in Fig. 5. Reduction processes for most other tetramer dianions show closely similar polarographic characteristics, but well-defined oxidations have been observed in only several instances (DePamphilis et al., 1974). One important advantage of synthetic species, derived from structural and analytical data, is knowledge of the net charge per complex unit, information not directly obtainable from proteins but required to specify the total oxidation level of the active site. With structural information in hand and anticipating the comparative physical properties in following sections, the known redox states in proteins can be unambiguously cor-

TABLE IV
OXIDATION–REDUCTION POTENTIALS OF ANALOGUES[a] AND PROTEINS[b]

Species	Couple 4−/3−	3−/2−	2−/1−	References[c]
1-Fe (DMF)				
$[Fe(S_2\text{-}o\text{-xyl})_2]^-$	—	—	−1.03	1
2-Fe (DMF)				
$[Fe_2S_2(S_2\text{-}o\text{-xyl})_2]^{2-}$	−1.73	−1.49	—	2
$[Fe_2S_2(SPh)_4]^{2-}$	−1.37	−1.09	—	2
4-Fe (DMF)				
$[Fe_4S_4(S\text{-}t\text{-Bu})_4]^{2-}$	−2.16	−1.42	−0.12	3
$[Fe_4S_4(SEt)_4]^{2-}$	−2.04	−1.33	—	3
$[Fe_4S_4(SCH_2Ph)_4]^{2-}$	−1.96	−1.25	—	3
$[Fe_4S_4(SPh)_4]^{2-}$	−1.75	−1.04	—	3
$[Fe_4S_4(SC_6H_4NMe_3)_4]^{2+}$	—	−0.79	—	3
$[Fe_4S_4(SC_6H_4NO_2)_4]^{2-}$	—	−0.70	—	3
(80% DMSO–H_2O)				
$[Fe_4S_4(SEt)_4]^{2-}$	—	−1.16	—	4
$[Fe_4S_4(S\text{-Cys(Ac)NHMe})_4]^{2-\ d}$	—	−0.91	—	4
$[Fe_4S_4(9\text{-peptide})(S\text{-Cys(Ac)-}$ $\text{NHMe})]^{2-\ e}$	—	−0.82	—	4
$[Fe_4S_4(12\text{-peptide})]^{2-\ e}$	—	−0.80	—	4
1-Fe proteins	—	—	−0.06	5
2-Fe proteins	—	−0.24 – −0.43	—	5, 6, 7
4-Fe proteins				
(HP)	—	≤ −0.64[f]	+0.35	5
(Fd)	—	−0.28 – −0.42	—	8–10
8-Fe proteins	—	−0.40 – −0.49	—	5

[a] $E_{1/2}(V)$ versus saturated calomel electrode, 25°C.
[b] $E_0'(E_m)(V)$ versus standard hydrogen electrode, pH ∼7, 25°C.
[c] *References*: (1) Lane *et al.*, 1975; (2) Mayerle *et al.*, 1975; (3) DePamphilis *et al.* 1974; (4) Que *et al.*, 1974b; (5) Hall *et al.*, 1974; (6) Estabrook *et al.*, 1973; (7) Gunsalus *et al.*, 1974; (8) Zubieta *et al.*, 1973; (9) Stombaugh *et al.*, 1973; (10) Mullinger *et al.*, 1975.
[d] DMSO solution.
[e] 9-peptide = **11**, 12-peptide = **12** (Table I).
[f] 80% DMSO–H_2O; Cammack, 1973.

related with analogue total oxidation levels as shown in columns (a–c) in series (5)–(7); protein potentials have been tabulated accordingly in Table IV. The indicated formal oxidation states of the metal follow directly and are such that species with an odd net charge contain an odd number of electrons. Several other useful points follow from Eqs. (5)–(7) and Table IV.

7. ACTIVE-SITE ANALOGUES OF IRON–SULFUR PROTEINS

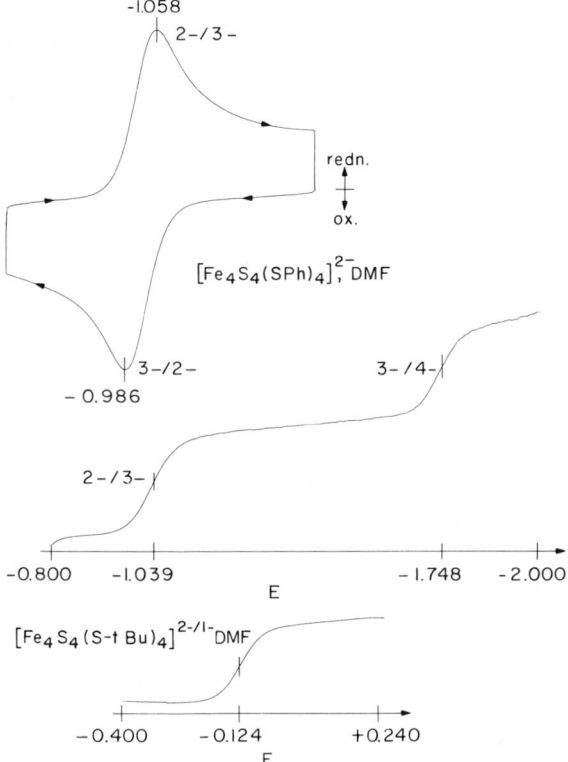

Fig. 5. Current–voltage curves for the redox processes of $[Fe_4S_4(SR)_4]^z$ in DMF: (*top*) cyclic voltammogram for the 2—/3— reaction of R = Ph at a stationary Pt electrode (Sweep rate 100 mV/second); (*middle*) conventional polarogram for $z = 2$—/3— and 3—/4— reductions of R = Ph at a rpe; (*bottom*) $z = 2$—/1— oxidation of R = t-Bu at a rpe. Reproduced with permission from DePamphilis *et al.* (1974). *J. Am. Chem. Soc.* **96**, 4159. Copyright by the American Chemical Society.

a. SCOPE OF ELECTION TRANSFER SERIES. Both analogue and protein series can be conceptualized in terms of terminal reduced and oxidized members that contain only Fe(II) and Fe(III), respectively, the stable oxidation states of iron when coordinated to conventional weak field ligands.† In this sense, the series (6) of three members and the trivial two-membered series (5) are complete; series (7) lacks the terminal neu-

---

† Among other types of Fe–S complexes, the Fe(IV) state, connected to Fe(III, II) by electron transfer, has been found but only with ligands (1,1-dithiolates, dithiolenes) capable of stabilizing high oxidation states (Martin *et al.*, 1974; Hollander *et al.*, 1974; Pignolet *et al.*, 1974; Cauquis and Lachenal, 1973; Pasek and Straub, 1972).

tral tetramer [4Fe(III)], which has not been detected. As for the unstable monoanion, it contains a highly oxidized core that may be reduced by thiolate ligands. Numerous attempts to isolate such species from oxidation reactions of tetramer dianions have failed.

b. DEPENDENCE OF ANALOGUE POTENTIALS ON LIGAND STRUCTURE. The data in Table IV, augmented by more extensive measurements (DePamphilis et al., 1974, Mayerle et al., 1975) show that half-wave potentials become more positive (increasing ease of reduction) as the electron-releasing tendency of R substituents decreases. For the 2-/3- reductions of 2-Fe and 4-Fe complexes, which achieve or approach more closely than other processes strict electrochemical reversibility, the trends can be described empirically in terms of linear free energy correlations with Taft σ^* (R = alkyl, Ph) and Hammett σ_p (R = Ph, p-C_6H_4X) constants. Least-squares fits of the data yield the relationships† (8), (9), and (10) for alkyl thiolate tetramers, aryl thiolate tetramers, and aryl thiolate dimers, respectively. The numerical similarity of the last two equations would appear to reflect restriction of redox events to the core structures,

$$E_{1/2}(V) = 0.411\sigma^* - 1.30 \tag{8}$$

$$E_{1/2}(V) = 0.298\sigma_p - 1.03 \tag{9}$$

$$E_{1/2}(V) = 0.328\sigma_p - 1.09 \tag{10}$$

whose intrinsic affinities for the first electron addition are regulated in nearly parallel fashion by substituent electronic properties.

c. PROTEIN OXIDATION LEVELS. In Eqs. (5)–(7) the nonbracketed forms of the proteins are doubtless those involved *in vivo* and have been isolated or produced *in vitro*. It is observed that the oxidation levels of the three types of synthetic analogues encompass all those currently known for the corresponding n-Fe proteins and, additionally, suggest the possible existence of other (bracketed) levels. One of these, the "super-reduced" form of *Chromatium* HP, $HP_{s\text{-}red}$, has been obtained by dithionite reduction of HP_{red} under denaturing conditions (Cammack, 1973; Dickson and Cammack, 1974). Some evidence has been presented for the formation of $Fd_{s\text{-}ox}$ by oxidation of *Clostridium acidi-urici* Fd_{ox} with ferricyanide (Sweeney et al., 1974). There is as yet no direct evidence bearing on the existence of reduced protein forms isoelectronic with

† As with other linear free energy correlations, these relationships do not hold for those groups X (e.g., NO_2, NMe_2) capable of resonance interaction with the phenyl ring.

$[Fe_2S_2(SR)_4]^{4-}$ or $[Fe_4S_4(SR)_4]^{4-}$. Under standard conditions of aqueous dithionite reduction 2-, 4-, and 8-Fe Fd_{ox} proteins are reduced to Fd_{red}. Reductions under denaturing conditions have not been reported. Lastly, analogue–protein oxidation level equivalencies provide a simple answer to earlier difficulties in understanding the approximately 0.8 V difference between the potentials for HP_{red}/HP_{ox} and 8-Fe Fd_{red}/Fd_{ox} (Table IV). From Eq. (7), Fd_{ox} and HP_{red} are isoelectronic, whereas Fd_{red} and HP_{ox} differ by two electrons. Consequently, these potentials do not refer to the same redox couple and are not directly comparable (Herskovitz et al., 1972b; DePamphilis et al., 1974). The independent "three-state hypothesis" of Carter et al. (1972) is entirely equivalent to the oxidation level relationships in series (7).

B. Electronic Features

Being colored and paramagnetic in accessible oxidation levels and containing a Mössbauer nucleus (^{57}Fe) that may be enriched in reconstitution reactions, the active centers of iron–sulfur proteins are nearly ideal targets for probing of electronic structural features by magnetic and spectroscopic methods. These include absorption, CD, MCD, NMR, EPR, ENDOR, ESCA, resonance Raman, and Mössbauer spectroscopy and magnetic susceptibility studies, and all have been applied to the proteins in one or more oxidation levels (Tsibris and Woody, 1970; Lovenberg, 1973; Orme-Johnson, 1973; Hall et al., 1973a,b, 1974; Sands and Dunham, 1975). Most of these methods have also been applied to the synthetic analogues, and the results are summarized in the following sections. With the exception of absorption, EPR, and Mössbauer spectra, results at present are limited to the oxidation levels $[Fe(SR)_4]^-$, $[Fe_2S_2(SR)_4]^{2-}$, and $[Fe_4S_4(SR)_4]^{2-}$ directly obtained by the synthetic methods in Section II,A. As indicated by the rather negative potentials in Table IV, lower oxidation states are very easily oxidized, and except for $[Fe(S_2\text{-}o\text{-}xyl)_2]^{2-}$, complexes in these states have not yet been isolated as salts of analytical purity.† Preliminary experiments suggest that $[Fe_2S_2(SR)_4]^{3-}$ converts spontaneously to $[Fe_4S_4(SR)_4]^{2-}$, although it may be initially formed at an electrode [series (5)] under diffusion-controlled conditions (Mayerle et al., 1975).

1. MAGNETIC SUSCEPTIBILITIES

The bulk magnetic properties of crystalline salts of representative 1-Fe, 2-Fe, and 4-Fe analogues have been examined over the approxi-

---

† Very recently, pure salts of $[Fe_4S_4(SR)_4]^{3-}$ have been isolated in these laboratories.

TABLE V

MAGNETIC AND PROTON CHEMICAL SHIFT DATA FOR ANALOGUES AND PROTEINS

Compound	μ_{Fe} (BM)	$-J$ (cm^{-1})	$-(\Delta H/H_0)^{obs}$ (ppm)^a	References^b
(Et$_4$N)[Fe(S$_2$-o-xyl)$_2$]	5.83–5.92 (60°–292°K)	—	—	1
C. pasteurianum Rd$_{ox}$	5.85 (280°–330°K)	—	—^{c}	2
(Et$_4$N)$_2$[Fe$_2$S$_2$(SPh)$_4$]	—	148	—	3
(Ph$_4$As)[Fe$_2$S$_2$(S$_2$-o-xyl)$_2$]	0.43–1.42 (100°–287°K)	183	33–42 (220°–350°K)	3
Spinach Fd$_{ox}$	—		~34.5–37 (278°–308°K)	4, 5
(Et$_4$N)$_2$[Fe$_4$S$_4$(SCH$_2$Ph)$_4$]	0.33–1.04 (100°–296°K)	—	—	6
	1.1 (DMSO, 299°K)			
(Ph$_4$As)$_2$[Fe$_4$S$_4$(SEt)$_4$]	—	—	12.2–14.9 (226°–355°K)	7
[Fe$_4$S$_4$(S-L-Cys(Ac(NHMe)$_4$]$^{2-}$	—	—	9.80–14.4 (195°–391°K)	7
[Fe$_4$S$_4$(9-peptide)(S-t-Bu)]$^{2-}$ ^{d}	—	—	12.4, 13.6 (304°K)	8
			9.9–13.6 (298°K)	8
Chromatium HP$_{red}$	~0.1–0.6 (100°–150°K)	—	11–18 (273°–353°K)	9, 10
C. pasteurianum Fd$_{ox}$	1.02–1.08 (278°–303°K)	—	10–19 (253°–363°K)	11
C. acidi-urici Fd$_{ox}$	1.20–1.26 (278°–303°K)	—	10–17 (278°–303°K)	12
B. polymyxa Fd$_{ox}$ I	0.86–0.91 (278°–303°K)	—	8–16 (277°–324°K)	13

^{a} CH$_2$S shifts.

^{b} References: (1) Lane et al., 1975; (2) Phillips et al., 1970a; (3) Gillum et al., 1976a; (4) Palmer et al., 1971; Palmer, 1973 (5) Salmeen and Palmer, 1972; (6) Herskovitz et al., 1972a; (7) Holm et al., 1974a; (8) Que et al., 1974b; (9) Cerdonio et al., 1974; (10) Phillips et al., 1970b; (11) Poe et al., 1970; McDonald et al., 1973; (12) Poe et al., 1971b; (13) Phillips et al., 1974.

^{c} Definite assignments not made.

^{d} 9-peptide = 11 (Table I).

mately 4.2°–300°K range. Magnetic moments per iron† (μ_{Fe}) of analogues and proteins calculated from Curie law using experimental susceptibilities at various temperatures are given in Table V. Results for [Fe(S_2-o-xyl$_2$)]$^-$ and Rd$_{ox}$ indicate simple Curie behavior for both. Magnetic moments are consistent with the $^6A_1$ ground state for tetrahedral Fe(III); to first order, the theoretical moment is 5.92 BM.

The magnetic behavior of 2-Fe and 4-Fe complexes is more complicated owing to exchange interactions between metal centers whose separations fall in the narrow range of 2.69–2.79 Å. Experimental susceptibility curves show that $\chi_{Fe} \to 0$ as $T \to 0$, indicating a singlet ground state and a positive temperature coefficient of paramagnetic susceptibility, characteristics of intramolecular antiferromagnetism. For dimeric dianions with $S_1 = S_2 = \frac{5}{2}$ [tetrahedral Fe(III)], theoretical susceptibility curves are readily generated from a model based on the Hamiltonian $\mathcal{H} = -2J\mathbf{S}_1 \cdot \mathbf{S}_2$ (Martin, 1968; Sands and Dunham, 1975). After corrections for paramagnetic impurities, the $\chi_{Fe}(T)$ data for the analogue (Ph$_4$As)$_2$-[Fe$_2$S$_2$(S$_2$-o-xyl)$_2$] best fit theory with $-J = 148$ cm^{-1} and $g = 2.00$ (Gillum et al., 1976a). Below about 100°K, plant, bacterial, and adrenal Fd$_{ox}$ proteins have been reported to show either no or a trace of paramagnetic component (Moss et al., 1969; Moleski et al., 1970; Kimura et al., 1970). Analysis of susceptibility results for spinach Fd$_{ox}$ at 77°–250°K has yielded $-J = 183$ cm^{-1} (Palmer et al., 1971).‡ Inasmuch as the observed antiferromagnetic interactions are presumably the net result of direct Fe · · · Fe exchange and superexchange through sulfide bridges, the similarity of J values (the moduli of the interactions) between the protein and analogue strongly implies that the geometries of at least the Fe$_2$S$_2$* cores of these species are similar. It remains to be seen if J values for a variety of proteins occur in a fairly narrow interval, as would be expected for a common core structure. Using J values obtained from [Fe$_2$(edt)$_4$]$^{2-}$ (23, Fe · · · Fe 3.41 Å, Herskovitz et al., 1975) and the μ-sulfido complex [Fe(salen)]$_2$S (Mitchell and Parker, 1973), the interaction sequence (Eq. 11) for the three known types of high-spin binuclear Fe(III)-S structural fragments emerges. Exchange coupling increases as Fe · · · Fe distances decrease and thiolate bridges are replaced by sulfide. That the large interactions found for the Fe(S*)$_2$Fe unit may be of general occurrence is supported by the extraction of $-J \sim 200$ cm^{-1} from

---

† $\mu_{Fe}^2 = 3kT\chi_m^p/nN$, where χ_m^p is the total paramagnetic susceptibility and n is the number of Fe atoms, all of which are assumed to contribute equivalently to the susceptibility.

‡ The value $-J = 143$ cm^{-1} has been estimated for spinach Fd$_{ox}$ in solution (Palmer et al., 1971). A value of $-J = 185$ cm^{-1} has recently been quoted for Synechoccus lividas Fd$_{ox}$ (Anderson et al., 1975b).

$$\text{Fe} \diamond_{S,R}^{S,R} \text{Fe} \quad < \quad \text{Fe}-\text{S}-\text{Fe} \quad < \quad \text{Fe} \diamond_{S}^{S} \text{Fe} \qquad (11)$$

$-J = 54 \text{ cm}^{-1} \qquad\qquad 75 \text{ cm}^{-1} \qquad\qquad 148 \text{ cm}^{-1}$

an analysis of the magnetism of the linear chain antiferromagnet $KFeS_2$ (Fe · · · Fe 2.70 Å, Sweeney and Coffman, 1972).

Magnetic data for 4-Fe analogues are currently limited to $[Fe_4S_4(SCH_2Ph)_4]^{2-}$. Moments of this complex (solid, solution) and proteins in aqueous solution are comparable (Table V). The small increases with temperature exhibited by the latter are consistent with those found for the analogue, which exhibits clear antiferromagnetic characteristics over a much wider temperature range (Herskovitz et al., 1972a). Particularly important are the recent results for HP_{red} obtained using a superconducting magnetometer (Cerdonio et al., 1974). Over a common temperature interval moments of this protein and the analogue differ by no more than about 0.1 BM. All of the comparative analogue–protein magnetic information supports fully the oxidation level equivalencies in columns (b), (c), and (b) of series (5), (6), and (7) respectively.

2. Proton Magnetic Resonance

The PMR spectra of 1-, 2-, 4-, and 8-Fe proteins have been examined in aqueous solution at or near ambient temperature in research pioneered by Phillips and co-workers, and many of the results and interpretations have been summarized (Palmer, 1973; Phillips, 1973; Phillips and Poe, 1973; Sands and Dunham, 1975). In both their oxidized and reduced forms the proteins exhibit, in addition to the complex absorption pattern at about 0–10 ppm resulting from nonexchangeable protons of the polypeptide chains, broadened and usually multiple resonances in D_2O solution about 8–20 ppm downfield of DSS internal reference (Table V). With the exception of Rd, which does not display signals in this region and whose spectra have not been otherwise assigned, the following tentative interpretations have been placed upon these resonances: (i) they arise from the α-methylene protons of the coordinated cysteinyl residues

$$\underset{\beta}{H-C}-\underset{\alpha}{\overset{H_a}{\underset{H_b}{C}}}-S-Fe$$

32; (ii) downfield (negative) shifts result from contact interactions generated by paramagnetic iron centers, with increasing downfield displacements with increasing temperature reflecting enhanced paramagnetism of the antiferromagnetically coupled protein cores; (iii) multiple resonances derive from inequivalent methylene groups and the angular dependence of hyperfine coupling constants in each, with the spatial distribution of α-H protons in **32** set by protein tertiary structure. The relative simplicity of the analogue spectra, the ease of incorporation of different thiolate substituents, and the thermal stability of the analogues allow conclusions (i), (ii), and, to a lesser extent, (iii) to be examined in some detail. Spectra of $[Fe_4S_4(SR)_4]^{2-}$ species are broadened owing to core paramagnetism, yet are sufficiently well resolved, as shown in Figs. 6 and 7, to permit signal assignments.

Comparison of chemical shifts of proteins containing 4-Fe centers with those of tetramer dianions having —CH_2S groups (Table V) confirms the assignment in (i) for these proteins. The spectrum of $[Fe_4S_4(S\text{-}L\text{-}Cys\text{-}(Ac)NHMe)_4]^{2-}$ (Fig. 6) clearly reveals the position of α-H resonances as well as that of β-H; in the proteins signals of the latter protons would be expected to be obscured by diamagnetic peptide absorptions. The very broad low-field resonance found for $[Fe_2S_2(S_2\text{-}o\text{-}xyl)_2]^{2-}$ substantiates the most recent assignment of α-H resonances in spinach Fd_{ox} (Salmeen and Palmer, 1972; Palmer, 1973), which occur as a single broad feature with a similar chemical shift (Table V). These results require revision of the α-H assignment to the feature observed at about -15 ppm for several plant-type Fd_{ox} proteins (Poe et al., 1971a; Glickson et al., 1971).

In paramagnetic molecules, observed resonances are generally shifted from their normal diamagnetic positions by isotropic magnetic interactions of the contact and/or dipolar (pseudocontact) type(s) (La Mar et al., 1973b). Thus

$$(\Delta H/H_0)^{iso} = (\Delta H/H_0)^{obs} - (\Delta H/H_0)^{dia} = (\Delta H/H_0)^{con} + (\Delta H/H_0)^{dip}$$

In testing conclusion (ii), it has been found that in the spectra of $[Fe_4S_4(SR)_4]^{2-}$ with R = n-alkyl, isotropic shifts attenuate down the carbon chain from —CH_2S and all shifts are to low field. With R = Ph, isotropic shifts of ring protons alternate in sign, being positive for o-H and p-H and negative for m-H. In the R = p-tolyl complex (Fig. 7), o-H and m-H shifts are virtually unchanged, but the p-CH_3 isotropic shift is negative and thus opposite in sign compared with p-H. These and related results (Holm et al., 1974a) satisfy qualitative diagnostic criteria for proton contact interactions (Horrocks, 1973). If tetramer isotropic shifts are wholly contact in origin, a corollary to conclusion (ii) is that their values should be proportional to the paramagnetic susceptibility.

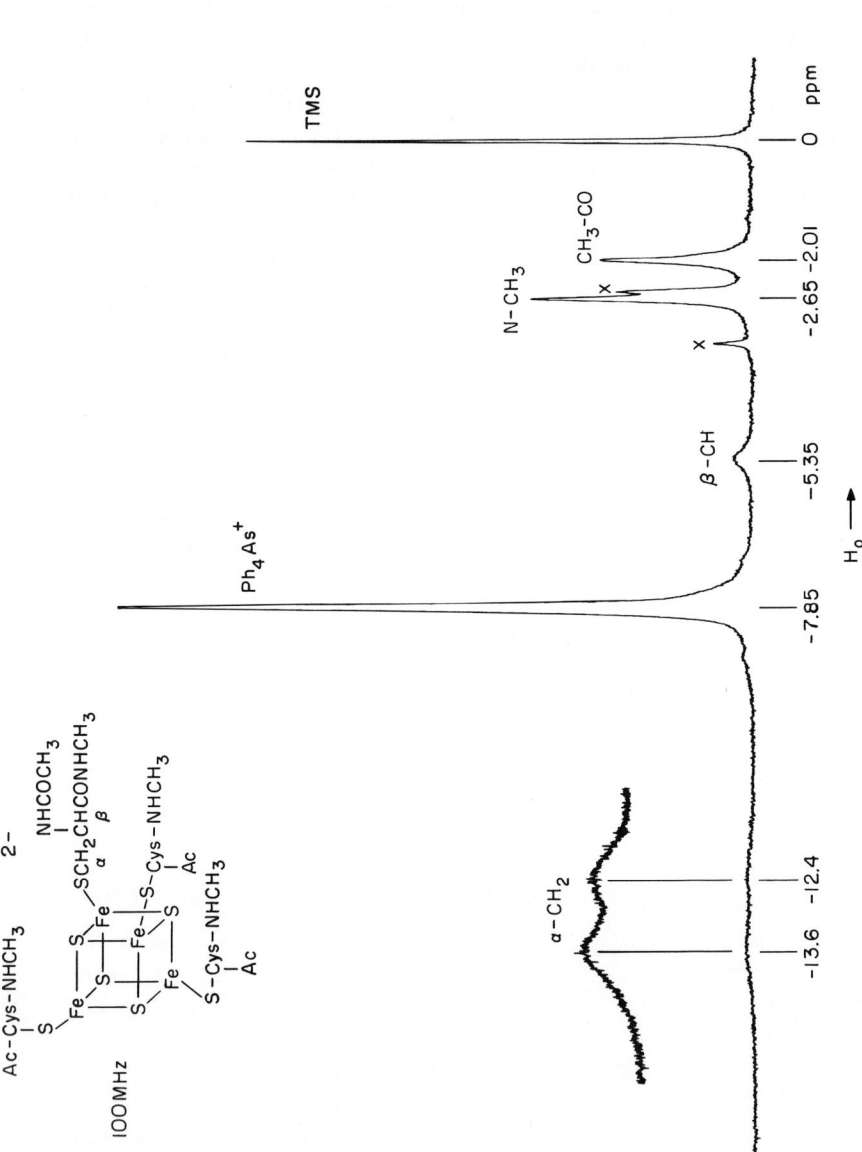

Fig. 6. PMR spectrum of $(Ph_4As)_2[Fe_4S_4(S\text{-}L\text{-}Cys(Ac)NHMe)_4]$ in DMSO-d_6 solution at 31°C. *Insert*: Methylene region recorded at higher gain. Reproduced with permission from Que *et al.* (1974b). *J. Am. Chem. Soc.* **96**, 6042. Copyright by the American Chemical Society.

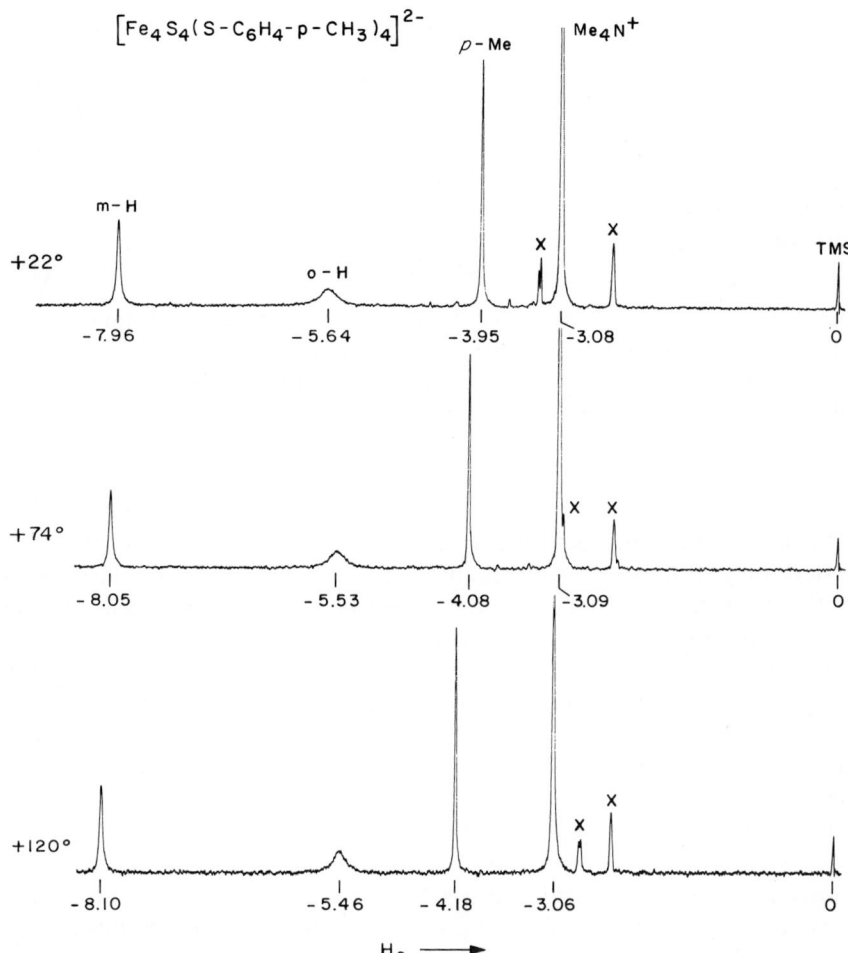

Fig. 7. PMR spectra (220 MHz) of $(Me_4N)_4[Fe_4S_4(S\text{-}p\text{-tol})_4]$ in DMSO-d_6 illustrating temperature dependent contact shifts (ppm) of o-H, m-H, and p-CH$_3$. Reproduced with permission from Holm et al. (1974a). J. Am. Chem. Soc. **96**, 2109. Copyright by the American Chemical Society.

In this case the net contact shift is given by Eq. (12) (La Mar et al., 1973a; Holm et al., 1974a)

$$\left(\frac{\Delta H}{H_0}\right)^{con} = \frac{P}{T} \frac{\Sigma_i A_i q_i S_i'(S_i' + 1) \exp(-E_i/kT)}{\Sigma_i q_i \exp(-E_i/kT)} \quad (12)$$

with $P = -g\beta/12\gamma_H \hbar k$. The S_i' are the various spin levels, $S' = 0, 1, 2, \ldots$ with energies E_i under the antiferromagnetic spin Hamiltonian,

$$\mathcal{H} = -2\Sigma J_{ij}\mathbf{S}_i \cdot \mathbf{S}_j$$

and the q_i are degeneracy factors appropriate to these levels. If the electron–nuclear hyperfine coupling constants A_i are independent of spin state, $A_i = A$ may be removed from the summation, and apart from a multiplicative constant, the quotient summed terms represent the temperature dependence of the magnetic susceptibility χ_{Fe}. Thus, at all temperatures,

$$\chi_{Fe} = R(\Delta H/H_0)^{con} \tag{13}$$

where the constant $(R = -g\beta\gamma_H\hbar/A)$ is independent of temperature. Contact shifts will then be directly proportional to magnetic susceptibility and their temperature dependencies will superimpose where scaled by R. Using solid state susceptibility data and CH_2 isotropic shifts in acetonitrile for $[Fe_4S_4(SCH_2Ph)_4]^{2-}$, it has been found that the two quantities have closely similar temperature dependencies in the interval (225°–295°K) common to both measurements (Holm et al., 1974a). The applicability of Eq. (13) indicates that A does not depend strongly on spin state or that only the lowest paramagnetic level ($S' = 1$) is populated and implies that the isotropic methylene shifts of $[Fe_4S_4(SCH_2Ph)_4]^{2-}$ are dominantly contact in nature. The similarity of magnetic and PMR properties between the proteins and this and other analogues having α-CH_2 groups suggests that Eq. (13) is likely to be appropriate to HP_{red} and Fd_{ox} as well, thereby supporting conclusion (ii) for the proteins. The same treatment has been successfully applied to the spectrum of $[Fe_2S_2(S_2\text{-}o\text{-xyl})_2]^{2-}$, and it has been shown that A_{CH_2} values for the analogue and spinach Fd_{ox} are virtually identical (Gillum et al., 1976a).

With $(\Delta H/H_0)^{iso} \cong (\Delta H/H_0)^{con}$ established, the signs of the shifts are readily interpreted. In arylthiolate tetramers ligand → metal (core) antiparallel spin transfer results in positive (parallel) spin on sulfur that is principally delocalized in the (approximately) orthogonal ring π-orbitals as indicated by VB structures **33–36**. [For a discussion of spin delocalization mechanisms, see La Mar (1973).] Negative spin density

33   34   35   36

arises at the meta positions through spin correlation effects. This mechanism results in the creation of net positive spin in the highest filled MO of the phenyl ring and imparts to the ligand the character of a π-radical.

From the McConnell equation $A_i = Q_{\rho C!}(Q_{CH^-}, Q_{CCH_3^+})$ and Eq. (12), the signs of shifts and coupling constants and the temperature dependencies of the former are accountable. With alkylthiolate substituents non-orthogonality of spin-containing sulfur and hydrogen 1s orbitals allows direct spin delocalization and leads to net parallel spin at the hydrogen nuclei and downfield shifts, as observed for 2-Fe and 4-Fe analogues.

Currently available 4-Fe analogues lack internally rigid R substituents, and consequently, the angular dependence of contact coupling constants predicted by $A_i = A_0 \cos^2 \theta$ is averaged to approximately or exactly zero. Conclusion (iii) for 4- and 8-Fe protein spectra follows from nonzero averaging of θ, the dihedral angle about the C_α-S bond axis. Multiple signal resolution of contact shifted resonances of the Fd_{ox} proteins in Table V undoubtedly arises in part from this effect, a point supported by the reduced resolution in denaturing solvents (McDonald et al., 1973). However, the α-protons H_a and H_b in **32** are intrinsically inequivalent (diastereotopic) owing to the adjacent chiral center. As may be seen in Fig. 6, this inequivalency is detectable in terms of a 1.2 ppm diastereotopic splitting. A similar effect (but of unknown magnitude) must obtain with the proteins even if θ values for the two α-H protons are equal.

3. ABSORPTION SPECTRA

Data obtained in the UV/visible region (300–800 nm) for proteins in aqueous solution and representative analogues in nonaqueous media are collected in Table VI. Not included are additional protein absorption bands detected by CD and MCD measurements, inasmuch as these techniques have not been applied to analogues. Spectra in this region identify the various n-Fe chromophores in most of the simpler proteins, a matter readily appreciated by inspection of typical spectra assembled by Palmer (1973) and Hall et al. (1974). Analogue spectra (Figs. 8–11) serve to establish the presence of chromophores electronically similar to those of the proteins and the oxidation level equivalencies in Eqs. (5–7). Thus, Rd_{ox} and $[Fe(S_2\text{-}o\text{-xyl})_2]^-$ (Fig. 8) both show two intense visible bands of comparable energies and intensities. Spectra of an algal 2-Fe Fd_{ox} protein and $[Fe_2S_2(S_2\text{-}o\text{-xyl})_2]^{2-}$ are directly compared in Fig. 9. Features of the latter at 338, 414, and 455 nm correlate well with the three protein bands, which are also observed in the spectra of adrenal Fd_{ox} in nonaqueous media (Kimura, 1971). The 590 nm band of the analogue is not readily detectable in protein solution spectra, but a corresponding feature is developed in low-temperature spectra (Palmer et al., 1967; Wilson, 1967; Kimura and Huang, 1970). Spectra of 4- and 8-Fe Fd_{ox} and HP_{red} above about 320 nm establish equivalent chromophores (Palmer, 1973). Counterparts of protein maxima at 388–400 nm are found at 413–421 nm

TABLE VI
UV–Visible Absorption Spectral Data for Proteins and Analogues

Species	Solvent	λ_{max}(nm)[ϵ_{mM}]	References[a]
Rd_{ox}	Water	~350 [sh]; 370–380 [10.9]; 490–497 [8.9]; 570 [sh]; 750 [360]	1–5
$[Fe(S_2\text{-}o\text{-}xyl)_2]^-$	DMF	354 [8.3]; 425 [sh]; 486 [sh]; 640 [1.6]; 684 [1.7]	6
Rd_{red}	Water	310–315 [10.8]; 333–335 [6.3]; 1600 [0.12]	3, 4
$[Fe(12\text{-peptide})]^{2-\,g}$	DMSO	1960 [0.11]	7
$[Fe(S_2\text{-}o\text{-}xyl)_2]^{2-}$	DMF, MeCN	322 [7.4], 355 [3.1], 1800 [0.12]	15
$[Fe_2(S_2\text{-}o\text{-}xyl)_3]^{2-}$	DMSO	360 [8.7]; 1930 [0.10]	6
$[Fe(SPh)_4]^{2-}$	—	1700 [0.098]	14
2-Fe Fd_{ox}	Water	325–333 [12–15]; 410–425 [9.0]; 455–470 [8.5–9.5]; 720 [0.8]	3, 5, 8
$[Fe_2S_2(S_2\text{-}o\text{-}xyl)_2]^{2-}$	DMF	294 [14.5]; 338 [16.2]; 414 [11.0]; ~455 [sh, 9.2]; 590 [4.8]	9
4, 8-Fe Fd_{ox}	Water	390–400 [15.3–16.8[b]]; 410[c] [80% DMSO]	5, 10
HP_{red}	Water	388 [15.3, 16.1]; 406[d] [80% DMSO]	11
$[Fe_4S_4(SEt)_4]^{2-}$	DMF	298 [23.3]; 420 [17.2]; 414[e] [15.7; 80% DMSO]	12
$[Fe_4S_4(S\text{-Cys}(Ac)NHMe)_4]^{2-}$	80% DMSO	294 [22.7]; 409 [16.6]; 413[f] [16.2; DMF]	13
$[Fe_4S_4(12\text{-peptide})]^{2-\,g}$	80% DMSO	290 [25.5]; 406 [17.0]	13

[a] *References*: (1) Lovenberg and Sobel, 1965; (2) Lovenberg and Williams, 1969; (3) Eaton et al., 1971; (4) Eaton and Lovenberg 1971, 1973; (5) numerous literature values; (6) Lane et al., 1975; (7) Anglin and Davison, 1975a; (8) Palmer et al., 1967; Rawlings et al., 1974; (9) Mayerle et al., 1975; (10) Hong and Rabinowitz, 1970a; Stombaugh et al., 1973; (11) Dus et al., 1967; (12) DePamphilis et al., 1974; (13) Que et al., 1974b; (14) Holah and Coucouvanis, 1975; (15) Lane et al., 1976.
[b] Per 4-Fe unit.
[c] Que et al., 1975.
[d] Cammack, 1973.
[e] Que et al., 1974b.
[f] DePamphilis et al., 1974.
[g] 12-Peptide = **12** (Table I).

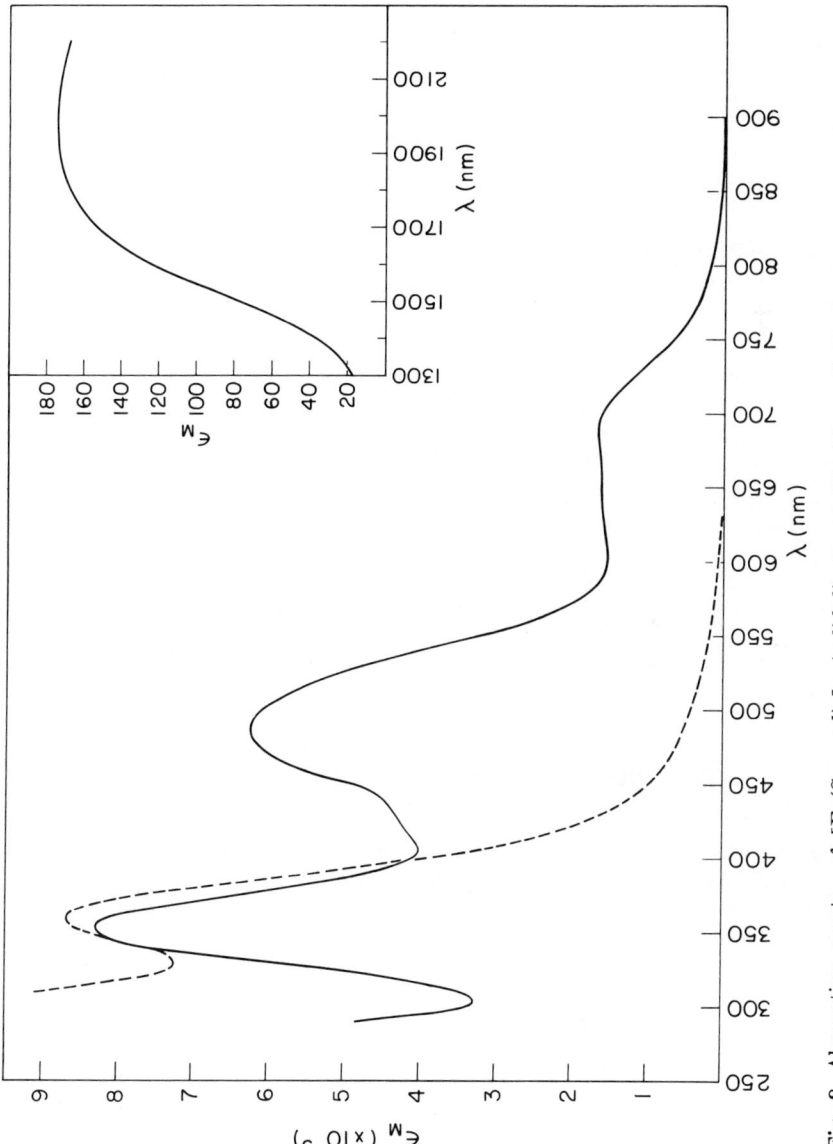

Fig. 8. Absorption spectra of [Fe(S$_2$-o-xyl)$_2$]$^-$ (solid line) and [Fe$_2$(S$_2$-o-xyl)$_3$]$^{2-}$ (broken line) in DMF solution. The insert shows the near-infrared spectrum of the latter complex in DMSO solution. From Lane et al. (1975).

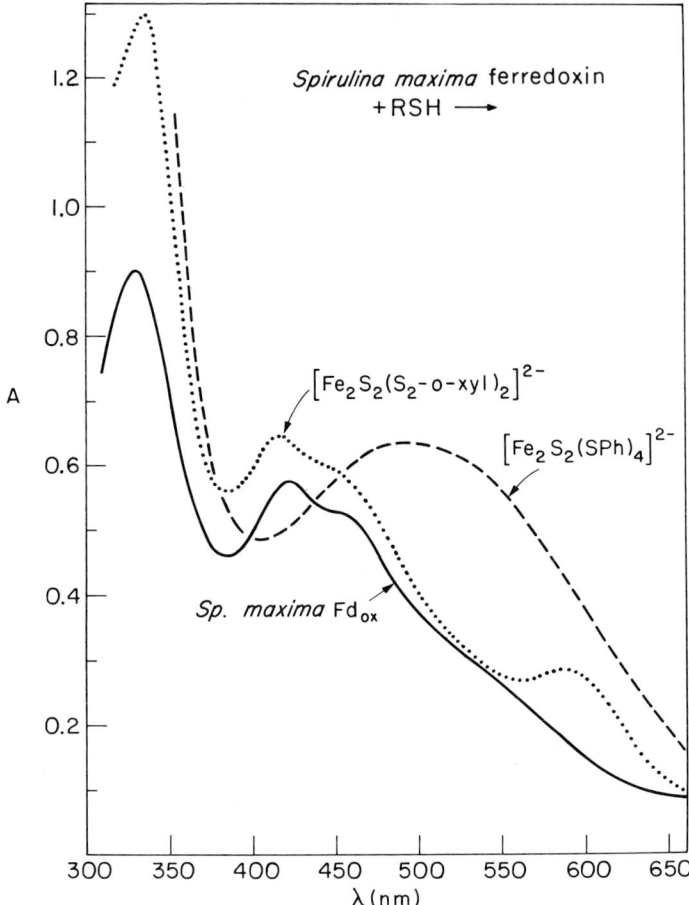

Fig. 9. Spectral comparison of *Spirulina maxima* Fd$_{ox}$, [Fe$_2$S$_2$(S$_2$-*o*-xyl)$_2$]$^{2-}$, and [Fe$_2$S$_2$(SPh)$_4$]$^{2-}$ in 80% DMSO–H$_2$O. The two complexes were obtained from the protein by extrusion reactions (see Section IV,B). Reproduced with permission from Que *et al.* (1975). *J. Am. Chem. Soc.* **97,** 463. Copyright by the American Chemical Society.

in the nonaqueous spectra of cysteinyl and alkylthiolate tetramer dianions (Fig. 10). The red shifts of analogue bands compared with those of proteins are diminished in 80% DMSO/H$_2$O. Indeed, in this denaturing medium, cysteinyl tetramers, HP$_{red}$, and Fd$_{ox}$ all have absorption maxima in the narrow 400–410 nm interval, indicating that disruption of tertiary structure renders protein and analogue chromophores nearly identical. Low-temperature (5°K) spectra of [Fe$_4$S$_4$(SR)$_4$]$^{2-}$, R = Et, CH$_2$Ph, re-

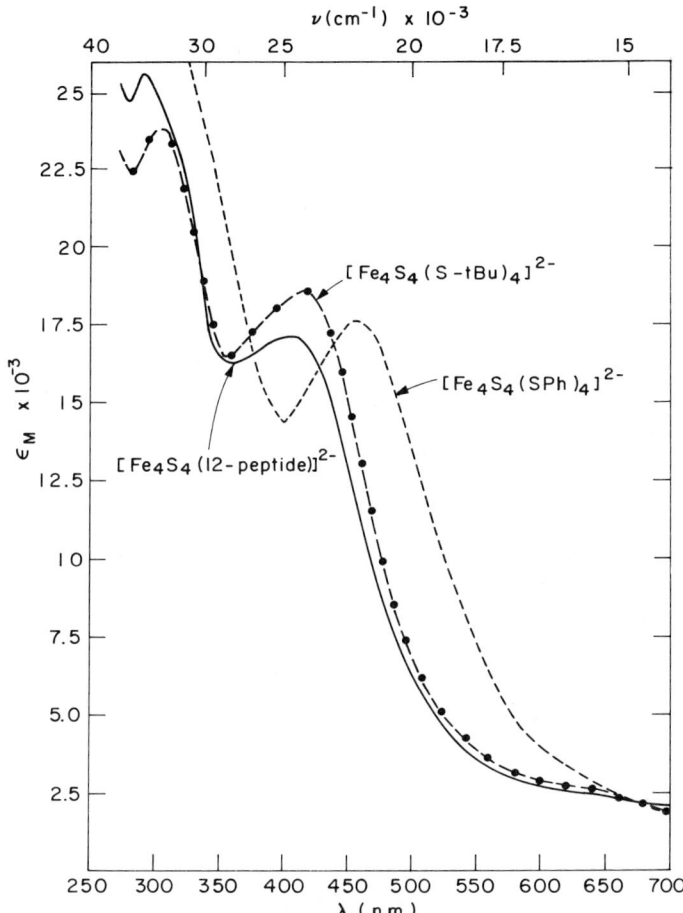

Fig. 10. Illustration of electronic spectral changes in the sequential ligand substitution reactions $[Fe_4S_4(S\text{-}t\text{-}Bu)_4]^{2-} \rightarrow [Fe_4S_4(12\text{-peptide})]^{2-} \rightarrow [Fe_4S_4(SPh)_4]^{2-}$ in DMSO solution (see Section IV,B). Reproduced with permission from Que et al. (1974b). *J. Am. Chem. Soc.* **96**, 6042. Copyright by the American Chemical Society.

veal additional features at ~500, ~600, and ~780 nm (Holm et al., 1974b), which appear to be present in the spectrum of HP_{red} measured at 77°K (Cerdonio et al., 1974). However, the latter displays a broad low-energy band at 1040 nm that appears to have no counterpart in the analogues, suggesting an electronic difference of as yet unspecified origin.

Spectra of Rd_{red} and all Fd_{red} proteins are characterized by substantial diminution of absorption in the visible region and consequent "bleaching" of color. When tetramer trianions are generated by chemical or elec-

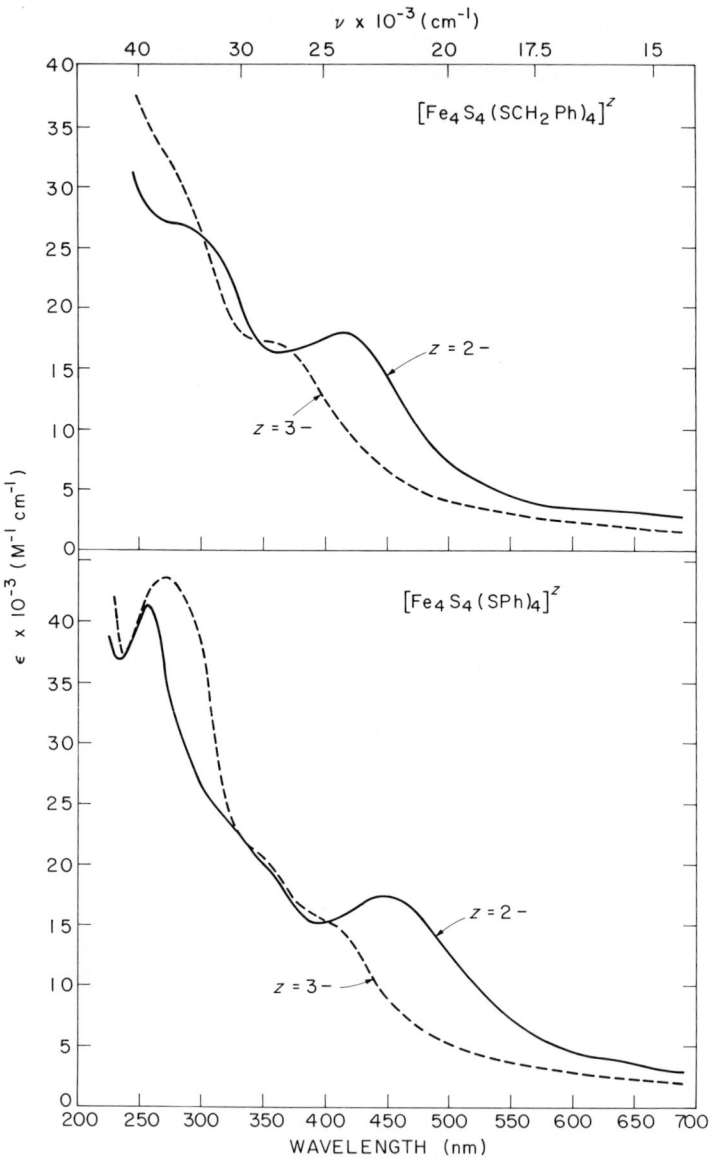

Fig. 11. Absorption spectra of dianions and electrochemically generated trianions of R = CH$_2$Ph and Ph tetramers in acetonitrile solution. From Frankel et al. (1974b).

trochemical reduction of dianions, their spectra show just this effect. In the two cases shown in Fig. 11, spectra in the visible revert to a nearly featureless form with an ~50% intensity reduction. The spectral changes concomitant with the one-electron reduction of $[Fe_4S_4(SCH_2Ph)_4]^{2-}$ are very similar to those observed upon dithionite reduction of *C. acidi-urici* Fd_{ox} (Mayhew *et al.*, 1969), except for the red shifts noted above, and support the equivalence $Fd_{red} \equiv [Fe_4S_4(SR)_4]^{3-}$.

Spectra of both analogues and proteins in the 300–800 nm region are dominated by relatively intense absorptions, most or all of which appear to arise from $S \to Fe$ charge-transfer excitations. As such, they do not readily convey information relevant to coordination sphere microsymmetries or the energetics of d-orbital splittings. Of the available protein–analogue chromophores, only the unit $Fe(II)-S_4$ is useful in this respect. In idealized T_d symmetry, a single spin-allowed, ligand-field transition $^5E(e^3t_2^3) \to {}^5T_2(e^2t_2^4)$ will arise whose energy is by definition the ligand-field splitting parameter Δ_t. In Rd_{red} a weak absorption is observed centered near 1600 nm (6250 cm^{-1}), a detailed analysis of which suggests the existence of three bands in this region (3700, 6000, and 7400 cm^{-1}) associated with transitions to the components of the $^5T_2$ state, whose orbital degeneracy is lifted by the distorted active-site structure (Eaton and Lovenberg, 1973). For the protein, $\Delta_t \sim 5000$ cm^{-1}, somewhat larger than values of 2800–3500 cm^{-1} (Low and Weger, 1960; Pappalardo and Dietz, 1961; Slack *et al.*, 1966) found for Fe(II) in tetrahedral sulfide lattice sites. Similarly, $\Delta_t \sim 3500$–3800 cm^{-1} for $Fe[(SPMe_2)_2N]_2$ (**18**) and $Fe[(SPPh_2)_2CH]_2$ (Davison and Switkes, 1971; Davison and Reger, 1971). More realistic chromophore models are provided by $[Fe(S_2\text{-}o\text{-}xyl)_2]^{2-}$ and $[Fe(12\text{-peptide})]^{2-}$ (Table VI) for which $\Delta_t \sim 5100$ to 5600 cm^{-1}, thereby tending to confirm the spectral assignment of Rd_{red} and also 2-Fd_{red} proteins, which exhibit similar weak absorption(s) in the near infrared (Eaton *et al.*, 1971). Although the ligand-field spectra of the synthetic complexes are broad and show actual or incipient band splitting, none suggests a total $^5T_{2g}$ term splitting as large as that in Rd_{red} (~4000 cm^{-1}, Eaton and Lovenberg, 1973). It appears, therefore, that Rd_{red}, as Rd_{ox}, probably has a distorted $[Fe(S\text{-Cys})_4]$ chromophore but one in which the "average" ligand-field strength is comparable with that generated by thiolate sulfur in synthetic species of apparently more regular coordination stereochemistry.

4. RESONANCE RAMAN SPECTRA

The presence of $S \to Fe$ charge-transfer absorption throughout the visible and near uv spectral regions of both analogues and proteins should selectively enhance the intensities of Fe–S vibrations when examined by resonance Raman spectroscopy. Although measurements of the vibrational properties of iron–sulfur proteins are not extensive, the available information indicates that Fe–S stretching motions absorb in the 400–250 cm^{-1} region. The resonance Raman spectra of Rd$_{ox}$ reveals bands at 365 and 311 cm^{-1} (Long and Loehr, 1970; Long et al., 1971). Similar measurements of adrenal Fd$_{ox}$ yield bands at 397, 350, and 297 cm^{-1} (Tang et al., 1973). Only the 350 cm^{-1} feature, also detectable in the infrared spectrum (Kimura and Huang, 1970), is unshifted upon S*/Se* replacement, suggesting its assignment as an Fe–S(Cys) stretch. Frequency tabulations given by Adams (1967) indicate that Fe–S vibrations should occur in this region. Recent work by Tang et al. (1975) has afforded the spectra of the biological and synthetic 4-Fe clusters shown in Fig. 12. All labeled peaks are polarized and arise from totally symmetric vibrations, which for idealized tetrahedral Fe$_4$S$_4$*S$_4$ cluster symmetry correspond to the breathing modes of (i) the Fe$_4$S$_4$* core (275, 250 cm^{-1}) and (ii) the terminal thiolate ligands (332–365 cm^{-1}). The indicated assignments of the three spectra are tentative, and that for (ii) is made primarily by analogy to adrenal Fd$_{ox}$. (The band expected at ~250 cm^{-1} for Fd$_{ox}$ is obscured by the fluorescent background.) While the frequency differences between [Fe$_4$S$_4$(SCH$_2$Ph)$_4$]$^{2-}$ and the proteins are appreciable, a more important difference seems to be the appearance of two polarized bands in the region assigned to (ii) in protein spectra. These bands presumably reflect a symmetry lowering of the protein clusters compared with that of the analogue, all species being compared

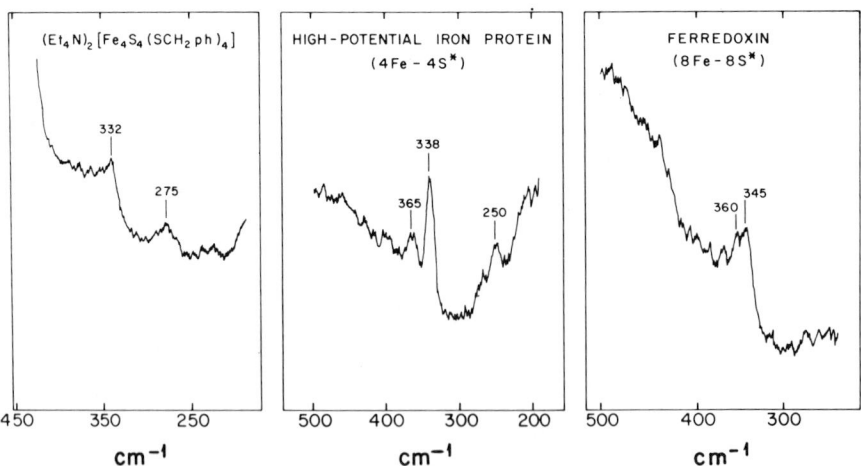

Fig. 12. Resonance Raman spectra of [Fe$_4$S$_4$(SCH$_2$Ph)$_4$]$^{2-}$ in acetonitrile solution and *Chromatium* HP$_{red}$ and *C. pasteurianum* Fd$_{ox}$ in aqueous solution (pH 7–8). From Tang et al. (1975).

in the solution state. The character and extent of the proposed symmetry lowering (Tang et al., 1975) cannot be determined from the present data nor are they evident from solid state structural information (Section II,B).

5. EPR SPECTRA

A highly characteristic feature of all 2-, 4-, and 8-Fe Fd_{red} proteins is the appearance of "$g = 1.94$" EPR spectra observable in frozen solutions at low temperatures (Orme-Johnson, 1973; Orme-Johnson and Sands, 1973; Hall et al., 1973a,b, 1974; Blumberg and Peisach, 1974). EPR spectra of several synthetic Fe–S complexes have shown a signal near $g \sim 1.94$ (Clare et al., 1970; Cotton and Gibson, 1971; Vergamini and Kubas, 1976). However, except for $[C_5H_5FeS(SEt)]_2^+$ (Vergamini and Kubas, 1976), these spectra have two g-values above 2, thereby not matching the anisotropy of protein spectra, and in some cases arise from mononuclear five- and six-coordinate complexes.

EPR spectra of 4-Fe Fd_{red} proteins exhibit rhombic symmetry with a signal at $g \sim 1.94$, $g_{av} < 2$, and highly temperature dependent signal intensities (Hall et al., 1974; Stombaugh et al., 1973; Zubieta et al., 1973). For *Bacillus polymyxa* Fd_{red}, $g = 1.88$, 1.92, 2.06 and $g_{av} = 1.95$, with signals rapidly becoming unobservable above 25°K (Stombaugh et al., 1973) owing to rapid electron spin relaxation. In contrast, spectra of 2-Fe Fd_{red} proteins are readily detectable at 77°K. Spectra of the reduction product of $[Fe_4S_4(SCH_2Ph)_4]^{2-}$, generated by controlled potential electrolysis (Fig. 11) or chemical reduction in nonaqueous media, reveal an axial spectrum with $g \sim 1.92$, 2.04 and $g_{av} = 1.96$, and unobservable signals above about 55°K (Frankel et al., 1974b). Both axial and rhombic spectra are found for the $[Fe_4S_4(SPh)_4]^{2-}$ reduction product depending upon the solvent and method of generation, typical values being 1.92, 2.06 and 1.86, 1.94, 2.06; $g_{av} = 1.95$ (Frankel et al., 1974b). The factors affording axial versus rhombic spectra of the analogues have not been thoroughly investigated. However, at least one protein center produces an axial spectrum; for HP_{s-red} (18°K, 80% DMSO) $g = 1.93$, 2.04; $g_{av} = 1.95$ (Cammack, 1973). The analogue spectra are associated with the doublet-state trianions $[Fe_4S_4(SR)_4]^{3-}$, whose EPR properties are obviously similar to those of the reduced proteins and offer further confirmation of their electronic equivalence with 4-Fe Fd_{red} and HP_{s-red} active centers. It is also clear that this equivalence extends to 8-Fe Fd_{red} centers, whose EPR spectra are, however, more complex owing to weak coupling between the two-spin doublet clusters (R. Mathews et al., 1974; Gersonde et al., 1974). The EPR fine structure from intramolecular magnetic interactions may be removed in denaturing solvents (Cammack,

1975), in which spectra of such proteins more closely resemble those of the analogues.

EPR spectra of Rd_{ox} proteins differ from those considered above in that no $g \sim 1.94$ signal is observed. The two most prominent spectral features are absorptions at $g \sim 4.3$ and 9.4. Such spectra, of which that of *P. oleovorans* Rd_{ox} has been examined in the most detail (Peisach *et al.*, 1971), are interpretable in terms of high-spin Fe(III) in a rhombically distorted ligand field (Orme-Johnson and Sands, 1973). The spectrum of $[Fe(S_2\text{-}o\text{-xyl})_2]^-$ at and below 77°K is complex (Lane *et al.*, 1975) and has not yet been satisfactorily interpreted. A strong $g = 4.3$ signal is observed, indicating that this complex does not assume a symmetry higher than rhombic in frozen solutions.

6. Mössbauer Spectra

The technique of Mössbauer spectroscopy has been applied extensively to the iron–sulfur proteins. Many of the experimental results, conclusions, and relevant theory, including the companion use of ENDOR, have been presented elsewhere (Dunham *et al.*, 1971a; Bearden and Dunham, 1973; Sands and Dunham, 1975). A model has also been elaborated to account for the zero field and magnetically perturbed spectra of 4-Fe centers (Eicher *et al.*, 1974). Further examination of the Mössbauer spectra of iron-sulfur centers is given in Chapter 10. Isomer shift (δ) and quadrupole splitting (ΔE_Q) data are summarized in Table VII for analogues and proteins. Both of these parameters are slightly temperature dependent, and not all isomer shifts were obtained with source and absorber at the same temperature. To facilitate internal comparison of δ values of proteins, data at 77°K summarized by Thompson *et al.* (1974) have been quoted (italicized values). Oxidation states are mean formal values from series (5)–(7).

All Mössbauer data are consistent with high-spin, distorted-tetrahedral $Fe\text{-}S_4$ units such as are found in numerous iron-sulfur compounds, minerals, and phases. Data recently obtained by Reiff *et al.* (1975), which refer to 2+, 3+, and intermediate oxidation states, are particularly useful for comparison. Isomer shifts show the expected trend of increasing values with decreasing formal oxidation state and approach $\delta \sim 0.6$ mm/second proposed as diagnostic of tetrahedral $Fe(II)\text{-}S_4$ (Reiff *et al.*, 1975). The substantial quadrupole splitting of $[Fe(S_2\text{-}o\text{-xyl})_2]^-$ in the solid state is retained in frozen solutions (Lane *et al.*, 1975), showing that upon release from a crystalline environment this complex does not adopt an effectively cubic $Fe\text{-}S_4$ structure. The internal hyperfine field at Fe in $[Fe(S_2\text{-}o\text{-xyl})_2]^-$ is −380 kOe, quite close to that (−370 kOe) in Rd_{ox} at low temperature (Phillips *et al.*, 1970a; Rao *et al.*, 1972) and sug-

TABLE VII
SELECTED MÖSSBAUER DATA FOR ANALOGUES AND PROTEINS

Species	Fe oxidation state	δ (mm/second)[a,b]	ΔE_Q (mm/second)[a]	Reference[c]
$[Fe(S_2\text{-}o\text{-xyl})_2]^-$	+3	0.13	0.57	1
Rd_{ox}	+3	*0.25*	0.74(?)	2, 9
$[Fe(S_2\text{-}o\text{-xyl})_2]^{2-}$	+2	0.61	3.28	12
Rd_{red}	+2	0.58, *0.65*	3.1–3.2	2
$[Fe_2S_2(S_2\text{-}o\text{-xyl})_2]^{2-}$	+3	0.17	0.36	3
2-Fe Fd_{ox}	+3	0.18–0.30 ⎫ *0.22*	0.60–0.66	4, 5
2-Fe Fd_{red}	+3	0.22–0.29 ⎭	0.60–0.80	4, 5
	+2	0.54–0.60, *0.56*	2.7–3.0	4, 5
HP_{ox}	+2.75	*0.32*	0.77	6
$[Fe_4S_4(SCH_2Ph)_4]^{2-}$	+2.5	*0.34*	1.15	7, 8
$[Fe_4S_4(SEt)_4]^{2-}$	+2.5	0.34	1.10	8
HP_{red}	+2.5	*0.42*	1.01	6
8-Fe Fd_{ox}	+2.5	*0.43*	0.75	9
$[Fe_4S_4(SCH_4Ph)_4]^{3-}$	+2.25	*0.46*	1.15	11
$HP_{s\text{-red}}$	+2.25	*0.59*	1.28	10
8-Fe Fd_{red}	+2.25	*0.57*	1.07	9

[a] Refers to 77°–300°K; italicized values were obtained at 77°K.
[b] Relative to Fe metal.
[c] *References*: (1) Lane *et al.*, 1975; (2) Phillips *et al.*, 1970a; Rao *et al.*, 1972; (3) Mayerle *et al.*, 1973; (4) Rao *et al.*, 1971; Cammack *et al.*, 1971; (5) Dunham *et al.* 1971a; Münck *et al.*, 1972; (6) Dickson *et al.*, 1974; (7) Holm *et al.*, 1974b; (8) Frankel *et al.*, 1974a; (9) Thompson *et al.*, 1974; see also Gersonde *et al.*, 1974; (10) Dickson and Cammack, 1974; (11) Frankel *et al.*, 1974b; (12) Lane *et al.*, 1976.

gestive of similar bond covalency effects. An equivalent comment applies to $[Fe(S_2\text{-}o\text{-xyl})_2]^{2-}$ [−150 kOe (Lane *et al.*, 1976)] and Rd_{red} [∼ −200 kOe (Rao *et al.*, 1972)].

Perhaps the greatest interest attends the Mössbauer properties of $[Fe_4S_4(SR)_4]^{2-,3-}$ for which formal oxidation state considerations imply different types of iron atoms (series 7). The spectra of $[Fe_4S_4(SCH_2Ph)_4]^{2-}$ in zero and applied magnetic fields at temperatures as low as 1.5°K (Fig. 13) reveal all iron sites to be strictly equivalent on the Mössbauer time scale. Hence, the lifetime τ of Fe(II,III) in the cluster must be $\lesssim 10^{-7}$ second. X-ray results together with other physical techniques spanning the τ range of about 10^{-4}–10^{-16} second also fail to detect distinct iron atoms (Holm *et al.*, 1974b). While linewidths in proteins appear to be about 0.1–0.2 mm/second larger than in the analogues, separate Fe(II,III) Mössbauer spectra are not seen in HP_{red} (Dickson *et al.*, 1974) and Fd_{ox} (Thompson *et al.*, 1974). Similarly,

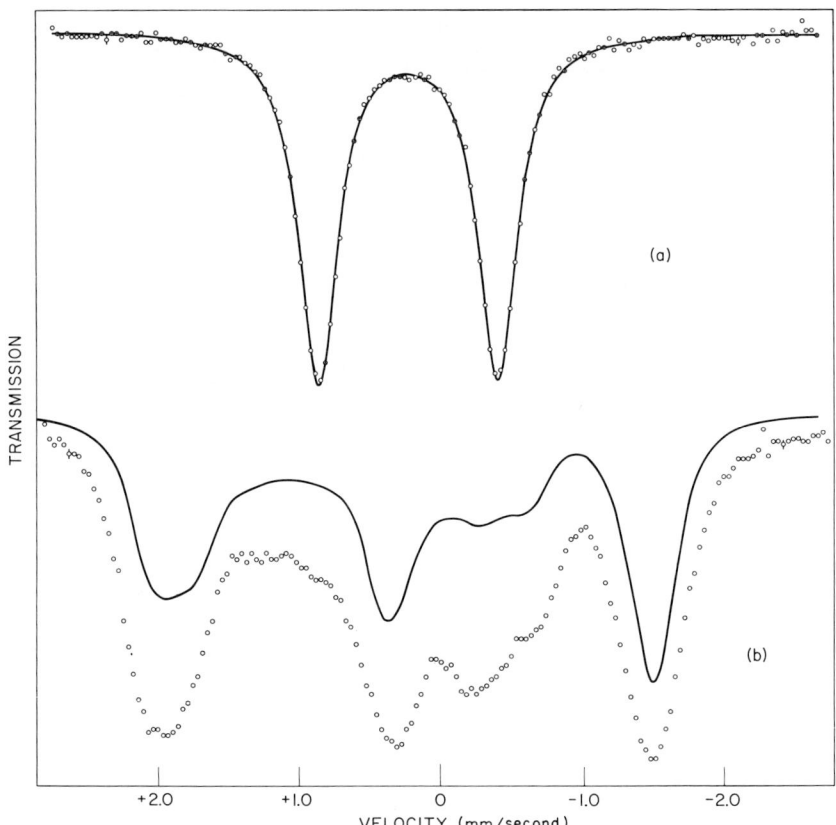

Fig. 13. Mössbauer spectra of $(Et_4N)_2[Fe_4S_4(SCH_2Ph)_4]$ at $1.5°K$, $H_0 = 0$; (b) $4.2°K$, $H_0 = 80$ kOe applied longitudinally. Solid lines in (a) and (b) are, respectively, theoretical least-squares fit to the data assuming Lorentzian line shape and theoretical computer-generated spectrum assuming $H_n = 80$ kOe, $\Delta E_Q = 1.26$ mm/second, $\eta = 0$, and the sign of the principal component of electric field gradient positive. Reproduced with permission from Holm et al. (1974b). *J. Am. Chem. Soc.* **96**, 2644. Copyright by the American Chemical Society.

Mössbauer spectra of several $[Fe_4S_4(SR)_4]^{3-}$ complexes generated in solution by reduction of dianions do not indicate localization of the electron added upon reduction, consistent with the findings for HP_{s-red} (Dickson and Cammack, 1974) and Fd_{red} (Thompson et al., 1974). The picture that emerges from these experiments is that of electronically delocalized clusters in both analogues and proteins rather than ones possessing trapped (integral) valence states. This situation stands in contrast to the 2-Fe Fd_{red} active sites for which both Mössbauer and ENDOR spectroscopy have clearly established the trapped valence Fe(II) +

Fe(III) situation (Sands and Dunham, 1975). At present, the tetranuclear clusters are the only biological active sites for which integral metal valence states are not adequate descriptions of oxidation levels.

7. Theoretical Electronic Structural Models

Based on iterative extended Hückel molecular orbital calculations performed for a variety of geometries, theoretical models have been described for oxidized and reduced Rd (Loew and Lo, 1974a,b; Loew et al., 1974a,b) and 2-Fe Fd sites (Loew and Steinberg, 1971, 1972). More recently, the SCF-Xα scattered wave method (Johnson, 1973, 1975), which has enjoyed considerable success in interpreting ground and excited state properties of metal-containing systems, has been applied to $[Fe(SR)_4]^-$ (Norman and Jaekels, 1975) and $[Fe_4S_4(SR)_4]^{2-}$ complexes (Slater and Johnson, 1974; Yang et al., 1975) with R = H, CH_3 as an initial approach to the development of electronic models for the Rd_{ox} and 4-Fe Fd_{ox}/HP_{red} sites, respectively. Space does not permit an examination of all of these models. Attention is directed to the tetranuclear dianion case, which, because of its delocalized nature and large number of orbital levels, is difficult to treat meaningfully in the absence of a calculated model.† Earlier considerations were based on a qualitative symmetry-factored MO model (Averill et al., 1973; Gall et al., 1974). Shown in Fig. 14a is the calculated energy level diagram for $[Fe_4S_4(SCH_3)_4]^{2-}$ in idealized T_d symmetry together with the dominant atomic contributions of each orbital. The bond distances employed are close to those in Fig. 3A,B and the actual $Fe_4S_4{}^*S_4$ cluster D_{2d} symmetry is considered a perturbation. The ground state electronic structure is $\ldots(4t_1)^6(10t_2)^4$, with the highest occupied orbital and those immediately above in energy predominantly tetrametal antibonding in character. The heavily tetrametal, lower energy $8t_2$ and $3e$ orbitals have net Fe–Fe bonding character, indicating that Fe atoms separated by about 2.7 Å generate bonding interactions stabilizing the core. All Fe–S and Fe–S* interactions have pronounced covalent character.

The model embodied in Fig. 14a provides a preliminary interpretation of certain electronic and structural features of alkylthiolate tetramer dianions and the isoelectronic sites of HP_{red} and Fd_{ox} (Yang et al., 1975). Analogue electronic spectra are dominated by intense features near 295 nm (4.20 eV) and 418 nm (2.97 eV), additional incompletely resolved bands at intermediate energies, and a low intensity shoulder at ~650 nm (1.91 eV) (Table VI in DePamphilis et al., 1974). The intense peaks and intermediate absorptions are assigned to S → Fe charge-transfer excita-

† An alternative theoretical treatment based on the extended Hückel molecular orbital model has recently been published in summary form (Thomson, 1975).

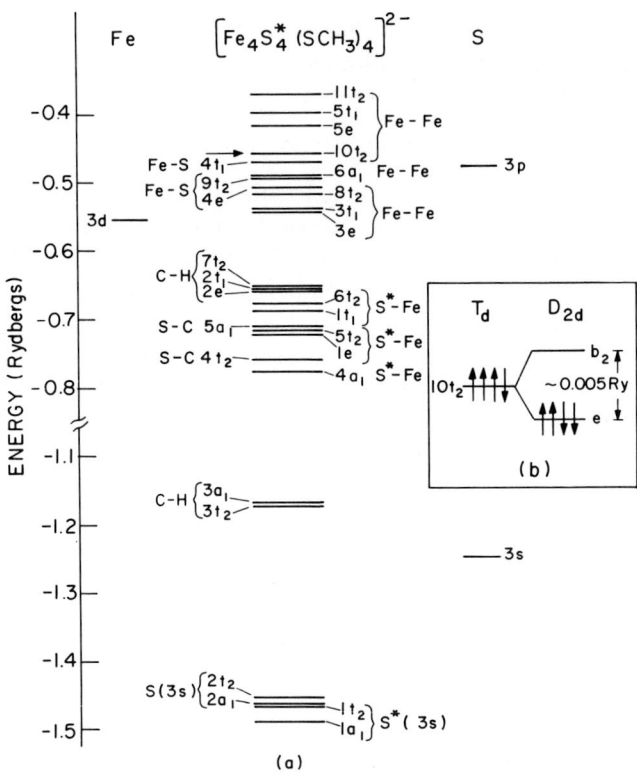

Fig. 14. (a) SCF-Xα-SW orbital energies of [Fe$_4$S$_4$(SCH$_3$)$_4$]$^{2-}$ in T$_d$ symmetry, and (b) splitting of the highest occupied orbital (10t$_2$) in D$_{2d}$ symmetry. Reproduced with permission from Yang *et al.* (1975). *J. Am. Chem. Soc.* **97**, 6595. Copyright by the American Chemical Society.

tions from the band of predominantly sulfurlike levels (—0.68 to —0.78 Ry) to the partially occupied 10t$_2$ tetrametal level. The top (6t$_2$) and bottom (4a$_1$) levels of this sulfur band are separated from 10t$_2$ by 2.98 and 4.32 eV, in excellent argeement with the energies of the principal features. The weaker shoulder is assigned to "d-d" transitions between occupied and unoccupied tetrametal orbitals.

The orbital degeneracy of the partially occupied 10t$_2$ level allows in principle Jahn–Teller distortion of a perfectly cubic cluster to a lower symmetry and is regarded as the probable cause of the analogue D$_{2d}$ structure. Under this symmetry, the 10t$_2$ orbital splits into e and b$_2$. The energy order in Fig. 14b results in zero net spin at low temperatures and paramagnetism at higher temperatures by excitation of electrons across the small e → b$_2$ energy gap, consistent with the magnetic properties of proteins and analogues (Table V). This model further suggests that the species $[Fe_4S_4(SR)_4]^{1-} \equiv HP_{ox}$ would have the $(10t_2)^3$ configuration and, in the simplest approximation, not be subject to Jahn–Teller distortions. The HP_{ox} site appears to be marginally more symmetric than the HP_{red} site, and the core volume contraction in the process $HP_{red} \to HP_{ox}$ is, as suggested (Carter et al., 1974b), partially accountable for in terms of electron removal from an Fe-Fe antibonding orbital (the $10t_2$ orbital or its components in lower symmetry). Isolation and characterization of a tetranuclear monoanion is a matter of increasing importance for several reasons. Among known types of Fe–S sites (uncoupled from other paramagnetic centers), only this oxidation level appears capable of sustaining an $S = \frac{3}{2}$ ground state. The presence of a cluster of iron atoms with this net electronic spin has been suggested from Mössbauer and EPR studies of the FeMo protein of *Azotobacter vinelandii* nitrogenase (Münck et al., 1975). If this magnetic center is of the $Fe_4S_4^*$ type, and not due to molybdenum, the theoretical model suggests that it possesses effective T_d symmetry. In HP_{ox} it now seems clear that an $S = \frac{1}{2}$ ground state exists and that not all iron atoms are equivalent (Antanaitis and Moss, 1975; Evans et al., 1970; Dickson et al., 1974; Phillips et al., 1970b; Anderson et al., 1975a). This raises the question whether a structurally unconstrained synthetic analogue $[Fe_4S_4(SR)_4]^{1-}$ will inherently adopt a cubic spin-quartet configuration or some lower symmetry, incompletely delocalized arrangement in which the imbalance of formal metal oxidation states [series (7)] will be manifested to a physically observable degree. The theoretical model further suggests that the species $[Fe_4S_4(SR)_4]^{3-} \equiv Fd_{red}$ with the $(10t_2)^5$ configuration would be expected to distort, with a possible slight elongation of Fe-Fe distances; no structural information is as yet available for any analogue or protein in this oxidation level. In this case, however, an $S = \frac{1}{2}$ ground state is probable regardless of the extent of distortion from T_d symmetry. While it is well recognized that Jahn–Teller distortions are not always found (or are not large enough to detect) in metal complexes with orbitally degenerate ground states, the work of Dahl and co-workers has produced a good correlation between the possible operation of the Jahn–Teller effect as deduced from a qualitative MO model and the M_4A_4 core symmetry in a variety of organometallic species (e.g., Gall et al., 1974; Simon and Dahl, 1973; Trinh–Toan et al., 1972).

8. Nonanalogue Complexes

The structurally defined coordination units in Table II possess one or more of the following features: presence of decidedly nonphysiological ligands, and in particular, the absence of terminal —CH_2S ligands potentially similar to cysteinyl residues; low formal metal oxidation states;

room temperature diamagnetism; absence of sulfide (S*) bridging components; and electronic spectra unrelated to those of the proteins. Thus, while many of these complexes are of much interest in their own right, they do not satisfy the requirements (1) and (2) for active site analogues given in Section I,A. Several additional observations are in order. Although derived from abiological ligands, $Fe[(SPMe_2)_2N]_2$ and $Fe[(SPPh_2)_2CH]_2$ (Section III,B,3) are useful approximate models for electronic properties of the tetrahedral $Fe(II)$-S_4 unit in proteins, and the former provides a measure of the Fe-S bond distance to high-spin $Fe(II)$. The same comment applies to $[Fe(SPh)_4]^{2-}$, which is only slightly less desirable as a Rd_{red} analogue than $[Fe(S_2$-o-xyl$)_2]^{2-}$ because of the presence of arylthiolate ligands. None of the first three complexes has been reported to be oxidizable to the $Fe(III)$ state. Other than $[Fe(Me_3AsS)_4](ClO_4)_2$ (Brodie et al., 1969), $(Ph_4P)_2[Fe(S_2C_4O_2)_2]$ (Coucouvanis et al., 1975), and, possibly, $Fe(S_2PF_2)_2$ (Tebbe and Muetterties, 1970), these are the only mononuclear tetrahedral $Fe(II)$-S_4 complexes yet isolated. The three tetranuclear complexes in Table II exhibit both similarities and differences compared to analogues and proteins. They contain the $Fe_4S_4{}^*$ core (30) with overall cubane-type stereochemistry and show electron-transfer behavior (Balch, 1969; Ferguson and Meyer, 1971; Gall et al., 1974). However, the detailed core structures of the three complexes and also of $[(C_5H_5)FeS]_4{}^{+,2+}$ (Gall et al., 1974) are not the same as those of proteins and analogues. It is readily seen, and has been shown using a qualitative MO model (Averill et al., 1973; Gall et al., 1974), that $[(C_5H_5)FeS]_4{}^{0,+,2+}$ and $[Fe_4S_4(NO)_4]$ are not isoelectronic with $[Fe_4S_4(SR)_4]^{3-,2-,1-}$ in terms of core and total numbers of valence electrons. This model has further been employed to interpret core structural differences. It is difficult to assign a core oxidation level in $[Fe_4S_4(tfd)_4]^{2-}$ owing to the ambiguity in oxidation state of the dithiolene ligand. In the present context, this is not a matter of importance, since Mössbauer studies have established the absence of a magnetic hyperfine field at the iron nuclei in each of the oxidation levels $[Fe_4S_4(tfd)_4]^{-2,1-,0}$ (Frankel et al., 1974c). Thus, the change in net charge is associated with predominant changes in ligand oxidation level, a situation not possible in analogues and proteins where redox events are restricted to the core.

IV. CHEMICAL REACTIVITY OF ANALOGUES

The simplest reaction of all analogues, electron transfer, has been described in Section III,A. Although the reaction chemistry of these complexes is in a developmental stage, certain reactivity properties of the 2-Fe and 4-Fe species have already emerged which allow manipulation of their $Fe_2S_2^*$ and $Fe_4S_4^*$ cores in chemically and biologically interesting and useful ways.

A. Thiolate Ligand Substitution

All analogues are paramagnetic and contain tetrahedrally coordinated metal centers, factors that in general promote lability of coordinated ligands. The premise that terminal thiolate ligands in bi- and tetranuclear analogue dianions would be substitution-labile has been demonstrated by PMR and spectrophotometric investigations (Bobrik et al., 1974; DePamphilis et al., 1974; Que et al., 1974a; Mayerle et al., 1975).

1. Types and Properties of Reactions

Reactions (14) and (15) occur upon the addition of an appropriate thiol to a solution of the initial complex in a nonaqueous solvent, such as acetonitrile or DMSO, at ambient temperature. Equilibrium is attained rapidly and the reactions at equilibrium may be conveniently monitored

$$[Fe_4S_4(SR)_4]^{2-} + n\ R'SH \rightleftharpoons [Fe_4S_4(SR)_{4-n}(SR')_n]^{2-} + n\ RSH \quad (14)$$

$$[Fe_2S_2(SR)_4]^{2-} + 4\ R'SH \rightleftharpoons [Fe_2S_2(SR')_4]^{2-} + 4\ RSH \quad (15)$$

by PMR, which allows detection of free thiol formed and the variously substituted tetramers owing to their differences in contact shifts, and by electronic spectra. The latter method is useful only when R = alkyl and R′ = aryl because of the red shifts of the prominent visible bands of the latter (Figs. 10 and 11). For the more thoroughly studied reaction (Eq. 14) under equilibrium conditions, the following characteristics have been established (Que et al., 1974a): (i) full substitution of one tetramer does not occur at the expense of mixed ligand tetramer formation, i.e., ratios of equilibrium constants for stepwise substitution approach statistical values; (ii) with R = t-Bu substitution tendencies of added thiols R′SH tend to parallel their acidities at least to $pK_a \lesssim 6.5$, leading to the ligand substitution series R′ = alkyl < Ac-L-Cys-NHMe $\lesssim$ aryl; (iii) addition of 4.5–6 equiv of aryl thiol results in complete formation of $[Fe_4S_4(SAr)_4]^{2-}$ from an alkylthiolate tetramer; (iv) because the structures of typical alkylthiolate (R = CH_2Ph) and arylthiolate (R = Ph) tetramers are nearly identical and evidence no intrinsic stability differences in the solid state, the position of Equilibrium (14) as expressed by properties (ii) and (iii) is assumed to be dominated by the acid–base characteristics of coordinated thiolate and R′SH; (v) no significant amount of core decomposition is observed under substitution conditions.

Properties (iii) and (v) extend to Reaction (15) also (Mayerle et al., 1975).

Reactions (14) and (15) and variants thereof are suitable for synthesis on a preparative scale. Arylthiolate tetramers and dimers can be obtained in good yield from $[Fe_4S_4(SR)_4]^{2-}$ (R = alkyl) and $[Fe_2S_2(S_2\text{-}o\text{-}xyl)_2]^{2-}$, respectively, in this manner (Que et al., 1974a; Mayerle et al., 1975). The first selenium-containing tetramer was synthesized by the substitution–redox reaction (16), which should be capable of extension to the preparation of other thio and seleno tetramers. Reaction (17) leads to

$$[Fe_4S_4(S\text{-}t\text{-}Bu)_4]^{2-} + 2\ PhSeSePh \rightarrow [Fe_4S_4(SePh)_4]^{2-} + 2\ t\text{-}BuSSBu\text{-}t \quad (16)$$

the replacement of thiolate with chloride and proceeds in good yield

$$[Fe_4S_4(SCH_2Ph)_4]^{2-} + 4\ PhCOCl \rightarrow [Fe_4S_4Cl_4]^{2-} + 4\ PhCH_2SCOPh \quad (17)$$

(Bobrik and Holm, 1975). Crystalline salts of the chloro complex have been isolated; bromide complexes may be obtained using benzoyl bromide. An x-ray structural study of $(Et_4N)_2[Fe_4S_4Cl_4]$ has verified retention of the cubane-type structure, whose core is virtually isometric with those of thiolate tetramer dianions in Fig. 3A and B (Bobrik and Holm, 1975). The fundamental core cation $[Fe_4S_4{}^*]^{2+}$ in solvated form should be accessible by solvolysis reactions of the labile halide species. Recently, in experiments employing the water-soluble complex $[Fe_4S_4(SCH_2CH_2CO_2)_4]^{6-}$, evidence has been presented for the multiple equili-

$$[Fe_4S_4(SR)_4]^{6-} + n\ L \rightleftharpoons [Fe_4S_4(SR)_{4-n}L_n]^{z-} + n\ (SR)^{2-} \quad (18)$$

bria (18) in basic aqueous media where apparently, $1 \leq n \leq 4$, and L, the lyate species, may be H_2O or OH^- (Job and Bruice, 1975). Further, the thiolate groups were found to undergo exchange with nucleophiles under these conditions, and a preliminary nucleophilicity order has been reported to be $Cl^- = Br^- < OH^- < CN^-$ (Job and Bruice, 1975). Processes (14–18) presage a rich reaction chemistry based on dimer and tetramer core structures, some of which, described in following sections, have definite biological utility.

2. Kinetics and Mechanism

The rates and mechanism of the initial thiolate substitution in reactions of R = Et and t-Bu tetramer dianions with arylthiols in acetonitrile solution have been investigated by stopped-flow spectrophotometry (Dukes and Holm, 1975). For Reaction (14) (R = t-Bu, $n = 1$), rates are overall second order, first order in tetramer and in thiol, and second order rate constants vary from about 2 to 3600 M^{-1} sec^{-1} at 25°C. The most interesting feature of the kinetic results is the dependence of rate

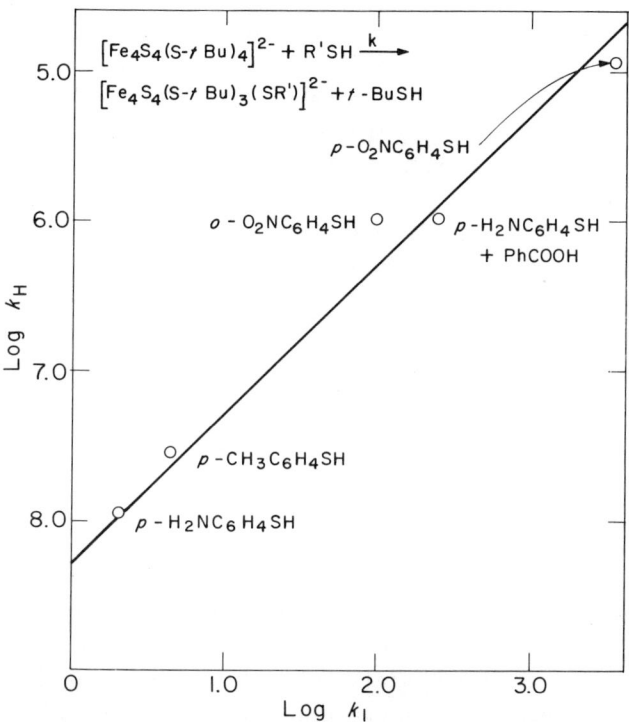

Fig. 15. Dependence of resolved second-order rate constants on thiol acidity (K_H) for the ligand substitution reaction between arythiols and $[Fe_4S_4(S\text{-}t\text{-}Bu)_4]^{2-}$ in acetonitrile solution. The solid line is constrained to a slope of unity. Reproduced with permission from Dukes and Holm (1975). *J. Am. Chem. Soc.* **97**, 528. Copyright by the American Chemical Society.

constants on thiol acidity shown in Fig. 15. This has led to the postulated mechanistic sequence (19). The slow protonation step, which is the simplest expression of the rate–acidity relation, is followed by fast separation of the alkylthiol from and rapid capture of the generated arylthiolate anion by the metal center. The rate law consistent with this scheme

$$\begin{array}{c}\text{S*}\\\text{S*—Fe—S̈}\\\text{S*}\quad\text{S}^{-}\!\!-\!\!\text{H}\\\quad\quad\text{R}'\end{array}\!\!\!\begin{array}{c}\text{R}\\\\\end{array}\longrightarrow\begin{array}{c}\text{S*}\\\text{S*—Fe}\overset{\oplus}{\underset{\ominus}{\rightleftarrows}}\text{S}\\\text{S*}\quad\text{S}\quad\text{H}\\\quad\text{R}'\end{array}\!\!\!\begin{array}{c}\text{R}\\\\\end{array}\longrightarrow\begin{array}{c}\text{S*}\\\text{S*—Fe—S}\\\text{S*}\quad\quad\text{R}'\end{array}+\text{RSH}\quad(19)$$

and other possible mechanisms are given elsewhere (Dukes and Holm, 1975). Examination of succeeding substitution reactions proved to be impractical, but this mechanism appears plausible for them as well, especially in view of the dependence of both initial substitution rates and equilibrium substitution tendencies on thiol acidity. Under basic aqueous conditions, a similar mechanism has been suggested in which bound thiolate is protonated, separates from the iron atom as thiol, and hydroxide is coordinated (Job and Bruice, 1975). The results of the aqueous solution studies indicate that, although the $Fe_4S_4^*$ core remains intact under anaerobic alkaline conditions, the same is not necessarily true of the entire cluster. Consequently, examination of intact clusters under these conditions should be conducted in the presence of excess thiolate. Solvolysis has not been observed for $[Fe_4S_4(SR)_4]^{2-}$ or $[Fe_2S_2(SR)_4]^{2-}$ complexes in dry organic media such as DMF, DMSO, or acetonitrile.

3. PEPTIDE SUBSTITUTION REACTIONS

A particular advantage of the thiolate replacement reactions is that they allow incorporation around a preformed core of relatively complicated ligand structure, affording complexes that may be difficult or inconvenient to prepare by the direct methods in Section II,A. The glycyl-L-cysteinylglycyl oligopeptide complexes required for the investigation of the dependence of 2-/3- tetramer potentials on ligand structure (Section V,B) were prepared from $[Fe_4S_4(S\text{-}tBu)_4]^{2-}$ and peptides 11 and 12 (Table I) by the substitution reactions in Fig. 16 (Que et al., 1974b). Here advantage was taken of the volatility of *tert*-butylthiol, removal of which from the DMSO reaction mixture allowed essentially quantitative product formation. By these means species of varying complexity can be obtained including, as noted earlier, the simple "monopeptide" complex $[Fe_4S_4(S\text{-}Cys(Ac)NHMe)_4]^{2-}$, species that reproduce the peptide spacing in 9 (e.g., $[Fe_4S_4(12\text{-peptide})]^{2-}$), and a mixed complex ($[Fe_4S_4(9\text{-peptide})(S\text{-}tBu)]^{2-}$) with this property. The latter is subject to further substitution giving $[Fe_4S_4(9\text{-peptide})(S\text{-}Cys(Ac)NHMe)]^{2-}$, which, as the Fd active site, has the Cys–X–X–Cys–X–X–Cys unit and an additional cysteinyl group roughly simulating the "distant" residue (18 or 45) in the protein structure. Some properties of these complexes are contained in Tables IV–VI. An important feature of the reaction chemistry of the peptide complexes is their quantitative conversion to $[Fe_4S_4(SPh)_4]^{2-}$ upon treatment with benzenethiol. The visible spectral changes pursuant to the reaction sequence are shown in Fig. 10.

$$[Fe_4S_4(S\text{-}t\text{-}Bu)_4]^{2-} \rightarrow [Fe_4S_4(12\text{-peptide})]^{2-} \rightarrow [Fe_4S_4(SPh)_4]^{2-}$$

B. Active-Site Core Extrusion Reactions

The binding of peptide structure around the $Fe_4S_4^*$ core and the conversion of peptide (and other alkylthiolate) tetramers and $[Fe_2S_2\text{-}$

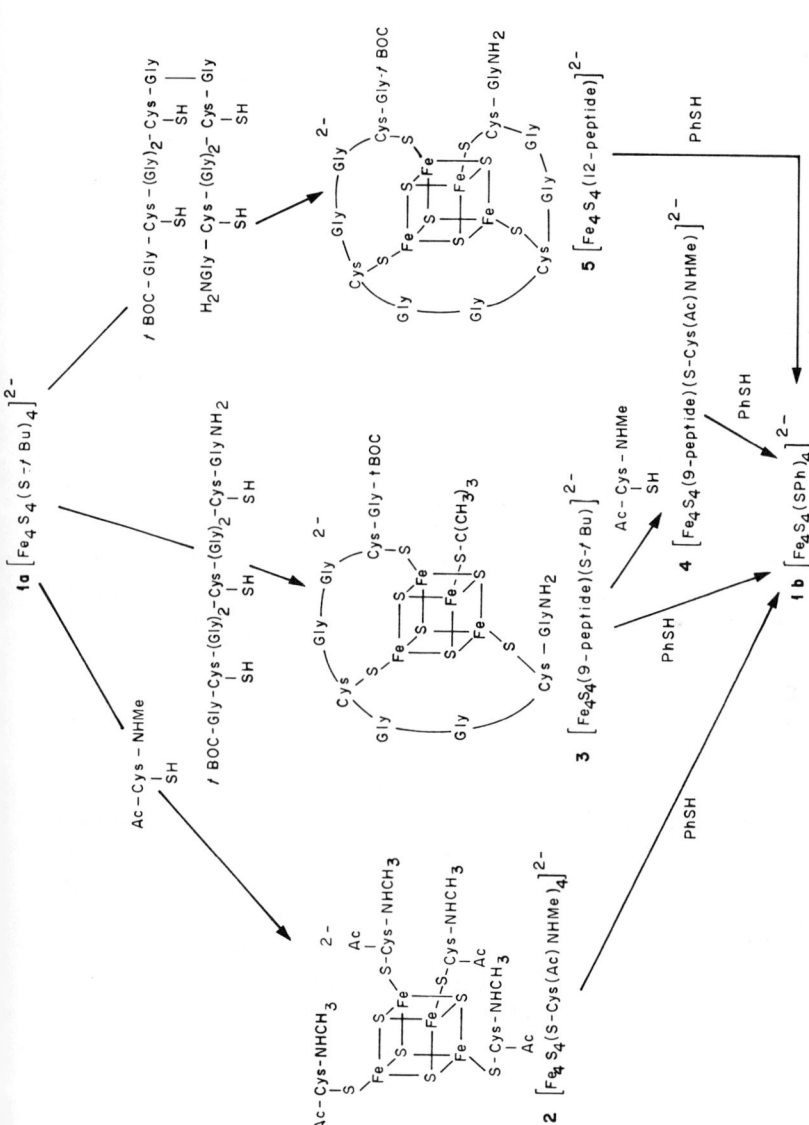

Fig. 16. Ligand substitution reactions of [Fe$_4$S$_4$(S-t-Bu)$_4$]$^{2-}$ with N-acetyl-L-cysteine-N-methylamide and glycyl-L-cysteinylglycyl peptides, also indicating conversion of peptide complexes to [Fe$_4$S$_4$(SPh)$_4$]$^{2-}$ by reaction with benzenethiol. Reproduced with permission from Que et al. (1974b). J. Am. Chem. Soc. **96**, 6042. Copyright by the American Chemical Society.

$(S_2\text{-}o\text{-xyl})_2]^{2-}$ to $[Fe_4S_4(SPh)_4]^{2-}$ and $[Fe_2S_2(SPh)_4]^{2-}$, respectively, are clear precedents for two types of thiolate substitution reactions of proteins. These are protein reconstitution, Reaction (20), and active center *core* extrusion, Reaction (21), in which the active centers formed or subject to extrusion of their cores are indicated. One reaction is the reverse of the other. The equilibrium substitution tendencies summarized above suggest that Reaction (20) is best conducted with R = alkyl and that R′ = aryl thiols are the preferable extrusion reagents, although these are

$$\left.\begin{array}{l}[Fe_2S_2(SR)_4]^{2-}\\ [Fe_4S_4(SR)_4]^{2-}\end{array}\right\} + \text{apoprotein} \rightarrow \text{holoprotein} \left\{\begin{array}{l}2\\3\end{array}\right\} + \text{RSH} \qquad (20)$$

$$\text{Holoprotein} \left\{\begin{array}{l}2\\3\end{array}\right\} + \text{R'SH} \rightarrow \left\{\begin{array}{l}[Fe_2S_2(SR')_4]^{2-}\ \text{and/or}\\ [Fe_4S_4(SR')_4]^{2-}\end{array}\right\} + \text{apoprotein} \qquad (21)$$

not necessary conditions. Both reactions as written involve transfer of $Fe_nS_n^*$ cores from one ligand environment to another, and both have been accomplished (Que *et al.*, 1975; Bale and Orme-Johnson, 1975; Erbes *et al.*, 1975; Gillum *et al.*, 1976b). Attention is restricted to extrusion, which is of potentially greater biological import.

Intact core extrusion from representative 2-Fe and 8-Fe Fd_{ox} proteins is demonstrated by the spectral changes in Figs. 9 and 17. These reactions were conducted in 80% DMSO–H_2O solution (pH ~ 8.5) and result in ~95% liberation of the active site cores from the proteins in the form of the characterized benzenethiolate derivatives using an about fortyfold excess of thiol. The spectra of the latter are identical with those of the separately prepared analogues. Under these experimental conditions, the proteins are expected to be substantially unfolded, thereby allowing easier access of thiol to the active site at which protonation of coordinated S-Cys presumably initiates the substitution process.

Extrusion reactions are possible only when the protein active site contains an integral substructure, in these cases an $Fe_nS_n^*$ core. The only other example of this type of reaction with metal-containing biological materials appears to be the release of iron protoporphyrin IX prosthetic groups from heme proteins by the conventional acetone–HCl treatment. An important application of Reaction (21) would involve extension to complex iron–sulfur proteins and enzymes, e.g., hydrogenase, nitrogenase components, mitochondrial proteins, in which organization of active sites into 2-Fe and 4-Fe types cannot be adequately defined by analytical and spectroscopic data. Prime requirements for the use of this unique approach appear to be (i) protein solubility and stability under denaturing conditions, (ii) structural integrity of the core(s) during and after extrusion, and (iii) distinctive spectral or other characteristics which serve to differentiate the extrusion product(s) from the holoprotein (and from each other). In a significant experiment Erbes *et al.* (1975) have

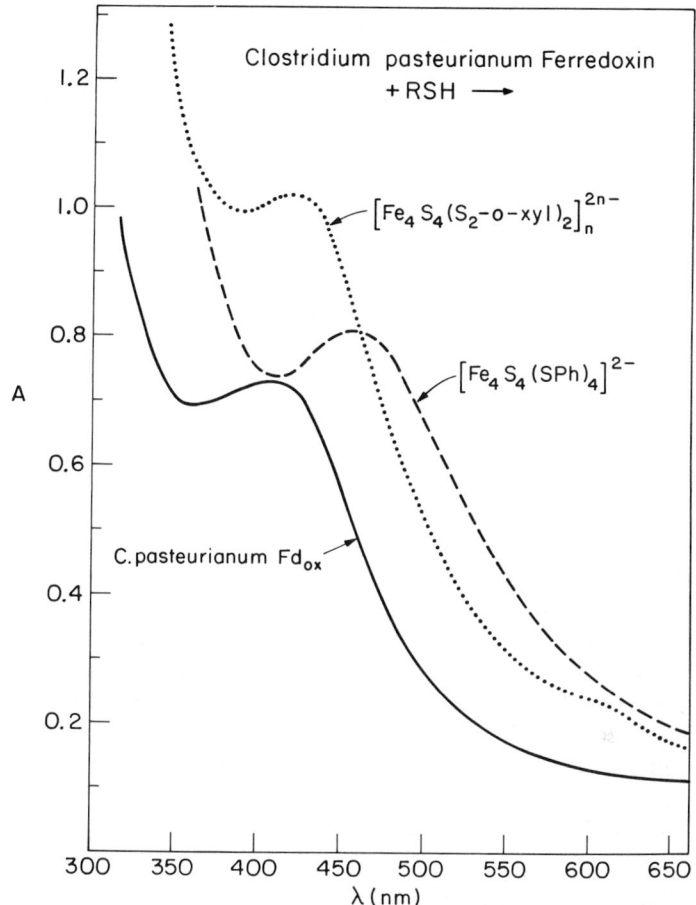

Fig. 17. Spectrophotometric demonstration of two active site extrusion reactions of 8-Fe Fd$_{ox}$ in DMSO solution using benzenethiol and o-xylyl-α,α'-dithiol. Reproduced with permission from Que et al. (1975). *J. Am. Chem. Soc.* **97**, 463. Copyright by the American Chemical Society.

reported the first application of the extrusion technique to enzymes. Treatment of *C. pasteurianum* hydrogenase with benzenethiol in 80% hexamethylphosphoramide–20% aqueous solution liberated [Fe$_4$S$_4$-(SPh)$_4$]$^{2-}$, indicating that the enzyme contains at least one 4-Fe site. The same observation has been made in another laboratory, where quantitation of the extrusion procedure has shown ~3 4-Fe sites in a different enzyme preparation from the same organism (Gillum et al., (1976b). Concerning property (ii), Reaction (22) has been found to occur in 80% DMSO–H$_2$O fairly rapidly below pH ~ 7.5 (Que et al., 1975).

$$2\,[\text{Fe}_2\text{S}_2(\text{SPh})_4]^{2-} \rightarrow [\text{Fe}_4\text{S}_4(\text{SPh})_4]^{2-} + \text{PhSSPh} + 2\,\text{PhS}^- \qquad (22)$$

This dimer–tetramer conversion is doubtless facilitated by the dimensional similarity of the dimer core and a tetramer face. Unless suppressed by higher pH values and excess of thiol, this reaction could lead to incorrect active site identification. Requirement (iii) is met to an at least adequate degree by the employment of benzenethiol as an extrusion reagent, inasmuch as its dimeric and tetrameric complexes possess characteristic red-shifted absorption spectra (Figs. 9 and 11). Particularly significant would be the availability of core-specific extrusion reagents, but, unfortunately, none has as yet been found (Que et al., 1975). As a case in point, o-xylyl-α,α'-dithiol extrudes the core of a 2-Fe Fd_{ox} protein in the form of $[Fe_2S_2(S_2\text{-}o\text{-xyl})_2]^{2-}$ (Fig. 9), resistant to dimerization under extrusion conditions. Yet this compound reacts cleanly with both 8-Fe Fd_{ox} and $[Fe_4S_4(SEt)_4]^{2-}$ to afford a product whose spectrum (Fig. 17) is that of an alkylthiolate tetramer, presumably an oligomer of tetramers, $[Fe_4S_4(S_2\text{-}o\text{-xyl})_2]_n{}^{2n-}$. Sufficient results exist to indicate that the extrusion method of active site identification has promise. Any further elaboration of this method, especially with regard to the development of thiol reagents that afford protein core derivatives with more distinct or intense spectral features, should be guided by the following observation. With all thiols thus far examined, the reactions of the synthetic analogues $[Fe_2S_2(SR)_4]^{2-}$ and $[Fe_4S_4(SR)_4]^{2-}$ exactly parallel those of 2-Fe and 8-Fe Fd_{ox} proteins, respectively. Lastly, it must be borne in mind that **2** and **3** merely are the two types of poly-Fe sites that have been recognized in the simpler proteins. The existence in higher molecular weight proteins and enzymes of more extensively oligomerized or otherwise structurally different active centers is a possibility that must be actively entertained.

C. Protonation and Acid Solvolysis

The reactivity properties described for tetranuclear anions in the preceding two sections refer to nonaqueous or partially or wholly basic aqueous media and are characterized by retention of core structure. Behavior under acidic conditions (pH $\lesssim 6.5$) is rather different. In a series of spectrophotometric experiments extrapolated to zero time, it has been reported that $[Fe_4S_4(S\text{-}s\text{-Bu})_4]^{2-}$ in 60% N-methylpyrrolidine–water and $[Fe_4S_4(SCH_2CH_2CO_2)_4]^{6-}$ in aqueous solution are the conjugate bases of weak acids with pK_a values of 3.9 and 7.5, respectively (Bruice et al., 1975; Job and Bruice, 1975). The site of protonation is unknown, but it has been proposed that the proton associates with a core face (Bruice et al., 1975). The protonated species is reported to be more susceptible to oxidation by O_2. An analysis of the irreversible acid solvolysis of $[Fe_4S_4(SR)_4]^{2-}$ requires two acid–base equilibria (Bruice et al., 1975). One involves the formation of $[HFe_4S_4(SR)_4]^-$ initially, followed a second protonation which ruptures Fe–S* core bonds. This species is proposed to undergo further core bond scission in rate-limiting steps affording (unidentified) products. Inasmuch as the rates are not appreciably dependent upon the presence of excess RSH, Fe–SR bonds are considered to be unbroken throughout the sequence. The relationship of this scheme to the well known acid-promoted decomposition of protein sites with liberation of H_2S* has not been established. In practical terms, these results emphasize the enhanced stability of the core structure under alkaline versus acidic conditions.

V. IRON–SULFUR UNITS AS REDOX CENTERS

A. Structural Aspects

The redox behavior of Fe–S proteins is examined in detail in Chapter 9. In this section, comments are restricted to certain structural features, brought into sharper focus by the availability of precise stereochemical determinations of analogues and related complexes, which may bear on the evolutionary selection of sites 1–3 as the most ubiquitous biological redox centers. The overriding structural feature of these sites is the approximately tetrahedral coordination at iron, which in the 1-Fe and 2-Fe cases is high-spin Fe(II,III). The one-electron redox events for each site reduce to the simple description of Eq. (23). In elementary ligand field

$$\text{Fe(III) } (e^2 t_2^3, S = \tfrac{5}{2}) \underset{-e^-}{\overset{+e^-}{\rightleftharpoons}} \text{Fe(II) } (e^3 t_2^3, S = 2) \tag{23}$$

terms this process involves one state [Fe(III)] with no d-electron stereochemical influence and interconversion between the two states accompanied by a change in population of the approximately nonbonding e orbitals in idealized T_d symmetry, with no d-electron rearrangement required. Two limiting views of coupled redox–structural effects may be entertained. In one the protein structure allows changes in Fe–S structural parameters, especially bond distances, such that the compromise geometry between the Fe(II,III) ground state energy minima that serves as a transition state for electron gain or loss (Reynolds and Lumry, 1966; Bennett, 1973; Sutin, 1973) can be achieved. The maximum spread in high-spin Fe(II)–S and Fe(III)–S nonbridging distances obtained from the structures of $Fe[(SPMe_2)_2N]_2$, thiolate complexes in Tables II and III, and an Fe(III) porphyrin thiolate (Koch et al., 1975) is 2.30–2.38 and 2.22–2.31 Å, respectively. Defining the bond distance difference $\Delta =$ (Fe(II)-L)−(Fe(III)-L) the most appropriate distance comparisons that can be made at present are between the mean values in $Fe[(SPMe_2)_2N]_2$ and $[Fe(S_2\text{-}o\text{-xyl})_2]^-$ ($\Delta = 0.09$ Å), and the mean Fe–S' distances in $[Fe(SCH_2CH_2NMe)_2(CH_2)_{2,3}]_2$ (**37**) and the corresponding distance ($\Delta = 0.07$ Å) or mean of nonbridging distances ($\Delta = 0.08$ Å) in $[Fe(edt)_2]_2^{2-}$ having the closely related centrosymmetric structure **23**. The correspond-

37

ing Δ value for tetrahedral $[FeCl_4]^{2-,-}$, using data from those precise structures that appear to be least complicated by intramolecular interactions (Lauher and Ibers, 1975; Kistenmacher and Stucky, 1968), is 0.11 Å. Thus, in the absence of protein constraints the metal–ligand distances in Rd and 2-Fe Fd proteins would be expected to decrease upon oxidation by about 0.07–0.11 Å.† This statement applies to the latter proteins inasmuch as they possess localized Fe(II)–Fe(III) oxidation states in the reduced form (Sands and Dunham, 1975). If 4-Fe sites are delocalized in their different total oxidation levels, the effect on terminal Fe–S bonds might be even smaller, since the changes can be attenuated over two or more metal centers and, from the model in Section III,B,7, the electron added to Fd_{ox} or removed from HP_{red} is described by an orbital of predominant antibonding tetrametal character. This model predicts that structural changes pursuant to electron transfer will mainly affect core dimensions, as observed for HP_{red} and HP_{ox} (Table III) (Freer et al., 1975). As for the 1-Fe and 2-Fe sites no electron rearrangement is required in traversing the oxidation levels of 4-Fe sites.

Electron transfer reactions of many metal complexes involve more substantial structural changes and electron rearrangement. The oxidation of octahedral high-spin $[Co(NH_3)_6]^{2+}(t_{2g}^5 e_g^2)$ to $[Co(NH_3)_6]^{3+}(t_{2g}^6)$ is accompanied by $\Delta = 0.18$ Å, owing to the removal of antibonding e_g electrons, and a formal two-step d-electron rearrangement to the product configuration. The rate of electron self-exchange is about 10^{15} times slower than that for $[Ru(NH_3)_6]^{2+,3+}$, where $\Delta = 0.04$ Å and electron re-

† The crystal structure of $Na(Ph_4As)[Fe(S_2\text{-}o\text{-xyl})_2]$ has recently been completed (Lane et al., 1976). Ranges and mean values of Fe–S distances and S–Fe–S angles are 2.324(5)–2.378(5), 2.356(13) Å and 103.5(2°)–114.9(2)°, 109.5(1.7)°, respectively. Thus the Fe(II)–S mean distance is 0.089(13) Å longer than the mean Fe(III)–S distance in $[Fe(S_2\text{-}o\text{-xyl})_2]^-$ (Table III), in good agreement with above estimate made prior to the completion of this structure.

arrangement is absent (Stynes and Ibers, 1971). In cases where the structural change is confined to less than the total number of metal–ligand bonds and antibonding electrons are involved, bond length alterations may be even larger. For the pair bis(3-picoline)Co(II)octaethylporphyrinate–bis(imidazole)Co(III)tetraphenylporphyrinate cation, $\Delta = 0.46$ Å for the axial Co–N bonds and is nearly zero for the equatorial Co–N bonds (Lauher and Ibers, 1974; Little and Ibers, 1974). Thus it is evident that the architecture and oxidation levels of protein sites 1–3 (and their analogues) are such as to provide relatively low local structure reorganizational and electron rearrangement barriers to outer-sphere electron transfer. While other factors also contribute to the total $\Delta G^{\ddagger}$ for such processes (Bennett, 1973; Sutin, 1973), it is difficult to imagine, given the weak-field nature of biological ligands and the requirement of a relatively abundant metal, how the natural construction of redox centers could have been more efficacious in minimizing these two types of barriers. The most comparable redox couple is Cu(I)–Cu(II) but unlike Fe(II)–Fe(III), in which both oxidation states can readily adopt a common unconstrained stereochemistry, Cu(II) has an intrinsic preference for tetragonal geometry and Cu(I) for nonplanar arrangements when four-coordinate.

The other limiting view of biological electron transfer holds that the redox site structure is constrained to be the same in both oxidation levels (Lippard, 1973; Bennett, 1973). Electron transfer to or from the site would not be accompanied by stereochemical adjustment. If the constant site geometry were intermediate between those inherently preferred in the absence of protein constraints, the efficiency of electron transfer in a kinetic sense would be enhanced on the basis of a small or zero structure reorganizational barrier. Such an arrangement would be a case of the entatic state. Prior to the availability of the structure of $[\text{Fe}(S_2\text{-}o\text{-xyl})_2]^-$, the Rd_{ox} site had been considered as an example of an entatic condition (Williams, 1971). This proposal is not inconsistent with the longer protein bond distances, which cover the range of Fe(III)–S and Fe(II)–S values determined for $[\text{Fe}(S_2\text{-}o\text{-xyl})_2]^-$ and $\text{Fe}[(\text{SPMe}_2)_2\text{N}]_2$, and the short distance, if real, might provide a favored path for the electron transfer itself. Further examination of structural and other contributions to electron transfer rates must await X-ray definition of other structurally uncharacterized protein and analogue oxidation levels and determination of rates and activation parameters of self-exchange reactions and redox processes involving outer-sphere reagents. A useful start in the latter direction has been made for Rd (Jacks et al., 1974) and HP (Mizrahi et al., 1976). For a more inclusive discussion of the kinetics of Fe–S and other metalloprotein redox reactions, see Chapter 9 and Bennett (1973).

B. Redox Potential Dependence on Ligand Structure

The physical properties of proteins localized essentially within sites 1–3 and (where known) the stereochemistry of these sites, when compared to those of 1-, 2-, and 4-Fe synthetic complexes, justify designation of the latter as analogues of the protein sites *in fixed oxidation levels*. Not unexpectedly, the analogy becomes somewhat less precise when redox potentials, which for proteins depend upon free energy differences between the entire molecule in two different oxidation states, are compared. As the data of Table IV reveal, analogues with simple alkylthiolate ligands do not in general closely approach what are perhaps the single most biologically significant protein parameters. Comparisons are given in the tabulation below. The differences are such that analogue potentials

Comparison	"$\Delta E_0'$" (mV)	Solvent
$[Fe_4S_4(SCH_2R')_4]^{2-,3-}/4$-, 8-Fe $Fd_{ox,red}$	$-(300–700)$	DMF, 80% DMSO–water
$[Fe_2S_2(S_2\text{-}o\text{-xyl})_2]^{2-,3-}/2$-Fe $Fd_{ox,red}$	$-(700–900)$	DMF–water
$[Fe(S_2\text{-}o\text{-xyl})_2]^{-,2-}/Rd_{ox,red}$	$-(600–700)$	DMF–water

are always more negative than those of proteins, i.e., reduced analogues are stronger reductants than are reduced proteins with their normal tertiary structure in aqueous solution. These differences cannot be accorded quantitative significance inasmuch as analogue potentials were determined polarographically in nonaqueous or partially aqueous media (versus saturated calomel electrode) whereas most of the protein values were obtained by other techniques in aqueous solution (versus standard hydrogen electrode). In the "$\Delta E_0'$" values above, analogue potentials have been converted to the hydrogen reference electrode; corrections for non-aqueous–aqueous junction potentials, previous values of which are only rough approximations (DePamphilis *et al.*, 1974), are not included. The few polarographic determinations of protein potentials (Weitzman *et al.*, 1971; Dalton and Zubieta, 1973) tend to support the cathodic shift of analogue compared with protein potentials.

The main utility in considering relative protein and analogue potentials is that the latter are those generated by the protein sites devoid of the perturbing effects of the protein on site structure and environment. In view of the relationships in Eqs. (8)–(10), this statement applies only at nominal parity of terminal ligands, namely, cysteinyl derivatives. With regard to localized active-site structural features, an electrostatic free energy model has been proposed to account for the less negative potentials of the Rd_{ox}/Rd_{red} compared with the 2-Fe Fd_{ox}/Fd_{red} couple (Kassner and Yang, 1973). Interestingly, this same trend holds for the analogue potentials $[Fe(S_2\text{-}o\text{-xyl})_2]^{-,2-}$ and $[Fe_2S_2(S_2\text{-}o\text{-xyl})_2]^{2-,3-}$ in nonaqueous media (Table IV). The rather negative potentials of 2-Fe proteins have been proposed to arise from the effects of a constant site geometry mentioned in the preceding section. Here it is speculated (Lippard, 1973) that the reduced form cannot adjust to a structure favorable to Fe(II) [implicitly, the site geometry is biased toward Fe(III)], making this oxidation level a strong reductant. Because potentials of 2-Fe analogues, which are unlikely to suffer structural constraints in either oxidation level, are far more negative than those of proteins, it seems unlikely that a localized structural effect of this sort plays a major role.

Until precise structures of both analogues and proteins in two redox states coupled by electron transfers are available,† it will be very difficult to evaluate the relationship of localized active site structural changes to redox potentials within or among classes of n-Fe proteins. Superimposed on these structural features are the environmental effects of the protein, as manifested by charge distribution of acidic and basic groups, medium polarity, and juxtaposition of aromatic residues and solvation properties at and near the active sites. Site structural and protein environmental effects are embodied in the considerations by Carter *et al.* (1974b) and Adman *et al.* (1975) as to the origin of the stability of HP_{ox} compared with 4,8-Fe Fd_{red}. It is our present conclusion that overall protein environmental effects are equally or more significant than stereochemical constraints localized at the active sites in "shifting" protein potentials to values less negative than those displayed by their analogues. Variations in potentials of proteins with a common prosthetic group, as with c-type cytochromes (Harbury and Marks, 1973), are well documented, and Kassner (1972) has demonstrated the effects of solvent polarity on the redox potentials of heme complexes. For Fe–S proteins, the effects of variant biological milieu on potentials is evident from the following observations, among others: the appreciable spread in potentials for proteins of the 2-Fe and 4- and 8-Fe types (Table IV); the approximately 160 mV difference for soluble and membrane-bound 2-Fe proteins in chloroplasts (Bearden and Malkin, 1975); the existence of two separate types of proteins with site 2, which from a common oxidation level are either reducible (Fd) or oxidizable (HP) [Eq. (7)]; and reduction of Hp_{red} to HP_{s-red} by dithionite in 80% DMSO–H_2O but not in pure aqueous medium (Cammack, 1973). Certainly there is no more dramatic evidence for the influence of protein structure on potentials than that described for *Azobacter vinelandii* Fd I (Sweeney *et al.*, 1975). This protein contains two 4-Fe clusters with the range of oxidation levels shown in the tabulation on p. 270. Potentials for conversion of the two different clusters between the

† See footnote on p. 266.

	Reducible cluster	Oxidizable cluster
Oxidized	-1	-1
Isolated form	-1 $\Big\} -420$ mV	-2 $\Big\} +340$ mV
Reduced	-2	-2

same oxidation levels differ by ~ 760 mV. While the oxidized cluster appears to behave like the HP site, there is no precedent for the potential of the reduced cluster. Indeed, all previous potentials near ~ -400 mV for $Fe_4S_4^*$ centers have been securely related to the $z = 3-/2-$ process of Eq. (7). At present, there is no ready interpretation for these large potential differences, which are all the more remarkable if each redox state involved retains the $[Fe_4S_4(S-Cys)_4]$ composition, with no substitution of terminal thiolate such as might alter cluster potentials.

It is evident from the foregoing considerations that interpretation of protein potentials at the level of a molecular description is a formidable yet necessary task if a satisfactory picture of the interplay between protein structure and function is to be attained (Holm, 1975). One approach to the problem, currently under investigation in these laboratories, is represented by the series (24) of compounds with increasing ligand complexity, whose redox potentials of interest are those for the process corresponding to Fd_{ox}/Fd_{red}. Taking $[Fe_4S_4(SEt)_4]^{2-}$ as the simplest analogue containing the $Fe-SCH_2-$ unit, the data for 80% $DMSO-H_2O$ solutions in Table IV reveal a 0.25 V anodic potential shift upon replacing the EtS⁻

$$[Fe_4S_4(SCH_2R')_4]^{-z} \cdots [Fe_4S_4(peptide)]^{2-} \cdots \text{holoprotein} \qquad (24)$$

group with acetylcysteinate-N-methylamide. A further positive shift of about 0.1 V is obtained by incorporation of glycyl-L-cysteinylglycyl oligopeptides around the core (Fig. 16). Using the highest and lowest values for the cysteinyl complexes and allowing from zero to $+0.20$ V liquid junction potential, apparent E_0' values of -0.66 to -0.86 and -0.55 to -0.75 V, respectively, are obtained after correction to the hydrogen electrode. The lowest of these values is 0.13 V more negative than the usual ~ -0.42 V Fd_{ox}/Fd_{red} potential. While these comparisons are hardly exact, it is clear that upon cysteinyl binding analogue potentials are shifted in the direction of protein values.† For the reduction $HP_{red} \to HP_{s-red}$ the midpoint potential has been estimated to be ≤ -0.64 V in 80% $DMSO-H_2O$ (Cammack, 1973), comparable to the potentials of

† A close approach to protein potentials is also shown by the couple $[Fe_4S_4(SCH_2CH_2CO_2)_4]^{6-,7-}$ in aqueous solution, for which $E_{1/2} = -0.83$ V, or -0.58 V versus the hydrogen electrode (Job and Bruice, 1975).

cysteinyl analogues. In this medium, the protein is expected to be substantially unfolded and $\lambda_{max} = 406$ nm, an 18 nm red shift compared with the aqueous solution spectrum. For *C. pasteurianum* Fd_{ox} (Que et al., 1975) and cysteinyl analogues (Table VI) (Que et al., 1974b) $\lambda_{max} = 410$ and 404–409 nm. These results suggest that environmental and other extrinsic effects on the active sites imposed by the normal protein conformation in aqueous solution are substantially removed in the unfolded (denatured) form of the protein. In this form, at least for *Chromatium* HP_{red}, features inherent to clusters of the essential $[Fe_4S_4(S\text{-Cys})_4]$ type, exemplified by the peptide complexes, begin to emerge. Further assessment of the influence of protein tertiary structure on potentials within a class of n-Fe proteins is best approached by controlled elaboration of analogue structures toward those present in proteins, ideally by the incorporation of cysteinyl-containing native protein fragments, and by potential measurements of these species and holoproteins using the same technique and under the same (denaturing, nondenaturing) conditions. Such experiments are in progress in these laboratories.

VI. PERSPECTIVES AND CONCLUSIONS

The results of the synthetic analogue approach as applied to Fe–S proteins are such as to demonstrate the viability of the method in achieving meaningful descriptions of the structural, electronic, and reactivity properties of active sites 1–3, particularly where these sites are free of the specific influences of the overall protein structure. The ease and demonstrated or probable flexibility of direct synthesis, augmented by the facility of incorporation of diverse ligand structures by substitution reactions, place synthetic analogues at ready access to any investigator. Consequently, these analogues are likely to attain enhanced utility as vehicles for examination of active site properties, as well as otherwise attractive objects for investigations in coordination chemistry. It appears likely that future research in these and other laboratories will focus in particular on structural and other characterizations of those analogues having oxidation levels ($[Fe(SR)_4]^{2-}$, $[Fe_2S_2(SR)_4]^{3-}$, $[Fe_4S_4(SR)_4]^{3-,-}$) corresponding to some of those incompletely defined in proteins, elucidation of factors controlling redox potentials, expansion of active site core extrusion reactions to complex proteins and enzymes whose active site structures are not securely established by spectroscopic methods, and further investigation of the reactivity of coordinated thiolate ligands, especially as pertinent to aqueous solution chemistry.

In addition to their role as electron transfer agents, there are a multitude of Fe–S proteins, usually of molecular weight $\gtrsim 60{,}000$, which are enzymes (Hall et al., 1974; Lovenberg, 1973; Orme-Johnson, 1973). Examples include the hydrogenases, nitrogenases, flavoprotein dehydrogenases and monoxygenases, and pyruvate dehydrogenase. All of these enzymes contain Fe–S centers possibly similar to 2 and 3 that may function as redox sites within the enzyme molecule and/or as the actual catalytic centers. In virtually all such enzymes the type(s) of Fe–S centers present are not known nor has the enzymatic mechanism been fully clarified at the molecular level. This problem is additionally compounded by the presence in these enzymes of other metals and potential prosthetic groups and by cofactor requirements. However, there are some cases in which Fe–S centers themselves appear to act as catalytic sites for substrate transformation. Prime among these is hydrogenase, containing only Fe–S centers, and, as embodied in a recent mechanistic proposal (Mortenson and Chen, 1974), these function as binding and catalytic sites for the reaction $2H^+ + 2e^- \rightleftharpoons H_2$. The terminal monooxygenase of the 4-methoxybenzoate-O-demethylase enzyme system from *Pseudomonas putida* has been isolated recently (Bernhardt et al., 1975). It is described as an Fe–S protein containing no heme, whose spectroscopic properties suggest that it contains a 2Fe–2S* site related to (but not identical with) those present in 2-Fe Fd proteins. Particularly intriguing are the nonheme dioxygenases, which catalyze the incorporation of both atoms of O_2 into substrate (Nozaki, 1973; Nozaki and Ishimura, 1974). These enzymes contain Fe(II,III) and cysteine but no other metals or prosthetic groups. Activity is lost by removal of iron, which appears to be present often in independent 1-Fe sites. While additional examples could be cited, these are perhaps sufficient to indicate that Fe–S centers may be the catalytic sites in certain enzymes. Consequently, in a second generation of experiments, the synthetic complexes (or variants thereof) described in this chapter may find profitable employment as initial test species for model studies of the reactivity properties of enzymatic Fe–S sites.

ACKNOWLEDGMENTS

The authors' research described herein has been supported by National Institutes of Health grants GM-19256 and GM-22352 (R.H.H., M.I.T., Stanford University) and HL-13157 (J.A.I., Northwestern University).

REFERENCES

Abel, E. A., and Crosse, B. C. (1967). *Organomet. Chem. Rev.* **2**, 443.
Adams, D. M. (1967). "Metal-Ligand and Related Vibrations," Chapter 7. Arnold, London.

Adman, E. T., Sieker, L. C., and Jensen, L. H. (1973). *J. Biol. Chem.* **248**, 3987.
Adman, E., Watenpaugh, K. D., and Jensen, L. H. (1975). *Proc. Natl. Acad. Sci. U.S.A.* **72**, 4854.
Ali, A., and Weinstein, B. (1971). *J. Org. Chem.* **36**, 3022.
Ali, A., Fahrenholz, F., Garing, J. C., and Weinstein, B. (1972). *J. Am. Chem. Soc.* **94**, 2556.
Ali, A., Fahrenholz, F., Garing, J. C., and Weinstein, B. (1973). *Int. J. Pept. Protein Res.* **5**, 91.
Anderson, R. E., Anger, G., Petersson, L., Ehrenberg, A., Cammack, R., Hall, D. O., Mullinger, R., and Rao, K. K. (1975a). *Biochim. Biophys. Acta* **376**, 63.
Anderson, R. E., Dunham, W. R., Sands, R. H., Bearden, A. J., and Crespi, H. L. (1975b). *Biochim. Biophys. Acta* **408**, 306.
Anglin, J. R., and Davison, A. (1975a). *Inorg. Chem.* **14**, 234.
Anglin, J. R., and Davison, A. (1975b). Unpublished results.
Antanaitis, B. C., and Moss, T. H. (1975). *Biochim. Biophys. Acta* **405**, 262.
Antonini, E., and Brunori, M. (1971). "Hemoglobin and Myoglobin in Their Reaction with Ligands," Chapter 4. North-Holland Publ., Amsterdam.
Averill, B. A., Herskovitz, T., Holm, R. H., and Ibers, J. A. (1973). *J. Am. Chem.* **95**, 3523.
Balch, A. L. (1969). *J. Am. Chem. Soc.* **91**, 6962.
Bale, J. R., and Orme-Johnson, W. H. (1975). To be published.
Bearden, A. J., and Dunham, W. R. (1973). In "Iron–Sulfur Proteins" (W. Lovenberg, ed.), Vol. 2, Chapter 6. Academic Press, New York.
Bearden, A. J., and Malkin, R. (1975). *Q. Rev. Biophys.* **7**, 131.
Bell, C. M., McKenzie, E. D., and Orton, J. (1971). *Inorg. Chim. Acta* **5**, 109.
Bennett, L. E. (1973). *Prog. Inorg. Chem.* **18**, 1.
Bernal, I., Davis, B. R., Good, M. L., and Chandra, S. (1972). *J. Coord. Chem.* **2**, 61.
Bernhardt, F. H., Pachowsky, H., and Staudinger, H. (1975). *Eur. J. Biochem.* **57**, 241.
Blumberg, W. E., and Peisach, J. (1974). *Arch. Biochem. Biophys.* **162**, 502.
Bobrik, M. A., and Holm, R. H. (1975). To be published.
Bobrik, M. A., Que, L., Jr., and Holm, R. H. (1974). *J. Am. Chem. Soc.* **96**, 285.
Boon, J. W., and MacGillavry, C. H. (1942). *Recl. Trav. Chim. Pays-Bas* **61**, 910.
Brintzinger, H., Palmer, G., and Sands, R. H. (1966). *Proc. Natl. Acad. Sci. U.S.A.* **55**, 397.
Brodie, A. M., Douglas, J. E., and Wilkins, C. J. (1969). *J. Chem. Soc. A* p. 1931.
Brostigen, G., and Kjekshus, A. (1969). *Acta Chem. Scand.* **23**, 2186.
Brostigen, G., and Kjekshus, A. (1970). *Acta Chem. Scand.* **24**, 1925.
Bruice, T. C., Maskiewicz, R., and Job, R. (1975). *Proc. Natl. Acad. Sci. U.S.A.* **72**, 231.
Burris, R. H., and Orme-Johnson, W. H. (1974). In "Microbial Iron Metabolism" (J. B. Neilands, ed.), Chapter 9. Academic Press, New York.
Cammack, R. (1973). *Biochem. Biophys. Res. Commun.* **54**, 548.
Cammack, R. (1975). *Biochem. Soc. Trans.* **3**, 482.
Cammack, R., Rao, K. K., Hall, D. O., and Johnson, C. E. (1971). *Biochem. J.* **125**, 849.
Carter, C. W., Jr., Freer, S. T., Xuong, Ng. H., Alden, R. A., and Kraut, J. (1971). *Cold Spring Harbor Symp. Quant. Biol.* **36**, 359.
Carter, C. W., Jr., Kraut, J., Freer, S. T., Alden, R. A., Sieker, L. C., Adman, E., and Jensen, L. H. (1972). *Proc. Natl. Acad. Sci. U.S.A.* **68**, 3526.
Carter, C. W., Jr., Kraut, J., Freer, S. T., Xuong, Ng. H., Alden, R. A., and Bartsch, R. G. (1974a). *J. Biol. Chem.* **249**, 4212.

Carter, C. W., Jr., Kraut, J., Freer, S. T., and Alden, R. A. (1974b). *J. Biol. Chem.* **249**, 6339.
Cauquis, G., and Lachenal, D. (1973). *Inorg. Nucl. Chem. Lett.* **9**, 1095.
Cerdonio, M., Wang, R.-H., Rawlings, J., and Gray, H. B. (1974). *J. Am. Chem. Soc.* **96**, 6534.
Churchill, M. R., and Wormald, J. (1971). *Inorg. Chem.* **10**, 1778.
Clare, M., Hill, H. A. O., Johnson, C. E., and Richards, R. (1970). *Chem. Commun.* p. 1376.
Coleman, J. M., Wojcicki, A., Pollick, P. J., and Dahl, L. F. (1967). *Inorg. Chem.* **6**, 1236.
Connelly, N. G., and Dahl, L. F. (1970). *J. Am. Chem. Soc.* **92**, 7472.
Cotton, S. A. (1972). *Coord. Chem. Rev.* **8**, 185.
Cotton, S. A., and Gibson, J. F. (1971). *J. Chem. Soc. A* p. 803.
Coucouvanis, D. (1970). *Prog. Inorg. Chem.* **11**, 233.
Coucouvanis, D., and Lippard, S. J. (1969). *J. Am. Chem. Soc.* **91**, 307.
Coucouvanis, D., Lippard, S. J., and Zubieta, J. A. (1970). *Inorg. Chem.* **9**, 2775.
Coucouvanis, D., Holah, D. G., and Hollander, F. J. (1975). *Inorg. Chem.* **14**, 2657.
Crichton, R. R. (1973). *Struct. Bonding (Berlin)* **17**, 67.
Dahl, L. F., and Wei, C. H. (1963). *Inorg. Chem.* **2**, 328.
Dalton, H., and Zubieta, J. A. (1973). *Biochim Biophys. Acta* **322**, 133.
Davison, A., and Reger, D. L. (1971). *Inorg. Chem.* **10**, 1967.
Davison, A., and Switkes, E. S. (1971). *Inorg. Chem.* **10**, 837.
de Médicis, R. (1970). *Rev. Chim. Miner.* **7**, 723.
DePamphilis, B. V., Averill, B. A., Herskovitz, T., Que, L., Jr., and Holm, R. H. (1974). *J. Am. Chem. Soc.* **96**, 4159.
Dickerson, R. E., Takano, T., Eisenberg, D., Kallai, O. B., Samson, L., Cooper, A., and Margoliash, E. (1971). *J. Biol. Chem.* **246**, 1511.
Dickson D. P. E., and Cammack, R. (1974). *Biochem. J.* **143**, 763.
Dickson, D. P. E., Johnson, C. E., Cammack, R., Evans, M. C. W., Hall, D. O., and Rao, K. K. (1974). *Biochem. J.* **139**, 105.
Dukes, G. R., and Holm, R. H. (1975). *Am. Chem. Soc.* **97**, 528.
Dunham, W. R., Bearden, A. J., Salmeen, I. T., Palmer, G., Sands, R. H., Orme-Johnson, W. H., and Beinert, H. (1971a). *Biochim. Biophys. Acta* **253**, 134.
Dunham, W. R., Palmer, G., Sands, R. H., and Bearden, A. J. (1971b). *Biochim. Biophys. Acta* **253**, 373.
Dus, K., De Klerk, H., Sletten, D., and Bartsch, R. G. (1967). *Biochim. Biophys. Acta* **140**, 291.
Eaton, W. A., and Lovenberg, W. (1971). *J. Am. Chem. Soc.* **92**, 7195.
Eaton, W. A., and Lovenberg, W. (1973). *In* "Iron–Sulfur Proteins" (W. Lovenberg, ed.), Chapter 3. Academic Press, New York.
Eaton, W. A., Palmer, G., Fee, J. A., Kimura, T., and Lovenberg, W. (1971). *Proc. Natl. Acad. Sci. U.S.A.* **68**, 3015.
Eicher, H., Parak, F., Bogner, L., and Gersonde, K. (1974). *Z. Naturforsch., Teil C* **29**, 683.
Eisenberg, R. (1970). *Prog. Inorg. Chem.* **12**, 295.
Eklund, H., Nordstrom, B., Zeppezauer, E., Söderlund, G., Ohlsson, I., Boiwe, T., and Brandén, I. (1974). *FEBS Lett.* **44**, 200.
Epstein, E. F., Bernal, I., and Balch, A. L. (1970). *Chem. Commun.* p. 136.
Erbes, D. L., Burris, R. H., and Orme-Johnson, W. H. (1975). *Proc. Natl. Acad. Sci. U.S.A.* **72**, 4795.

Estabrook, R. W., Suzuki, K., Mason, J. I., Baron, J., Taylor, W. E., Simpson, E. R., Purvis, J., and McCarthy, J. (1973). In "Iron–Sulfur Proteins" (W. Lovenberg, ed.), Vol. 1, Chapter 3. Academic Press, New York.
Evans, M. C. W., Hall, D. O., and Johnson, C. E. (1970). *Biochem. J.* **119**, 289.
Fee, J. A., and Palmer, G. (1971). *Biochim. Biophys. Acta* **245**, 175.
Ferguson, G., Hannaway, C., and Islam, K. M. S. (1968). *Chem. Commun.* p. 1165.
Ferguson, J. A., and Meyer, T. J. (1971). *Chem. Commun.* p. 623.
Flahaut, J. (1972). *MTP Int. Rev. Sci., Inorg. Chem., Ser. One* **10**, 189.
Fleet, M. E. (1970). *Z. Kristallogr., Kristallgeom., Kristallphys., Kristallchem.* **132**, 276.
Frankel, R. B., Averill, B. A., and Holm, R. H. (1974a). *J. Phys. (Paris)* **35**, C6-107.
Frankel, R. B., Herskovitz, T., Averill, B. A., Holm, R. H., Krusic, P. J., and Phillips, W. D. (1974b). *Biochem. Biophys. Res. Commun.* **58**, 974.
Frankel, R. B., Reiff, W. M., Bernal, I., and Good, M. L. (1974c). *Inorg. Chem.* **13**, 493.
Freer, S. T., Alden, R. A., Carter, C. W., Jr., and Kraut, J. (1975). *J. Biol. Chem.* **250**, 46.
Fritchie, C. J., Jr. (1975). *Acta Crystallogr., Sec. B* **31**, 802.
Gall, R. S., Chu, C. T.-W., and Dahl, L. F. (1974). *J. Am. Chem. Soc.* **96**, 4019.
Gersonde, K., Schlaak, H. E., Breitenbach, M., Parak, F., Eicher, H., Zgorzalla, W., Kalvius, M. G., and Mayer, A. (1974). *Eur. J. Biochem.* **43**, 307.
Gibson, J. F., Hall, D. O., Thornley, J. H. M., and Whatley, F. R. (1966). *Proc. Natl. Acad. Sci. U.S.A.* **56**, 987.
Gillum, W. O., Frankel, R. B., Foner, S., and Holm, R. H. (1976a). *Inorg. Chem.* **15**, 1095.
Gillum, W. O., Mortenson, L. E., Chen, J. S., and Holm, R. H. (1976b). To be published.
Gilmour, A. D., and McAuley, A. (1970). *J. Chem. Soc. A* p. 1006.
Glickson, J. D., Phillips, W. D., McDonald, C. C., and Poe, M. (1971). *Biochem. Biophys. Res. Commun.* **42**, 271.
Gunsalus, I. C., and Lipscomb, J. D. (1973). In "Iron–Sulfur Proteins" (W. Lovenberg, ed.), Vol. 1, Chapter 6. Academic Press, New York.
Gunsalus, I. C., Meeks, J. R., Lipscomb, J. D., Debrunner, P., and Munck, E. (1974). In "Molecular Mechanisms of Oxygen Activation" (O. Hayaishi, ed.), Chapter 14. Academic Press, New York.
Hall, D. O., Cammack, R., and Rao, K. K. (1973a). *Pure Appl. Chem.* **34**, 553.
Hall, D. O., Cammack, R., and Rao, K. K. (1973b). *Space Life Sci.* **4**, 445.
Hall, D. O., Cammack, R., and Rao, K. K. (1974). In "Iron in Biochemistry and Medicine" (A. Jacobs and M. Worwood, eds.), Chapter 8. Academic Press, New York.
Hall, S. R., and Stewart, J. M. (1973). *Acta Crystallogr. Sect. B* **29**, 579.
Hamilton, W. C., and Bernal, I. (1967). *Inorg. Chem.* **6**, 2003.
Harbury, H. L., and Marks, R. H. L. (1973). In "Inorganic Biochemistry" (G. Eichhorn, ed.), Vol. 2, Chapter 26. Elsevier, Amsterdam.
Harrison, P. M., Hoare, R. J., Hoy, T. G., and Macara, I. G. (1974). In "Iron in Biochemistry and Medicine" (A. Jacobs and M. Worwood, eds.), Chapter 3. Academic Press, New York.
Healy, P. C., and White, A. H. (1972). *J. Chem. Soc., Dalton Trans.* p. 1163.
Herriott, J. R., Sieker, L. C., Jensen, L. H., and Lovenberg, W. (1970). *J. Mol. Biol.* **50**, 391.

Herskovitz, T., Averill, B. A., Holm, R. H., Ibers, J. A., Phillips, W. D., and Weiher, J. F. (1972a). *Proc. Natl. Acad. Sci. U.S.A.* **69**, 2437.

Herskovitz, T., Averill, B. A., Holm, R. H., and Ibers, J. A. (1972b). *Abstr. Int. Conf. Coord. Chem., 15th, 1972* p. 1.

Herskovitz, T., DePamphilis, B. V., Gillum, W. O., and Holm, R. H. (1975) *Inorg. Chem.* **14**, 1426.

Holah, D. G., and Coucouvanis, D. (1975). *J. Am. Chem. Soc.* **97**, 6917.

Hollander, F. J., Pedelty, R., and Coucouvanis, D. (1974). *J. Am. Chem. Soc.* **96**, 4032.

Holm, R. H. (1975). *Endeavour* **34**, 38.

Holm, R. H., Phillips, W. D., Averill, B. A., Mayerle, J. J., and Herskovitz, T. (1974a). *J. Am. Chem. Soc.* **96**, 2109.

Holm, R. H., Averill, B. A., Herskovitz, T., Frankel, R. B., Gray, H. B., Siiman, O., and Grunthaner, F. J. (1974b). *J. Am. Chem. Soc.* **96**, 2644.

Hong, J.-S., and Rabinowitz, J. C. (1970a). *J. Biol. Chem.* **245**, 4982.

Hong, J.-S., and Rabinowitz, J. C. (1970b). *J. Biol. Chem.* **245**, 6574.

Horrocks, W. D. (1973). *In* "NMR of Paramagnetic Molecules" (G. N. La Mar, W. D. Horrocks, Jr., and R. H. Holm, eds.), Chapter 4. Academic Press, New York.

Hoskins, B. F., and Pannan, C. D. (1975). *Inorg. Nucl. Chem. Lett.* **11**, 409.

Hu, W., and Lippard, S. J. (1974). *J. Am. Chem. Soc.* **96**, 2366.

Jacks, C. A., Bennett, L. E., Raymond, W. N., and Lovenberg, W. (1974). *Proc. Natl. Acad. Sci. U.S.A.* **71**, 1118.

Jeannin, S., Jeannin, Y., and Lavinge, G. (1972). *J. Organomet. Chem.* **40**, 187.

Jensen, L. H. (1973). *In* "Iron–Sulfur Proteins" (W. Lovenberg, ed.), Vol. 2, Chapter 4. Academic Press, New York.

Jensen, L. H. (1974). *Annu. Rev. Biochem.* **43**, 471.

Job, R. C., and Bruice, T. C. (1975). *Proc. Natl. Acad. Sci. U.S.A.* **72**, 2478.

Johansson, G., and Lipscomb, W. N. (1958). *Acta Crystallogr.* **11**, 594.

Johnson, K. H. (1973). *Adv. Quantum Chem.* **7**, 143.

Johnson, K. H. (1975). *Annu. Rev. Phys. Chem.* **26**, 39.

Jørgensen, C. K. (1968). *Inorg. Chim. Acta, Rev.* **2**, 65.

Kagamiyama, H., Rao, K. K., Hall, D. O., Cammack, R., and Matsubara, H. (1975). *Biochem. J.* **145**, 121.

Kassner, R. J. (1972). *Proc. Natl. Acad. Sci. U.S.A.* **69**, 2263.

Kassner, R. J., and Yang, W. (1973). *Biochem. J.* **133**, 283.

Kimura, T. (1968). *Struct. Bonding (Berlin)* **5**, 1.

Kimura, T. (1971). *Biochem. Biophys. Res. Commun.* **43**, 1145.

Kimura, T., and Huang, J. J. (1970). *Arch. Biochem. Biophys.* **137**, 357.

Kimura, T., Tasaki, A., and Watari, H. (1970). *J. Biol. Chem.* **234**, 4450.

Kistenmacher, T. J., and Stucky, G. D. (1968). *Inorg. Chem.* **7**, 2150.

Kjekshus, A., Nicholson, D. C., and Mukherjee, A. D. (1972). *Acta Chem. Scand.* **25**, 1105.

Koch, S., Tang, S. C., Holm, R. H., Frankel, R. B., and Ibers, J. A. (1975). *J. Am. Chem. Soc.* **97**, 916.

La Mar, G. N. (1973). *In* "NMR of Paramagnetic Molecules" (G. N. La Mar, W. D. Horrocks, Jr., and R. H. Holm, eds.), Chapter 3. Academic Press, New York.

La Mar, G. N., Eaton, G. R., Holm, R. H., and Walker, F. A. (1973a). *J. Am. Chem. Soc.* **95**, 63.

7. ACTIVE-SITE ANALOGUES OF IRON–SULFUR PROTEINS

La Mar, G. N., Horrocks, W. D., Jr., and Holm, R. H., eds. (1973b). "NMR of Paramagnetic Molecules." Academic Press, New York.
Lane, R. W., Ibers, J. A., Frankel, R. B., and Holm, R. H. (1975). *Proc. Natl. Acad. Sci. U.S.A.* **72**, 2868.
Lane, R. W., Ibers, J. A., Frankel, R. B., Papaefthymiou, G. C., and Holm, R. H. 1976). *J. Am. Chem. Soc.,* in press.
Lappin, A. G., and McAuley, A. (1975). *J. Chem. Soc., Dalton Trans.* p. 1560.
Lauher, J. W., and Ibers, J. A. (1974). *J. Am. Chem. Soc.* **96**, 4447.
Lauher, J. W., and Ibers, J. A. (1975). *Inorg. Chem.* **14**, 348.
Leipoldt, J. G., and Coppens, P. (1973). *Inorg. Chem.* **12**, 2269.
Leussing, D. L., and Jayne, J. (1962). *J. Phys. Chem.* **66**, 426.
Leussing, D. L., and Kolthoff, I. M. (1953). *J. Am. Chem. Soc.* **75**, 3904.
Leussing, D. L., and Mislan, J. P. (1960). *J. Phys. Chem.* **64**, 1908.
Leussing, D. L., and Newman, L. (1956). *J. Am. Chem. Soc.* **78**, 552.
Leussing, D. L., and Tischer, T. N. (1963). *Adv. Chem. Ser.* **37**, 216.
Leussing, D. L., Mislan, J. P., and Coll, R. J. (1960). *J. Phys. Chem.* **64**, 1070.
Lippard, S. J. (1973). *Acc. Chem. Res.* **6**, 282.
Lipscomb, W. N. (1973). *Proc. Natl. Acad. Sci. U.S.A.* **70**, 3797.
Little, R. G., and Ibers, J. A. (1974). *J. Am. Chem. Soc.* **96**, 4440.
Livingstone, S. E. (1965). *Q. Rev., Chem. Soc.* **19**, 386.
Llinás, M. (1973). *Struct. Bonding (Berlin)* **17**, 135.
Lode, E. T., and Coon, M. J. (1973). In "Iron–Sulfur Proteins" (W. Lovenberg, ed.), Vol. 1, Chapter 7. Academic Press, New York.
Loew, G. H., and Lo, D. (1974a). *Theor. Chim. Acta* **32**, 217.
Loew, G. H., and Lo, D. (1974b). *Theor. Chim. Acta* **33**, 137.
Loew, G. H., and Steinberg, D. A. (1971). *Theor. Chim. Acta* **23**, 239.
Loew, G. H., and Steinberg, D. A. (1972). *Theor. Chim. Acta* **26**, 107.
Loew, G. H., Chadwick, M., and Steinberg, D. A. (1974a). *Theor. Chim. Acta* **33**, 125.
Loew, G. H., Chadwick, M., and Lo, D. (1974b). *Theor. Chim. Acta* **33**, 147.
Long, T. V., and Loehr, T. M. (1970). *J. Am. Chem. Soc.* **92**, 6384.
Long, T. V., Loehr, T. M., Allkins, J. R., and Lovenberg, W. (1971). *J. Am. Chem. Soc.* **93**, 1809.
Lovenberg, W., ed. (1973). "Iron–Sulfur Proteins," Vols. 1 and 2. Academic Press, New York.
Lovenberg, W., and Sobel, B. E. (1965). *Proc. Natl. Acad. Sci. U.S.A.* **54**, 193.
Lovenberg, W., and Williams, W. M. (1969). *Biochemistry* **8**, 141.
Low, W., and Weger, M. (1960). *Phys. Rev.* **118**, 1130.
McAuliffe, C. A., and Murray, S. G. (1972). *Inorg. Chim. Acta. Rev.* **6**, 103.
McDonald, C. C., Phillips, W. D., Lovenberg, W., and Holm, R. H. (1973). *Ann. N.Y. Acad. Sci.* **222**, 789.
Malkin, R., and Rabinowitz, J. C. (1966). *Biochem. Biophys. Res. Commun.* **23**, 822.
Malkin, R., Aparicio, P. J., and Arnon, D. I. (1974). *Proc. Natl. Acad. Sci. U.S.A.* **71**, 2362.
Martin, R. L. (1968). In "New Pathways in Inorganic Chemistry" (E. A. V. Ebsworth, A. G. Maddock, and A. G. Sharpe, eds.), Chapter 9. Cambridge Univ. Press, London and New York.
Martin, R. L., Rohde, N. M., Robertson, G. B., and Taylor, D. (1974). *J. Am. Chem. Soc.* **96**, 3647.

Mason, R., and Zubieta, J. A. (1973). *Angew. Chem., Int. Ed. Engl.* **12**, 390.
Mathews, B. W., Wearer, L. H., and Kester, W. R. (1974). *J. Biol. Chem.* **249**, 8030.
Mathews, F. S., Levine, M., and Argos, P. (1972). *J. Mol. Biol.* **64**, 449.
Mathews, R., Charlton, S., Sands, R. H., and Palmer, G. (1974). *J. Biol. Chem.* **249**, 4326.
Mathur, H. B., Gupta, M. P., and Kavedia, C. V. (1966). *Indian J. Chem.* **4**, 337.
Mayerle, J. J., Frankel, R. B., Holm, R. H., Ibers, J. A., Phillips, W. D., and Weiher, J. F. (1973). *Proc. Natl. Acad. Sci. U.S.A.* **70**, 2429.
Mayerle, J. J., Denmark, S. E., DePamphilis, B. V., Ibers, J. A., and Holm, R. H. (1975). *J. Am. Chem. Soc.* **97**, 1032.
Mayhew, S. G., Petering, D., Palmer, G., and Foust, G. P. (1969). *J. Biol. Chem.* **244**, 2830.
Mehrotra, R. C., Gutpa, V. D., and Sukhani, D. (1968). *Inorg. Chim. Acta, Rev.* **2**, 111.
Mitchell, P. C. H., and Parker, D. A. (1973). *J. Inorg. Nucl. Chem.* **35**, 1385.
Mizrahi, F. A., Wood, F. E., and Cusanovich, M. A. (1976). Biochemistry **15**, 343.
Moleski, C., Moss, T. H., Orme-Johnson, W. H., and Tsibris, J. C. M. (1970). *Biochim. Biophys. Acta* **214**, 548.
Mortenson, L. E., and Chen, J.-S. (1974). *In* "Microbial Iron Metabolism" (J. B. Neilands, ed.), Chapter 4. Academic Press, New York.
Moss, T. H., Petering, D., and Palmer, G. (1969). *J. Biol. Chem.* **244**, 2275.
Mullinger, R. N., Cammack, R., Rao, K. K., Hall, D. O., Dickson, D. P. E., Johnson, C. E., Rush, J. D., and Simopoulos, A. (1975). *Biochem. J.* **151**, 75.
Münck, E., Debrunner, P. G., Tsibris, J. C. M., and Gunsalus, I. C. (1972). *Biochemistry* **11**, 855.
Münck, E., Rhodes, H., Orme-Johnson, W. H., Davis, L. C., Brill, W. J., and Shah, V. K. (1975). *Biochim. Biophys. Acta* **400**, 32.
Murray, K. S., and Newman, P. J. (1975). *Australian J. Chem.* **28**, 773.
Norman, J. G., Jr., and Jackels, S. C. (1975). *J. Am. Chem. Soc.* **97**, 3833.
Nozaki, M. (1973). *In* "Molecular Mechanisms of Oxygen Activation" (O. Hayaishi, ed.), Chapter 4. Academic Press, New York.
Nozaki, M., and Ishimura, Y. (1974). *In* "Microbial Iron Metabolism" (J. B. Neilands, ed.), Chapter 16. Academic Press, New York.
Orme-Johnson, W. H. (1973). *Annu. Rev. Biochem.* **42**, 159.
Orme-Johnson, W. H., and Sands, R. H. (1973). *In* "Iron–Sulfur Proteins" (W. Lovenberg, ed.), Vol. 2, Chapter 5. Academic Press, New York.
Page, F. M. (1955). *Trans. Faraday Soc.* **51**, 919.
Palmer, G. (1973). *In* "Iron–Sulfur Proteins" (W. Lovenberg, ed.), Vol. 2, Chapter 8. Academic Press, New York.
Palmer, G., Brintzinger, H., and Estabrook, R. W. (1967). *Biochemistry* **6**, 1658.
Palmer, G., Dunham, W. R., Fee, J. A., Sands, R. H., Iizuka, T., and Yonetani, T. (1971). *Biochim. Biophys. Acta* **245**, 201.
Pappalardo, P., and Dietz, R. E. (1961). *Phys. Rev.* **123**, 1188.
Pasek, E. A., and Straub, D. R. (1972). *Inorg. Chem.* **11**, 2285.
Peisach, J., Blumberg, W. E., Lode, E. T., and Coon, M. J. (1971). *J. Biol. Chem.* **246**, 5877.
Phillips, W. D. (1973). *In* "NMR of Paramagnetic Molecules" (G. N. La Mar, W. D. Horrocks, Jr., and R. H. Holm, eds.), Chapter 11. Academic Press, New York.

Phillips, W. D., and Poe, M. (1973). *In* "Iron–Sulfur Proteins" (W. Lovenberg, ed.), Vol. 2, Chapter 7. Academic Press, New York.
Phillips, W. D., Poe, M., Weiher, J. F., McDonald, C. C., and Lovenberg, W. (1970a). *Nature (London)* **227,** 574.
Phillips, W. D., Poe, M., McDonald, C. C., and Bartsch, R. G. (1970b). *Proc. Natl. Acad. Sci. U.S.A.* **67,** 682.
Phillips, W. D., McDonald, C. C., Stombaugh, N. A., and Orme-Johnson, W. H. (1974). *Proc. Natl. Acad. Sci. U.S.A.* **71,** 140.
Pignolet, L. H., Patterson, G. S., Weiher, J. F., and Holm, R. H. (1974). *Inorg. Chem.* **13,** 1263.
Poe, M., Phillips, W. D., McDonald, C. C., and Lovenberg, W. (1970). *Proc. Natl. Acad. Sci. U.S.A.* **65,** 797.
Poe, M., Phillips, W. D., Glickson, J. D., McDonald, C. C., and San Pietro, A. (1971a). *Proc. Natl. Acad. Sci. U.S.A.* **68,** 68.
Poe, M., Phillips, W. D., McDonald, C. C., and Orme-Johnson, W. H. (1971b). *Biochem. Biophys. Res. Commun.* **42,** 705.
Que, L., Jr., Bobrik, M. A., Ibers, J. A., and Holm, R. H. (1974a). *J. Am. Chem. Soc.* **96,** 4168.
Que, L., Jr., Anglin, J. R., Bobrik, M. A., Davison, A., and Holm, R. H. (1974b). *J. Am. Chem. Soc.* **96,** 6042.
Que, L., Jr., Holm, R. H., and Mortenson, L. E. (1975). *J. Am. Chem. Soc.* **97,** 463.
Quiocho, F. A., and Lipscomb, W. N. (1971). *Adv. Protein Chem.* **25,** 1.
Rae, A. I. M. (1967). *Chem. Commun.* p. 1245.
Rao, K. K., Cammack, R., Hall, D. O., and Johnson, C. E. (1971). *Biochem. J.* **122,** 257.
Rao, K. K., Evans, M. C. W., Cammack, R., Hall, D. O., Thompson, C. L., Jackson, P. J., and Johnson, C. E. (1972). *Biochem. J.* **129,** 1063.
Rawlings J., Siiman, O., and Gray, H. B. (1974). *Proc. Natl. Acad. Sci. U.S.A.* **71,** 125.
Reiff, W. M., Grey, I. E., Fan, A., Eliezer, Z., and Steinfink, H. (1975). *J. Solid State Chem.* **13,** 42.
Reynolds, W. L., and Lumry, R. W. (1966). "Mechanisms of Electron Transfer." Ronald Press, New York.
Salemme, F. R., Freer, S. T., Xuong, Ng. H., Alden, R. A., and Kraut, J. (1973a). *J. Biol. Chem.* **248,** 3910.
Salemme, F. R., Kraut, J., and Kamen, M. D. (1973b). *J. Biol. Chem.* **248,** 7701.
Salmeen, I. T., and Palmer, G. (1972). *Arch. Biochem. Biophys.* **150,** 767.
Sands, R. H., and Dunham, W. R. (1975). *Q. Rev. Biophys.* **7,** 443.
Sayers, D. E., Stern, E. A., and Herriott, J. R. (1976). *J. Chem. Phys.* **64,** 427.
Schöberl, A., Rimpler, M., and Dethlefsen, U. (1973a). *Justus Liebigs Ann. Chem.* **1973,** 1372.
Schöberl, A., Rimpler, M., and Dethlefsen, U. (1973b). *Justus Liebigs Ann. Chem.* **1973,** 1612.
Schrauzer, G. N., Kiefer, G. W., Tano, K., and Doemeny, P. A. (1974). *J. Am. Chem. Soc.* **96,** 641.
Schubert, M. (1932). *J. Am. Chem. Soc.* **54,** 4077.
Schubert, M. (1933). *J. Am. Chem. Soc.* **55,** 4563.
Schultz, A. J., and Eisenberg, R. (1973). *Inorg. Chem.* **12,** 518.
Schunn, R. A., Fritchie, C. J., Jr., and Prewitt, C. T. (1966). *Inorg. Chem.* **5,** 892.

Shulman, R. G., Eisenberger, P., Blumberg, W. E., and Stombaugh, N. A. (1975). *Proc. Natl. Acad. Sci. U.S.A.* **72**, 4003.
Sieker, L. C., Adman, E., and Jensen, L. H. (1972). *Nature (London)* **235**, 40.
Simon, G. L., and Dahl, L. F. (1973). *J. Am. Chem. Soc.* **95**, 2164, 2175.
Slack, G. N., Ham, N. S., and Chrenko, R. M. (1966). *Phys. Rev.* **152**, 376.
Slater, J. C., and Johnson, K. H. (1974). *Phys. Today* **27**, 34.
Snow, M. R., and Ibers, J. A. (1973). *Inorg. Chem.* **12**, 249.
Stadtherr, L. G., and Martin, R. B. (1972). *Inorg. Chem.* **11**, 92.
Stombaugh, N. A., Burris, R. H., and Orme-Johnson, W. H. (1973). *J. Biol. Chem.* **248**, 7951.
Stynes, H. C., and Ibers, J. A. (1971). *Inorg. Chem.* **10**, 2304.
Sugiura, Y., and Tanaka, H. (1972). *Biochem. Biophys. Res. Commun.* **46**, 335.
Sugiura, Y., Kunishima, M., and Tanaka, H. (1972). *Biochem. Biophys. Res. Commun.* **48**, 1400 and 1518.
Sugiura, Y., Kunishima, M., Tanaka, H., and Dearman, H. H. (1975). *J. Inorg. Nucl. Chem.* **37**, 1511.
Sutin, N. (1973). *In* "Inorganic Biochemistry" (G. L. Eichhorn, ed.), Vol. 2, Chapter 19. Elsevier, Amsterdam.
Sweeney, W. V. and Coffman, R. E. (1972). *Biochim. Biophys. Acta* **286**, 26.
Sweeney, W. V., Bearden, A. J., and Rabinowitz, J. C. (1974). *Biochem. Biophys. Res. Commun.* **59**, 188.
Sweeney, W. V., Rabinowitz, J. C., and Yoch, D. C. (1975). *J. Biol. Chem.* **250**, 7842.
Takano, T., Kallai, O. B., Swanson, S., and Dickerson, R. E. (1973). *J. Biol. Chem.* 5234.
Tanaka, M., Haniu, M., Yasunobu, K. T., and Kimura, T. (1973). *J. Biol. Chem.* **248**, 1141.
Tanaka, M., Haniu, M., Yasunobu, K. T., Dus, K., and Gunsalus, I. C. (1974). *J. Biol. Chem.* **249**, 3689.
Tanaka, N., Kolthoff, I. M., and Stricks, W. (1955). *J. Am. Chem. Soc.* **77**, 1996 and 2004.
Tang, S.-P. W., Spiro, T. G., Mukai, K., and Kimura, T. (1973). *Biochem. Biophys. Res. Commun.* **53**, 869.
Tang, S.-P. W., Spiro, T. G., Antanaitis, C., Moss, T. H., Holm, R. H., Herskovitz, T., and Mortenson, L. E. (1975). *Biochem. Biophys. Res. Commun.* **62**, 1.
Tano, K., and Schrauzer, G. N. (1975). *J. Amer. Chem. Soc.* **97**, 5404.
Taylor, J. E., Yan, J. F., and Wang, J. (1966). *J. Am. Chem. Soc.* **88**, 1663.
Tebbe, F. N., and Muetterties, E. L. (1970). *Inorg. Chem.* **9**, 629.
Terzis, A., and Rivest, R. (1973). *Inorg. Chem.* **12**, 2132.
Thomas, J. T., Robertson, J. H., and Cox, E. G. (1958). *Acta Crystallogr.* **11**, 599.
Thompson, C. L., Johnson, C. E., Dickson, D. P. E., Cammack, R., Hall, D. O., Weser, U., and Rao, K. K. (1974). *Biochem. J.* **139**, 97.
Thomson, A. J. (1975). *Biochem. Soc. Trans.* **3**, 468.
Tomita, A., Hirai, H., and Makashima, S. (1967). *Inorg. Chem.* **6**, 1746.
Tomita, A., Hirai, H., and Makashima, S. (1968a). *Inorg. Chem.* **7**, 760.
Tomita, A., Hirai, H., and Makashima, S. (1968b). *Inorg. Nucl. Chem. Lett.* **4**, 715.
Trinh-Toan, Fehlhammer, W. P., and Dahl, L. F. (1972). *J. Am. Chem. Soc.* **94**, 3389.
Tsibris, J. C. M., and Woody, R. W. (1970). *Coord. Chem. Rev.* **5**, 417.

Tsibris, J. C. M., Tsai, R. L., Gunsalus, I. C., Orme-Johnson, W. H., Hansen, R. E., and Beinert, H. (1968a). *Proc. Natl. Acad. Sci. U.S.A.* **59**, 959.
Tsibris, J. C. M., Nantvedt, M. J., and Gunsalus, I. C. (1968b). *Biochem. Biophys. Res. Commun.* **30**, 323.
Vahrenkamp, H. (1975). *Angew. Chem., Int. Ed. Engl.* **14**, 322.
Vallee, B. L., and Wacker, W. E. C. (1970). *In* "The Proteins" (H. Neurath, ed.), 2nd ed., Vol. 5, Chapter 4. Academic Press, New York.
Vallee, B. L., and Williams, R. J. P. (1968). *Proc. Natl. Acad. Sci. U.S.A.* **59**, 498.
Vergamini, P. J., and Kubas, G. J. (1976). *Prog. Inorg. Chem.* **21**, 261.
Ward, J. C. (1970). *Rev. Pure Appl. Chem.* **20**, 175.
Watenpaugh, K. D., Sieker, L. C., Herriott, J. R., and Jensen, L. H. (1971). *Cold Spring Harbor Symp. Quant. Biol.* **36**, 359.
Watenpaugh, K. D., Sieker, L. C., Herriott, J. R., and Jensen, L. H. (1973). *Acta Crystallogr., Sect. B* **29**, 943.
Wei, C. H., and Dahl, L. F. (1965a). *Inorg. Chem.* **4**, 1.
Wei, C. H., and Dahl, L. F. (1965b). *Inorg. Chem.* **4**, 493.
Wei, C. H., Wilkes, G. R., Treichel, P. M., and Dahl, L. F. (1966). *Inorg. Chem.* **5**, 900.
Weitzman, P. D. J., Kennedy, I. R., and Caldwell, R. A. (1971). *FEBS Lett.* **17**, 241.
Williams, R. J. P. (1971). *Inorg. Chim. Acta, Rev.* **5**, 137.
Wilson, D. F. (1967). *Arch. Biochem. Biophys.* **122**, 254.
Yajima, H., Shirai, N., and Kiso, Y. (1971). *Chem. Pharm. Bull.* **19**, 1900.
Yang, C. S., and Huennekens, F. M. (1970). *Biochemistry* **9**, 2127.
Yang, C. Y., Johnson, K. H., Holm, R. H., and Norman, J. G., Jr. (1975). *J. Am. Chem. Soc.* **97**, 6596.
Yasunobu, K. T., and Tanaka, M. (1973). *In* "Iron-Sulfur Proteins" (W. Lovenberg, ed.), Vol. 2, Chapter 2. Academic Press, New York.
Yoch, D. C., and Valentine, R. C. (1972). *Annu. Rev. Microbiol.* **26**, 139.
Zubieta, J. A., Mason, R., and Postgate, J. R. (1973). *Biochem. J.* **133**, 851.

CHAPTER 8

Evidence from Mössbauer Spectroscopy and Magnetic Resonance on the Active Centers of the Iron–Sulfur Proteins

R. CAMMACK, D. P. E. DICKSON, and C. E. JOHNSON

I. Introduction	283
II. Spectroscopy of the Active Center	289
A. Magnetic Moment of Iron in Molecules	290
B. Electron Spin Relaxation	292
C. Magnetic Hyperfine Interactions	293
D. Transferred Hyperfine Interactions	296
E. Mössbauer Chemical Shifts	297
F. Electric Quadrupole Interactions	298
III. Proteins with 1 Fe Centers	298
IV. Proteins with 2 Fe–2 S* Centers	301
A. General Properties	301
B. Mössbauer Data	302
V. Proteins with 4 Fe–4 S* Centers	305
A. General Properties	305
B. Mössbauer Data	310
C. Similarity of the 4 Fe–4 S* Centers	315
D. Equivalence and Valence of the Iron Atoms	317
E. Spin Coupling	322
VI. Conclusions	326
References	327

I. INTRODUCTION

In the understanding of the structures and electronic states of atoms in large biological molecules, both diffraction and spectroscopic methods have played important and often complementary parts. Each has its own advantages and limitations. X-ray diffraction has given us detailed models of the atomic positions in many biological molecules, but it requires single crystals, and it is a difficult and lengthy process to derive the structure from the diffraction pattern unambiguously. Spectroscopic methods are more generally applicable and are quicker to use and easier to interpret, but they usually give data in a limited region of the molecule, e.g., magnetic methods give data on those atoms (or free radicals) with unpaired electrons. Nevertheless, since the magnetic atoms are often rather special ones (e.g., they occur at the active sites of the iron–sulfur proteins), a knowledge of their state often tells us a lot about the biological properties of the molecule. The iron atom plays a central role in the biological activity of the molecule, and the changes that occur in a biochemical reaction are localized near it. A knowledge of the state of the iron as well as the overall molecular structure might enable one to understand the mechanism by which the molecule works. The structural data obtained from spectroscopic methods are often crude compared with those available from X-ray diffraction, but spectroscopy also gives information about the electronic state and structure of atoms that cannot be obtained from diffraction.

The iron–sulfur proteins form a very important and widely occurring class of biological molecule, as other chapters in this and the preceding volumes testify. The molecular structure and amino acid sequence of several of them have now been determined, and the classification of them into proteins with different active centers is now becoming clearer. It appears there are active centers containing one, two, and four iron atoms, though some molecules may contain more than one of these centers. The structure of one of the proteins with a 1 Fe center and two of the proteins containing 4 Fe–4 S* centers have been determined by X-ray diffraction, but as yet it has not been possible to grow good enough crystals of any of the proteins with 2 Fe–2 S* centers to measure their structure. On the other hand, the spectroscopic methods, notably electron paramagnetic resonance (EPR), electron nuclear double resonance (ENDOR), nuclear magnetic resonance (NMR), and Mössbauer spectroscopy have been very successful in elucidating the electronic structure of the 2 Fe–2 S* centers, as well as the comparatively simple 1 Fe centers. Reviews of the use of these techniques have been given in Volume II by Orme-Johnson and Sands (EPR and ENDOR), Phillips and Poe (NMR), Bearden and Dunham (Mössbauer spectroscopy), and Palmer (a general survey of all the techniques as applied to the 2 Fe–2 S* center in particular). Current work is beginning to make advances in the understanding of proteins containing the more complex 4 Fe–4 S* centers.

The aim of this chapter is to describe the physical background to the behavior of the iron atoms in the 1 Fe and 2 Fe–2 S* centers, which are now reasonably well understood, and then to summarize and discuss the experimental evidence relating to the structure of the 4 Fe–4 S* centers. However, before going on to do this, it is useful to consider briefly the general properties, structural information, and classification of the iron–sulfur proteins that are relevant to the discussion of the data.

Spectroscopic methods have been important in the discovery and understanding of the iron–sulfur proteins, and in particular, the EPR signal centered around $g = 1.94$ (Palmer and Sands, 1966; Hall et al., 1966) in the reduced ferredoxins is one of their most characteristic properties. This signal was one of the first indications of the highly unusual chemistry of the iron–sulfur center, as it is very rare for iron compounds to exhibit average g values less than the free electron value of 2.0023. It is also unusual that oxidized iron (normally Fe^{3+}) does not give an EPR signal while reduced iron (normally Fe^{2+}, a non-Kramers ion) does. Hyperfine effects seen in the EPR spectra when the proteins are enriched with ^{57}Fe and ^{33}S show that the iron and labile sulfur atoms are at the biologically active center of the proteins at which the electron transfer takes place (Chapter 5 of Volume II).

Initially, many of the iron–sulfur proteins were called ferredoxins, while others were named according to their function (e.g., xanthine oxidase) or according to the organism from which they were extracted (e.g., putidaredoxin). For the present purposes, it appears more logical to classify them in terms of their iron–sulfur active centers. This form of classification is particularly relevant when discussing the spectroscopic techniques that yield direct information on the active centers of the proteins. The proteins that can at present be fitted into this scheme are the proteins with 1 Fe centers (rubredoxins), which contain one iron atom and no labile sulfur atoms at their active center; the proteins with 2 Fe–2 S* centers (plant ferredoxins, adrenal ferredoxin, *Pseudomonas putida* ferredoxin, etc.), which have a center containing two iron atoms and two labile sulfur atoms, and the proteins with 4 Fe–4 S* centers. The proteins in this last group contain one (HiPIP's and four-iron bacterial ferredoxins) or two (eight-iron bacterial ferredoxins) active centers each containing four iron atoms and four labile sulfur atoms. There are also more complex iron–sulfur proteins, which may contain flavin groups as well as metal atoms other than iron. Some of these are now thought to contain active centers like those of the simpler proteins described above, but the understanding of these conjugated iron–sulfur proteins is still at an elementary level. However, insight into the properties of the basic types of center, and the ability to recognize them by spectroscopic methods should prove helpful in elucidating the nature of these complex iron–sulfur proteins.

So far, the structures of three iron–sulfur proteins have been determined by X-ray diffraction techniques. They are the rubredoxin from *Clostridium pasteurianum* (Rasse et al., 1974), the HiPIP from *Chromatium* (Carter et al., 1974a), and the eight-iron bacterial ferredoxin from *Peptococcus aerogenes* (Adman et al., 1973). An obvious gap in the series at present is the absence of a structure of a two-iron ferredoxin.

The determination of the rubredoxin structure is described by Jensen in Chapter 4 of Volume II in which the structures of both the rubredoxin and the eight-iron bacterial ferredoxin are illustrated. The rubredoxin center has a single iron atom in a near tetrahedral coordination to four cysteinyl sulfur atoms.

The X-ray analyses of the structures of *P. aerogenes* ferredoxin and *Chromatium* HiPIP show that both contain a center that approximates to four iron atoms at four points of a cube of side 0.2 nm with the four labile sulfur atoms at the alternate points of a concentric cube of side 0.25 nm. The iron atoms are coordinated out along the diagonals of the cube to cysteinyl sulfur atoms in the amino acid chain of the protein. Each iron atom is thus in a near tetrahedral environment of four sulfur atoms. The ferredoxin has two such centers, and so far, the resolution of the X-ray data does not show any significant differences between the centers in the two proteins. Models of these centers are shown in Chapter 8 of Volume II. The latest X-ray studies of *Chromatium* HiPIP at 0.2 nm resolution (Freer et al., 1975) show that the structure of the 4 Fe–4 S* center in the reduced protein is not significantly different from that in the analog compounds containing the anion $[Fe_4S_4(SR)_4]^{2-}$. In these compounds, the cubic structure is distorted from tetrahedral symmetry so that the point group is more nearly tetragonal (Averill et al., 1973). On oxidation of the HiPIP molecule, there is a significant decrease in the Fe–S* bond lengths along one axis of the cube, with the result that the symmetry becomes closer to tetrahedral (Carter et al., 1974b). This effect could result from the removal of an electron from an antibonding orbital. The iron atoms in the protein might also be expected to show differences that are not observed in the low molecular weight analogue compounds and that arise from small distortions to the center caused by the asymmetry of the protein chain.

The occurrence of apparently identical active centers in two proteins with opposite redox properties (HiPIP has a redox potential of $+350$ mV compared with -400 mV for the ferredoxin) has been rationalized by the three-state scheme of Carter et al. (1972). This hypothesis proposes that in oxidized eight-iron ferredoxin and reduced HiPIP, the centers are in an equivalent nonmagnetic state C; the ferredoxin can be reduced to give a paramagnetic state C^-, whereas the HiPIP can be oxidized to give a different paramagnetic state C⁺. On this scheme, it might be expected to be possible to prepare a "super-oxidized" ferredoxin with a center in a state C⁺ and a "super-reduced" HiPIP with a center in a state C⁻. The latter state has been observed by Cammack (1973) under conditions in which the protein structure is modified (Section V,A). The mechanism by which the protein environment of the active centers dictates the normal redox properties of these proteins is not yet understood. The analogue compounds also exhibit three oxidation states (Frankel et al., 1974a).

Although there is no direct X-ray evidence, it seems reasonable on the basis of amino acid sequences and other criteria that all the iron–sulfur proteins containing either four or eight iron and labile sulfur atoms have similar centers to those described above and may also be fitted into the three-state scheme for the different redox states of the centers. This general assumption is further justified by data obtained in various spectroscopic investigations carried out on this family of iron–sulfur proteins with 4 Fe–4 S* centers.

Although as yet, no iron–sulfur protein with a 2 Fe–2 S* center has been analyzed by X-ray diffraction, it has been possible to arrive at a consistent model of the active center from a variety of other data. Gibson et al. (1966) proposed a model for the center of spinach ferredoxin in which the iron atoms in the oxidized protein are high-spin Fe^{3+} antiferromagnetically coupled to give no net spin at low temperatures. In reduced ferredoxin, one of the iron atoms (the Mössbauer spectra suggest that it may be a specific one) is reduced to high-spin Fe^{2+}; antiferromagnetic coupling then gives a net spin $S = \frac{1}{2}$. Gibson et al. (1966) and Thornley et al. (1966) pointed out several consequences of this model, and it is considered in detail by Palmer in Chapter 8 of Volume II. A diagrammatic representation of this model is shown in Fig. 1. The proposed structure of the center involves two labile sulfur atoms providing bridging ligands between the two iron atoms, which are also each coordinated to two cysteinyl sulfur atoms. Thus, the iron atoms are in a tetrahedral or near tetrahedral coordination to four sulfur atoms, as first suggested by Brintzinger et al. (1966). The near tetrahedral arrangement of four sulfur atoms around each iron atom appears, therefore, to be a common feature of the iron–sulfur proteins.

In Section II, some of the theoretical background to the behavior of iron atoms in the active centers of these proteins is given, particularly as it relates to Mössbauer and magnetic resonance spectroscopies. The data on the proteins with 1 Fe and 2 Fe–2 S* centers, which are now reasonably well understood, are described in Sections III and IV. The treatment shows how this information can be fitted into the theoretical framework and also how it provides the necessary background for

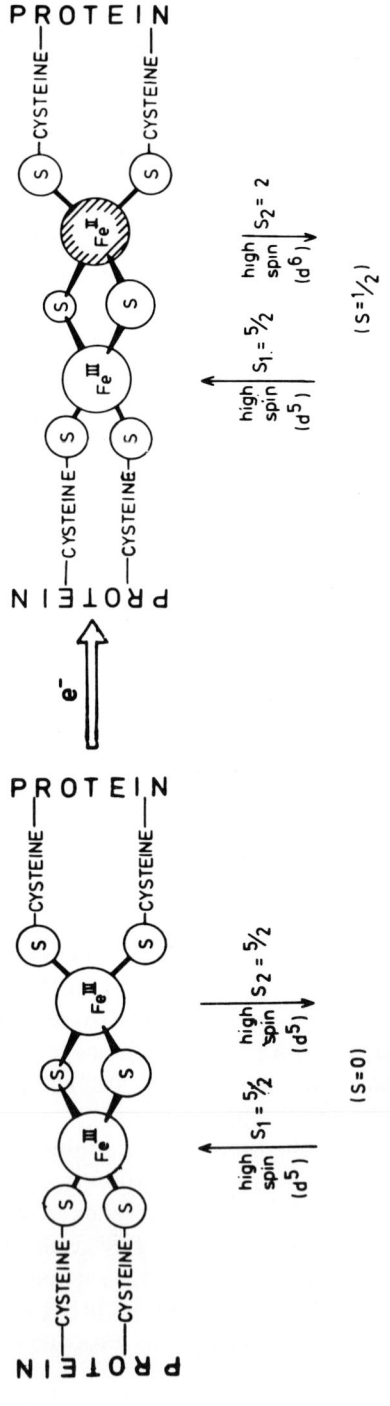

Fig. 1. Proposed model for the 2 Fe–2 S* center of the two-iron ferredoxins. (From Rao et al., 1971.)

understanding the work on the proteins with 4 Fe–4 S* centers discussed in Section V.

II. SPECTROSCOPY OF THE ACTIVE CENTER

Spectroscopic data essentially give information local to the iron atoms in a protein. Conventional macroscopic physical methods for studying the state of iron atoms are often difficult to apply to biological molecules because of their small number compared with the total number of atoms Wet chemical methods are also frequently ruled out because of the inherent instability of the isolated protein.

The measurement of the magnetic properties of transition metal elements in biological molecules has for many years been recognized as an important way of finding the electronic state of the metal atom and hence of providing a clue to the structure and function of the molecule. NMR in biochemistry applies to both protons (PMR) and ^{13}C atoms and is mainly used in determining details of the structure of protein molecules. It may also provide information on a magnetic atom by measuring the field it produces at neighboring nuclei. EPR and ^{57}Fe Mössbauer spectroscopy are very different techniques but are related to each other in that they measure the paramagnetic transition metal atoms themselves and may be used to study their chemical state and bonding, and to obtain qualitative data on the local structure and symmetry in their neighborhood.

In practice, EPR is seldom observed in ions with an even number of electrons (e.g., Fe^{2+}) because Kramers' theorem (which says that in atoms with an odd number of electrons all the levels are at least twofold degenerate) does not apply, and hence the ground state is usually a nonmagnetic singlet. The Mössbauer spectrum may be observed equally well for diamagnetic or paramagnetic ions, since it is not a magnetic resonance technique. This is a great advantage for the iron–sulfur proteins, since both the oxidized and reduced forms of the molecule may be studied.

The magnetic hyperfine coupling between the spins of the electrons and nuclei provides a particularly sensitive and detailed way of measuring magnetic properties. This essentially uses the ^{57}Fe nuclei, which have a spin $I = \frac{1}{2}$, as a probe of the electronic magnetism. The magnetic hyperfine splitting in EPR, ENDOR, or Mössbauer spectra is strongly dependent on and characteristic of the state of the iron atom and hence can be a very powerful method of investigating the iron atoms in a molecule. Furthermore, the magnetic hyperfine interaction is a tensor quantity (i.e., its value is different when measured in different axes in the molecule), and when its anisotropy can be measured, it enables the orbital wavefunction of the iron d electrons to be determined.

For Fe^{3+} atoms where the electron spin relaxation times are long, hyperfine splitting may be observed at low temperatures and by using ^{57}Fe Mössbauer spectroscopy even in zero applied magnetic field. For such slowly relaxing paramagnetic atoms, the application of a small external magnetic field usually produces a simpler Mössbauer spectrum, which is often easier to interpret than the zero-field spectrum. For Fe^{2+} atoms, the electron spin–lattice relaxation time is short and the magnetic resonance technique cannot be applied. Using Mössbauer spectroscopy, it is necessary to apply a strong magnetic field at low temperatures in order to observe magnetic splittings.

The Mössbauer spectrum provides further useful information through both the chemical isomer shift and the electric quadrupole splitting. The shift is sensitive to the chemical state of the iron atom (whether it is ferrous or ferric and high or low spin) and on the degree of covalency. The quadrupole splitting is principally a probe of the local stereochemistry and is a qualitative measure of the distortion from cubic symmetry. It may also provide information on the orbital wave function of the iron d electrons and may help in confirming the chemical state of the iron that is assigned from chemical shift data.

In this section we review the basic ideas that describe how the state of iron atoms in molecules and molecular complexes is determined by their interactions with their surroundings and show how their properties influence the various spectroscopic measurements.

A. Magnetic Moment of Iron in Molecules

The magnetic moment of an atom arises both from the orbital and the spin angular momentum of electrons. However, in compounds of iron (and other atoms containing an incomplete 3d shell), the orbital moment is generally quenched by the electrostatic field of the ligands. A small amount of orbital moment is restored by the spin–orbit coupling, and the magnetic moment becomes a tensor quantity. The energy levels of the ground state of the atom in an applied magnetic field H, which has components H_x, H_y, and H_z along the principal axes x, y, and z of the ligand field may be described by the spin Hamiltonian (Abragam and Bleaney, 1970)

$$\mathcal{H} = g_x \beta H_x S_x + g_y \beta H_y S_y + g_z \beta H_z S_z \\ + D[S_z^2 - \tfrac{1}{3}S(S+1)] + E(S_x^2 - S_y^2)$$

where D and E are the splittings produced by the combined effects of the ligand field and the spin–orbit coupling, and β is the Bohr magneton. For a free electron, g is isotropic and very close to 2 (2.0023). Deviations from this value arise from contributions from the orbital magnetic moment.

For a ferric atom, the electron configuration $3d^5$ gives a ground state ^6S; i.e., there is no orbital moment $(L = 0)$, and so g is isotropic and equal to 2 and $S = \frac{5}{2}$. In typical inorganic Fe^{3+} salts, D lies between 0.01 and 0.1 cm^{-1}, but with highly covalent sulfur ligands (as in reduced rubredoxin), it may be much larger.

For a ferrous atom, the configuration $3d^6$ gives a ground state ^5D. The action of the ligand field is to remove the orbital degeneracy so that the lowest-lying state is an orbital singlet with orbital angular momentum quenched. However, spin–orbit coupling mixes in excited orbital states, which restores some orbital moment and removes some or all of the five-fold spin degeneracy. Since the spin–orbit coupling parameter is negative, corresponding to the spin and orbital moments being coupled parallel to each other, the g values of Fe^{2+} are all greater than or equal to the free electron value. Typically they range between 2.1 and 2.4.

Data on the g values and the ligand field splittings of iron atoms may be obtained from measurements of magnetic susceptibility. For the Fe^{3+} atom, they may more accurately be found from EPR spectra.

It is found from the EPR spectra of the 2 Fe–2 S* centers of reduced two-iron ferredoxins that the spin is $\frac{1}{2}$ and the g values are on average less than 2. Typical values are given in Chapter 5 of Volume II. These low g values are now accepted as originating from the antiparallel coupling between the magnetic moments of the iron atoms. The center contains a ferric atom with spin $S_1 = \frac{5}{2}$ and a ferrous atom with spin $S_2 = 2$.

We consider next, therefore, a system of two atoms with different spins S_1 and S_2 coupled together antiferromagnetically. This is covered in some detail by Palmer in Chapter 8 of Volume II. The spin Hamiltonian for the system is of the form

$$\mathcal{H} = \beta \mathbf{H} \cdot (\mathbf{g}_1 \cdot \mathbf{S}_1 + \mathbf{g}_2 \cdot \mathbf{S}_2) + J \mathbf{S}_1 \cdot \mathbf{S}_2 + D_1 S_{1z}^2 + D_2 S_{2z}^2 + \cdots$$

where J is the exchange coupling and its positive sign corresponds to antiparallel alignment of the two spins. It is assumed that J is large compared with any other term in the Hamiltonian. The coupling produces states of resultant spin S, where S is the vector sum of the spin of the individual atoms, i.e., $\mathbf{S} = \mathbf{S}_1 + \mathbf{S}_2$. For the reduced ferredoxins, the ground state of the center has $S = \frac{1}{2}$, and it lies 100 cm^{-1} or more below the other states, so that it is the only one that needs to be considered at the temperatures where experiments are carried out. The terms in D_1

and D_2 in the Hamiltonian arise from ligand field interactions with the electronic spins, but their resultant will be zero for the state with $S = \frac{1}{2}$. The effective spin Hamiltonian of the ground state may then be written as

$$\mathcal{H} = \beta \mathbf{H} \cdot \mathbf{g} \cdot \mathbf{S}$$

In zero field, the spins S_1 and S_2 precess about their resultant S. In an external magnetic field H, S precesses about the field direction, and hence so do S_1 and S_2. It has been shown (Gibson et al., 1966; Thornley et al., 1966) that this leads to the characteristic g values with an average of less than 2.

B. Electron Spin Relaxation

In order to observe EPR, the electron spin relaxation times τ must be such that their contribution $\hbar/\tau$ to the width of the level is smaller than the separation $g\beta H$ of the levels. Similarly, magnetic hyperfine interactions will only be resolved in either EPR, ENDOR, or Mössbauer spectra if $\hbar/\tau$ is less than the separation of the nuclear levels. The electron spin relaxation times result from the spin–lattice relaxation time T_1 and the spin–spin relaxation time T_2.

Spin–lattice relaxation processes take place via the electron orbit, since there is no direct interaction between an electron spin and an oscillating electric field (i.e., the lattice). The orbital magnetic moment behaves like a classical electric current loop and so can interact with the lattice. In this direct relaxation process, which is important at low temperatures, and for a system with one electron spin ($S = \frac{1}{2}$), the electron spin–lattice relaxation rate $1/T_1$ is proportional to the orbital magnetic moment, i.e., to $(g - 2)$. For the Fe^{3+} ion which is in an S state, i.e., $L = 0$, T_1 is long, so that EPR is often observed at room temperature. For Fe^{2+}, there is a considerable amount of orbital magnetic moment $[(g - 2) \sim 0.1$ to $0.4]$ and T_1 is usually very short even at liquid helium temperatures.

Gayda et al. (1976) have investigated the electron spin relaxation in reduced spinach ferredoxin at relatively high temperatures (above 77°K). The variation of $1/T_1$ with temperature as determined from the linewidth of the EPR signal indicates that the relaxation takes place via an Orbach process (Orbach, 1961; Gibson et al., 1966), which involves electron spin transitions to the excited states of the antiferromagnetically coupled system.

The condition for observing the magnetic hyperfine interaction is that $\tau \gg \hbar/A$, where A is the strength of the hyperfine interaction (see Section II,C). This essentially means that the electronic magnetic moment must remain stationary for a time long compared with the precession period of the nuclei in the field of the electrons (or the electrons in the field of the nuclei). Otherwise, if $\tau \sim \hbar/A$, both the EPR and Mössbauer lines broaden and overlap and the magnetic hyperfine interaction cannot be resolved. If the electron spins flip rapidly compared with $\hbar/A$, no magnetic effects are seen in the Mössbauer spectrum, which then looks similar to that of a diamagnetic substance. Finally, when $\tau \ll \hbar/A$, it becomes impossible for the spins to come into equilibrium with the lattice so that the EPR signal may be unobservable (i.e., it saturates). However, the Mössbauer spectrum is still observable, provided the sample remains in the solid state. So magnetic hyperfine effects are seen in Mössbauer and EPR spectra when the relaxation times are long. This may be achieved by cooling the specimen to low temperatures, which increases the spin–lattice relaxation time T_1, and by using magnetically dilute samples, which increases the spin–spin relaxation time T_2. Iron–sulfur proteins are generally magnetically dilute (except possibly in the case of proteins containing more than one paramagnetic center in each molecule), so that hyperfine interactions are observable in their ^{57}Fe Mössbauer spectra at low temperatures.

C. Magnetic Hyperfine Interactions

The magnetic interactions between the electronic and nuclear magnetic moments give rise to a splitting of the energy levels of an atom, which is known as hyperfine splitting, and which may be observed in EPR, ENDOR, and Mössbauer spectra. These kinds of measurement are complementary; in EPR and ENDOR, the magnetic field produced by the nuclei at the electrons is measured, while with Mössbauer spectroscopy, it is the field produced by the electrons at the nuclei that is measured. We shall therefore consider the relation between the two kinds of spectra by describing the splitting of the energy levels of an atom produced by a magnetic field.

We consider as a simple example an atom with electron spin $S = \frac{1}{2}$ with isotropic g values and with nuclear spin I. The energy levels in a magnetic field H are eigenvalues of the spin-Hamiltonian

$$\mathcal{H} = g\beta\mathbf{H}\cdot\mathbf{S} + A\mathbf{S}\cdot\mathbf{I} - g_n\beta_n\mathbf{H}\cdot\mathbf{I}$$

where g_n is the nuclear g factor and β_n is the nuclear magneton. The energy levels of this atom are shown in Fig. 2 as a function of the magnetic field. Two nuclear states are shown, one with spin $I_g = \frac{1}{2}$ and an isotropic negative hyperfine coupling constant A representing the ground state of ^{57}Fe, and the other with spin $I_e = \frac{3}{2}$ and hyperfine coupling constant

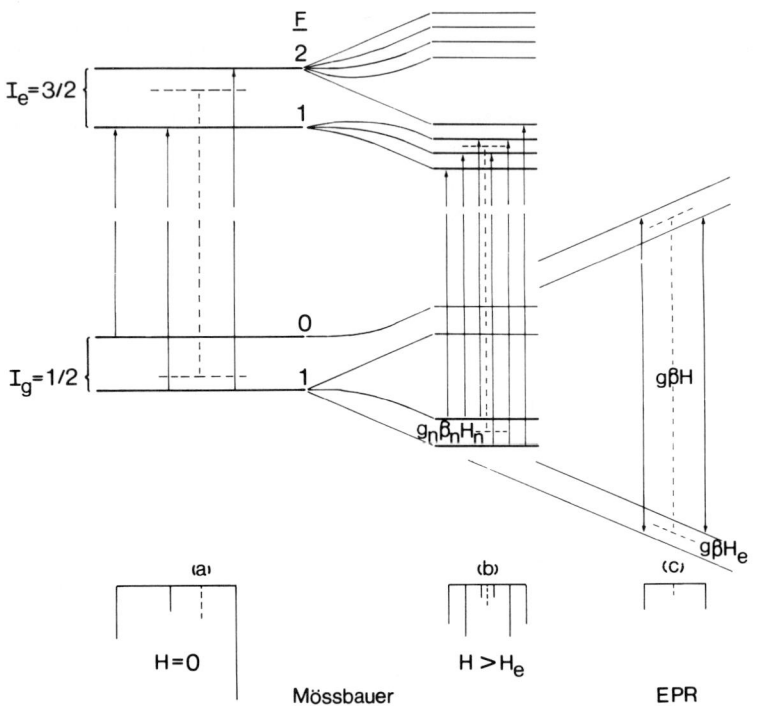

Fig. 2. Energy levels of a paramagnetic atom with $S = \frac{1}{2}$ and $I = \frac{1}{2}$ and $\frac{3}{2}$. The Mössbauer transitions are shown in (a) for $H = 0$ and (b) for $H > H_e$, where $H_e = A/2g\beta$. The EPR transitions are shown in (c). (From Johnson, 1971).

$-0.562A$, corresponding to the first excited state of ^{57}Fe. The Mössbauer spectrum involves transitions between these states with the selection rules $\Delta m_I = 0, \pm 1$, $\Delta m_S = 0$, while the EPR spectrum involves transitions within the ground nuclear state only, and with selection rules $\Delta m_I = 0$, $\Delta m_S = \pm 1$; m_I and m_S are the projections of the nuclear and electron spins along the z direction.

In zero magnetic field, the hyperfine interaction couples S and I to give a total angular momentum $\mathbf{F} = \mathbf{S} + \mathbf{I}$, similar to the situation in a free atom. For slow electron-spin relaxation, the Mössbauer spectrum shows the asymmetrical three-line pattern of Fig. 2a, as seen, for example, in the low-temperature spectra of the reduced two-iron ferredoxin from *Euglena* (Johnson et al., 1968).

When the applied field is larger than $A/2g\beta$, the fluctuating components of the hyperfine interaction tensor are suppressed, and the splitting of the hyperfine levels is $A/2$. The Mössbauer spectrum becomes symmetrical and shows a six-line Zeeman pattern (Fig. 2b) with an effective field at the nuclei $H_n = A/2g_n\beta_n$. Effectively, the nuclear and electronic spins have become decoupled from each other and precess independently about the external field. In the EPR spectrum, the line is split by $2H_e$, where $H_e = A/2g\beta$ is the field at the electrons produced by the nuclei (Fig. 2c). For ^{57}Fe, $g_n = 0.18$, and since $\beta/\beta_n = 1840$, H_n is about 10^4 times $A/g\beta$ (or $2H_e$). Typical values for $A/g\beta$ from EPR measurements on iron–sulfur proteins are about 20 G and for H_n from Mössbauer data are 200 kG.

When the external magnetic field becomes large (tens of kilogauss), the magnitude of the Zeeman splitting changes, since the term $-g_n\beta_n\mathbf{H}\cdot\mathbf{I}$ in the Hamiltonian becomes important and the effective field at the nuclei is then $\mathbf{H}_n + \mathbf{H}$. A simple isotropic magnetic hyperfine interaction gives a Mössbauer spectrum of six lines with intensities in the ratio 6:0:2:2:0:6 for a field applied parallel to the γ-ray direction and 3:4:1:1:4:3 for a field applied perpendicular to the γ-ray direction. Thus observation of changes in intensity of the lines in a magnetically split spectrum with different orientations of the applied field can assist in the interpretation of complex spectra. The separation of the lines is proportional to the effective magnetic field at the nucleus, which equals the hyperfine field plus the applied field. Thus, if the outer lines move out as the applied field is increased, the hyperfine field is positive, i.e., parallel to the applied field, whereas if the lines move in as the applied field is increased, the hyperfine field is negative, i.e., antiparallel to the applied field. The hyperfine field at the nucleus is directed antiparallel to the magnetic moment of the atom, and therefore positive hyperfine fields are usually only found in systems with antiferromagnetic coupling. Although the iron–sulfur proteins do not usually exhibit such a simple isotropic magnetic hyperfine interaction, the above features can frequently be observed in their Mössbauer spectra.

For anisotropic g and A values, the splittings become a function of the angle that the field makes with the symmetry axes of the atom, and in a specimen where these axes are randomly oriented, the lines will become broadened.

We now consider the case of a system consisting of a pair of antiferromagnetically coupled atoms. The magnetic hyperfine interaction can be described by the spin Hamiltonian derived by Johnson et al. (1971).

$$\mathcal{H} = \beta\mathbf{H}\cdot\mathbf{g}\cdot\mathbf{S} + \tfrac{7}{3}\mathbf{S}\cdot\mathbf{A}_1\cdot\mathbf{I}_1 - \tfrac{4}{3}\mathbf{S}\cdot\mathbf{A}_2\cdot\mathbf{I}_2 - \beta_n\mathbf{H}\cdot\mathbf{g}_n\cdot(\mathbf{I}_1+\mathbf{I}_2)$$

The EPR spectrum results mainly from the first term, and thus a single spectrum is observed resulting from the spin and g values of the coupled system. However, the two atoms will have different magnetic hyperfine interactions arising from the second and third term of the above Hamiltonian, and thus two components are seen in the Mössbauer spectrum.

In an external magnetic field, only the components of the g and A tensors parallel to the field are important. With randomly oriented molecular axes, the field has a maximum probability of being perpendicular to the axis of highest symmetry. Referred to the axes x, y, and z of the ligand field, the effective spin Hamiltonian then approximates to

$$\mathcal{H}_{\text{eff}} = g_x \beta H S_z + \tfrac{7}{3} A_{1x} S_z I_{1z} - \tfrac{4}{3} A_{2x} S_z I_{2z} - g_n \beta_n H (I_{1z} + I_{2z})$$

where z is the direction of the magnetic field. The two atoms give separate Mössbauer spectra with hyperfine fields given by

$$H_{1z} = \frac{7}{3} \frac{A_{1x} S_z}{g_n \beta_n} \quad \text{and} \quad H_{2z} = -\frac{4}{3} \frac{A_{2x} S_z}{g_n \beta_n}$$

Since S_z, the component of S along the field direction, can take either of the two values $\pm \tfrac{1}{2}$, there are two possible hyperfine field values for each atom, one corresponding to the electron spin pointing up, and the other for spin down. The lower state has $S_z = -\tfrac{1}{2}$, and it can be seen that for Fe^{3+} the hyperfine field is negative, i.e., directed antiparallel to the electronic magnetic moment, so that the effective field at the nuclei becomes smaller in the presence of an applied field. For Fe^{2+}, on the other hand, the hyperfine field is positive. The spectra for the excited electronic state with $S_z = +\tfrac{1}{2}$ can also be observed, and they have the opposite signs of hyperfine field. The intensity of these spectra is small because of the lower population of the $S_z = +\tfrac{1}{2}$ state relative to $S_z = -\tfrac{1}{2}$. In large fields at low temperatures (e.g., 60 kG at 1°K), their population is practically negligible.

The hyperfine field at the nucleus of an atom is made up of several contributions (Marshall and Johnson, 1962):

$$H_n = H_s + H_L + H_d$$

H_s is the contribution from the polarization of the core s electrons caused by the distortion of the s electron orbitals by the 3d electrons, which carry the magnetic moment of the atom. This is frequently known as the Fermi contact term; it is large, negative, and isotropic and is usually the dominant contribution to the hyperfine field. H_L arises from the orbital magnetic moment of the 3d electrons and H_d is the dipolar contribution from their spin magnetic moment.

D. Transferred Hyperfine Interactions

In the region of a paramagnetic atom, there will be a hyperfine magnetic field at neighboring nuclei as well as at the nucleus of the atom itself. If a nuclear magnetic resonance measurement is performed on these neighboring nuclei (protons in PMR and ^{13}C nuclei in ^{13}C NMR), the transferred hyperfine field will add to the static external magnetic field H and will shift the position of the resonance signal by an amount ΔH.

The PMR measurements are usually made at about room temperature on the proteins in liquid solution. The electronic relaxation times are then very short and the transferred hyperfine field is a time average. The shift $\Delta H/H$ is proportional to χ the susceptibility of the paramagnetic atom times H_n^t the transferred hyperfine field at the proton, and it can provide a probe of the magnetism in the region of the paramagnetic atom. It is found for solutions of paramagnetic atoms that $\Delta H/H$ varies inversely with the temperature T, as expected for the Curie law ($\chi \propto 1/T$).

The behavior of contact-shifted β-CH$_2$ proton resonances in iron–sulfur proteins has been discussed in Chapters 7 and 8 in Volume II. For the 2 Fe–2 S* and 4 Fe–4 S* centers of the iron–sulfur proteins, $\Delta H/H$ is found for some protons to increase and for others to decrease linearly with $1/T$. This provides indirect evidence for the antiferromagnetic coupling between the iron atoms. Similarly, Packer et al. (1972) have observed contact shifted resonances in the ^{13}C NMR spectra of *Clostridium acid-urici* ferredoxin. These have been assigned to the phenyl carbon atoms of the two tyrosine residues which are situated close to the iron–sulfur centers of the protein. The position of these resonances depends on the redox state of the protein.

E. Mössbauer Chemical Shifts

The Mössbauer chemical shift has become well established as a powerful tool for probing the electronic charge density at nuclei in solids. It enables the oxidation state and degree of covalency to be characterized. It is not, however, an absolute measurement, and it requires an empirical calibration in order to make use of its rather low sensitivity, at any rate as far as iron is concerned.

It is the effects of the covalency of the sulfur ligands that makes the chemical shifts observed in the iron–sulfur proteins rather different from those found in typical inorganic complexes of iron. There is as yet no general theory of the chemical shift, but it is found that the chemical shift tends to decrease (i.e., moves to more negative velocities) as the degree of covalency of the ligands increases (i.e., in the order –H$_2$O, –Cl$^-$, –O^{2-}, –S^{2-}, –CN$^-$, etc.). Also, it is systematically less by about 0.2 mm/second for tetrahedral coordination compared with octahedral coordination to the same ligands. The only case for which the value of the chemical shift can lead to a positive identification of the chemical state without calibration is Fe^{2+}, which normally has a large positive value. In inorganic salts, the Fe^{2+} shift is of the order of 1.5 mm/second, but in biological compounds with tetrahedral coordination (e.g., reduced rubredoxin), it can be as low as 0.6 mm/second, which overlaps with the values found for Fe^{3+} in many other compounds.

F. Electric Quadrupole Interactions

In addition to magnetic hyperfine splitting, the energy levels of the nuclei may be split by the electric quadrupole interaction between the electric quadrupole moment of the nucleus eQ (where e is the electronic charge) and the electric field gradient eq ($=V_{zz}$) produced by the atom itself and surrounding atoms at the nucleus. For ^{57}Fe, a quadrupole moment exists only for the excited state of the nucleus that has a spin I of $\frac{3}{2}$ and a quadrupole moment Q of about 0.18 b. The interaction is then described by the Hamiltonian

$$\mathcal{H}_Q = \tfrac{1}{4}e^2qQ[I_z^2 - \tfrac{5}{4} + \tfrac{1}{3}\eta(I_x^2 - I_y^2)]$$

where the asymmetry parameter $\eta = (V_{xx} - V_{yy})/V_{zz}$, and V_{xx} ($=\partial^2 V/\partial x^2$), etc., are the components of the electric field gradient tensor. When there is no magnetic hyperfine interaction or applied magnetic field, the resulting Mössbauer spectrum is split into two lines separated in energy by

$$\Delta E_Q = \tfrac{1}{2}e^2qQ(1 + \tfrac{1}{3}\eta^2)^{1/2}$$

Hence, the quadrupole splitting ΔE_Q is mainly a measure of the asymmetry of the local structure in the region of the iron atom. It may also help to confirm the oxidation state of the atom. A larger value (~ 3 mm/second) is usually associated with Fe^{2+} compounds, compared with a much smaller value (<0.6 mm/second) found in Fe^{3+} compounds. The sign of the splitting may be determined by the application of an external magnetic field, or by measurements on single crystals. This enables the orbital wave-function of the d electrons to be determined. This, in turn, may be used to deduce the nature of the distortion of the neighboring atoms from cubic symmetry.

III. PROTEINS WITH 1 Fe CENTERS

The simplest active center in the iron–sulfur proteins is that of the rubredoxins. Rubredoxins contain only one center per molecule, and that center consists of an iron atom surrounded by a distorted tetrahedron of cysteinyl (i.e., nonlabile) sulfur atoms. Mössbauer data has been obtained on the rubredoxins from *C. pasteurianum* and *Chloropseudomonas ethylica* (Phillips et al., 1970a; Rao et al., 1972a). From measurements on the rubredoxins in the oxidized and reduced states, the chemical shifts and magnetic hyperfine interactions of Fe^{2+} and Fe^{3+} in tetrahedral sulfur coordination have been measured. This effectively calibrates these quantities by allowing for the effects of covalency on the iron atom in this environment. The data on the proteins with 1 Fe centers can then be used in interpretation of the data on the proteins with 2 Fe–2 S* and 4 Fe–4 S* centers, where the antiferromagnetic coupling between the iron atoms produces a behavior very different from that with single atoms.

The rubredoxins give an EPR signal when oxidized (Fe^{3+}), but not when reduced (Fe^{2+}) (Atherton et al., 1966), in contrast to the ferredoxins, where the opposite behavior is found. The EPR spectra of oxidized rubredoxin show a strong signal at $g \sim 4.3$ and a weaker one at $g \sim 9.4$. These are characteristic of a nonaxial ligand field (Wickman et al., 1965; Blumberg, 1967; Dowsing and Gibson, 1969) and can be described by a spin Hamiltonian of the form

$$\mathcal{H} = D[S_z^2 - \tfrac{1}{3}S(S+1)] + E(S_x^2 - S_y^2)$$
$$+ \tfrac{1}{6}a[S_x^4 + S_y^4 + S_z^4 - \tfrac{1}{5}S(S+1)(3S^2 + 3S - 1)]$$

with axial (D), nonaxial (E), and possibly fourth-order cubic (a) terms. E is of the order of a third of D.

The Mössbauer spectrum of oxidized rubredoxin at 4.2°K is shown in Fig. 3a and is a six-line pattern characteristic of a Fe^{3+} atom with an effective field at the nuclei of -370 kG. The hyperfine splitting is observed because the electron spin relaxation times are long, and a single field arises because the distorted tetrahedral ligand field splits the energy levels of the atom so that the ground state is approximately $S_z = \pm\tfrac{5}{2}$ and is the only state with an appreciable population at this temperature. The value of the hyperfine field is -395 kG (Johnson, 1974) and may be compared with -550 kG found in Fe_2O_3, with octahedral oxygen coordination, and -475 kG in ferric trispyrollidyl dithiocarbamate, which has octahedral sulfur coordination. The low magnitude in rubredoxin is partly a result of the covalent bonds between the iron and the sulfur ligands and partly due to the tetrahedral coordination. The sign of the quadrupole splitting is positive. It arises from the asymmetry of the charges of the sulfur ligands, and shows that they are distorted by a compression along the symmetry (z) axis.

The Mössbauer spectra of reduced rubredoxin at 4.2°K (Fig. 3b and c) are characteristic of high spin Fe^{2+}. In zero field, there is a large quadrupole splitting (3.3 mm/second) and a shift (0.65 mm/second) that is

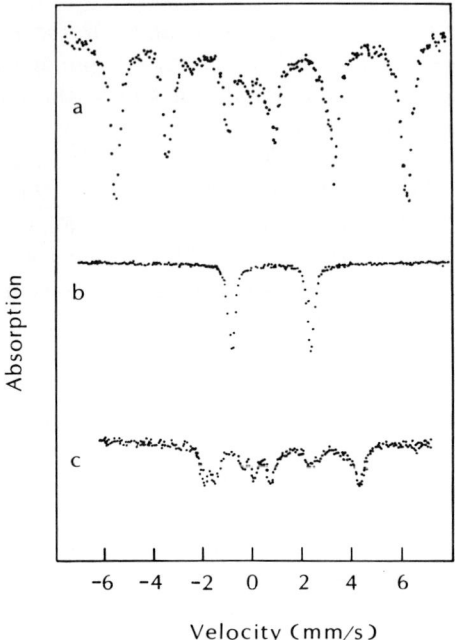

Fig. 3. Mössbauer spectra of *C. pasteurianum* rubredoxin at 4. 2°K (a) oxidized, (b) reduced, and (c) reduced with a field of 30 kG applied perpendicular to the direction of the γ-ray beam. (After Rao et al., 1972a.)

low for Fe^{2+} but that is entirely consistent with tetrahedral sulfur ligands. Magnetic hyperfine splitting is observed when large fields (up to 60 kG) are applied and the effective field at the nuclei is predominantly perpendicular to the symmetry axis of the ligand field, so that

$$\mathbf{H}_{eff} = \mathbf{H} + \mathbf{H}_{nx}$$

The internal field H_{nx} is a function of the applied field H, and its saturated value is estimated to be -210 kG (Johnson, 1974).

The sign of the quadrupole splitting is seen to be negative, and its value of 3.3 mm/second at 4.2°K is a typical value for Fe^{2+}. The splitting arises from the asymmetrical charge distribution of the electrons of the Fe^{2+} atom and shows that the ground state is d_{z^2}, so that once again the sulfur tetrahedron is distorted by a compression along the z axis.

A value for the hyperfine field has been obtained by Johnson (1974) using the approach of Marshall and Johnson (1962). The core polarization field for Fe^{2+} ($S = 2$) may be estimated to be $\frac{4}{5}$ of the Fe^{3+} ($S = \frac{5}{2}$) field, i.e. $H_s = -315$ kG. The dipolar field is about $+64$ kG, and the orbital field H_{Lx} is estimated to be $+36$ kG. Adding these values gives a total of -215 kG, which agrees well with the measured value of -210 kG.

IV. PROTEINS WITH 2 Fe–2 S* CENTERS

A. General Properties

After the rubredoxins, the iron–sulfur proteins with 2 Fe–2 S* centers are the next simplest molecule. The generally accepted model for the active center of these proteins is shown in Fig. 1.

The plant two-iron ferredoxins were the first proteins of this type to be isolated (Davenport et al., 1952). Their biological role and many of their properties have already been dealt with in other chapters of this treatise and will not be described here. The EPR spectrum of the reduced form (Palmer and Sands, 1966; Hall et al., 1966) was the first indication of the unusual chemistry of these proteins.

A second type of two-iron ferredoxin comprises those such as adrenal ferredoxin (adrenodoxin) and P. putida ferredoxin (putidaredoxin) that are involved in hydroxylation reactions. Proteins with similar EPR spectra have been isolated from a number of aerobic bacteria (see Hall et al., 1974). Although their amino acid sequences are very different from those of the plant ferredoxins, they appear to contain the same type of iron–sulfur center, as indicated by the similarity of their ENDOR (Fritz et al., 1971) and Mössbauer spectra (Cammack et al., 1971; Dunham et al., 1971; Münck et al., 1972).

The EPR spectra of reduced adrenal ferredoxin and other proteins of this type are nearly axial (Chapter 5 of Volume II), in contrast to the rhombic spectra of reduced plant ferredoxins. The electron spin relaxation in the former is slower, as indicated by the fact that the EPR spectra are observed at higher temperatures and are more readily saturated by microwave power at low temperatures. The slower relaxation rate is consistent with the smaller g value anisotropy of the signal, i.e., a smaller value of $(g - 2)$, resulting from a smaller contribution from the orbital magnetic moment.

The Mössbauer spectra of reduced adrenodoxin (Cammack et al., 1971) show magnetic hyperfine structure at all temperatures from $1.7°K$ to $244°K$, in contrast with the reduced plant ferredoxins, which show it only at the lower temperatures. This is again a consequence of a longer electron spin relaxation time in reduced adrenodoxin.

The two-iron ferredoxins obtained from anaerobic bacteria such as C. pasteurianum appear to be of a third type, although relatively little is known of their structure or function. Their EPR signal shows rhombic symmetry but has less g value anisotropy than the plant ferredoxins (Chapter 5 of Volume II), and the temperature dependence of the signal is intermediate between that of plant and adrenal ferredoxins. Mössbauer spectra indicate a close similarity to the other two-iron ferredoxins (Dunham et al., 1971).

Xanthine oxidase is a complex protein that behaves as a dimer, each half of the molecule containing a molybdenum atom, a flavin group and four atoms each of iron and labile sulfur. The latter are now believed to be arranged as two 2 Fe–2 S* centers that have redox potentials of −345 and −303 mV and give rise to different EPR signals (Bray et al., 1976). In xanthine oxidase from milk, one center has an EPR spectrum very like that of the plant ferredoxins, but the other center has a signal that is only observed at low temperatures and has $g_{av} = 2.01$. There is evidence that this center is spin-coupled to Mo^{5+} in the reduced protein (Lowe et al., 1972). In spite of the anomalous g values of the second center, it is believed that both types of center are similar to those in the two-iron ferredoxins.

Xanthine oxidase was one of the first iron–sulfur proteins to be investigated by Mössbauer spectroscopy (Johnson et al., 1967). The spectra of the protein over a range of temperatures and magnetic fields were found to resemble those of the plant ferredoxins (Johnson et al., 1969).

B. Mössbauer Data

There have been a number of Mössbauer investigations of the proteins with 2 Fe–2 S* centers, mainly using samples enriched in ^{57}Fe either by chemical exchange or growing on an enriched medium (e.g., Johnson et al., 1967, 1968; Johnson and Hall, 1968; Moss et al., 1968; Cammack et al., 1971; Dunham et al., 1971; Rao et al., 1971, 1972b; Münck et al., 1972).

At 195°K (where $\tau \ll \hbar/A$), the Mössbauer spectra of the oxidized plant ferredoxins consist of a Fe^{3+} quadrupole-split doublet, with some broadening and structure that can be attributed to a slight inequivalence of the two iron atoms. On reduction, two well-separated doublets are observed (Fig. 4). The doublet with the smaller splitting is very similar to that of the oxidized protein and arises from a Fe^{3+} atom that is unchanged on reduction. The doublet with the larger splitting also has an increased chemical shift, indicating that the other atom is reduced to Fe^{2+} (see Fig. 1).

At low temperatures (e.g., 4.2°K), the Mössbauer spectra of the reduced proteins show magnetic hyperfine splitting because the electron spin–lattice relaxation time has become long compared with $\hbar/A$. The spins S_1 and S_2 precess about their resultant S, and the observed spectrum is asymmetrical and has broad and partly resolved lines. When a small magnetic field ($H > A/2g\beta$) is applied, a more symmetrical spectrum

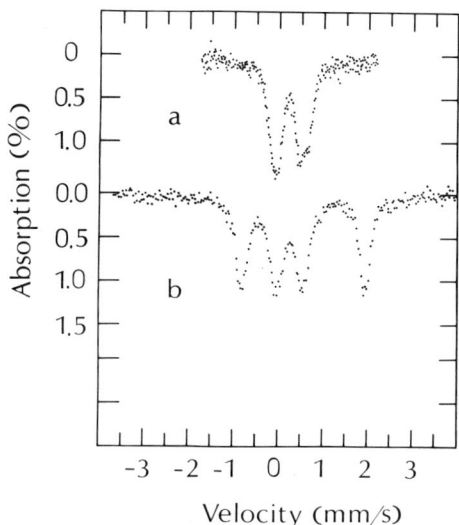

Fig. 4. Mössbauer spectra of *Scenedesmus* ferredoxin at 195°K (a) oxidized and (b) reduced. (From Johnson, 1971.)

with sharper lines results. The effect of the field is to cause S_1 and S_2 to precess about it, and this results in two separate effective magnetic fields at the nuclei. The resulting spectrum is a superposition of two contributions, one from the Fe^{3+} atom and the other from the Fe^{2+} atom, and the lines from each are strongly overlapping.

On increasing the size of the applied field, the spectrum splits into two components, showing that H_{1z} and H_{2z} (see Section II,C) have different signs owing to the antiferromagnetic coupling between the two iron atoms. However, a problem can arise in interpreting these spectra, because lines arise from the $S_z = +\frac{1}{2}$ excited state as well as from the $S_z = -\frac{1}{2}$ ground state. A comparison of the spectra taken in large applied magnetic fields at 4.2°K and at lower temperatures reveals the presence of lines arising from the excited state; the excited state population decreases significantly as the temperature is lowered, and these lines virtually disappear at 1.7°K (Cammack *et al.*, 1971).

The separation of the spectrum due to the ground state from that due to the excited state has been carried out by Johnson (1974) by subtracting the 1.7° and 4.2°K spectra of reduced adrenal ferredoxin taken in a perpendicular field of 30 kG, suitably weighting each according to the population of the two states at these temperatures. The separated spectra for the two states are shown in Fig. 5 and are interpreted in terms of stick spectra. The Fe^{3+} atom gives a spectrum that approximates to the six-line pattern of an isotropic magnetic atom with a small quadrupole

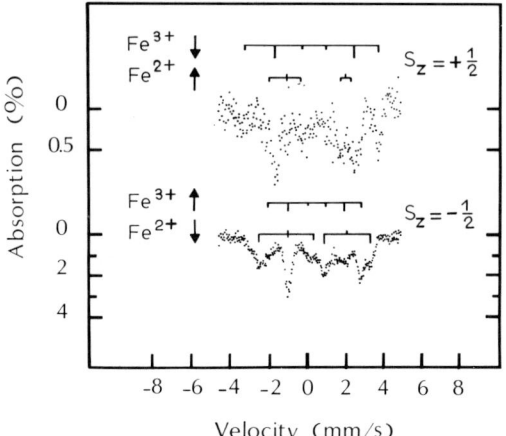

Fig. 5. Mössbauer spectra of the $S_z = -\frac{1}{2}$ and $S_z = +\frac{1}{2}$ states of reduced adrenal ferredoxin in a perpendicular field of 30 kG derived from spectra taken at 4.2° and 1.7°K. The stick spectra show proposed assignments for the principal lines contributed by the Fe^{3+} and Fe^{2+} atoms. (From Johnson, 1974.)

interaction, while the spectrum due to the Fe^{2+} atom is more closely described by the action of an effective magnetic field on a large quadrupole splitting.

It has been remarked (Dunham et al., 1971) that conclusions based on line positions are not valid in such complicated spectra and that only a detailed computer fit gives meaningful results. However, in a large applied field, the effective field approximation gives a good description of the hyperfine interaction, and thus the essential physics of the spectra can be pictured as described here. This is particularly true in the case of adrenal ferredoxin, where the overlapping of the outermost lines is not so severe as in the case of the plant ferredoxins.

The Mössbauer spectra give confirmation of the antiferromagnetic coupling between the Fe^{2+} and Fe^{3+} atoms in the reduced 2 Fe–2 S* centers. Atoms with spins pointing up (parallel to the resultant spin S) have the effective field at the nucleus reduced by the external field (since their hyperfine field is negative), so that the spectrum is compressed by the field. On the other hand, atoms with spins down have their effective field increased, and the spectrum appears to expand in the external field. In general, it is not easy to identify all the lines, especially those due to Fe^{2+}, from the Mössbauer spectrum alone. Fritz et al. (1971) have measured the A values for the iron atoms in a number of two-iron ferredoxins, and using these data, Dunham et al., (1971) have performed a detailed interpretation of the Mössbauer spectra of reduced spinach ferredoxin at low temperatures in applied magnetic fields using computer simulation techniques.

From Fig. 5, the hyperfine fields at the nuclei are estimated to be -180 kG for the Fe^{3+} atom and very approximately $+120$ kG for the Fe^{2+} atom. Because of the coupling between the two atoms, the field is reduced from its value in the free atoms, and this may be understood in terms of the discussion in Chapter 8 of Volume II for coupled spins S_1 and S_2. The hyperfine fields of the Fe^{3+} and Fe^{2+} atoms are proportional to the projections $\langle S_1 \rangle$ and $\langle S_2 \rangle$ of their spins along the total spin S. Thus,

$$H_{1z} = (\langle S_1 \rangle / S_1) H_{nz}(Fe^{3+}) \quad \text{and} \quad H_{2z} = (\langle S_2 \rangle / S_2) H_{nz}(Fe^{2+})$$

where H_{nz} (Fe^{3+}) and H_{nz} (Fe^{2+}) are z components of the hyperfine field for the uncoupled ferric and ferrous atoms, respectively. Following the approach adopted for rubredoxin in Section III, and using values found for that protein, Johnson (1974) has obtained values of -185 and $+135$ kG for the z components of the hyperfine fields of the coupled Fe^{3+} and Fe^{2+} atoms, in good agreement with the experimental values.

V. PROTEINS WITH 4 Fe–4 S* CENTERS

A. General Properties

The most complicated of the iron–sulfur proteins that have so far been systematically studied are those containing 4 Fe–4 S* centers. These include the HiPIP's and four-iron bacterial ferredoxins, as well as the eight-iron bacterial ferredoxins, which contain two centers per molecule. As discussed earlier, it is useful to consider these proteins together, as it is their similar active centers that are investigated by spectroscopic techniques.

The simplest members of this class are the four-iron ferredoxins, which have been extracted from a number of microorganisms, in particular *Bacillus* and *Desulfovibrio*. Although no X-ray studies have yet been reported on these proteins, it seems very probable that they contain a center similar to those in *P. aerogenes* ferredoxin and *Chromatium* HiPIP. The ferredoxins from *Desulfovibrio* have molecular weights of about 6000. The amino acid sequence of *Desulfovibrio gigas* ferredoxin is homologous with the N-terminal half of the sequence of the eight-iron ferredoxins; in particular, the four cysteines that bind to one of the iron–sulfur centers in *P. aerogenes* ferredoxin are conserved in *D. gigas* ferredoxin (Travis et al., 1971). The ferredoxins from *Bacillus* have higher molecular weights, about 8500, but are otherwise similar in properties.

The proteins are reduced with redox potentials between −280 mV and −380 mV (Mullinger et al., 1975; Zubieta et al., 1973).

The four-iron ferredoxins show only a small EPR signal at $g = 2.02$ in the oxidized state as normally prepared, but on reduction, they show a strong signal (accounting for one unpaired electron per protein molecule), with $g_{av} = 1.96$ (Shethna et al., 1971). The EPR spectrum of reduced *B. stearothermophilus* ferredoxin is shown in Fig. 6a. When the ferredoxin

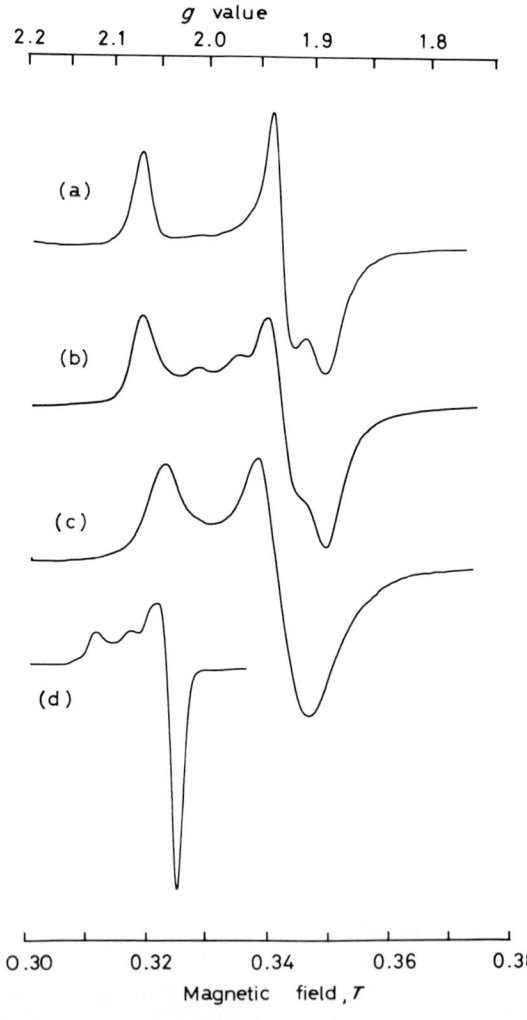

Fig. 6. EPR spectra of proteins with 4 Fe–4 S* centers: (a) reduced *B. stearothermophilus* ferredoxin. (b) partially reduced *C. pasteurianum* ferredoxin, (c) superreduced *Chromatium* HiPIP in 80% DMSO, and (d) oxidized *Chromatium* HiPIP.

is reconstituted with ^{57}Fe, a broadening is observed in the EPR spectra that arises from magnetic hyperfine interactions (Mullinger et al., 1975). The optical absorption, optical rotatory dispersion, and circular dichroism spectra of B. stearothermophilus ferredoxin are indicative of a 4 Fe–4 S* center similar to those in the eight-iron ferredoxins (Mullinger et al., 1975).

The temperature dependence of the EPR spectra of the four-iron ferredoxins is comparable with that of the eight-iron ferredoxins. The signal is extremely broad at 77°K and clearly seen at temperatures below about 35°K. However, at about 10°K, the signal is more easily saturated with microwave power than that of the eight-iron ferredoxins. This suggests that the eight-iron ferredoxins have an additional mechanism of electron spin relaxation at low temperatures that may be associated with the presence of two centers in one molecule (Mullinger et al., 1975).

High-potential iron-sulfur proteins (HiPIP's) have been isolated from a number of photosynthetic bacteria. The HiPIP from *Chromatium* has a molecular weight of about 10,000, and its role appears to be as an electron transport carrier in chromatophores (Evans et al., 1974). Its amino acid sequence shows no homology with those of the four-iron and eight-iron ferredoxins, in spite of the fact that it has an almost identical 4 Fe–4 S* center. As normally isolated, the protein is in the reduced state, which is nonmagnetic. On the scheme of Carter et al. (1972), this is the C state, equivalent to oxidized ferredoxin. It is oxidized with a midpoint potential of +350 mV to a C$^+$ state. The oxidized form shows an EPR signal that is probably axial (Fig. 6d) with $g_\parallel = 2.12$ and $g_\perp = 2.04$, but shows additional structure that may arise from heterogeneity of paramagnetic species (Palmer et al., 1967).

Reduction of HiPIP to the C$^-$ state equivalent to reduced ferredoxin has not been observed in aqueous solution, but in the presence of 80% dimethyl sulfoxide (DMSO), which causes a structural change in the protein, the 4 Fe–4 S* center can be reduced to a C$^-$ (so-called superreduced) state (Cammack, 1973). The midpoint potential for this reduction has been estimated to be —600 mV or lower. The EPR spectrum of this state is shown in Fig. 6c.

Clostridium pasteurianum and *Chromatium* ferredoxins are of the eight-iron type, containing two 4 Fe–4 S* centers per molecule. *C. pasteurianum* ferredoxin probably has a very similar structure to that determined for *P. aerogenes* ferredoxin, as the amino acid sequences show close homology. *Chromatium* ferredoxin has a higher molecular weight, about 9500, but the amino acid sequence is homologous with that of *C. pasteurianum* ferredoxin, if allowance is made for two extra loops of peptide chain in the sequence (Matsubara et al., 1970). *Chromatium* ferredoxin also has a more negative redox potential (—490 mV) (Bachofen and Arnon, 1966) than those of the clostridial ferredoxins (about —390 mV) (Eisenstein and Wang, 1969).

Both proteins are in a nonmagnetic C state as prepared and are reduced to a C⁻ state. The two centers in clostridial ferredoxin can be reduced one at a time by careful titration with dithionite, and the EPR spectrum of molecules with only one center reduced is similar to those of reduced four-iron ferredoxins (Orme-Johnson and Beinert, 1969) and is shown in Fig. 6b. The more complex spectrum of the fully reduced form (Fig. 7a) has been associated (Mathews et al., 1974; Gersonde et al., 1974) with a spin coupling between the two paramagnetic centers. This is confirmed by observation of a weak $\Delta m_S = 2$ transition at $g = 3.88$. In agreement with this, no such signal is observed in the EPR spectrum of the reduced four-iron B. stearothermophilus ferredoxin (Mullinger et al., 1975). The lineshape of the EPR spectrum of reduced clostridial ferredoxin is somewhat variable, depending on the salt concentration (Cammack, 1975). The spectra in low salt or in the presence of 15% DMSO (Fig. 7b), are somewhat different from those in 0.8 M NaCl (Fig. 7a). These differences may reflect a decreased degree of spin coupling between the centers in the presence of NaCl. A small effect is observed in the corresponding Mössbauer spectra (Fig. 14a is in the presence of 0.8 M NaCl; Fig. 14b is with 15% DMSO).

In the presence of strong denaturing agents such as 70–90% DMSO, C. pasteurianum ferredoxin becomes unfolded, as shown by PMR measurements (McDonald et al., 1973). The presence of contact-shifted resonances means that the 4 Fe–4 S* centers are not destroyed but remain attached to the cysteine residues. The ^{57}Fe hyperfine coupling constants in the protein reduced in this state are similar to those observed in the native state (Anderson et al., 1975). Similar unfolding has been observed in a wide range of ferredoxins and HiPIP (Cammack, 1975). Provided that precautions are taken to exclude oxygen, the proteins are restored to their native state if the concentration of the denaturing agent falls below a critical level.

When the protein is unfolded, the 4 Fe–4 S* centers all show similar behavior. Their EPR spectra in the reduced state show nearly axial symmetry (Figs. 6c and 7c) and are rather similar to those of the reduced four-iron ferredoxins and partly reduced eight-iron ferredoxins (Fig. 6a and b). They are only detected below 30°K. These spectra resemble those of the analog compounds $[Fe_4S_4(SR)_4]^{3-}$ that are in the equivalent oxidation state (Frankel et al., 1974a). Ferredoxins that contain 2 Fe–2 S* centers show a distinctive, isotropic EPR signal in the unfolded state (Cammack, 1975) that can be readily detected at 77°K.

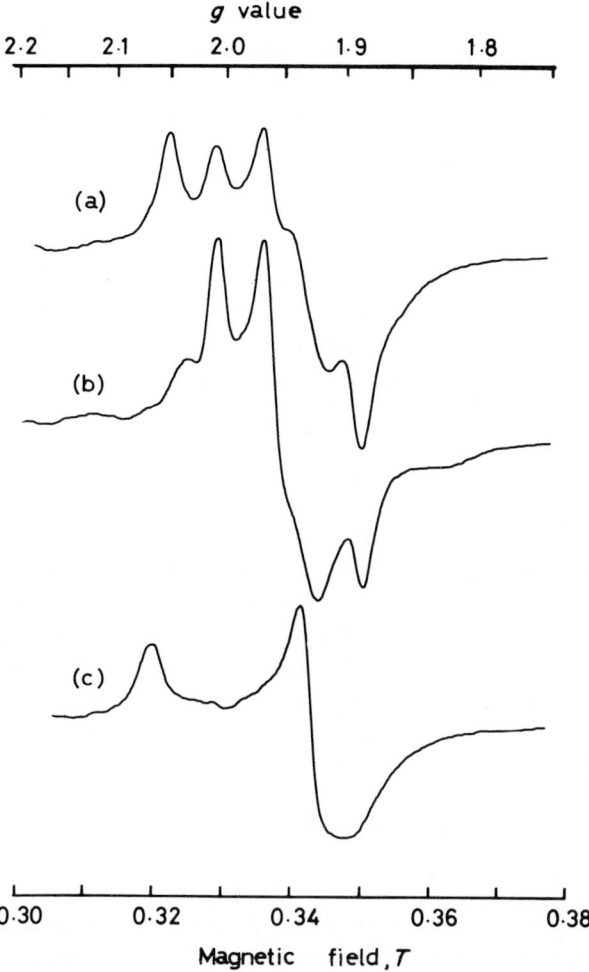

Fig. 7. EPR spectra of reduced *C. pasteurianum* ferredoxin showing the effect of reversible protein denaturation: (a) as prepared in 0.8 M NaCl, (b) in 15% DMSO, and (c) in 80% DMSO.

This difference suggests a method of distinguishing the two types of iron–sulfur centers, by examining the line shape and temperature dependence of the reduced protein in 80% DMSO. This method has been applied to show that the membrane-bound iron–sulfur proteins of photosystem I in chloroplasts are of the 4 Fe–4 S* type (Cammack and Evans, 1975).

As already noted, *Chromatium* HiPIP in concentrated DMSO behaves similarly to the ferredoxins and is readily reduced to a super-reduced state (Cammack, 1973). The complex EPR spectrum of the eight-iron ferredoxins is simplified in high concentrations of DMSO, indicating that spin coupling between the centers has been removed. These changes can be considered as a relaxation of the constraints imposed on the 4 Fe–4 S* centers by the protein, which produce the redox behavior of HiPIP and the spin coupling in *C. pasteurianum* ferredoxin.

B. Mössbauer Data

Mössbauer studies have been carried out on the four-iron ferredoxin from *B. stearothermophilus* (Mullinger *et al.*, 1975), the HiPIP from *Chromatium* (Moss *et al.*, 1968; Evans *et al.*, 1970; Dickson *et al.*, 1974a; Dickson and Cammack, 1974), and the eight-iron ferredoxins from *C. pasteurianum* (Blomstrom *et al.*, 1964; Thompson *et al.*, 1974; Gersonde *et al.*, 1974) and *Chromatium* (Moss *et al.*, 1968; R. Cammack, D. P. E. Dickson, and C. E. Johnson, unpublished work). A problem in some of these investigations has been the presence of adventitious iron-containing species and denaturation products. However, now that a number of the proteins in this group have been studied in both redox states and in a wide range of temperature and magnetic field environments, a consistent pattern of behavior can be observed, and the presence of contaminants is unlikely to go unnoticed and lead to an incorrect interpretation.

In this section, the Mössbauer data on these proteins will be summarized, and in the subsequent sections, the conclusions from the Mössbauer data concerning various aspects of the properties of the 4 Fe–4 S* centers will be discussed and compared with information obtained in other ways.

1. FOUR-IRON FERREDOXIN

The Mössbauer spectra of oxidized *B. stearothermophilus* ferredoxin in the absence of an applied field consist of quadrupole-split doublets. The broadening and asymmetry of the lines can be attributed to the overlapping of at least two slightly differing spectra, indicating that the iron atoms in the 4 Fe–4 S* center are not precisely equivalent. The spectra in strong applied magnetic fields show no internal hyperfine field, evidence that the center is nonmagnetic in the oxidized state.

The spectra of the reduced ferredoxin at various temperatures are shown in Fig. 8. At both 195° and 77°K, they consist of two significantly different quadrupole-split doublets. At 77°K, the right-hand lines of these doublets are well resolved. Computer fitting shows that the two component doublets have nearly equal areas, which implies that the 4 Fe–4 S* center contains two pairs of nearly equivalent iron atoms.

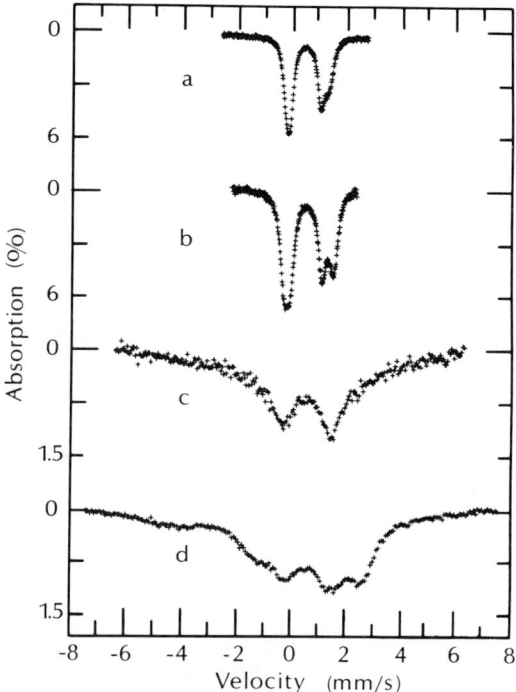

Fig. 8. Mössbauer spectra of reduced *B. stearothermophilus* ferredoxin at (a) 195°K, (b) 77°K, (c) 35°K, and (d) 4.2°K. (From Mullinger *et al.*, 1975.)

At 35°K, the spectrum consists of a single quadrupole-split doublet with broad lines superimposed on a wide band of absorption characteristic of a magnetic spectrum with intermediate electron spin relaxation ($\tau \sim \hbar/A$). At temperatures below 20°K, the spectra show a broad asymmetric pattern of absorption resulting from magnetic hyperfine interaction (Fig. 8d). A number of features of these spectra and in particular the broad line at approximately -4 mm/second indicate that at least part of the spectrum may result from a hyperfine interaction coupling S and I as described in Section II,C, and similar to that observed in the proteins with 2 Fe–2 S* centers. The application of a small magnetic field suppresses this coupling and gives spectra more akin to a symmetrical six-line Zeeman pattern, with line intensities being dependent on the applied field orientation (Fig. 9b and c).

In the spectra in applied magnetic fields (Fig. 9), there is evidence for lines (shown arrowed in the low-field spectra Fig. 9b and c) that move in as the applied field is increased, as well as lines shown arrowed in the

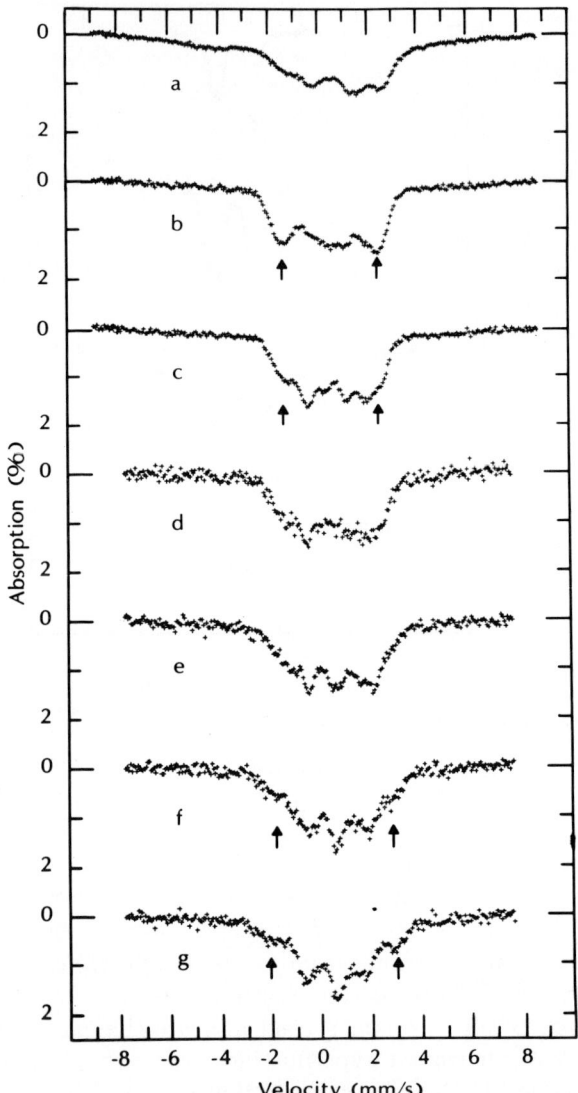

Fig. 9. Mössbauer spectra of reduced *B. stearothermophilus* ferredoxin at 4.2°K showing the effect of applied magnetic fields: (a) zero field, (b) ~500 G parallel, (c) ~500 G perpendicular, (d) 15 kG perpendicular, (e) 30 kG perpendicular, (f) 45 kG perpendicular, and (g) 60 kG perpendicular. (From Mullinger *et al.*, 1975.)

high-field spectra that move out as the applied field is increased (Fig. 9f and g). The presence of the outer lines in the spectra taken in large fields is confirmed by spectra taken in parallel applied fields in which their intensity is increased. Spectra taken in large fields at 2°K show no significant differences to those obtained at 4.2°K, which means that these outer lines cannot arise from an excited spin state. There is therefore good evidence for both positive and negative hyperfine fields and hence antiferromagnetic coupling within the 4 Fe–4 S* center. This is discussed more fully in Section V,E.

2. HiPIP

Mössbauer spectra of reduced *Chromatium* HiPIP show quadrupole-split doublets in the absence of an applied magnetic field and no internal hyperfine field when a magnetic field is applied at low temperatures. This confirms the EPR and magnetic susceptibility evidence that this state is nonmagnetic. The spectra of reduced *Chromatium* HiPIP are very closely similar to the corresponding spectra of oxidized *B. stearothermophilus* ferredoxin.

The oxidized form of *Chromatium* HiPIP shows a quadrupole-split doublet spectrum at 77° and 195°K, with some broadening and asymmetry, implying that all four iron atoms are not exactly equivalent. At 4.2°K (Fig. 10), the spectrum becomes a broad band of absorption consistent with a magnetic hyperfine interaction in the intermediate spin relaxation region. Further cooling to 1.3°K has very little effect on this spectrum, but the application of small magnetic fields sharpens it up to show a number of absorption lines. These have been interpreted (Evans *et al.*, 1970) as arising from inequivalent pairs of iron atoms with hyperfine fields of 90 and 121 kG, although these values should be regarded as only approximate.

Because of the particular conditions associated with its production, super-reduced *Chromatium* HiPIP cannot be completely isolated. Samples, therefore, always contain some reduced protein. This is unimportant in EPR measurements, as the reduced protein, being nonmagnetic, does not give a signal. Mössbauer spectroscopy, however, is sensitive to all the iron atoms present, and therefore the spectra obtained consist of superimposed contributions from both the reduced and super-reduced proteins. (This sensitivity of Mössbauer spectroscopy to all the iron atoms present in a sample is often a problem, but it means that this technique can be an excellent monitor of the purity of samples.) At 195° and 77°K, the spectra of super-reduced HiPIP samples consist of quadrupole-split doublets. Because of the multiple components, these spectra are difficult to deconvolute, but more than two doublet components are required to obtain a reasonable fit, which implies that the super-reduced protein must itself contain inequivalent iron atoms. The spectra at 4.2°K and with small applied magnetic fields show broad magnetic patterns similar to

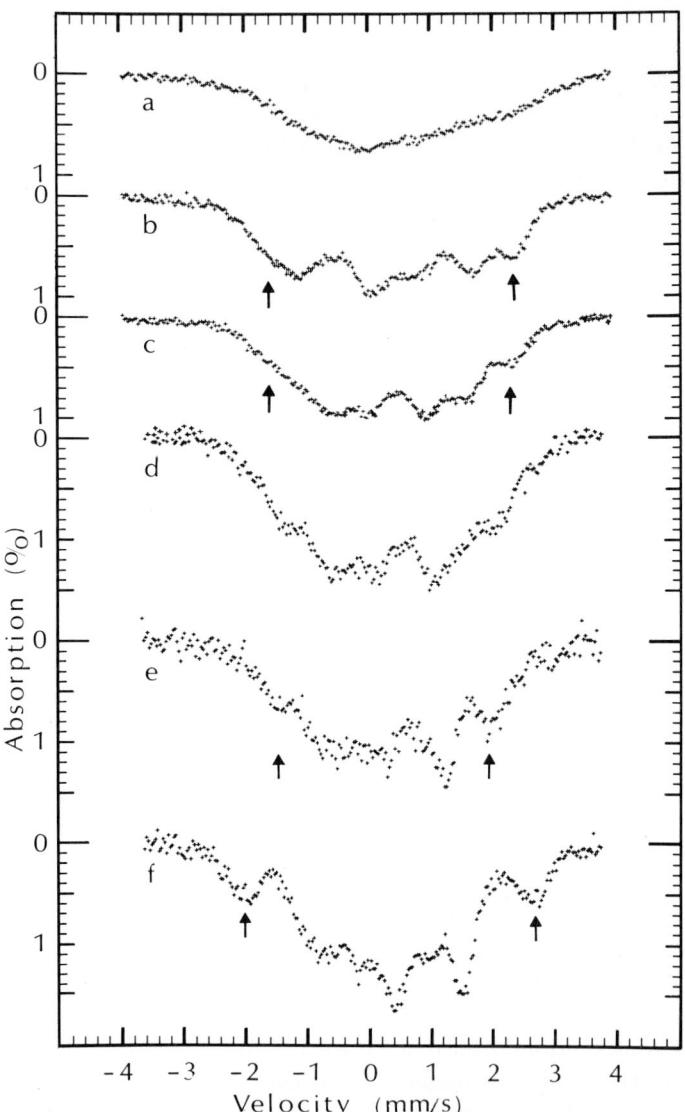

Fig. 10. Mössbauer spectra of oxidized *Chromatium* HiPIP at 4.2°K showing the effect of applied magnetic fields: (a) zero field, (b) ~500 G parallel, (c) ~500 G perpendicular, (d) 15 kG perpendicular, (e) 30 kG perpendicular, and (f) 60 kG perpendicular. (From Dickson *et al.*, 1974a.)

those seen in reduced *B. stearothermophilus* ferredoxin, as well as the doublet due to the contaminating reduced (nonmagnetic) protein. The Mössbauer spectra of the analog compounds in the equivalent (trianion) oxidation state are very similar, as contamination with the dianion state is also experienced (Frankel et al., 1974b).

3. EIGHT-IRON FERREDOXINS

The Mössbauer spectra of *C. pasteurianum* and *Chromatium* ferredoxins are very closely similar under all conditions. The spectra of the oxidized ferredoxins in the absence of an applied field show quadrupole-split doublets with some broadening and asymmetry arising from small differences between the four iron atoms within each center and possibly some difference between the two centers. The structure seen in the spectra is somewhat greater in the case of *Chromatium* ferredoxin, suggesting that there is a slightly greater inequivalence of the iron atoms in this protein. The spectra of the oxidized ferredoxins at 4.2°K in applied magnetic fields show no internal hyperfine field, thus confirming the EPR evidence that the oxidized state of these proteins is nonmagnetic. These spectra are very similar to the corresponding spectra of oxidized *B. stearothermophilus* ferredoxin and reduced *Chromatium* HiPIP.

The reduced eight-iron ferredoxins show quadrupole-split doublet spectra at all temperatures between 1.3° and 195°K. The 77° and 195°K spectra show some asymmetry but do not exhibit resolved components as seen in the spectra of reduced *B. stearothermophilus* ferredoxin. In all cases, the whole spectrum is changed on reduction, which means that all the iron atoms are affected. This confirms that one of the reducing electrons must go to each center.

At 4.2°K, one would expect the electron spin relaxation time to be long and the magnetic hyperfine structure to be resolved. However both ferredoxins show spectra consisting of a doublet but with broad lines (Fig. 14a), quite different from the broad magnetic spectra seen in the magnetic forms of the other iron–sulfur proteins. Some magnetic interaction is apparent when small fields ($\sim$500 G) are applied, as these produce lines in the wings of the broadened doublet, behavior not seen in the nonmagnetic oxidized forms. The absence of large magnetic hyperfine splittings in the spectra of the reduced eight-iron ferredoxins at low temperatures is discussed in Section V,E.

The spectra of the reduced eight-iron ferredoxins in large applied magnetic fields are very closely similar to those of *B. stearothermophilus* ferredoxin and exhibit the same evidence for antiferromagnetic coupling.

C. Similarity of the 4 Fe–4 S* Centers

Mössbauer spectroscopy and EPR are sensitive probes of the iron atoms and hence of the active centers of these proteins. According to the three-state scheme, the oxidized ferredoxins and reduced HiPIP contain 4 Fe–4 S* centers in the same redox state. This is nicely confirmed by the Mössbauer spectra, which are closely similar for all the proteins in these states. The spectra in applied magnetic fields in particular are very similar (Fig. 11). These spectra have been fitted using a computer program suitable for diamagnetic iron atoms. The Mössbauer parameters obtained from these fits are close for all these proteins. In large applied magnetic fields, the proteins with C⁻ state centers show very similar spectra. At low temperatures in the absence of an applied field, the proteins having one and two centers give different spectra, but as discussed in Section V,E, this is not related to differences in the active centers.

The Mössbauer spectroscopic data are therefore consistent with the assumption made in Section I that this group of proteins all contain basically similar 4 Fe–4 S* centers. This is also confirmed by a comparison of the EPR spectra of the proteins with C⁻ state centers (Figs. 6 and 7), where under suitable conditions all the proteins exhibit the characteristic signal of the 4 Fe–4 S* center (see Section V,A).

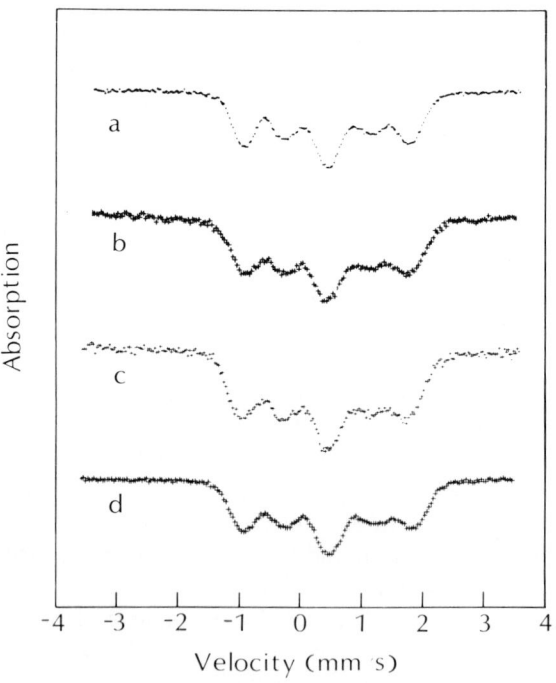

Fig. 11. Mössbauer spectra of proteins with 4 Fe–4 S* centers in the C state, taken at 4.2°K in a perpendicularly applied field of 60 kG: (a) reduced *Chromatium* HiPIP, (b) oxidized *B. stearothermophilus* ferredoxin, (c) oxidized *C. pasteurianum* ferredoxin, and (d) oxidized *Chromatium* ferredoxin.

D. Equivalence and Valence of the Iron Atoms

In a molecule containing a number of iron atoms, each one contributes a component spectrum to the Mössbauer spectrum of the sample. The Mössbauer spectra of the iron–sulfur proteins will therefore only exhibit one component if all the iron atoms in each molecule are completely equivalent. In the absence of magnetic effects, the iron atoms in the iron–sulfur proteins exhibit quadrupole-split doublet spectra that can be interpreted to give useful information about the equivalence of the iron atoms.

In rubredoxins with only one iron atom per molecule, the spectra of both redox states show only one component as expected. In the case of the two-iron ferredoxins, the spectra of the oxidized form show two near identical components that change on reduction to give two very different components, one being like that of the oxidized protein. Thus, the oxidized protein contains two closely equivalent iron atoms, one of which changes completely on reduction. This is seen in Fig. 4.

In the proteins with 4 Fe–4 S* centers, the situation is less simple. The spectra do not in general show either a single component (four equivalent iron atoms) or four separate resolved components.

Quadrupole-split doublet spectra can be defined by the chemical shift and the quadrupole splitting. These parameters give information on the valence state (predominately the chemical shift) and the environment (predominately the quadrupole splitting) of the iron atoms. In principle, multicomponent spectra can be computer fitted to give the chemical shift and quadrupole splitting of the individual components. This procedure is reliable when the components are well resolved but becomes less reliable in the situation of potentially four (or even eight) overlapping components that cannot be resolved by eye. The latter situation occurs frequently with these proteins, and it may then be difficult to say more than that the iron atoms are similar.

The chemical shift is a measure of the electronic charge density at the ^{57}Fe nucleus and relates mainly to the valence of the iron atoms and hence to the distribution of the 3d electrons between them; it is therefore a relevant parameter from the biological viewpoint. The Mössbauer chemical shift (often backed up by the quadrupole splitting) is frequently used to characterize the valence of ionic compounds. In compounds with considerable covalency, this is more difficult; theoretical calculations are unreliable, but a semi-empirical calibration of the chemical shift is possible for a series of similar compounds. In the iron–sulfur proteins we have a series of molecules of increasing complexity, in all of which the iron atom is in a near tetrahedral environment of four sulfur atoms. The effects of covalency will therefore be similar in all cases, and being a local probe, the chemical shift in different members of the series should be comparable. The quadrupole interaction contains contributions from atoms outside the iron–sulfur center and the quadrupole splitting is therefore sensitive to the whole protein as well as to the valence state of the iron atoms. However, the quadrupole splitting and its temperature dependence can provide useful confirmatory evidence for valence assignments.

In the proteins with 1 Fe and 2 Fe–2 S* centers, the valence of the iron atoms in the various redox states are well defined, both from Mössbauer data by comparison with data on inorganic compounds and from other measurements. Thus, the chemical shifts in these proteins can be used to calibrate the chemical shift as an indicator of the valence state of the iron atoms in the 4 Fe–4 S* proteins. Table I shows the values of the chemical shifts obtained from least-squares computer fitting of the spectra. Except in the case of reduced *B. stearothermophilus* ferredoxin, where the two components of the spectrum are well resolved, only average values of the chemical shift are quoted for the proteins with 4 Fe–4 S* centers. The errors are ±0.01 mm/second in all cases except super-reduced *Chromatium* HiPIP, where because of the presence of contaminating reduced protein the errors are ±0.04 mm/second. Taking average values of the chemical shifts is justified in that, to within the errors, the same values are obtained whether the spectra are fitted to one component or to multiple components and the weighted mean taken. Table I shows the valence assignments based on the chemical shifts in terms of average valences of equivalent iron atoms, giving fractional values as well as formal valences. These assignments are consistent with the three-state scheme, and the C^+, C, C^- designations are also given in the table.

All the C state centers seem to consist of four closely equivalent iron atoms of average valence $Fe^{2.5+}$ (i.e., twenty-two 3d electrons over the four iron atoms with nearly equivalent electronic charge distributions). The C^+ state of oxidized *Chromatium* HiPIP also appears to have four iron atoms with closely equivalent electron charge distributions and 21 3d electrons. The C^- state centers have 23 3d electrons, but there are differences as to how these electrons are arranged. In the case of reduced *B. stearothermophilus* ferredoxin, the spectrum at 77°K consists of two components with nearly equal areas, suggesting two pairs of equivalent iron atoms with average valence assignments of 2 $Fe^{2.5+}$ and 2 Fe^{2+}, although all the iron atoms are changed on reduction. In the reduced eight-iron ferredoxins, however, the four iron atoms within each center appear to have a more nearly equivalent electronic charge distribution. The difference in the behavior of the two-iron, four-iron, and eight-iron

TABLE I
MÖSSBAUER CHEMICAL SHIFTS OF THE IRON–SULFUR PROTEINS[a]

Protein	Chemical shift	Average valences	State	Formal valences
Fe^{3+} in rubredoxin[b]	~0.25	—	—	—
Fe^{3+} in adrenal ferredoxin[c]	0.26	—	—	—
Fe^{3+} in spinach ferredoxin[d]	0.26	—	—	—
Oxidized *Chromatium* HiPIP[e]	0.31	4 $Fe^{2.75+}$	C^+	3 Fe^{3+}, 1 Fe^{2+}
Reduced *Chromatium* HiPIP[e]	0.42	4 $Fe^{2.5+}$		
Oxidized *B. stearothermophilus* ferredoxin[f]	0.42	4 $Fe^{2.5+}$	C	2 Fe^{3+}, 2 Fe^{2+}
Oxidized *C. pasteurianum* ferredoxin[g]	0.43	4 $Fe^{2.5+}$		
Oxidized *Chromatium* ferredoxin[h]	0.41	4 $Fe^{2.5+}$		
Super-reduced *Chromatium* HiPIP[i]	0.59	4 $Fe^{2.25+}$		
Reduced *B. stearothermophilus* ferredoxin[f]	0.50, 0.60	2 $Fe^{2.5+}$, 2 Fe^{2+}	C^-	1 Fe^{3+}, 3 Fe^{2+}
Reduced *C. pasteurianum* ferredoxin[g]	0.57	4 $Fe^{2.25+}$		
Reduced *Chromatium* ferredoxin[h]	0.54	4 $Fe^{2.25+}$		
Fe^{2+} in rubredoxin[b]	0.65	—	—	—
Fe^{2+} in spinach ferredoxin[d]	0.60	—	—	—

[a] At 77°K, relative to pure iron metal at room temperature (quoted in millimeters per second).
[b] From Rao et al. (1972a).
[c] From Cammack et al. (1971).
[d] From Rao et al. (1971). 195°K value adjusted for second-order Doppler shift.
[e] From Dickson et al. (1974a).
[f] From Mullinger et al. (1975).
[g] From Thompson et al. (1974).
[h] R. Cammack, D. P. E. Dickson, and C. E. Johnson, unpublished results.
[i] From Dickson and Cammack (1974).

ferredoxins as regards the localization of the extra electron that goes to the active centers on reduction is a striking feature that is well illustrated by comparison of Fig. 12 and Fig. 4.

The valence assignments based on the chemical shift data are supported by the quadrupole splittings obtained for these proteins. These are intermediate in value between those obtained for ferric and ferrous atoms in the proteins with 1 Fe and 2 Fe–2 S* centers. Although the individual proteins have somewhat different quadrupole splittings, there is a general

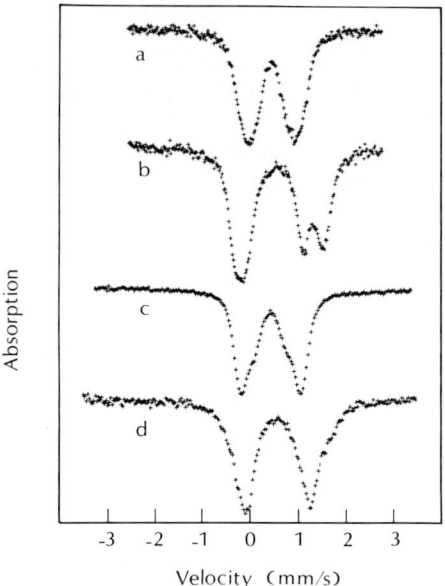

Fig. 12. Mössbauer spectra at 77°K of (a) oxidized and (b) reduced *B. stearothermophilus* ferredoxin and (c) oxidized and (d) reduced *Chromatium* ferredoxin.

trend of increasing quadrupole splitting and temperature dependence of quadrupole splitting with increasing ferrous character consistent with the assignments given in Table I.

Further confirmatory evidence for the valence assignments comes from the spectra of reduced *B. stearothermophilus* ferredoxin at various temperatures (Fig. 8). The two resolved doublet components at 77°K are interpreted as arising from two iron atoms of ferrous character and two iron atoms with more ferric character. At 35°K, the more ferric component (i.e., the component with the smaller chemical shift and quadrupole splitting) has broadened into a wide band of absorption resulting from magnetic hyperfine interaction, while the ferrous component remains as a doublet. This is in keeping with the assigned valences, as the electron spin–lattice relaxation is much faster for ferrous atoms.

Information on the formal valences of the iron atoms can also be inferred indirectly. EPR and magnetic susceptibility data show that the proteins with C state centers have no unpaired spins, while the C⁺ and C⁻ state centers have $S = \frac{1}{2}$. As there are four iron atoms in a center that gains or loses one electron on reduction or oxidation, then unless some of the iron atoms are monovalent or tetravalent, the most probable scheme is that shown in Table I. Arguments similar to these have been used by Palmer in Chapter 8 of Volume II.

The dianion form of the analog compounds (which is equivalent to the proteins in the C state) also contains formally 2 Fe^{3+} and 2 Fe^{2+} iron atoms, although the Mössbauer data (Frankel et al., 1974b) as well as information from other spectroscopic techniques (Holm et al., 1974) suggest complete equivalence of the four iron atoms.

The near equivalence of electron charge distribution between the four iron atoms in the centers of the proteins cannot readily be explained by a model in which the 3d electrons are in molecular orbitals completely delocalized over all four iron atoms, as this is incompatible with the direct evidence of iron atoms with both positive and negative hyperfine fields. The explanation must lie in partial spatial delocalization or fast hopping of the 3d electrons between certain of the iron atoms. Molecular orbital calculations for a system of completely equivalent iron atoms have been successful in explaining the different g values of oxidized HiPIP and reduced ferredoxin as well as some of the features of the optical spectra (Thomson, 1975). However it is difficult to extend this delocalized model to the condition where the iron atoms are not exactly equivalent. Eicher et al. (1974) have computer fitted the Mössbauer spectra of reduced C. pasteurianum ferredoxin using a Hamiltonian implying four equivalent iron atoms, although this approach does not give good simulations of the spectra obtained in large magnetic fields.

A model for the C state center that is compatible with the Mössbauer data is shown in Fig. 13 (Dickson et al., 1974b). The antiferromagnetic coupling between the iron atoms can occur via 90° superexchange through the sulfur atoms. There is fast hopping (relative to the lifetime of the ^{57}Fe excited state) or delocalization of the sixth 3d electrons between the two spin-up iron atoms and between the two spin-down iron atoms. Effectively, each iron atom is $Fe^{2.5+}$ giving the observed chemical shift and the

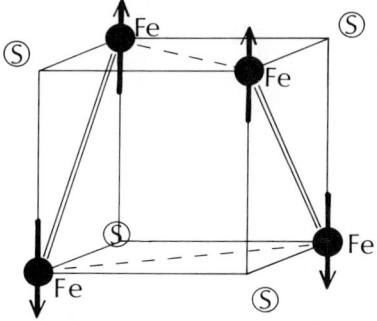

Fig. 13. Model for the 4 Fe–4 S* center in the C state. The parallel lines represent antiferromagnetic coupling, while the dashed lines represent electron delocalization or fast hopping. (From Dickson et al., 1974b.)

center has the required net spin of zero. Movement or delocalization of the sixth 3d electrons between spin-up and spin-down iron atoms is inhibited by the Pauli exclusion principle. (It is the operation of the exclusion principle that ensures the localization of the extra 3d electron in the case of the reduced two-iron ferredoxins.) This model for the C state center is compatible with the observed behavior of *B. stearothermophilus* ferredoxin on reduction, when the extra electron presumably goes to one pair of iron atoms that are then Fe^{2+}. On the other hand, the continued near equivalence of the iron atoms on reduction of the eight-iron ferredoxins or oxidation of HiPIP is difficult to reconcile with this simple ionic model, although the effects of covalency could blur the distinction between the iron atoms.

Recently, Antanaitis and Moss (1975) have studied oxidized *Chromatium* HiPIP by magnetic susceptibility, and computer simulation of the EPR spectrum (which, as already noted, appears to consist of more than one component). They proposed a model analogous to that of Fig. 13, in which three of the iron atoms at any time are ferric, and the other is ferrous. The extra electron hops between two iron atoms at a rate that is slow on the EPR time scale ($>10^{-8}$ second), but rapid on the Mössbauer time scale ($<10^{-7}$ second).

E. Spin Coupling

The proteins with C state centers show non-Curie law temperature dependence of susceptibility at low temperatures and no EPR signal. The Mössbauer spectra show zero hyperfine field and no magnetic splitting at low temperatures in the absence of an applied field. There is thus conclusive evidence that the C state center has a nonmagnetic ground state with total spin zero. This implies antiferromagnetic coupling between the four iron atoms with two atoms spin-up and two atoms spin-down. At higher temperatures, excited spin states will become populated, and evidence for these is seen in susceptibility and PMR measurements (Chapter 7 of Volume II). However in EPR and Mössbauer measurements the presence of fast electron spin relaxation at higher temperatures means that those higher spin states are not observed.

In the C^+ and C^- state centers, which have one less or one more electron, antiferromagnetic coupling within the center gives a ground state with total spin $S = \frac{1}{2}$, and magnetic susceptibility and EPR measurements are consistent with this.

When magnetic hyperfine interactions are present, the Mössbauer spectra of these proteins are considerably more complex than the simple case of a single isotropic magnetic site, which gives the familiar six-line

Zeeman pattern described in Section II,C. A complete understanding of these spectra would require computer simulation and fitting of the spectra starting from a suitable Hamiltonian for the system, however, a qualitative view, embodying the essential physics, can be obtained by comparing the behavior seen in the spectra of these proteins with that in a simple magnetic spectrum. The features described in Section II,C for a simple magnetic spectrum are all observed in the applied field spectra of the proteins with 4 Fe–4 S* centers. This is well illustrated by the spectra of oxidized *Chromatium* HiPIP taken in applied fields (Fig. 10). In the spectra in low magnetic fields there are outer lines, shown arrowed in Fig. 10b and c, that are greater in intensity for a parallel applied field as compared to a perpendicular field. As the field is increased, these outer lines are no longer apparent, which means that they must have moved inward. In the high-field spectra, other lines have appeared at the outside of the spectrum, shown arrowed in Fig. 10e and f, that clearly move out as the applied field increases from 30 to 60 kG. These lines increase in intensity when the large magnetic fields are applied parallel to the γ-ray direction, as would be expected if they can be related to the outer lines of a magnetic six-line pattern. Thus, there is direct evidence for both positive and negative hyperfine fields and hence antiferromagnetic coupling between the iron atoms. The area of the spectrum due to the iron atoms with the positive hyperfine field (as defined by the area of the outermost lines) is half the total spectral intensity, indicating that the iron atoms are equally divided between those with positive and negative hyperfine fields. These features are also seen in the applied field spectra of all the other proteins with magnetic state centers, although in the reduced ferredoxins the magnetic hyperfine interaction is less isotropic as a result of the increased ferrous character of the iron atoms so that some of the lines are less clearly observed.

Reduced *B. stearothermophilus* ferredoxin shows clear evidence in its zero field Mössbauer spectra (e.g., Fig. 12b) for two distinct types of iron atom, and it therefore presents the best possibility for computer fitting of its applied field spectra. Preliminary work suggests that good fits can be obtained with two component spectra yielding two sets of A values which are consistent with the approximate values of hyperfine field (-90 and $+120$ kG) obtained from a more qualitative interpretation of the spectra in the effective field approximation.

The Mössbauer spectroscopic data on these proteins show iron atoms with spin-up and with spin-down; i.e., there is an inequivalence of the iron atoms as regards the distribution of electron spin between them. ENDOR and PMR can provide complementary data.

Because of the greater complexity of the 4 Fe–4 S* centers, ENDOR measurements (Anderson et al., 1975) have so far given a less clear picture than in the case of the proteins with 2 Fe–2 S* centers (Chapter 5 of Volume II). Measurements on *Chromatium* HiPIP and *C. pasteurianum* ferredoxin indicate that the resonances are nearly isotropic; no highly anisotropic A values like those found for the ferrous atom in the reduced two-iron ferredoxins have been detected. The couplings observed are also smaller in magnitude, indicating greater electron delocalization onto the sulfur ligands. Only one type of resonance is observed in reduced *C. pasteurianum* ferredoxin in the native state or unfolded in 80% DMSO and in super-reduced HiPIP in 80% DMSO. In oxidized HiPIP, however, two types of hyperfine couplings were detected, with components ranging from 117 to 105 kG for one, and from 83 to 76 kG for the other. These values are in reasonably good agreement with the values for two hyperfine fields of approximately -120 and $+90$ kG observed in the Mössbauer spectra (Dickson et al., 1974a). The results are also in agreement with the observation of two classes of contact-shifted β-CH_2 proton resonances in the PMR spectra (Phillips et al., 1970b). These observations all support the proposal that the iron atoms in oxidized HiPIP are inequivalent in pairs.

In the case of reduced *C. pasteurianum* ferredoxin, only one set of A values was detected. This seems difficult to reconcile with the observation in the Mössbauer spectra of iron atoms with both positive and negative hyperfine fields, although it must be remembered that the ENDOR technique cannot give the sign of the hyperfine field.

Magnetic hyperfine structure would be expected at low temperatures in the Mössbauer spectra of all the proteins with C^+ and C^- state centers. However, the eight-iron ferredoxins from *C. pasteurianum* and *Chromatium*, that have two 4 Fe–4 S* centers show broadened quadrupole-split doublets at 4.2°K. The broad lines may indicate the onset of some magnetic effects, although cooling to 1.3°K does not produce any detectable change in the spectra. This anomalous behavior has been attributed by Gersonde et al. (1974) to a spin coupling between the two centers in the molecule to give a state with total spin zero and the apparently diamagnetic Mössbauer spectra. A spin–spin interaction is indicated by the lack of any change of the spectra with temperature. As already mentioned, Mathews et al. (1974) have postulated a spin coupling between the two centers to explain the EPR spectra of the eight-iron ferredoxins.

At first sight, a coupling between the two centers appears unlikely, as they are approximately 1.2 nm apart. Although the two nearest iron atoms are appreciably closer together than this, there are no direct bridging ligands between the two centers for the superexchange inter-

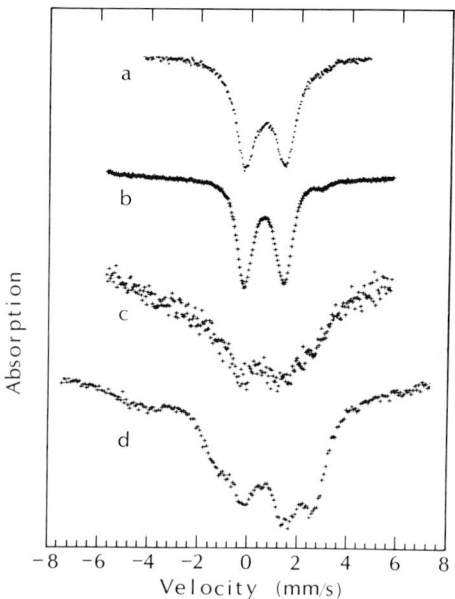

Fig. 14. Mössbauer spectra of reduced ferredoxins at 4.2°K showing the effect of DMSO: *C. pasteurianum* ferredoxin (a) native in 0.8 *M* NaCl, (b) in 15% DMSO, and (c) in 80% DMSO and (d) native *B. stearothermophilus* ferredoxin. (These spectra compare with the EPR spectra Figs. 7a–c and 6a.)

action that usually gives risk to antiferromagnetic coupling, and therefore the interaction is relatively weak.

In order to examine this coupling theory, a denaturing agent (DMSO) has been applied to reduced *C. pasteurianum* ferredoxin. At large concentrations, this unfolds the protein and thus modifies and presumably decreases any interaction between the centers. The effect of this treatment on the EPR spectrum is to remove the additional features seen in the spectrum of the native reduced eight-iron ferredoxin (Fig. 7a) and produce an axial signal like that of a reduced four-iron ferredoxin (Fig. 7c). The effect on the Mössbauer spectrum is similarly to change the anomalous behavior of the native reduced eight-iron ferredoxin (Fig. 14a) and to produce a broad asymmetric spectrum (Fig. 14c) like that seen in the reduced four-iron ferredoxin from *B. stearothermophilus* (Fig. 14d). There is, therefore, clear evidence for an interaction between the two centers of the eight-iron ferredoxins, although the exact mechanism of this interaction is not yet understood. In high applied magnetic fields (15 kG and above), Mössbauer spectra of reduced *Chromatium* and *C. pasteurianum* ferredoxins are virtually identical to the equivalent spectra of *B.*

stearothermophilus ferredoxin, suggesting that large magnetic fields break the coupling between the two centers in the eight-iron ferredoxins and give a behavior characteristic of the 4 Fe–4 S* centers.

VI. CONCLUSIONS

The spectroscopic techniques such as EPR and Mössbauer spectroscopy have made some important contributions to our understanding of the iron–sulfur proteins. This understanding is fairly well developed in the case of the proteins with 1 Fe and 2 Fe–2 S* centers, although an X-ray crystallographic determination of the structure of a protein with a 2 Fe–2 S* center would be extremely useful to resolve this system more completely.

Chemical shift and quadrupole splitting data from Mössbauer spectroscopy have provided valuable evidence for the chemical state of the iron in these molecules. In the 2 Fe–2 S* centers, both iron atoms are ferric in the oxidized state, while on reduction one becomes ferrous and the other is unchanged. In the oxidized four-iron ferredoxin, it appears that there is delocalization of the iron valence electrons within the 4 Fe–4 S* center and that the average electronic state of the iron corresponds to 2 Fe^{3+} and 2 Fe^{2+} atoms. In the reduced state, two quadrupole doublets are observed; one with a chemical shift only slightly larger than in the oxidized state, which suggests that it arises from a Fe^{3+}–Fe^{2+} pair in which the d electrons are delocalized, while the other has a larger shift corresponding to two Fe^{2+} atoms. The eight-iron ferredoxins contain two 4 Fe–4 S* centers in each molecule, and the chemical shifts indicate that the iron atoms are in a similar state to those in the four-iron ferredoxin, although distinct spectra are not resolved.

The 4 Fe–4 S* center in HiPIP shows only one quadrupole-split doublet in both the oxidized and reduced forms. The average chemical shift suggests that the reduced protein contains 2 Fe^{3+} and 2 Fe^{2+} atoms with strong electron delocalization as in the oxidized four-iron and eight-iron ferredoxins, while the oxidized center corresponds to 3 Fe^{3+} and 1 Fe^{2+} atoms.

Xanthine oxidase, with eight iron atoms per molecule, gives Mössbauer spectra similar to those of the two-iron ferredoxins and unlike those of the proteins with 4 Fe–4 S* centers. This suggests that the molecule contains four 2 Fe–2 S* centers.

In all these proteins that contain two or more iron atoms, they are antiferromagnetically coupled together. This was first suggested to account for the $g = 1.94$ EPR signal from the reduced two-iron ferredoxins.

In oxidized ferredoxins and reduced HiPIP, where there are pairs of identical atoms, the resultant magnetic moment for the molecule is zero. In the reduced ferredoxins and oxidized HiPIP, there is a net spin of $\frac{1}{2}$, since the spins of the iron atoms can no longer pair off. Direct evidence for this comes from the observation of the magnetic hyperfine splitting of Mössbauer spectra in large magnetic fields, where effective magnetic fields at the nuclei are observed that are both parallel and antiparallel to the external field. Indirect evidence for this coupling comes from (i) the average g values of less than 2 for the reduced ferredoxins, (ii) the low values (about half that of uncoupled ions) of the magnetic hyperfine fields, (iii) the absence of Curie law paramagnetism in oxidized ferredoxins, and (iv) PMR data.

Significant progress has been made in the study of the proteins with 4 Fe–4 S* centers, and a reasonably complete and consistent pattern has emerged. However, much remains to be done in order to obtain a complete theoretical understanding of the electronic structure of this more complex system, with particular regard to the effects of the protein environment.

With our present knowledge of the simpler members of the family of iron–sulfur proteins, gleaned from a whole range of spectroscopic techniques, we can now move on to apply these techniques to give us information on more complex iron–sulfur proteins, membrane-bound proteins, and complete enzyme systems. Such work is already well advanced with the more sensitive technique of EPR (see Chapter 2 of this volume). Mössbauer spectroscopy, too, is beginning to be applied with some success to systems such as nitrogenase (Smith and Lang, 1974; Münck et al., 1975). As this work progresses, the results will surely prove to be of great biological interest.

REFERENCES

Abragam, A., and Bleaney, B. (1970). "Electron Paramagnetic Resonance of Transition Ions," p. 131. Oxford Univ. Press, London and New York.
Adman, E. T., Sieker, L. C., and Jensen, L. H. (1973). *J. Biol. Chem.* **248**, 3987.
Anderson, R. E., Anger, G., Petersson, L., Ehrenberg, A., Cammack, R., Hall, D. O., Mullinger R. N., and Rao, K. K. (1975). *Biochim. Biophys. Acta* **376**, 63.
Antanaitis, B. C., and Moss, T. H. (1975). *Biochim. Biophys. Acta* **405**, 262.
Atherton, N. M., Garbett, K., Gillard, R. D., Mason, R., Mayhew, S. J., Peel, J. L., and Stangroom, J. E. (1966). *Nature (London)* **212**, 590.
Averill, B. A., Herskovitz, T., Holm, R. H., and Ibers, J. A. (1973). *J. Am. Chem. Soc.* **95**, 3523.
Bachofen, R., and Arnon, D. I. (1966). *Biochim. Biophys. Acta* **120**, 259.
Blomstrom, D. C., Knight, E., Phillips, W. D., and Weiher, J. F. (1964). *Proc. Natl. Acad. Sci. U.S.A.* **51**, 1085.

Blumberg, W. E. (1967). *In* "Magnetic Resonance in Biological Systems" (A. Ehrenberg, B. G. Malmstrom, and T. Vanngard, eds.), p. 119. Pergamon, Oxford.
Bray, R. C., Barber, M. J., Lowe, D. J., Fox, R., and Cammack, R. (1976). *Proc. FEBS Meet., 10th,* (in press).
Brintzinger, H., Palmer, G., and Sands, R. H. (1966). *Proc. Natl. Acad. Sci. U.S.A.* **55**, 397.
Cammack, R. (1973). *Biochem. Biophys. Res. Commun.* **54**, 548.
Cammack, R. (1975). *Biochem. Soc. Trans.* **3**, 482.
Cammack, R., and Evans, M. C. W. (1975). *Biochem. Biophys. Res. Commun.* **67**, 544.
Cammack, R., Rao, K. K., Hall, D. O., and Johnson, C. E. (1971). *Biochem. J.* **125**, 849.
Carter, C. W., Jr., Kraut, J., Freer, S. T., Alden, R. A., Sieker, L. C., Adman, E. T., and Jensen, L. H. (1972). *Proc. Natl. Acad. Sci. U.S.A.* **69**, 3526.
Carter, C. W., Jr., Kraut, J., Freer, S. T., Xuong, Ng. H., Alden, R. A., and Bartsch, R. G. (1974a). *J. Biol. Chem.* **249**, 4212.
Carter, C. W., Jr., Kraut, J., Freer, S. T., and Alden, R. A. (1974b). *J. Biol. Chem.* **249**, 6339.
Davenport, H. E., Hill, R., and Whatley, F. R. (1952). *Proc. R. Soc. London, Ser. B* **139**, 346.
Dickson, D. P. E., and Cammack, R. (1974). *Biochem. J.* **143**, 763.
Dickson, D. P. E., Johnson, C. E., Cammack, R., Evans, M. C. W., Hall, D. O., and Rao, K. K. (1974a). *Biochem. J.* **139**, 105.
Dickson, D. P. E., Johnson, C. E., Thompson, C. L., Cammack, R., Evans, M. C. W., Hall, D. O., Rao, K. K., and Weser, U. (1974b). *J. Phys. (Paris)* **35**, C6-343.
Dowsing, R. D., and Gibson, J. F. (1969). *J. Chem. Phys.* **50**, 294.
Dunham, W. R., Bearden, A. J., Salmeen, I. T., Palmer, G., Sands, R. H., Orme-Johnson, W. H., and Beinert, H. (1971). *Biochim. Biophys. Acta* **253**, 134.
Eicher, H., Parak, F., Bogner, L., and Gersonde, K. (1974). *Z. Naturforsch., Teil B* **29**, 683.
Eisenstein, K. K., and Wang, J. H. (1969). *J. Biol. Chem.* **244**, 1720.
Evans, M. C. W., Hall, D. O., and Johnson, C. E. (1970). *Biochem. J.* **119**, 289.
Evans, M. C. W., Lord, A. V., and Reeves, S. G. (1974). *Biochem. J.* **138**, 177.
Frankel, R. B., Herskovitz, T., Averill, B. A., Holm, R. H., Krusic, P. J., and Phillips, W. D. (1974a). *Biochem. Biophys. Res. Commun.* **58**, 974.
Frankel, R. B., Averill, B. A., and Holm, R. H., (1974b). *J. Phys. (Paris)* **35**, C6-107.
Freer, S. T., Alden, R. A., Carter, C. W., Jr., and Kraut, J. (1975). *J. Biol. Chem.* **250**, 46.
Fritz, J., Anderson, R. E., Fee, J., Palmer, G., Sands, R. H., Tsibris, J. C. M., Gunsalus, I. C., Orme-Johnson, W. H., and Beinert, H. (1971). *Biochim. Biophys. Acta* **253**, 110.
Gayda, J. P., Gibson, J. F., Cammack, R., Hall, D. O., and Mullinger, R. N. (1976). *Biochim. Biophys. Acta* **434**, 154.
Gersonde, K., Schlaak, H. E., Breitenbach, M., Parak, F., Eicher, H., Zgorzalla, W., Kalvius, M. G., and Mayer, A. (1974). *Eur. J. Biochem.* **43**, 307.
Gibson, J. F., Hall, D. O., Thornley, J. H. M., and Whatley, F. R. (1966). *Proc. Natl. Acad. Sci. U.S.A.* **56**, 987.
Hall, D. O., Gibson, J. F., and Whatley, F. R. (1966). *Biochem. Biophys. Res. Commun.* **23**, 81.
Hall, D. O., Cammack, R., and Rao, K. K. (1974). *In* "Iron in Biochemistry and

Medicine" (A. Jacobs and M. Worwood, eds.), p. 279. Academic Press, New York.
Holm, R. H., Averill, B. A., Herskovitz, T., Frankel, R. B., Gray, H. B., Siiman, O., and Grunthaner, F. J. (1974). *J. Am. Chem. Soc.* **96**, 2644.
Johnson, C. E. (1971). *J. Appl. Phys.* **42**, 1325.
Johnson, C. E. (1974). *J. Phys. (Paris)* **35**, C1–57.
Johnson, C. E., and Hall, D. O. (1968). *Nature (London)* **217**, 446.
Johnson, C. E., Knowles, P. F., and Bray, R. C. (1967). *Biochem. J.* **103**, 10C.
Johnson, C. E., Elstner, E., Gibson, J. F., Benfield, G. L., Evans, M. C. W., and Hall, D. O. (1968). *Nature (London)* **220**, 1291.
Johnson C. E., Bray, R. C., Cammack, R., and Hall, D. O. (1969). *Proc. Natl. Acad. Sci. U.S.A.* **63**, 1234.
Johnson, C. E., Cammack, R., Rao, K. K., and Hall, D. O. (1971). *Biochem. Biophys. Res. Commun.* **43**, 564.
Lowe, D. J., Lynden-Bell, R. M., and Bray, R. C. (1972). *Biochem. J.* **130**, 239.
McDonald, C. C., Phillips, W. D., Lovenberg, W., and Holm, R. H. (1973). *Ann. N.Y. Acad. Sci.* **222**, 789.
Marshall, W., and Johnson, C. E. (1962). *J. Phys. Radium* **23**, 733.
Mathews, R., Charlton, S., Sands, R. H., and Palmer, G. (1974). *J. Biol. Chem.* **249**, 4326.
Matsubara, H., Sasaki, R. M., Tsuchiya, D. K., and Evans, M. C. W. (1970). *J. Biol. Chem.* **245**, 2121.
Moss, T. H., Bearden, A. J., Bartsch, R. G., Cusanovich, M. A., and San Pietro, A. (1968). *Biochemistry* **7**, 1591.
Mullinger, R. N., Cammack, R., Rao, K. K., Hall, D. O., Dickson, D. P. E., Johnson, C. E., Rush, J. D., and Simopoulos, A. (1975). *Biochem. J.* **151**, 75.
Münck, E., Debrunner, P. G., Tsibris, J. C. M., and Gunsalus, I. C. (1972). *Biochemistry* **11**, 855.
Münck, E., Rhodes, H., Orme-Johnson, W. H., Davis, L. C., Brill, W. J., and Shah, V. K. (1975). *Biochim. Biophys. Acta* **400**, 32.
Orbach, R. (1961). *Proc. R. Soc. London, Ser. A* **264**, 458.
Orme-Johnson, W. H., and Beinert, H. (1969). *Biochem. Biophys. Res. Commun.* **36**, 337.
Packer, E. L., Sternlicht, H., and Rabinowitz, J. C. (1972). *Proc. Natl. Acad. Sci. U.S.A.* **69**, 3278.
Palmer, G., and Sands, R. H. (1966). *J. Biol. Chem.* **241**, 253.
Palmer, G., Brintzinger, H., Estabrook, R. W., and Sands, R. H. (1967). In "Magnetic Resonance in Biological Systems" (A. Ehrenberg, B. G. Malmström, and T. Vänngård, eds.), p. 159. Pergamon, Oxford.
Phillips, W. D., Poe, M., Weiher, J. F., McDonald, C. C., and Lovenberg, W. (1970a). *Nature (London)* **227**, 574.
Phillips, W. D., Poe, M., McDonald, C. C., and Bartsch, R. G. (1970b). *Proc. Natl. Acad. Sci. U.S.A.* **67**, 682.
Rao, K. K., Cammack, R., Hall, D. O., and Johnson, C. E. (1971). *Biochem. J.* **122**, 257.
Rao, K. K., Evans, M. C. W., Cammack, R., Hall, D. O., Thompson, C. L., Jackson, P. J., and Johnson, C. E. (1972a). *Biochem. J.* **129**, 1063.
Rao, K. K., Smith, R. V., Cammack, R., Evans, M. C. W., Hall, D. O., and Johnson, C. E. (1972b). *Biochem. J.* **129**, 1159.
Rasse, D., Warme, P. K., and Sheraga, H. A. (1974). *Proc. Natl. Acad. Sci. U.S.A.* **71**, 3736.

Shethna, Y. I., Stombaugh, N. A., and Burris, R. H. (1971). *Biochem. Biophys. Res. Commun.* **42**, 1108.
Smith, B. E., and Lang, G. (1974). *Biochem. J.* **137**, 169.
Thompson, C. L., Johnson, C. E., Dickson, D. P. E., Cammack, R., Hall, D. O., Weser, U., and Rao, K. K. (1974). *Biochem. J.* **139**, 97.
Thomson, A. J. (1975). *Biochem. Soc. Trans.* **3**, 468.
Thornley, J. H. M., Gibson, J. F., Whatley, F. R., and Hall, D. O. (1966). *Biochem. Biophys. Res. Commun.* **24**, 877.
Travis, J., Newman, D. J., LeGall, J., and Peck, H. D. (1971). *Biochem. Biophys. Res. Commun.* **45**, 452.
Wickman, H. H., Klein, M. P., and Shirley, D. A. (1965). *J. Chem. Phys.* **42**, 2113.
Zubieta, J. A., Mason, R., and Postgate, J. R. (1973). *Biochem. J.* **133**, 851.

CHAPTER 9

Redox Mechanisms of Iron–Sulfur Proteins

LARRY E. BENNETT

   I. Introduction.. 331
 II. Recent Developments in Simple Redox Chemistry................ 332
     A. Outer-Sphere Reactions..................................... 332
     B. Inner-Sphere Reactions..................................... 345
     C. Other Relevant Redox Transformations in Simple Systems..... 348
III. Redox Dynamics of Structurally Characterized Iron–Sulfur
     Proteins.. 352
     A. Reactions with Simple Reductants and Oxidants.............. 352
     B. Protein–Protein Reactions Involving Iron–Sulfur Species....... 369
IV. Survey of the Redox Behavior of Structurally Less Characterized
     Iron–Sulfur Proteins... 372
     References.. 375

I. INTRODUCTION

    Over the little more than a decade since the first iron–sulfur proteins were discovered, these intriguing and ubiquitous biological redox agents have challenged scientists to a remarkable extent. Testimony to this can be found both in the substantial volume of imaginative and productive research directed toward understanding these proteins and in the variety of perspectives from which it has been pursued. The resulting knowledge of the biological origins and function and the structure of the metal sites of these unique molecules is considerable, although by no means complete. Much less is known in any detail of their kinetic properties, which, we shall infer, are intrinsic to their function.
    It is the purpose of this chapter to consider, to the limited extent presently possible, the mechanisms by which iron–sulfur proteins undergo redox transformations. The principal, although not exclusive, focus will be on the smaller proteins (the biological function of which appears

to be electron transfer) for which X-ray structures are available to aid in the interpretation of kinetic results. This approach is based on the philosophy that these redox agents are exceedingly complex, relative to the nonbiological agents for which there is increasing theoretical understanding, and that the details of mechanism are most likely to be revealed with the simpler proteins of known structure.

With the exception of ferredoxins in photosynthetic organisms, little research is available on the detailed mechanistic behavior of iron–sulfur proteins in their biological environments. Thus, much of this report will deal with reactions in aqueous media. In fact, it can be fairly stated, at least with regard to our own work, that the proteins are simply being studied as coordination complexes, albeit rather fancy ones. This may not prove completely satisfying to those whose primary concern is biological function. However, we hope to be at least partially persuasive that the kinetic characteristics of the proteins can be identified and that at least some of those characteristics can reasonably be expected to be operative during physiological function.

We cannot hope to be exhaustive in our coverage of all iron–sulfur protein behavior that may reflect on mechanistic issues. For more comprehensive coverage the reader is referred to the other chapters in this volume, Volumes I and II of this treatise, and recent review articles (Palmer and Brintzinger, 1972; Orme-Johnson, 1973; Bennett, 1973), overlap with which we shall attempt to minimize.*

II. RECENT DEVELOPMENT IN SIMPLE REDOX CHEMISTRY

In a previous review of metalloprotein redox reactions (Bennett, 1973), a brief summary was provided of the redox reactions of simple metal complexes and of the theoretical foundations on which such reactions seemed best understood at the time. There is no demonstrated need for a major revision of that treatment as it may relate to our consideration of iron–sulfur protein behavior. Only an outline will be presented here in an effort to summarize relevant recent developments as an extension of that or other previously referenced reviews of the subject. Mention should be made of the somewhat more theoretical and typically excellent chapter along related lines of inquiry by Sutin (1973), which appeared after our previous writing.

A. Outer-Sphere Reactions

There is no convincing evidence at present that electron-transfer metalloproteins function physiologically via mechanisms involving the

* While this chapter was written, a related review by Palmer (1975) appeared.

substitution of a bridging ligand from one reactant into the first-coordination sphere of a metal center on the other reactant, although such a possibility should not be universally dismissed. (This category of proteins is not intended to include the more complex redox metalloenzymes of the iron–sulfur or other classes, where substitution by substrate often seems essential.) The outer-sphere reaction category, within which presently studied proteins of this type appear most likely to be placed, has proven particularly susceptible to theoretical treatment, at least with simple complexes, with that of Marcus (fully cited by Marcus and Sutin, 1975), having received the widest critical acceptance.

In aqueous solution, $\Delta G_{ii}^{\ddagger}$ for the outer-sphere, self-exchange reaction between the two components of a redox couple, e.g., $Ru(NH_3)_6^{2+/3+}$, is theoretically attributed to four contributions:

1. An association term $(RT \ln k_b T/hZ)$, which reflects the losses in translational and rotational free energy on bringing the hypothetically uncharged reactants together along a consummative electron transfer pathway. The contribution to $\Delta G_{ii}^{\ddagger}$ lies predominantly in the accompanying entropy loss, $\Delta S_a^{\ddagger}$, which is calculated to be ~ -10 to -13 eu for simple complexes at 25°C, depending on the value chosen for Z, the collision frequency. This frequency is calculated to be somewhat smaller for larger protein molecules in solution, but it should be emphasized that the important feature is the collision frequency of their redox sites, i.e., along a consummative electron-transfer pathway. For example, the frequency of collision of $Ru(NH_3)_6^{2+}$ with the redox site of rubredoxin may be diminished primarily only by blockage of approach on those sides of the site from which there is an extension of protein material and, therefore, not be greatly reduced from that with simple complexes. The association term is, of course, subject to considerable reevaluation for physiological protein function where different considerations for consummative association of reactants may often be necessary. (For the moment, an indulgence is requested of the reader for whom the physiological protein counterpart of the self-exchange reaction might understandably be unimaginable.) The associative term in aqueous solution appears at present to be susceptible to being factored out of the total $\Delta S^{\ddagger}$, at least approximately, based on this theory.

2. A coulombic term $\Delta G_c^{\ddagger}$, which reflects the electrostatic interaction between appropriately charged reactants at their separation distance in the activated complex. This contribution becomes negligible in aqueous solutions of sufficient ionic strength to shield the reactant charges, at least for simple complexes. (We shall consider the nature of this contribution at lower ionic strengths below and with small proteins in Section III.) In view of presently available testimony to the importance of specific

electrostatic interactions at the consummative interfaces between bioselective metalloproteins, e.g., with cytochrome c and its oxidase (Dickerson and Timkovich, 1975), it would be foolhardy to expect any such conclusion to be applicable to physiological function. However, it is not our present objective to explore this aspect of iron–sulfur protein behavior in detail, since with a few exceptions, the necessary experiments to examine such specific and possibly multisite interactions have yet to be done, particularly within the electron-transfer subcategory.

3. An inner-sphere rearrangement term $\Delta G_i^{\ddagger}$, which arises due to the widely-discussed Franck–Condon activation of the first coordination sphere. Such requirements are almost certainly operative, in principle, for metalloproteins even during physiological function, although the contribution may be small as it is with some, but not all, simple complexes, e.g., $Ru(NH_3)_6^{2+/3+}$. Intuitively and in practice, the effect is attributable primarily to an enthalpy component, $\Delta H_i^{\ddagger}$.

4. A corresponding outer-sphere rearrangement term $\Delta G_o^{\ddagger}$, reflecting Franck–Condon activation of the surrounding medium. In media of high dielectric constant, such as water, there is little contribution to this in the entropy $\Delta S_o^{\ddagger}$ on a theoretical basis. Experimental reasons for questioning this conclusion will be discussed shortly. For aquo or ammine complexes exchanging one electron, a value of $\Delta H_o^{\ddagger} \sim 6$ kcal mole^{-1} can be calculated from the theoretical formulations. (The 3 kcal mole^{-1} value quoted in our previous review arose from a misinterpretation.)

As we have previously stated, it seems particularly desirable to consider carefully the nature of outer-sphere contributions to protein activation, since the material exterior to the first coordination sphere is nonhomogeneous, which compounds the problem theoretically, and is substantially of low dielectric constant. The latter parameter enters in the denominators of expressions for $\Delta G_c^{\ddagger}$, $\Delta S_c^{\ddagger}$, $\Delta G_o^{\ddagger}$ and $\Delta S_o^{\ddagger}$ (Reynolds and Lumry, 1966) and may significantly influence kinetic behavior. Substantial effects of dielectric changes have been reported (Kassner, 1972) on ground-state thermodynamic parameters and modeled electrostatically for cytochrome c (Kassner, 1973) and the two-iron ferredoxins (Kassner and Yang, 1973). It would not be at all surprising if these effects operated to influence kinetic behavior as well and may, for this reason, be responsible for the evolution of certain structural features.

Returning attention to simple complexes, the $Ru(NH_3)_6^{2+/3+}$ self-exchange is characterized (Meyer and Taube, 1968) by a $\Delta S^{\ddagger} = -11 \pm 3$ eu, in seemingly good agreement with theory, and $\Delta H^{\ddagger} = 10.3 \pm 1$ kcal mole^{-1}, of which perhaps 4 kcal lie in $\Delta H_c^{\ddagger}$ or remains unaccounted for, since $\Delta H_i^{\ddagger}$ is expected to be small [Stynes and Ibers (1971) to whom we apologize for our previous miscalculation of $\Delta H_o^{\ddagger}$]. Several $M(H_2O)_6{}^{2+/3+}$ self-exchanges have values of $\Delta H^{\ddagger} \sim 10\text{--}13$ kcal mole^{-1} and $\Delta S^{\ddagger} \sim -20$ to -25 eu, for example (Krishnamurty and Wahl, 1948). If these reactions are outer sphere, as now seems probable for at least some of them, the significant discrepancies with theory may be resolvable in terms of specific interactions, not taken into account by the theory, of ligand molecules with the solvent (e.g., hydrogen bonding), which may be related to the poorly understood but substantial decreases in partial molar entropy of transition states with increasing charge, referenced previously (Bennett, 1973).

1. Progress in Modeling Nonprotein Redox Behavior

If there are no complications (e.g., spin state changes, excessively negative $\Delta G°$), outer-sphere cross-reactions between two components of two redox couples frequently obey the Marcus relationship (Bennett, 1973),

$$\Delta G_{12}^{\ddagger} = m(\Delta G_{11}^{\ddagger} + \Delta G_{22}^{\ddagger} + \Delta G_{12}^{\circ}) - 1.15 RT \log f \qquad (1)$$

where the last term is often negligible, and m is nominally 0.5, although it may vary when ΔG_{12}° is very positive or very negative (Reynolds and Lumry, 1966). The applicability of Eq. (1) across a variety of cross-reactions is especially remarkable in view of the emphasis by Marcus (cited by Marcus and Sutin, 1975; and R. A. Marcus, personal communication) that the relationship is theoretically justified only when the parameters involved are available *in the same medium*, e.g., ionic strength, a circumstance not always available. Moreover, the derivation of the cross-reaction relationship requires the fulfillment of one of three conditions: (1) the work terms in bringing the reactants together must be negligible, (2) these terms must cancel, or (3) the $\Delta G^{\ddagger}$ values must be corrected for the work terms if they are sizable and different, which is not usually or readily attempted.

While the widespread applicability of Eq. (1) may be due to fortuitous cancellation of effects in some cases, there is a strong empirical and theoretical indication of the contribution to the cross-reaction rate constant from the "standard" free-energy change (that which is operative under the particular experimental conditions studied). Thus, the self-exchange parameters can be considered as more reflective of the inherent reactivity of a redox couple than any particular cross-reaction parameters that are dependent additionally on both the inherent reactivity and the reduction potential of the second couple. Evidence that similar considerations related to Eq. (1) are apparently applicable, at least empirically, to the reactions of small electron-transfer proteins in aqueous solution of moderate ionic strength will be presented in the next subsection.

Prior to that, it is appropriate to devote some attention, using simple complexes, to the way in which the effects that seem so pervasive in the Marcus free-energy correlation [Eq. (1)] are partitioned between enthalpy and entropy contributions. Since these are two important parameters among a very limited number obtainable by kinetic measurements, and since they have proven useful in organic and inorganic systems for interpretation of events at the molecular level, it is important to understand how they arise for cross-reactions.

Being somewhat naive in theoretical matters, we derived straightforwardly from Eq. (1), neglecting small f terms, the cross-reaction relationships

$$\Delta H_{12}^\ddagger = m(\Delta H_{11}^\ddagger + \Delta H_{22}^\ddagger + \Delta H_{12}^\circ) \qquad (2)$$
$$\Delta S_{12}^\ddagger = m(\Delta S_{11}^\ddagger + \Delta S_{22}^\ddagger + \Delta S_{12}^\circ) \qquad (3)$$

where m is nominally 0.5 also, in an attempt to understand our activation parameters (Jacks and Bennett, 1974) for the $V(H_2O)_6{}^{2+}$–$Ru(NH_3)_6{}^{3+}$ cross-reaction involving simply an outer-sphere electron transfer between t_{2g} orbitals. All of the necessary parameters are independently available, although not universally under the same conditions. [The previous conditions for the theoretical applicability of Eq. (1) are also in force with Eqs. (2) and (3).]

The results are summarized in Table I. The agreement between calculated and observed values of $\Delta H^\ddagger$ and $\Delta S^\ddagger$, as well as $\Delta G^\ddagger$, is almost certainly too good to be true in the sense of empirically validating Eqs. (2) and (3). It can be noted that all independent parameters are available at or near 0.10 M ionic strength except the activation parameters for the $V(H_2O)_6{}^{2+/3+}$ self-exchange. In view of our calculation from Eq. (1) of only a $\sim$1 kcal increase in $\Delta G^\ddagger$ for that reaction at the lower ionic strength, it does not seem likely that the calculated $\Delta H^\ddagger$ and $\Delta S^\ddagger$ values would be very different if the appropriate self-exchange parameters were available.

However semiquantitative these calculations may be, it seems improbable that the very low $\Delta H^\ddagger$ and $\Delta S^\ddagger$ values for the cross-reaction can be understood except in terms of substantial contributions from ΔH_{12}° and ΔS_{12}°. Our molecular interpretation of the entropy effect is essentially that, in process of activation, the increase in the vibration-orientation polarization of the surrounding medium required by vanadium is not compensated for by the smaller decrease in polarization around $Ru(NH_3)_6{}^{3+\to 2+}$

TABLE I
Summary of Parameters for the $V(H_2O)_6^{2+}$–$Ru(NH_3)_6^{3+}$ Reaction[a]

System	ΔS[b]	$\Delta S°$	$\Delta H°$	$\Delta G°$	$\Delta S^\ddagger$ Obs	$\Delta S^\ddagger$ Calc	$\Delta H^\ddagger$ Obs	$\Delta H^\ddagger$ Calc	$\Delta G^\ddagger$ Obs	$\Delta G^\ddagger$ Calc
$V(H_2O)_6^{2+}$ to $V(H_2O)_6^{3+}$	−42	−26	−13.6	−5.9	−25	—	12.6	—	20.2	21.0[c]
$Ru(NH_3)_6^{3+}$ to $Ru(NH_3)_6^{2+}$	−1	−17	−6.9	−1.8	−11	—	10.3	—	13.6	—
$V(H_2O)_6^{2+}$ + $Ru(NH_3)_6^{3+}$ to $V(H_2O)_6^{3+}$ + $Ru(NH_3)_6^{2+}$	—	−43	−20.5	−7.7	−42	−40	0.6	1.2	13.2	13.0

[a] S values are in entropy units (cal mole^{-1} deg^{-1}), H and G values are in kcal mole^{-1}, all rounded and reported or referenced by Jacks and Bennett (1974).
[b] Absolute entropy changes referenced to a scale where $S°_{H^+} = 0$ eu (rather than −5.5 eu previously used) and $S°_{H_2} = 31.2$ eu.
[c] Calculated from Eq. (1) for 0.10 M ionic strength using our cross-reaction results.

suggested by the absolute entropy differences (Table I).* The molecular source of the $\Delta H°$ contribution appeared to us to result from a diminished rearrangement barrier (outer and inner sphere) relative to the self-exchange reactions as the result of the ground-state difference in redox orbital energies (Jacks and Bennett, 1974), a conclusion that appears consistent with the related treatment described next.

Prior to the appearance of our article, Marcus and Sutin (1975) submitted an elegant article in which relationships were derived that are equivalent to Eqs. (2) and (3) except for a correction term α, which is small ($\sim$0.05) for our reaction. They have clearly described important assumptions underlying the derivation of all three cross-reaction relationships. Satisfactory agreement was found for the calculated $\Delta S^{\ddagger}$ and, to a lesser extent, the $\Delta H^{\ddagger}$ and $\Delta G^{\ddagger}$ values for the reduction of $Fe(bipy)_3^{3+}$ and $Ru(bipy)_3^{3+}$ by $Fe(H_2O)_6^{2+}$, with a discussion of the discrepancies observed that may be partially applicable to our system with a highly negative $\Delta H_{12}°$. A particularly important point is that these relationships rationalize observed values of $\Delta H_{12}^{\ddagger}$ that are negative. We shall summarize the eminently readable molecular interpretation and reaction coordinate diagrams of Marcus and Sutin (1975) only briefly, believing that this article is mandatory study for any serious investigator of electron transfer kinetics.

The essential points include: (1) When $\Delta S_{12}°$ is very negative, the quantum states of the products are more widely spaced than for the reactants, resulting in intersections for the activated complexes that are quantum state dependent and must be Boltzmann weighted. (2) The result, when $\Delta H_{12}°$ is very negative, is that the less energetic activated complexes are more reactive than those with higher energy, and the mean thermal energy of the reactants is higher than the mean energy of the species that do react.†

---

* In addition to the $\sim -43/2$ eu contribution from this factor, there exists a -10 to -13 eu association contribution $\Delta S_a^{\ddagger}$ from the self-exchange terms. Also from the self-exchange terms, a coulombic contribution $\Delta S_c^{\ddagger}$ of about -9 eu can be calculated from the available equations (Reynolds and Lumry, 1966) for a theoretical charge product of $+6$ at $I = 0.10$ M. The actual contribution from the latter may be closer to -4 eu, since an effective charge product of $+2.4$ seems indicated by a Debye–Hückel treatment of the $Ru(NH_3)_6^{2+/3+}$ self-exchange data (Meyer and Taube, 1968). In any event, the observed $\Delta S^{\ddagger}$ seems largely accounted for on a rational molecular basis.

† It can be mentioned that negative activation energies are established for certain gas phase combinations involving radicals and find a related interpretation (e.g., Benson, 1960). At our request, an expert in this area, H. E. O'Neal of San Diego State University, who pointed this out, has further considered the problem and personally communicated that a large negative $\Delta S°$ is not a required condition in transition state theory for negative activation energies; the rate constant will always decrease with

9. REDOX MECHANISMS OF IRON–SULFUR PROTEINS

Considerable further study is necessary before the application of Marcus cross-reaction enthalpy and entropy relationships can be made blindly. One disturbing serious apparent discrepancy already exists between observed and our (calculated) values of $\Delta G^{\ddagger} = 9.8$ (7.5), $\Delta H^{\ddagger} = 3.2$ (9.3), and $\Delta S^{\ddagger} = -22$ (6.0) for the necessarily outer-sphere $Ru(NH_3)_6{}^{2+}$–$Fe(H_2O)_6{}^{3+}$ reaction (Meyer and Taube, 1968), which has a small negative ΔH_{12}° (-2.2) and a very positive ΔS_{12}° ($+44$). Two transparent copies of Figure 1 in Marcus and Sutin (1975) can be overlaid to obtain a composite picture of this case, but it is difficult to understand the discrepancy in $\Delta S_{12}^{\ddagger}$ particularly. It may be that for reactions with very negative ΔG_{12}°, the activated complex closely resembles the reactants and requires little inner- or outer-sphere rearrangements prior to electron transfer (Reynolds and Lumry, 1966, p. 135). Thus, the observed $\Delta S^{\ddagger}$ may be primarily due to a $\Delta S_a{}^{\ddagger}$ contribution of -10 to -13 eu and a $\Delta S_c{}^{\ddagger}$ of -4 to -9 eu with little effect of the net ΔS_{12}° on the reactant-like activated complex. Similarly, $\Delta H^{\ddagger}$ might be lower than anticipated using Eq. (2) for reactions with less negative ΔG_{12}°.

It should be clear, from this discussion and other available evidence not pertinent to our immediate objectives, that even simple electron-transfer reactions are not as well understood as the recently deficient federal support of demonstrably talented investigators in this fundamental area would appear to suggest. It has become obvious (in fairness, largely due to federal support) that the failure to consider possible contributions from ΔH° and ΔS° to $\Delta H^{\ddagger}$ and $\Delta S^{\ddagger}$, respectively, is potentially as hazardous as the previously frequent failure to consider the contribution from ΔG° to $\Delta G^{\ddagger}$. The possible implications for protein behavior of this most important recent development in the outer-sphere reaction category will be considered below. It is appropriate here to mention the latest article we have seen concerning possible compensatory changes in $\Delta H^{\ddagger}$ and $\Delta S^{\ddagger}$ (Kemeny and Mahanti, 1972) and note that their quadratic polaron model, which, of the two discussed therein, seems most similar to usual models of outer-sphere reactions, always yields positive energies and negative entropies of activation.

2. Potential Implications for Modeling Metalloprotein Redox Behavior

In attempting to develop rational models in terms of which the redox behavior of metalloproteins can be at least partially understood, it can be stated unequivocably that their redox sites are coordination com-

---

increasing temperature if the partition function ratio $(Q^{\ddagger}/Q)$ decreases significantly with temperature. From statistical thermodynamics, this condition can be met in the case of a large negative $\Delta S^{\ddagger}$.

plexes, however exotic. We have tentatively adopted the hypothesis that the portion of the protein that is exterior to the first coordination shell(s) of the metal(s), while unquestionably being essential to physiological function, will fulfill that role as the result of modifications it imposes on the coordination complex, including the development of a shape and surface topography that can selectively favor particular interactions with physiological reactants. One important challenge confronting chemists of biological, structural, thermodynamic, or kinetic persuasions is the characterization of those modifications and their purposes. This challenge is not likely to be met overnight.

Since the characteristics of any particular protein will arise from a combination of the coordination complex characteristics of the redox site and the modifications characteristic of the surrounding material, one reasonable approach seemed to us to be an evaluation of behavior in comparison to that of simpler coordination complexes. Therefore, we suggested rather strongly (Bennett, 1973) that the pervasive adherence to Eq. (1) in many instances of simple cross-reactions might extend to cross-reactions involving relatively small electron-transfer proteins. That suggestion can now be considered in more detail.

Mammalian cytochrome c is the only protein in this class for which the self-exchange activation parameters have been determined (Gupta *et al.*, 1972; Gupta, 1973). Its rate of reduction by $Ru(NH_3)_6^{2+}$, well characterized as a straightforward outer-sphere reactant, is in agreement with the independently available predictions of Eq. (1) (Ewall and Bennett, 1974). Readily available recent reports from other laboratories, particularly those of Harry Gray and Norman Sutin, point to a similar free-energy correlation with other simple redox agents, e.g., $Fe(EDTA)^{2-}$ and $Co(phen)_3^{3+}$.

Such a correlation is also supported by as yet unpublished work of Joanne Popa in the author's laboratory, carried out in collaboration with Robert Bartsch, on the $Ru(NH_3)_6^{2+}$ reduction of cytochrome c_2 from *Rhodospirillum rubrum*. The rate at pH 7 is faster than with cytochrome c, although not so much so as would be predicted from the ~ 1.4 kcal $\Delta G°$ increase alone, perhaps as the result of a slightly smaller protein self-exchange contribution. More significantly, over a pH range where the cytochrome c reduction potential is constant and the rate changes negligibly, the cytochrome c_2 potential varies and the rate changes in good correlation with the predictions of Eq. (1) for the changes in net driving force contribution alone (provided that minor variations in the pH rate profile are considered as second-order perturbations on the $\Delta G°$ influence).

In Section III,A,2 kinetic results for the reaction of high potential iron

9. REDOX MECHANISMS OF IRON–SULFUR PROTEINS

protein (HiPIP) from *Chromatium* with two oxidants and two reductants will be summarized. The results are reasonably consistent in the predictions they afford of the unknown self-exchange rate of the HiPIP couple in spite of considerable variation in the nature of the four simple reactants.

While no conclusive statement can be made at this time, these lines of evidence are consistent with Eq. (1) being applicable to cross-reactions of electron transfer proteins with simple reagents, at least on a semi-quantitative basis. Until sufficient evidence to the contrary is presented, it seems advisable to evaluate cross-reactions with proteins in this light and to factor the self-exchange and reduction potential characteristics of the simple reagent out of the results, thereby revealing, potentially, the inherent reactivity characteristics of the protein.

The next step, of course, is to examine cross-reactions involving proteins for enthalpy and entropy correlations within Eqs. (2) and (3) similar to those discussed above with simple reagents. Such correlations cannot be convincingly predicted on the basis of any present or future free-energy correlations, since cancellations responsible for apparent agreement in the latter may not be fortuitously partitioned among the enthalpy and entropy separately.

The only protein for which even approximately appropriate data sufficient to examine the question are available is cytochrome c. We have used the thermodynamic data of George *et al.* (1968), summarized with comparable other results by Margalit and Schejter (1973), and have estimated $\Delta H^{\ddagger} = E_a - RT = 9.4$ kcal mole^{-1} and $k = 1.4 \times 10^3$ M^{-1} sec^{-1} at 25°C from Gupta *et al.* (1972) as being closest to appropriate for the cytochrome self-exchange under the conditions ($I = 0.10$ M, tris–tris HCl) of our Ru(NH$_3$)$_6^{2+}$ study (Ewall and Bennett, 1974). However, no strictly comparable conditions are available, and it is particularly evident in several of these references, as well as Gupta (1973), that the enthalpy and entropy characteristics of the protein couple are remarkably (and often compensatingly) sensitive to media, especially to our chloride media.

Independently measured and calculated parameters are summarized in Table II. The agreement between observed and calculated activation parameters is excellent for $\Delta G^{\ddagger}$ and probably adequate for present purposes with $\Delta H^{\ddagger}$ and $\Delta S^{\ddagger}$. Note should be taken that: (1) the latter values would be very difficult to account for in terms of $\Delta H^{\ddagger}$ and $\Delta S^{\ddagger}$ values for the self-exchanges alone, (2) $\Delta H°$ and $\Delta S°$ appear to contribute significantly in the *direction* of agreement with observation, and (3) compensatory decreases of $\Delta H° - \Delta S°$ and $\Delta H^{\ddagger} - \Delta S^{\ddagger}$ values for the protein couple that are worth only ~2 kcal within each pair would result in

TABLE II
SUMMARY OF PARAMETERS FOR THE $Ru(NH_3)_6^{2+}$–CYTOCHROME c REACTION[a]

System	$\Delta H°$	$\Delta S°$	$\Delta G°$	$\Delta H^‡$	$\Delta S^‡$	$\Delta G^‡$
$Ru(NH_3)_6^{2+}$ to $Ru(NH_3)_6^{3+}$	6.9	17	1.8	10.3	−11	13.6
Cyt c^{III} to cyt c^{II}	−14.4	−28	−6.0	9.4	−13	13.2
$Ru(NH_3)_6^{2+}$ + cyt c^{III} to $Ru(NH_3)_6^{3+}$ + cyt c^{II}	−7.5	−11	−4.2	2.9	−28	11.2 (obs)
				6.1	−18	11.3 (calc)

[a] As for footnotes a and b in Table I with additional references given in the paragraph preceding first mention of Table II in the text.

acceptable agreement. The latter variations seem within the realm of possibility, given the media sensitivity and differences previously mentioned.

Alternatively, it may be that the cross-reaction differs in the details of net activation sufficiently from the self-exchanges that cancellation of certain enthalpy and entropy components does not occur. If data were available with a sufficient variety of reactants, such differences might be illuminated and prove interesting, from a fundamental if not a physiological perspective.

Unfortunately, the only ground-state thermodynamic parameter that seems available for the $Fe(EDTA)^{2-/1-}$ couple is $\Delta G°$ [from a reduction potential of 0.117 V at $I = 0.10$ M (Schwarzenbach and Heller, 1951)]. Using $k \sim 2 \times 10^4$ M^{-1} sec^{-1} which can be estimated for this couple's self-exchange at $I = 0.10$ M from data on closely related couples at slightly higher ionic strengths (Wilkins and Yelin, 1968; Grossman and Wilkins, 1967), we can directly calculate only a $k = 5.5 \times 10^4$ M^{-1} sec^{-1} for the $Fe(EDTA)^{2-}$–cytochrome c^{III} reaction, in good agreement with the observed $k = 2.5 \times 10^4$ M^{-1} sec^{-1} (Hodges et al., 1974). Further, it seems clear from this paper that changes in buffer identity alone (at pH 7 and $I = 0.10$ M (NaCl)) can result in largely compensatory $\Delta H^‡$ and $\Delta S^‡$ changes of $\sim$1 kcal and $\sim$3 eu, respectively, emphasizing again substantial media sensitivity arising most probably from the protein.

The $\Delta S°$ for this favorable reaction seems likely to us, in view of the negative $\Delta S_{1/2}$ for cytochrome $c^{III/II}$ (Table II) and the size and charge type of the $Fe(EDTA)^{2-/1-}$ couple, to be small and quite possibly negative, which would require that $\Delta H°$ be negative. Values of $\Delta H^‡ \sim 4.0$ kcal mole^{-1} and $\Delta S^‡ \sim -25$ eu can be estimated for the latter couple from Wilkins and Yelin (1968). A completely hypothetical fit of the observed cross-reaction parameters, $\Delta H^‡ = 5.1$ kcal mole^{-1} and $\Delta S^‡ = -21$ eu (Hodges et al., 1974), to Eqs. (2) and (3) would require $\Delta H_{12}° \sim -3.2$

kcal mole^{-1} and $\Delta S_{12}^\circ \sim -4$ eu, which values are not greatly out of line with expectation since $\Delta G_{12}^\circ = -3.3$ kcal mole^{-1}, although they must certainly remain hypothetical until measured directly.

The widely studied reaction

$$\text{Fe(CN)}_6{}^{3-} + \text{cytochrome } c^{II} \rightleftharpoons \text{Fe(CN)}_6{}^{4-} + \text{cytochrome } c^{III}$$

now seems likely to involve stable precursor and successor complexes (Stellwagen and Shulman, 1973; Miller and Cusanovich, 1975; Stellwagen and Cass, 1975), which, according to the appropriate kinetic expression (Marcus and Sutin, 1975), will result in an increase of the calculated cross-reaction rate constant above the value calculated by the usual expression, perhaps sufficiently to find agreement with observed values. In view of these complications, it does not seem possible at present to evaluate the enthalpy and entropy effects separately, although a contribution from $\Delta H^\circ = -12.3$ kcal mole^{-1} seems possible in view of $\Delta H^\ddagger = 0$ for the cross-reaction (Morton et al., 1970), compared to significantly positive $\Delta H^\ddagger$ values for both self-exchanges (see Table II and Section III,A,2).

The oxidation of cytochrome c^{II} by Co(phen)$_3{}^{3+}$ ($k = 1.50 \times 10^3$, $\Delta H^\ddagger = 11.3$, $\Delta S^\ddagger = -6.2$) has been reported as conforming to Eq. (1) (McArdle et al., 1974). Our opportunity to evaluate the applicability of Eqs. (2) and (3) is severely limited by the absence of necessary data under comparable conditions. Values of $\Delta H^\ddagger = 5.1$ kcal mole^{-1} and $\Delta S^\ddagger = -34$ eu for the Co(phen)$_3{}^{3+/2+}$ exchange at $I = 0.10$ (Farina and Wilkins, 1968, footnote 2) seem in reasonable agreement with values for related reactions reported therein under similar conditions, but are in substantial contrast to $\Delta H^\ddagger = 16.4 \pm 5$ and $\Delta S^\ddagger = +4$ eu obtained with no added electrolyte (Baker et al., 1959). The use of Neumann's data appears to require, for agreement with $\Delta S^\ddagger = -6.2$ eu, a substantial positive ΔS° contribution, which seems unlikely to be sufficiently positive in view of a ΔS° value of -17.2 ± 2 eu, which can be calculated for the Fe(phen)$_3{}^{3+/2+}$ couple from data at $I = 0.10$ M (George et al., 1959).

It seems advisable in attempting to evaluate the applicability of Eqs. (2) and (3) to protein-simple reagent cross-reactions to pay particular attention to media effects, not only on the protein, but also on the simple reagent couple. As is also the case with the Fe(CN)$_6{}^{3-/4-}$ couple (Eaton et al., 1967; Campion et al., 1967), the M(phen)$_3{}^{3+/2+}$ and related nonspherical, hydrophobic couples seem especially sensitive to the identity of the counterion, most notably when oxyanions are compared to other anions (Baker et al., 1959; Bennett and Taube, 1968). Other possible complications with such couples have been discussed by Marcus and Sutin (1975).

With these considerations in mind, it presently seems prudent to rely only on results with the presumably uncomplicated reductants, $Ru(NH_3)_6^{2+}$ and $Fe(EDTA)^{2-}$, in judging the applicability of Eqs. (2) and (3) to cytochrome c reduction. These limited results seem consistent with the tentative and admittedly speculative hypothesis that the overall kinetic behavior of a cross-reaction involving a small protein may, in favorable cases, be factorable into enthalpy and entropy contributions from both the ground state and transition state characteristics of the two couples involved, at least semiquantitatively. It seems reasonable from the data in Table II, for example, that the $\Delta H°$, $\Delta S°$, $\Delta H^{\ddagger}$, and $\Delta S^{\ddagger}$ characteristics of the cytochrome c couple alone appear to contribute significantly and, to a considerable extent, identifiably to $\Delta H^{\ddagger}$ and $\Delta S^{\ddagger}$ of the cross-reaction. If this hypothesis proves true, the possible portent for other electron-transfer protein couples (for which the inherent enthalpy and entropy characteristics for both ground and transition states of the self-exchange are usually unknown) is that suitable cross-reaction studies may reveal aspects of those characteristics that could provide insight into the molecular details of kinetic behavior. (The ramifications of this hypothesis for iron–sulfur proteins will be considered in Section III.)

One intriguing set of observations can be made in this context on the basis of what is known about cytochrome c and the photosynthetic and respiratory redox chains. It is well known that the transfer of redox equivalents within these chains is frequently between agents differing little in reduction potential [see Section V by Bennett (1973)].* However, it is also known that the potential of cytochrome c remains essentially constant from its physiological environment (Bennett, 1973, Table IV) to a variety of aqueous media of different ionic strengths (Margalit and Schejter, 1973). The same cannot be said of substantial and opposing values of $\Delta H°$ and $\Delta S°$ in aqueous media, which are therein reported to vary compensatingly by the equivalent of ~ 6.8 kcal mole^{-1}! Similarly, the effect on the self-exchange of an ionic strength change that affects $\Delta G^{\ddagger}$ by only ~ 1.4 kcal mole^{-1} is to change $\Delta H^{\ddagger}$ by ~ 6 kcal mole^{-1} (Gupta, 1973). It almost appears as if, for this ancient protein whose functional demands are presumably at constant mean temperature, some principle of evolution has been invoked to develop a structure that maintains ground-state and transition-state free-energy characteristics relatively independent of media while the corresponding enthalpy and en-

*The physiologically oriented person may initially question our interest in Eq. (1) for this reason. Our reply is that, under circumstances of small $\Delta G°$, the conceptual value of Eq. (1) lies in its factoring of the net behavior into inherent reactivities ($\Delta G_{11}^{\ddagger}$ and $\Delta G_{22}^{\ddagger}$) to a first approximation, which we believe will carry over to and probably be modified by the physiological environment.

9. REDOX MECHANISMS OF IRON–SULFUR PROTEINS

tropy characteristics (having molecular sources) fluctuate substantially and, to a large extent, compensatingly. (This behavior has not been tested physiologically to the author's knowledge.)

The lesson we derive from these observations is that $\Delta H^{\ddagger}$ and $\Delta S^{\ddagger}$ values for a cross-reaction (physiological or nonphysiological) and any contribution to them that can be expressed as being from molecularly-based and possibly substantial values of $\Delta H°$ and $\Delta S°$ must be very carefully evaluated if the measurement of these few accessible kinetic parameters is to lead to insight at the molecular level. Moreover, to the extent that the ground state protein structure is mostly unchanged so long as it remains undenatured, the contributions to its inherent electron-transfer reactivity from inner- and outer-sphere rearrangement barriers seem likely to remain largely in effect for aqueous and physiological reactions alike (unless some uniquely different electron transfer pathway, as yet undemonstrated, is operative physiologically).

Thus, one major objective of studies such as those described herein is to provide information on these intrinsic activation characteristics. Modifications arising from differences that are operative in the physiological reactions, e.g., in the association and coulombic terms, are·virtually certain to be significant but in most cases insufficient information is available at present to do more in any quantitative way than speculate about them.

We are aware of only one study of an outer-sphere reaction involving a simple complex containing coordinated sulfur. The reaction between $Ru(NH_3)_6^{2+}$ and $Co(en)_2(SCH_2COO)^+$ is somewhat faster, $k \sim 0.15\ M^{-1}$ sec^{-1} ($I = 0.10$) than is the analogous reduction of $Co(en)_2(OCH_2COO)^+$, $k \sim 0.001\ M^{-1}$ sec^{-1} ($I = 0.50$) (Lane et al., 1976a). This result was attributed to a ground-state elongation of the Co–N bond that is trans to sulfur in $Co(en)_2(SCH_2COO)^+$ (Elder et al., 1973), and a greater ease of activation of the Co–S versus the Co–O bond. It was concluded that reasonable outer-sphere reactivity is conferred on cobalt(III) by the thiolate function once allowances are made for its basicity and the kinetic retardation that seems to arise with chelate ligands. It seems reasonable to expect, consistently with the results described in Section III,A,1 for rubredoxin, that thiolate functions (coordinated to mononuclear centers, at least) will not result in particularly difficult redox activation on other metal centers as well.

B. Inner-Sphere Reactions

No major extensions of our previous summary (Bennett, 1973) of this class seems necessary in the present context. Since inner-sphere behavior has yet to be demonstrated for iron–sulfur proteins, it seems appropriate to restrict our attention, briefly, to recent results with thiolate-coordinated complexes and other ligands of possible biological significance.

1. REACTIONS OF THIOLATE-COORDINATED METAL CENTERS

Results that were mostly obtained in author's laboratory when he was at the University of Florida are expected to appear shortly (Lane et al., 1976a,b). In the first of these reports, the reductions by $Cr(H_2O)_6^{2+}$ of $Co(en)_2(SCH_2COO)^+$, $Co(en)_2(SCH(CH_3)COO)^+$, $Co(en)_2(SCH_2CH_2-NH_2)^{2+}$, $Co(en)_2(SCH_2CH_2OH)^{2+}$, and their oxygen analogues are characterized by rate constants and, for accessible systems, activation parameters. While the patterns emerging for the latter are not completely understood, the major conclusion of potential biological relevance seems clear; one possible uniqueness of divalent sulfur as a donor atom is that it can bear a substituent larger than a proton and still allow a high level of inner-sphere reactivity, at least on cobalt(III). Since some donor substituent larger than hydrogen, e.g., —CH_2R, is required in order to attach a redox metal center to its protein, the conclusion may become relevant if inner-sphere mechanisms are discovered for thiolate-coordinated metalloproteins.

Substitutions of the thiolate-coordinated chromium(III) products of these reactions appeared in certain instances to be redox-catalyzed by $Cr(H_2O)_6^{2+}$ (Lane et al., 1976b). The high bridging efficiency of the thiolate function appeared to be retained, but the results were obtained under restrictive conditions, and acceptance of our interpretation of the relatively rapid rates reported must await more exhaustive study in view of the work to be described next.

The rates of reduction by $Cr(H_2O)_6^{2+}$ of several chromium(III) analogs of the cobalt(III) complexes have been found to be unusually slow in comparison to rates of the latter and to available chromium(III) rates (Weschler and Deutsch, 1976). The conclusion was reached that thiolate sulfur does not serve as a very efficient bridging function when coordinated to chromium(III). While this appears to be true in comparison to bridging functions without sterically-demanding substituents, we must hasten to add that comparisons with appropriately substituted bridging functions (a potentially important requirement in relation to possible inner-sphere biological function) are not available. Thus, our conclusion may not be at variance with theirs, and the information afforded by the comparisons must be regarded as moot in regard to this point.

Three aspects of chromium(III) substitution chemistry involving thiolate functions can be mentioned in passing. Strong labilization toward substitution, apparently in the trans position, is effected by thiolate ligands on aquo chromium(III) centers (Asher and Deutsch, 1976). Such an influence might contribute to lower Franck–Condon barriers for the redox catalysis described above. Second, a number of studies from Deutsch's laboratory and our own [described or referenced by Lane et al. (1976b)] have now characterized one path for the loss of a variety of thiolate functions from chromium(III) as involving the protonation of the sulfur prior to achieving the transition state. This seems somewhat related to the effects of acid on the oxidation and solvolysis of one of Holm's $Fe_4S_4(SR)_4^{2-}$ model complexes, studied by Bruice et al. (1975). Third, there is increasing evidence [presented or referenced by Lane et al. (1976b)] that substitution on chromium(III) centers, which until recently was quite widely believed to be a dissociatively dominated process (and, therefore, largely independent of the nature of the incoming ligand), may be associative in character, at least with a certain class of incoming ligands. This class appears to include those ligands with an appropriate combination of basicity and nucleophilicity, on which points thiolate and sulfide functions qualify abundantly. Other donor functions of proteins may also fall into this category. Thus, it can no longer be assumed that chromium(III) will retain its traditional inertness under conditions of low acidity and exposure to potentially nucleophilic substituents such as might be found within proteins near neutral pH, e.g., in the imaginative effort to locate the site of $Cr(H_2O)_6^{2+}$ attack on cytochrome c (Grimes et al., 1974).

2. Simple Reaction Systems of Potential Biological Significance

While it is not particularly relevant to the present state of knowledge of iron–sulfur protein redox behavior, brief mention seems warranted of developments in the important area of intramolecular electron transfer for possible future reference. In important continuing work from Taube's laboratory, the rates of electron transfer from Ru(II) to Co(III) have been compared when both centers are bound to remote sites of nicotinate, isonicotinate, and related intervening bridging ligands (Isied and Taube, 1973). Work also in progress in Hurst's laboratory promises to elucidate the details of electron transfer from Cu(I) to Ru(III) when the former is remotely bound to an unsaturated carbon–carbon bond of a carboxylate ligand attached to the latter (see Hurst and Lane, 1973). While there are potential complications with both systems, they hold forth the potential of better defining the barriers to electron transfer that are imposed by the electronic properties of the bridging ligand and the metal–metal separation distance. The nature of such barriers may some day prove useful for understanding certain metalloproteins with multiple redox sites such as exist in the iron–sulfur category (Bennett, 1973; Bray, 1975; Palmer, 1975).

C. Other Relevant Redox Transformations in Simple Systems

In this subsection, we summarize a sort of potpourri of possibly relevant simple redox transformations that do not readily fit into the two previous traditional categories. Several particularly interesting discoveries have come from Deutsch's laboratory.

When $Cr(en)_2(SCH_2COO)^+$ in 0.01 M $HClO_4$ was exposed to excesses of either of the one-equivalent oxidants, $Np(VI)$ or $Ce(IV)$, the expected oxidation of sulfur was found to be masked by its coordination, relative to the ease with which the adjacent methylene carbon was oxidized to a carbonyl function (Weschler et al., 1973). Consistency was found with reports that the autooxidations of certain organic substrates are catalyzed by copper and iron thiolate complexes without the net oxidation of sulfur (see reference 11 of their article).

The potentially significant proposal was made that the detailed function of hepatic aldehyde dehydrogenase and xanthine oxidase, each of which is a complex iron–sulfur flavoprotein containing molybdenum and catalyzes the conversion of aldehydes to carboxylic acids, may be related to these observations. Thus, it was suggested that this conversion might involve addition of the aldehyde substrate to a metal-coordinated hydrosulfide function to yield a hemimercaptal intermediate, Enz–M–S–CH(R)OH analogous to that previously proposed for coenzyme A-dependent bacterial aldehyde dehydrogenase conversions. The intermediate was proposed to be oxidized by two equivalents at the presumably similarly activated carbon to yield a coordinated thioacid which is then proposed to hydrolyze, releasing carboxylate and regenerating the metal hydrosulfide catalyst. The question of whether some adjacent function similar to carboxylate is also necessary in order to activate the methylene carbon seems to be raised by the second set of observations, made, however, under slightly different conditions.

When $Np(VI)$ or $Co(III)$ is injected into a 0.10 M $HClO_4$ solution containing excess $Co(en)_2(SCH_2CH_2NH_2)^{2+}$ (1), net reaction (4) occurs

$$NpO_2^{2+} + 5\,H^+ + 2\,(1) \longrightarrow (en)_2Co\begin{pmatrix} S-CH_2 \\ \ \ \ \vert \\ N-CH_2 \\ H_2 \end{pmatrix}^{SCH_2CH_2NH_3^{4+}} + Co^{2+} + NpO_2^+ + 2\,enH_2^{2+} \quad (4)$$

2

according to Woods et al. (1975). It was proposed that the one-equivalent oxidation of the coordinated thiolate on **1** produces a radical ligand complex (**3**) which dimerizes with **1** to give a relatively stable radical intermediate, 4. This intermediate was presumed to decay via the

$$\left[\begin{array}{c}(en)_2Co-S-S-Co(en)_2\\ |||\\ H_2N\diagdown_{C}\diagup^{C} {}^{C}\diagdown_{C}\diagup NH_2\\ H_2H_2H_2H_2\end{array}\right]^{5+}$$

4

internal induced reduction (see Bennett, 1973) of one of the Co(III) centers to account for the net stoichiometry, including the production of a coordinated disulfide complex (2), the only one known to the author containing a biologically relevant ligand. [X-ray studies of disulfide complexes with multidentate ligands have been reported by Warner et al. (1974) and Riley and Seff (1972).] Part of the support for the formulation of (2) is its reaction with excess free cysteamine to regenerate (1) along with free cystamine, which can be viewed as a thiol–disulfide interchange or a two-equivalent reduction of the coordinated disulfide (Woods et al., 1975, 1976).

These studies significantly expand the number of diverse pathways (Weschler et al., 1974) by which coordinated thiols can be oxidized; those pathways also include oxidation of sulfur to yield a coordinated sulfinic acid (Gillard and Maskill, 1968; Schubert, 1933) and metal–sulfur bond fission leading to the free disulfide (Asher and Deutsch, 1972). To this we can add what appear to be closely related but very preliminary unpublished observations by J. P. Bennett, Jr. in the author's laboratory. At the relatively high concentrations required for study by PMR, solutions of $Co(en)_2(SCH_2CH_2OH)^{2+}$ (Lane et al., 1976a), which were spectrally stable for considerable periods under more dilute conditions, developed a paramagnetic component as indicated by progressively broadened PMR signals. This was interpreted as most likely being due to Co(II), the development of which at high concentrations might be understood in terms of a bimolecular, two-equivalent oxidation by cobalt(III) to the disulfide according to a rate law of the form, $k[Co(en)_2(SCH_2CH_2OH)^{2+}]^2$, which is reminiscent of that previously referenced and discussed (Bennett, 1973) for the oxidation of water by Co_{aq}^{3+}. The reaction is under further study. Also noteworthy is the oxidation of thiols to disulfides by sulfoxides, including DMSO (Snow et al., 1975).

While the author is unaware of any presently demonstrated relevance of these observations to iron–sulfur protein behavior other than those already mentioned, it seems worthwhile to keep these and anticipated future developments in the area in mind when considering the behavior of thiolate- (and probably sulfide-) coordinated metalloproteins and model compounds in oxidizing environments (e.g., Bruice et al., 1975).

Because they exhibit charge-transfer absorption in the near ultraviolet, it is worth noting the photosensitivity of thiolate-coordinated Cr(III)–ammine complexes (Weschler and Deutsch, 1973), which may extend to certain metal–sulfur proteins. Also, the quenching of $Cr(H_2O)_6^{2+}$ with O_2, but not p-aminophenyl disulfide, is reported to catalyze the decomposition of $Cr(H_2O)_4(SCH_2COO)^+$ with release of H_2S (Weschler and Deutsch, 1975). Persons interested in iron–sulfur compounds may wish to learn of the synthesis and characterization of $[Co(en)_2(SeCH_2CH_2NH_2)](NO_3)_2$ for which the elongation of the Co–N bond trans to selenium is 0.063 Å compared to 0.041 Å for the sulfur analogue (Stein et al., 1976).

We conclude this section with four brief observations. The first concerns the remarkable "inorganic cubane" and related iron–sulfur model compounds developed in Holm's laboratory. Since they will be expertly discussed elsewhere in this volume, we shall simply cite the latest in a long line of contributions (Lane et al., 1975) in which a model for the $Fe-S_4$ unit in rubredoxin is described. Kinetic studies of the redox behavior of these models will be invaluable, in comparison to the protein results, for attempting to assess the modifications in kinetic characteristics that the proteins impose on the coordination complex (Section II,A,2).

Second, since we previously discussed the possible kinetic consequences of structuring a metalloprotein redox site with two appreciably populated spin states of possibly different kinetic responses to external redox agents (Bennett, 1973), it is appropriate to note a report on the rate of spin-state change for an Fe(II) complex in solution (Beattie et al., 1973). The reversible changes between singlet and quintet states occur at rates of $\sim 10^7$ sec^{-1} from which a tentative estimate of 10^{-5} for the transmission coefficient κ (see Section III,A,2) was calculated.

Their conclusion that spin state changes are not likely to be rate determining for most electron-transfer reactions in aqueous solution seems well taken. Whether it applies as expansively to physiological electron-transfer where reactants may be sterically juxtaposed remains to be examined. (It can be noted that few net observable redox processes for physiological systems have been reported to approach the rate of spin-state change observed.) At the same time, the thermodynamic advantages of such fine spin-state adjustments, where they occur, are not clear (Bennett, 1973), and the question of their kinetic effects in physiological systems appears open until further mechanistic information is available.

Third, flavins and iron–sulfur clusters, which are frequently strongly reducing, are sometimes found in the same enzyme, e.g., the aldehyde–

9. REDOX MECHANISMS OF IRON–SULFUR PROTEINS

carboxylate oxidoreductases mentioned above (see also Bray, 1975; Beinert et al., 1975, Bennett, 1973). We therefore mention, without any specific implications, a study of the interaction of flavins with electron-rich metals by Yu and Fritchie (1975).

Finally, since many iron–sulfur proteins are of interest to investigators of photosynthesis, it seems appropriate to describe recent developments that have led to a simple model system that converts light energy into chemical energy (Young et al., 1975). While the components differ substantially, of course, conceptual parallels with certain principles underlying photosynthesis are evident.

In a series of studies referenced by Young et al. (1975), evidence has been obtained that the weak reductant $Ru(bipy)_3^{2+}$ can absorb a photon to become a correspondingly more energetic reductant, $Ru(bipy)_3^{2+*}$, which can be quenched by electron transfer to a variety of oxidants, O, according to

$$Ru(bipy)_3^{2+*} + O \rightarrow Ru(bipy)_3^{3+} + O^-$$

While this behavior is not surprising in principle, studies with similar objectives in the author's laboratory in 1971–1972 and the paucity of related reports suggest that the choice of excited state system (and perhaps the oxidant system) is critical, there being a variety of potentially competitive alternate quenching pathways.

In their intriguing system, Young et al. (1975) included a reductant R, which reacted with the $Ru(bipy)_3^{3+}$ product

$$Ru(bipy)_3^{3+} + R \rightarrow Ru(bipy)_3^{2+} + R^+$$

regenerating the ruthenium starting material. The occurrence of these reactions requires, among other things, that the reduction potential of the O/O^- and R^+/R couples lie in the 2.1 V range between the $Ru(bipy)_3^{3+/2+}$ and $Ru(bipy)_3^{3+/2+*}$ couples. If, in addition, the R^+/R potential is more positive than for the O/O^- couple, the sum of the above conversions amounts to the unfavorable reaction

$$O + R \xrightarrow{h\nu} O^- + R^+$$

being catalytically driven by part of the light energy absorbed (69% in this case). Young et al. (1975) report evidence for this occurring and also for the spontaneous reverse reaction. Creutz and Sutin (1975) have studied the related oxidation of hydroxide to O_2 by $Ru(bipy)_3^{3+}$ and its application in a solar energy storage system. Such model systems offer much promise for better understanding and utilizing the conversion of light energy to chemical energy.

III. REDOX DYNAMICS OF STRUCTURALLY CHARACTERIZED IRON–SULFUR PROTEINS

A. Reactions with Simple Reductants and Oxidants

Detailed mechanistic studies of iron–sulfur proteins are decidedly in their infancy and much remains to be learned. Most of the few studies available are with simple reactants and are described in this subsection according to protein category.

1. Rubredoxins

The simplest known iron–sulfur protein is the rubredoxin first isolated from *Clostridium pasteurianum* and described by Lovenberg and Sobel (1965). Further attraction to its kinetic study is provided by an elegant X-ray structure of unusually high resolution (Jensen, 1974) and low R index (Watenpaugh et al., 1973) for a protein.

Of possibly significant note with regard to the tetrahedrally-distorted $Fe^{III}(SR)_4^-$ redox site are a wide range of S–Fe–S bond angles (101–115°) and one particularly short Fe–S bond (2.05 Å) in comparison to the results of Kistenmacher and Stucky (1968) for the $Fe^{III}Cl_4^-$ ion.* According to Jensen (1974), the angular distortions are in the same range as has been reported for a model structure (Churchill and Wormald, 1971) and not much greater than those arising from packing forces with small molecules.

On the basis of comparisons made possible by their determination of the $Fe^{II}Cl_4^{2-}$ ion structure, Lauher and Ibers (1975) expect a substantial extension of about 0.11–0.12 Å in the average Fe–S bond length on reduction to Fe(II), which could result in significant conformational changes. Evidence exists for small conformational differences between reduced and oxidized forms in solution (Lovenberg and Williams, 1969), where approximately tetrahedral coordination is retained (Eaton and Lovenberg, 1970; Long et al., 1971). Jensen's earlier personal communication (Bennett, 1973; Jacks et al., 1974) that, from a preliminary difference map, it appears that the dimensions of the immediate metal environments in the crystal change by no more than about 0.1 Å in the reduced form seems consistent with the proposed expansion of the FeS_4 core on reduction if the distortions of the oxidized form are largely retained. In any event, if the sorts of considerations entertained in Section

---

*Since this chapter was written, studies by X-ray absorption spectroscopy of rubredoxin (Schulman et al., 1975) have appeared which suggest that these structure anomalies may be significantly smaller (see also Lane et al., 1975).

II,A,2 prove applicable even partially to protein kinetic studies such as ours, these ground-state features will deserve attention, particularly as they contribute to $\Delta H°$ and $\Delta S°$ of a reaction.

For reasons detailed elsewhere (Bennett, 1973; Jacks et al., 1974), the latter authors undertook the study of electron transfer to oxidized clostridial rubredoxin [Rd^{III}] from $Ru(NH_3)_6^{2+}$, $V(H_2O)_6^{2+}$, and $Cr(H_2O)_6^{2+}$. Outer-sphere reaction is ensured for the first reductant and strongly indicated for the latter two with the results summarized in Table III where parentheses indicate limited experiments with $Cr(H_2O)_6^{2+}$ and a necessarily limited temperature range with $Ru(NH_3)_6^{2+}$ (13.8–25.0°C).

The variation in the rate of reaction with $V(H_2O)_6^{2+}$ at the lower ionic strengths is suggestive of a direct approach by the reductants to the uninegative $Fe(SR)_4^-$ redox site. Further, it is clear that outer-sphere agents can oxidize or reduce the iron site of clostridial rubredoxin quite rapidly, as indicated by these results and a calculated $k = 3.3 \times 10^7\ M^{-1}$ sec^{-1} for the only slightly favorable ($\Delta G° = -1.8$ kcal mole^{-1}) $Ru(NH_3)_6^{3+}$–Rd^{II} reaction (Jacks et al., 1974).

A second indication of the high inherent reactivity of rubredoxin is provided by our calculation, from an equation related to Eq. (1), of self-exchange rates of 1.0×10^9 and $1.7 \times 10^8\ M^{-1}$ sec^{-1} from the $Ru(NH_3)_6^{2+}$ and $V(H_2O)_6^{2+}$ data, respectively (Jacks et al., 1974). In view of the low $\Delta H^\ddagger$ and $\Delta S^\ddagger$ values for all three cross-reactions, it is of interest to examine the source of rubredoxin's high reactivity in the context of Eqs. (2) and (3) discussed in Section II,A.

Unfortunately, being less aware of the possible significance of these relationships at the time and being hampered by described experimental difficulties, we obtained only an indication of the ground-state thermodynamic parameters necessary for such an examination. Within our ex-

TABLE III
SUMMARY OF KINETIC DATA FOR THE REDUCTION OF CLOSTRIDIAL RUBREDOXIN[a]

Reductant	k (25°C) (M^{-1} sec^{-1})	I (M)	pH	$\Delta H^\ddagger$ (kcal/mole)	$\Delta S^\ddagger$ (eu)
$Ru(NH_3)_6^{2+}$	9.5×10^4	0.10	6.3–7.0	(1.4)	(−31)
$V(H_2O)_6^{2+}$	1.1×10^4	0.50	3.5–4.5	0.1	−40
	1.6×10^4	0.10	4.0	—	—
	4.0×10^4	0.010	4.0	—	—
$Cr(H_2O)_6^{2+}$	1.2×10^3	0.10	3.5–4.0	∼(0)	∼(−44)

[a] From Table 4 of Jacks et al. (1974); parentheses denote values obtained under limited experimental conditions, see text.

perimental uncertainties, no appreciable variation was found in the equilibrium constant of the $Ru(NH_3)_6{}^{2+}$–Rd^{III} reaction over an 11.2°C change in temperature (Jacks et al., 1974). This suggests $\Delta H° \sim 0$ (and $\Delta S° \sim -12$) for the reaction, but examination of the limited data base suggests that this indication might be no more accurate than to within ~ 5 kcal mole^{-1}. Accurate measurements of the Rd^{III}/Rd^{II} reduction potential as a function of temperature are now clearly desirable.

After examining the more reliable of those data contained in Table IV and, having no expectation that $\Delta H^{\ddagger}$ for the Rd^{II}/Rd^{III} self-exchange could be negative, we have quite arbitrarily obtained an acceptable hypothetical fit to Eqs. (1)–(3) in the following way and explored it for reasonableness. To the extent that our model is even approximately applicable and that $\Delta H^{\ddagger}_{RuRu}$ approaches its theoretical contribution, $\Delta H^{\ddagger}_{RdRd}$ would have to be low. We have assigned it a value of zero and then, from the apparent $\Delta G^{\ddagger}_{RdRd}$ available through Eq. (1) from the cross-reaction, calculated $\Delta S^{\ddagger}_{RdRd}$ to be ~ -18 eu (on the order of -10 eu of which is expected from losses in rotational and translational entropy on forming a collision complex along a consummative path). In order to obtain internal consistency, we have taken $\Delta H°$ to be -5 kcal mole^{-1}, seemingly within reach of our limited equilibrium results, and calculated $\Delta S°$ to be -28 eu using our $\Delta G°$ (3.5), which is in reasonable agreement with independent measurements (see Jacks et al., 1974).

It must be emphasized that in order to find reasonably good agreement between calculated and observed values of $\Delta H^{\ddagger}_{RuRd}$ and $\Delta S^{\ddagger}_{RuRd}$, extremes in $\Delta H°_{12}$ and $\Delta S°_{12}$ (which, nevertheless, seem consistent with available data) have been necessary and should not be taken as other than approximate in the absence of more concrete evidence. In particular, for the conversion of Rd^{III} to Rd^{II}, the changes in enthalpy of -11.9 kcal

TABLE IV
SUMMARY OF PARAMETERS FOR THE $Ru(NH_3)_6{}^{2+}$–Rd^{III} REACTION[a]

System	$\Delta H°$	$\Delta S°$	$\Delta G°$	$\Delta H^{\ddagger}$	$\Delta S^{\ddagger}$	$\Delta G^{\ddagger}$
$Ru(NH_3)_6{}^{2+}$ to $Ru(NH_3)_6{}^{3+}$	6.9	17	1.8	10.3	-11	13.6, 12.7[b]
Rd^{III} to Rd^{II}	[-11.9]	[-45]	(1.7)	[0]	[-18]	[5.2]
$Ru(NH_3)_6{}^{2+} + Rd^{III}$ to $Ru(NH_3)_6{}^{3+} + Rd^{II}$	[-5]	[-28]	(3.5)	(1.4)	(-31)	10.7 (obs)
				[2.6]	[-28]	— (calc)

[a] As for footnotes a and b in Table I with additional data provided or referenced by Jacks et al. (1974). Parentheses denote values obtained under limited experimental conditions; square brackets denote values that have been calculated on the basis of Eqs. (1), (2), or (3) or estimated, see text.

[b] Calculated at $I = 0.10$ M from the data of Meyer and Taube (1968).

mole^{-1} and absolute entropy of ~-23 eu (on a scale where $S^\circ_{H^+}$ = -5.5 eu and $S^\circ_{H_2}$ = 31.2 eu for this consideration) required by our estimates seem surprisingly negative [although being probably in the right direction given the greater expected polarization of the environment around the more highly charged Fe^{II}(SR)$_4^{2-}$ site (Jacks et al., 1974)].

Somewhat more positive estimates of $\Delta H^\ddagger$ and $\Delta S^\ddagger$ (RdRd) may turn out to be more realistic (and, in fact, would be required to fit the V(H$_2$O)$_6^{2+}$ data similarly, although media differences here do not encourage such an attempt). But, if so, the agreement would be poorer. It might be noted that the disagreements such as those for cytochrome c (Table II) result from calculated enthalpy and entropy barriers that are higher and lower, respectively, than those actually observed. This may turn out to be fortuitous or the result of appreciable work terms that do not cancel. For example, if a portion of the observed $\Delta H^\ddagger_{RuRu}$ were due to unshielded coulombic effects (perhaps the $\sim$4 kcal mentioned earlier), that portion would not be expected to carry over to the $\Delta H^\ddagger$ of a cross-reaction with a negatively charged species except, perhaps, to lower the barrier. Also, the outer-sphere rearrangements may be modified in cross-reactions involving proteins if the redox sites must be closely juxtaposed. In any event, as more data is obtained disagreements such as these may be better understood and prove enlightening.

However objectionable such relatively simple relationships may prove to be for proteins on a quantitative basis, they seem, for at least the two examples presented thus far, to have qualitative and possibly semiquantitative significance. We find it hard to believe from our results that the inherent enthalpy barrier to the activation of rubredoxin for electron transfer is more than a few kilocalories. Similarly, the inherent entropy barrier does not appear to be much more than that arising from encounter complex formation. Thus, rubredoxin is inherently an extremely reactive electron-transfer protein whose cross-reactions may be relatively fast or slow depending on the enthalpy and entropy characteristics of its reaction partner(s), any specific interactions between them, and probably the way in which the ground-state characteristics combine to determine ΔH°, ΔS°, and ΔG° (Jacks et al., 1974). In fact, as has been suggested, the physiological function of rubredoxin, presently unknown, may be to serve as a catalyst for redox reactions between proteins that are not sufficiently reactive with each other (perhaps analogously to cytochrome c).

At a molecular level the low apparent $\Delta H^\ddagger_{RdRd}$, which we rationalized on an electronic basis (Jacks et al., 1974), now seems a little surprising in view of the substantial bond-length differences between FeCl$_4^-$ and FeCl$_4^{2-}$ (Lauher and Ibers, 1975). Perhaps the ground-state vibrational levels overlap or little thermal excitation is required in order for overlap to occur (see Section II,A and Fig. 1 in Bennett, 1973). Alternatively, if $\Delta H_i^{\ddagger}$ is appreciable, some compensation from $\Delta H_o^{\ddagger}$ would seem necessary. The apparent value of $\Delta S_{RdRd}^{\ddagger}$ seems consistent with little conformational change being required to attain the activated complex.

It would, of course, be highly desirable to compare these estimated values with direct measurements of them. However, if the self-exchange is as rapid as our results consistently suggest, it may be possible to make these measurements only with EPR techniques (see Peisach et al., 1971), if then. The next best thing appears to be the study of this and other rubredoxins with carefully selected and characterized oxidants and reductants under conditions where all other necessary parameters are available and comparable. We hope to pursue this as resources permit. Since we are unaware of other kinetic studies of simple reactants with rubredoxins, we shall defer their further consideration to Section III,B,2.

2. High Potential Iron Proteins (HiPIP)

The intriguing structures of these $Fe_4S_4(SR)_4$ proteins, related ferredoxins, and model compounds are discussed in Chapters 6 and 7. The question of how the details of molecular structure affect their unique redox potential characteristics should prove more accessible in light of those discussions. In addition, the elegant structural determinations of both $HiPIP_{ox}$ and $HiPIP_{red}$ from *Chromatium vinosum* strain D in Kraut's laboratory (Carter et al., 1974a,b) reveal several distinctive features of potential mechanistic significance.

The iron–sulfur cluster shrinks significantly upon oxidation by about 0.15 Å in one direction and 0.08 Å in another, which could result in significant inner-sphere rearrangement barriers. The cluster is buried in a large interior cavity and surrounded by a number of aromatic and other nonpolar side chains. These interpenetrate the hydrogen-bonded secondary structure creating the cavity and are believed to protect the cluster from contact with the solvent as well as influencing the reduction potential characteristics. It can be expected that they will also affect the outer-sphere rearrangement barrier.

Perhaps most notably from a mechanistic standpoint, there is no evident connection of the redox site with a bond system to the protein surface that could delocalize the redox orbital for interaction with external agents. The closest distance of approach from any protein surface to a van der Waals surface of the cluster can be estimated from models to be ~ 3.5 Å to a sulfur atom from the "bottom" surface, all other approach distances being significantly longer, e.g., $\lesssim 10$ Å. In this respect, HiPIP is unique in comparison to all other proteins and coordination complexes

9. REDOX MECHANISMS OF IRON–SULFUR PROTEINS

for which there is reasonably definitive structural and kinetic information (providing that the ground-state structures are not significantly altered in solution). Based on the crystal structure and the analysis of the chemical foundations for it (Carter *et al.*, 1974a,b), there is little reason to believe that dissolution of HiPIP crystals, which already contain about 50% water (Carter *et al*, 1974a), in an aqueous medium is likely to result in very substantial structural change. Until evidence to the contrary is presented, we adopt the position that the structures in the highly hydrated crystal and in aqueous solution are probably quite similar. Support for this position is provided by the observations (Cammack, 1973; Dickson and Cammack, 1974) that DMSO concentrations in excess of 70% are required to shift the native HiPIP$_{red}$ absorption spectrum, under which conditions the protein can be superreduced to the oxidation state level of structurally similar reduced ferredoxins. This behavior is consistent with the three oxidation state hypothesis of Carter *et al.* (1972).

Given this structure as a starting point from which to proceed to the geometry of the transition state for electron transfer, an essential mechanistic question must be asked. How does the electron gain access to or egress from the buried redox site?

One obvious possibility is that the protein changes conformation so as to allow sufficiently close approach for "normal" electron transfer. This would seem to be against the structural character that has evolved for HiPIP and appears to require rather severe treatment, e.g., the DMSO example just mentioned. In any event, such an activation pathway would be expected to receive relatively positive contributions to $\Delta S^{\ddagger}$ and, given the structure, probably $\Delta H^{\ddagger}$ as well. A physiological reactant might be structured so as to compensate for this while possibly excluding access of solvent to the cluster, but few simple reagents seem likely to be so structured.

An interesting second proposal has been made by Carter *et al.* (1974b) on the basis of a consistent but weak thread of evidence that, on oxidation of the cluster, Tyr 19, which is in contact with it, moves further away. It was suggested that physiological oxidation of HiPIP$_{red}$ might involve the oxidation of this tyrosine by some group on the oxidase acting at $C_{\epsilon}2$ or the hydroxyl oxygen, which are accessible to solvent, to produce a protein-stabilized phenoxy radical intermediate or transition state, which is presumably then reduced by the cluster. They proposed analogously that reduction of HiPIP$_{ox}$ could take place by oxidation of Tyr 19 by the cluster, presumably followed by reduction of the stabilized phenoxy radical by the reductase. (The energy of the reduced cluster–phenoxy radical species must be at least 2 kcal mole^{-1} higher than that of $HiPIP_{ox}$, since it would seem that a few percent would be spectroscopically detectable.)

This mechanism is energetically distinct from the reduction of aromatic groups, which has sometimes been proposed (see Ewall and Bennett, 1974), and does not seem objectionable on these grounds in view of presently available evidence. Ferricyanide, of comparable reduction potential, is known to oxidize phenoxide ions, albeit in alkaline media (Stewart, 1964; McDonald and Hamilton, 1973). This author is not aware of sufficient kinetic data on phenoxide oxidations to warrant any predictions of the activation parameter characteristics of this mechanism. It will probably best be tested by comparative studies of other HiPIP's yet to be found, that lack this tyrosine or carefully modified proteins. The history of such approaches in cytochrome c studies (Dickerson and Timkovich, 1975) emphasizes the importance of characterizing the effect of such modifications on the reduction potential and the integrity of the redox site.

In case this proposed mechanism does not survive such tests, other roles can be envisaged for the effects on electron transfer dynamics of tyrosine and other aromatic or nonpolar groups. They can affect (a) the thermodynamics of the couple's two ground states, which in view of previous discussion, may contribute to the kinetic characteristics of cross-reactions, (b) the outer-sphere rearrangement barrier contributing to inherent reactivity (Section II,A), and possibly (c) the tunneling barrier discussed below. Further evidence regarding the roles of such groups will be discussed in Section IV.

The third, and perhaps not the last, possibility involves the quantum-mechanical tunneling of an electron to or from a reactant at the protein surface through the intervening protein material (or possibly unoccupied space). Abundant reports, too numerous to be completely cited here, provide evidence for electron tunneling over substantial distances between electrodes under an applied voltage (e.g., Kurtin *et al.*, 1971) and between aromatic residues in rigid glasses (e.g., Miller, 1975). To our knowledge, the first proposal that tunneling might be important physiologically came from DeVault and Chance (1966; see also Chance, 1972). Related results have recently appeared (McElroy *et al.*, 1974).

We first began to consider the possibility of tunneling pathways within proteins in connection with our kinetic studies of HiPIP, being aware of its structural features through the courtesy of Carter and Kraut (Bennett *et al.*, 1974). Under the subsequent and appreciated encouragement of Max Delbrück, we attempted to develop a model within the framework of the Marcus theory and within which the feasibility of tunneling within proteins could be examined (L. E. Bennett and W. D. Jones, unpublished work, 1974). Shortly after the conclusion of our work, a theoretically more complex treatment of thermally activated tunneling between biological molecules appeared (Hopfield, 1974). Our model will be summarized briefly here in the hope that its simplicity might be helpful and in order to examine the possible kinetic ramifications of tunneling for subsequent consideration.

Elaborating slightly on Marcus' original model (Marcus, 1956, 1964), the reaction between two proteins is envisaged to occur via four idealized steps: (a) encounter of reduced protein I, Red_I, with oxidized protein II, Ox_{II}, to form a reactant complex, R; (b) Franck–Condon activation of this complex to R^* (for optimal tunneling efficiency the two redox orbitals are expected to be made effectively equal in energy); (c) electron transfer between redox sites to create an activated product complex, P^*, which differs from R^* only in the electron location; (d) Frank–Condon relaxation to form P in the products' vibronic ground states, perhaps accompanied or followed by dissociation. Schematically represented (Marcus, 1956), this corresponds to

$$Red_I + Ox_{II} \underset{k_{-a}}{\overset{k_a}{\rightleftarrows}} R \qquad K_a = k_a/k_{-a} \qquad (5a)$$

$$R \underset{k_{-b}}{\overset{k_b}{\rightleftarrows}} R^* \qquad K_b = k_b/k_{-b} \qquad (5b)$$

$$R^* \underset{k_{-c}}{\overset{k_c}{\rightleftarrows}} P^* \qquad (5c)$$

$$P^* \underset{k_{-d}}{\overset{k_d}{\rightleftarrows}} P \rightleftharpoons Ox_I + Red_{II} \qquad (5d)$$

The energetics of the Franck–Condon rearrangements during steps (5b) and (5d) can be represented by the usual potential energy versus nuclear configuration diagram for reactants and products (Marcus, 1964; Reynolds and Lumry, 1966). Interaction between the redox orbitals of the two reacting species provides adiabatic reactions with sufficient splitting at the intersection region so that reactants are converted to products on the lower surface with a transmission coefficient κ of unity (Marcus, 1964). Thus, in the Eyring formulation,

$$k_{rate} = (\kappa k_B T/h) [\exp(-\Delta H^{\ddagger}/RT)] [\exp(\Delta S^{\ddagger}/R)] \qquad (6)$$

experimental $\Delta H^{\ddagger}$ and $\Delta S^{\ddagger}$ values are associated with those events up to and including Franck–Condon activation.

When the interaction between redox orbitals is quite small, such as seems possible with proteins like conformationally native HiPIP, the splitting at the interaction region is small, and Franck–Condon activated reactants frequently remain on the reactant surface rather than always going to products (Marcus, 1964). For such reactions, described as nonadiabatic, the transmission coefficient can be significantly less than unity.

If activation parameters are determined under the usual assumption of $\kappa = 1$, any real deviations from this will show up as an excessively negative apparent $\Delta S^{\ddagger}$ (by -4.58 eu for each factor of ten by which κ is less than one).

In terms of this model, any rate diminution that results from inefficient electron tunneling through a classically impenetrable barrier arises from a value for k_c that is significantly less than the rate at which the electron bombards the barrier in the Franck–Condon activated complex R*. From the third ionization energy of iron and Planck's relationship, $E = h\nu$, a typical bombardment frequency ν_e was estimated to be 7.4×10^{15} sec^{-1}. Thus, $k_c = p\nu_e = p(7.4 \times 10^{15}$ sec$^{-1})$, where p is the probability of a bombarding electron arriving at the other side of the barrier (Marcus, 1964).

In attempting to evaluate p, we noted that the Franck–Condon relaxation rates k_{-b} and k_d, are of the order of vibrational relaxation rates, i.e., $\sim 10^{13}$ sec^{-1} (Marcus, 1956). In any tunneling process that is net exoergic and very much slower than this, the tunneled electron is efficiently captured in the more stable product configuration and has a substantially diminished opportunity to tunnel in the reverse direction. Therefore, such a process can be approximated as a one-dimensional tunneling of a bound particle from R* to a free particle in P* $\to$ P. This is analogous to treating radioactive α-decay as the tunneling of a particle in a finite-walled box. The equations can be found in most quantum mechanics texts (e.g., Rapp, 1971).

Under circumstances where these limiting conditions are satisfied, the expected dependence on the barrier height B and width x is mathematically straightforward, providing that B is modeled as constant. This amounts to an effective averaging of the real barrier. The probability, then, is

$$p = \exp[-(2/\hbar)(2mB)^{1/2}x] \qquad (7)$$

where m is the effective mass of the tunneled electron, taken here to be unity. Thus,

$$k_c = [7.4 \times 10^{15} \text{ sec}^{-1}][\exp(-1.025B^{1/2}x)] \qquad (8)$$

when B is in electron volts and x is in angtroms. The barrier widths for reaction of structurally characterized species with HiPIP, at least, seem definable in the absence of conformational change. The barrier heights present a more difficult problem that cannot be rigorously solved at present. Our approach has been to estimate the energetics of discretely transferring the electron to intervening residues, since this is presumably the classical path.

In one sense, this may overestimate the effective real barrier, since it neglects interactions with filled orbitals. In the same sense, the tunneling of an electron hole, corresponding classically to the electron moving in a reversed sequence of events, may present a lower barrier. Third, the classical barrier to transfer between residues could increase the effective barrier height by an amount difficult to estimate from available data, although it can be noted that all electron transfers are tunneled through such classic barriers (Reynolds and Lumry, 1966, p. 12), many quite readily. Unfortunately, reduction potentials (or oxidation potentials in the case of hole tunneling) of the various intervening residues are required for our approach, but except for the disulfide function, they appear to be outside the range of electrochemical accessibility.

Only in the case of benzene (as a model for phenylalanine) can a reasonable extrapolation be made from linear relationships between the reduction potentials of related hydrocarbons and their lowest-energy electronic absorptions (Bergman, 1954) or their electron affinities (Dewar et al., 1968, 1970). The results suggest a reduction potential of -2.8 ± 0.3 V for benzene. The effect of the medium being different in a protein than for the electrochemical measurements could well be significant, but is not now assessable and its impact may not exceed other present uncertainties.

No such extrapolation seems possible at present for imidazole (histidine), indole (tryptophan), or phenol (tyrosine). However, like benzene, these species react with the hydrated electron ($E^{\circ}_{1/2} = -2.8$ V) at rates (His H^+ > Trp > Phe $\gtrsim$ Tyr > His), which in general for such reactions, are influenced by the reducibility of the substrate (Hart and Anbar, 1970). Also, reduction potentials are available for related azaheterocycles in DMF (Tabner and Yandle, 1968, 1971). It does not seem likely that any of these residues are less oxidizing than -3.5 V.

One-electron oxidation potentials of aromatic compounds for estimating hole tunneling barriers are even less available because of the high reactivity of the immediate products, but some have been derived for polycyclic aromatic hydrocarbons in acetonitrile (Peover, 1971) that suggest a moderate barrier for the oxidation of benzene. The oxidizability of tyrosine has been mentioned and seems likely to extend to histidine and tryptophan.

We shall assume a model in which such relatively reducible or oxidizable residues [including possibly arginine (Dewar et al., 1968) or disulfide linkages] intervene between a protein surface and the redox site such as seems to be the case for the "left channel" of cytochrome c (Dickerson and Timkovich, 1975), and, with HiPIP, for the previously mentioned Tyr 19 "channel" or possibly the aromatic ring between the cluster and the "bottom" face. Any electron tunneling through such material will confront an effective barrier that is modified by the protein environment from the above estimates (which are rough anyway) and, in addition, is almost certainly increased by the inability of the nuclei in the vicinity of the tunneling pathway to reorient themselves during the time of electron traversal. That this second factor may not be excessively large is suggested by the reversibility of many of the electrode reactions studied and the magnitude of usual Franck–Condon barriers. In view of these uncertainties, we have attempted to be conservative in calculating representative values of k_c [from Eq. (8) and summarized in Table V] by using heights that seem rather high for reactants with reduction potentials in the neighborhood of 0.1–0.4 V (such as cytochrome c or HiPIP).

Even after allowance is made for the approximate nature of our model and barrier height estimates, it seems possible that the real values of k_c would be compatible with many observed rates. In fact, relatively unstable encounter complexes (K_a) and/or sizable Franck–Condon barriers (K_b) could be accommodated in addition without the calculated overall rate constant [k_{obs} (M^{1-} sec^{-1}) $= k_c K_a K_b$ in our model] being driven out of line with observation.

With regard to the effect of tunneling on the apparent $\Delta S^\ddagger$, it has been emphasized that the probabilities used in the calculations ($p = k_c/7.4 \times 10^{15}$) cannot be employed directly as transmission coefficients in the Eyring theory (Marcus, 1964). Nevertheless, at least a qualitative indication of κ should be given by p. It does not appear unreasonable at, say a width of $\sim$6 Å and $B \sim 4.2$ eV, to expect a κ of the order of 10^{-6}, the equivalent of ~ -25 eu. Since such $\Delta S^\ddagger$ values are not unknown, and since a tunneling model is apparently consistent with observed rates, we conclude that tunneling pathways may be feasible and deserve detailed experimental consideration. At $x = 4$ Å, one assumption of our model seems marginal, but a negative contribution to $\Delta S^\ddagger$ is still expected.

It must be emphasized that our simple model is intended to be more

TABLE V
Representative Calculated Values of k_c (sec^{-1})

B (eV) \ x (Å)	4	6	8	10
3.6	3.1×10^{12}	6.2×10^{10}	1.3×10^9	2.6×10^7
4.2	1.7×10^{12}	2.5×10^{10}	3.7×10^8	5.6×10^6
4.8	9.3×10^{11}	1.0×10^{10}	1.2×10^8	1.3×10^6

provocative than rigorous. It is not the first or likely to be the last word on the subject, and readers may wish to consult a related treatment by Sutin (1973) and considerations of tunneling in bridged systems (Halpern and Orgel, 1960; Dogonadze et al., 1973).

It would seem most desirable theoretically to obtain the interaction operator H_{if} (or the coulomb exchange integral) between the two redox sites as it is affected by the intervening material in the nuclear configuration of the activated complex, since the resonance frequency, $2|H_{if}|/h$, can be equated with pv_e (Sutin, 1973). This is beyond the theoretical talents of this author, at least, but it can be anticipated that the molecular orientations of the intervening material will affect in possibly critical ways the resonance or tunneling frequencies. This feature is not taken into account in our model, perhaps at its jeopardy. Should tunneling prove operative, a critical component of Franck–Condon activation may, in some cases, be a reorientation of the intervening residues that could conceivably be facilitated particularly well by specific physiological reactants relative to structurally less complex agents. Also, the formation of charge transfer complexes could serve to facilitate tunneling.

The essentially nonadiabatic theory of Levich and Dogonadze, discussed by Reynolds and Lumry (1966), may prove useful in better treating these matters should rate-influencing tunneling processes become more widely and convincingly demonstrated than is currently the case. A particularly interesting system is the heme a of cytochrome oxidase with a triplyunsaturated C_{17} side chain which has been proposed (Caughey, 1971) to stack in a way that might provide a tunneling channel to the iron center.

Before turning to experimental studies of HiPIP, a brief word is in order regarding the possible influence of tunneling or radical–ion pathways on biological specificity. If such "channels" were collaborated with tailored protein–protein surface interactions, facile energy matching of donor and acceptor orbitals (B. Chance, personal communication), (and, possibly, physiologically unique molecular reorientations during activation), advantages in the direction of specificity are evident, even if several proteins are bound together. In fact, one practicable evolutionary means of imposing redox specificity on the internal coordination complex might have been to insulate it effectively against electron transfer in most directions while providing pathways in one or several specific directions to the surface where charge distributions, for example, could provide for alignment of the reaction partner. In the light of recent observations, if tunneling pathways are only occasionally utilized by proteins, the ways in which they are avoided posses important fundamental questions to the understanding of redox behavior.

A number of intriguing experimental studies involving simple reagents and *Chromatium* HiPIP have recently been completed and should appear shortly. A partial summary of the results is provided in Table VI. From substantial evidence, all of the reactions are first order in each reactant with no present evidence for rate saturation. The results on more than three separately prepared samples (our samples courtesy of R. G. Bartsch, University of California) from three different laboratories are in remarkably good agreement wherever comparisons are possible, particularly in view of media differences to which the reactions seem relatively insensitive.

The available ionic strength variations for reductions of the uninegative cluster of HiPIP$_{ox}$ suggest reaction at a protein site(s) of quite low and possibly slightly negative charge, the remaining negative charge on the protein [totalling ~ -3 according to Mizrahi et al. (1976)] apparently being well shielded from the reductants at $I > 0.01$. (In this connection, see results above and below this ionic strength for reaction with cytochrome c^{II} in Section III,B,1 below.) This would be consistent with reductant approach to the "bottom" protein surface affording the closet approach distance to the cluster in a structure which is largely intact conformationally. This surface is uncharged except for positive Arg 33 (Carter et al., 1974a), which is somewhat removed from a reasonable site of attack. A similar approach by Fe(CN)$_6^{3-}$ to HiPIP$_{red}$ (supported also by subsequent arguments) seems consistent with the direction and magnitude of that ionic strength variation since the cluster in this case bears a -2 charge (see this volume, Chapter 7), which is expected to be felt somewhat through the intervening 3–4 Å of low dielectric constant material.

Since the redox orbitals of the Fe(CN)$_6^{4-/3-}$ and Ru(NH$_3$)$_6^{2+/3+}$ couples are both t$_{2g}$ and there is no appreciable evident difference in their electrostatic interaction with HiPIP, there is little current reason to expect that the electron transfer pathway of lowest $\Delta G^{\ddagger}$ will be significantly different for their reactions with HiPIP. Circumstantial support for this being the case is provided by a remarkably consistent calculated prediction by our group and Rawlings et al. (1976), based on reactions 1–3 in Table VI and a derivative of Eq. (1), of $k \sim 5$–$13 \times 10^2 \, M^{-1}$ sec^{-1} for the "apparent" HiPIP self exchange at $I = 0.10 \, M$. [We have attempted to take appropriate precautions with the medium-sensitive Fe(CN)$_6^{4-/3-}$ couple (Campion et al., 1967).] This suggests a markedly lower inherent reactivity for HiPIP than that suggested earlier for rubredoxin. We can now examine in more detail our proposal (Bennett, 1973) that this might be due to the buried nature of the HiPIP redox site. We shall not discuss reactions 4–6 in Table VI here and will postpone

TABLE VI
Summary of Observed Kinetic Parameters with High Potential Iron Protein (Chromatium)[a]

	Reaction	pH	I (M)	k (M^{-1} sec^{-1})	$\Delta H^{\ddagger}$	$\Delta S^{\ddagger}$	Reference[b]
1.	$Ru(NH_3)_6^{2+}$ + $HiPIP_{ox}$	7	0.024	4.0×10^5	—	—	1
		5–8	0.10	3.5×10^5	1.5 ± 0.3	-28 ± 1	1
2.	$Fe(CN)_6^{4-}$ + $HiPIP_{ox}$	7.3	0.01	1.5×10^2	4.2	-35	2
		7.3	0.11	1.5×10^2	—	—	2
		7.0	0.10	1.8×10^2	—	—	3
3.	$Fe(CN)_6^{3-}$ + $HiPIP_{red}$	7.3	0.01	1.2×10^3	0	-45	2
		7.3	0.11	2.5×10^3	—	—	2
		7.0	0.10	2.4×10^3	-0.2 ± 0.5	-44 ± 1	4
		7.0	0.10	4.2×10^3	0	-42	3
		7.0	0.01	1.1×10^3	—	—	3
4.	Ascorbate + $HiPIP_{ox}$	7.3	0.01–0.11	3.9 ± 1.4	—	—	2
5.	SO_2^- + $HiPIP_{ox}$	7.3	0.01–0.11	$2.1 \pm 0.9 \times 10^6$	—	—	2
6.	$S_2O_4^{2-}$ + $HiPIP_{ox}$	7.3	0.01–0.11	$1.2 \pm 0.3 \times 10^3$	—	—	2
7.	$Co(phen)_3^{3+}$ + $HiPIP_{red}$	7.0	0.10	2.8×10^3	14.0 ± 0.5	4 ± 1	4
		8.0	0.10	5.3×10^2	—	—	3
8.	$Co(bipy)_3^{3+}$ + $HiPIP_{red}$	7.0	0.10	4.5×10^2	—	—	3
		6.0	0.10	3.5×10^2	—	—	3
		5.3	0.10	3.0×10^2	—	—	3

[a] Results in Tris–Tris HCl medium unless otherwise specified subsequently and with rate constants at 25°C except for reference 2, which are at 20°C.

[b] References: (1) work of T. S. Roemer (M.S. Thesis, San Diego State University) (to be submitted); (2) Mizrahi et al. (1976), preprint kindly provided by Mike Cusanovich; at $I = 0.11$ M [NaCl] = 0.10; the rates of reactions 2 and 3 are independent of pH over the ranges 7–9 and 6–10, respectively; increase at lower pH values, and appear free of specific ion effects; (3) work of R. X Ewall at San Diego State (to be submitted); (4) Rawlings et al. (1976), preprint kindly provided by Harry Gray; phosphate buffer; small pH variation with reaction 7. The data in the paper as published on the $Fe(EDTA)^{2-}$–$HiPIP_{ox}$ reaction was added subsequent to this writing.

consideration of reactions 7–8 in Table VI (with markedly different activation parameters).

Reasonably satisfactory data seem available for a tentative analysis of the $Fe(CN)_6^{4-/3-}$ reactions in the context of Eqs. (2) and (3), particularly at $I = 0.01$ M. At pH 7.3, all available evidence (Mizrahi et al., 1976) is consistent with both the oxidation and reduction reactions with HiPIP being microscopically reversible (Frost and Pearson, 1961; see also Krupka et al., 1966; Burwell and Pearson, 1966). Assuming this to be the case, we have used the activation parameters at $I = 0.01$ M (Mizrahi et al., 1976) and ground-state thermodynamic parameters for the $Fe(CN)_6^{4-/3-}$ that appear to be little different at two slightly higher ionic strengths (Hanania et al., 1967; O'Reilly, 1973) to calculate the ground state parameters contained in Table VII for $I = 0.01$ M. In addition, reasonable estimates of activation parameters for the $Fe(CN)_6^{4-/3-}$ self-exchange can be made from Campion et al. (1967) under these conditions enabling a hypothetical calculation of the HiPIP self-exchange parameters (Table VII). For comparison, at $I = 0.10$ values of $\Delta H^\ddagger = [0]$, $\Delta S^\ddagger = [-46]$ and $\Delta G^\ddagger = [13.2]$ can be calculated for HiPIP using the data of Rawlings et al. (1976), if it is assumed, somewhat less reliably, that $\Delta H^\ddagger = 4.0$ and $\Delta S^\ddagger = -32$ for the $Fe(CN)_6^{4-/3-}$ self-exchange under approximately this set of conditions (Stasiw and Wilkins, 1969) and that the ground-state HiPIP parameters are relatively insensitive to this change in ionic strength. Satisfaction of the latter condition is implied but not established by the close similarity in $\Delta H^\ddagger$ and $\Delta S^\ddagger$ for the slightly ionic strength-dependent $Fe(CN)_6^{3-}$–$HiPIP_{red}$ reaction at the two ionic strengths and other insensitivities already mentioned. While these results are unsurprisingly not in complete quantitative agreement, they seem consistent in qualitatively characterizing the HiPIP self-exchange relevant to these cross-reactions as having low enthalpy and high entropy barriers to be discussed further below.

TABLE VII
SUMMARY OF PARAMETERS FOR THE $Fe(CN)_6^{4-}$–$HiPIP_{ox}$ REACTION[a]

System	$\Delta H°$	$\Delta S°$	$\Delta G°$	$\Delta H^\ddagger$	$\Delta S^\ddagger$	$\Delta G^\ddagger$
$Fe(CN)_6^{4-}$ to $Fe(CN)_6^{3-}$	27	60	9.1	0	−45	13.5
$HiPIP_{ox}$ to $HiPIP_{red}$	−22.8	−50	−8.0	[4.2]	[−35]	[14.4]
$Fe(CN)_6^{4-} + HiPIP_{ox}$ to $Fe(CN)_6^{3-} + HiPIP_{red}$	4.2	10	1.1	4.2	−35	14.5

[a] As for footnote a in Table IV, with the exception that data have been obtained or deemed approximately appropriate at $I = 0.01$ M. See text for references.

A seemingly different picture emerges for the oxidation of HiPIP$_{red}$ by Co(phen)$_3^{3+}$, an oxidant we approach with some reluctance for previously stated reasons regarding its self-exchange behavior. Data in the present references and those given in Section II,A,2 can be used to estimate very tentatively certain desired parameters of Table VIII. However much one wants to argue about the quantitative significance of these calculations, it seems unlikely that more appropriate data or theoretical treatment would revise the qualitative indication that the HiPIP self-exchange characteristics appropriate to this cross-reaction are dramatically different than those for the previous cross-reaction. The enthalpy barrier now seems high whereas $\Delta S_{HH}^{\ddagger}$ is apparently favorable. Thus, the proposal of Rawlings et al. (1976), based largely on rate parameters (Table VI), that HiPIP can be activated via at least two significantly different routes seems to be supported when Eq. (2) and (3) are taken approximately into account. The possible significance of the calculated $\Delta G_{HH}^{\ddagger}$ being lower (but remarkably similar to those previously calculated in view of the large but apparently compensatory changes in $\Delta H^{\ddagger}$ and $\Delta S^{\ddagger}$) will be further considered in Section III,B,1.

Quite a tidy portrait could be presented were it not for the fact that data available in Tables I, VI, and VIII can be used to calculate values of $\Delta H_{HH}^{\ddagger} = [8.6]$, $\Delta S_{HH}^{\ddagger} = [-12]$, and $\Delta G_{HH}^{\ddagger} = [13.3]$ for the HiPIP self-exchange characteristics that are hypothetically appropriate to its cross-reaction with Ru(NH$_3$)$_6^{2+}$. It is certainly conceivable that HiPIP behavior intermediate between the previous extremes is operative here, but it seems at least as reasonable that the present state of the art is inadequate to draw any such conclusion convincingly. Intuitively and for reasons previously mentioned, we are inclined to include this cross-reaction within the idealized category of low $\Delta H^{\ddagger}$ and negative $\Delta S^{\ddagger}$ HiPIP "inherent reactivity" characteristics and to tentatively attri-

TABLE VIII
Summary of Parameters for the Co(phen)$_3^{3+}$–HiPIP$_{red}$ Reaction[a]

System	$\Delta H°$	$\Delta S°$	$\Delta G°$	$\Delta H^{\ddagger}$	$\Delta S^{\ddagger}$	$\Delta G^{\ddagger}$
Co(phen)$_3^{3+}$ to Co(phen)$_3^{2+}$	(−14.8)	(−17)	−9.7	5.1	−34	15.2
HiPIP$_{red}$ to HiPIP$_{ox}$	(22.8)	(50)	8.0	[14.9]	[9]	[12.1]
Co(phen)$_3^{3+}$ + HiPIP$_{red}$ to Co(phen)$_3^{2+}$ + HiPIP$_{ox}$	(8.0)	(33)	−1.7	14.0	4	12.8

[a] As for footnote a in Table IV, with data obtained or approximated as being appropriate at $I = 0.10$ M. See text for references.

bute discrepancies to data or theoretical applications that may not be appropriate.

With the simple reactants studied so far, the patterns of HiPIP reactivity behavior seem to be describable in terms of two extreme kinetic models or, possibly, some combination thereof. One seems to be characterized by rather negative $\Delta S_{HH}^{\ddagger}$ values, which are consistent with, but not necessarily demonstrative of, a nonadiabatic tunneling pathway involving either Tyr 19 or the aromatic residue interfaced between the cluster and the "bottom" surface. The apparent $\Delta H_{HH}^{\ddagger}$ values associated with this pathway are not yet sufficiently well defined to discriminate between Franck–Condon rearrangements associated with the cluster [perhaps modulated by the protein (Carter et al., 1974b)] and its environs, which are probably required by the tunneling model, and any additional heat input that might be required for activation via a phenoxy radical pathway. The combined characteristics do seem incompatible with a "conformational unwrapping" pathway but it seems advisable to retain the tunneling and phenoxy radical pathways, which may be closely related if hole tunneling is operative, as viable possibilities for this kinetic mode.

On the other hand, providing that we have not been led astray by oversimplifications that may arise in applying Eqs. (2) and (3) to these systems, the high enthalpy and low entropy barrier mode seems compatible with a conformational unwrapping of the cluster that might be assisted by the hydrophobic nature of the Co(phen)$_3^{3+}$ ligands. The activation characteristics of the phenoxy radical pathways are sufficiently unpredictable to retain it as a contender for this mode but any nonadiabatic pathway seems considerably less likely in view of the entropy characteristic.

3. FERREDOXINS

We are unaware of any kinetic studies of the reactions of ferredoxins with simple reagents except the dithionite reductions of spinach, *Micrococcus lactilyticus*, and *Clostridium pasteurianum* ferredoxins, which under pseudo first order conditions, follow a rate law $k_{obs}[Fd][S_2O_4^{2-}]^{1/2}$ with $k = 8.6$, 11, and 19 $M^{-1/2}$ sec^{-1}, respectively, which seems to implicate SO_2^- as the reductant (Lambeth and Palmer, 1973). In contrast to three half-life conformity with spinach ferredoxins, slight kinetic deviations were evident after one half-life with the other two, possibly due to their having two redox sites. The similarity in rate constants was remarked upon in view of differences in their redox sites but seems to us not necessarily surprising considering their similarities in reduction potential.

9. REDOX MECHANISMS OF IRON–SULFUR PROTEINS

Several eight-iron ferredoxins, which are presumably related to the structurally characterized *Peptococcus aerogenes* ferredoxin (Adman et al., 1973), have been studied by ESR (Mathews et al., 1974). The two four-iron clusters each have $S = \frac{1}{2}$ in the reduced state and are spin coupled. The mechanism of the total interaction remains in question, particularly in regard to the possibility of intramolecular electron transfer over the ~ 10 Å separation between redox sites. Should such transfer be established, it might bear testimony on the tunneling question posed earlier, particularly if $\Delta S^{\ddagger}$ for the process is negative. However, in the X-ray-based model we have seen (courtesy of J. Kraut), there is an "empty" space between the redox sites that might well be occupied by a water molecule. Therefore, a hydrogen atom transfer pathway (Bennett, 1973) between redox sites also should be considered.

B. Protein–Protein Reactions Involving Iron–Sulfur Species

1. HIGH POTENTIAL IRON PROTEINS

Very recently kinetic data has been obtained for the second-order reactions of *Chromatium* HiPIP with horse heart cytochrome c^{II} and *R. rubrum* cytochrome c_2, all of which have been characterized by X-ray structures (see previous references and those given by Dickerson and Timkovich, 1975). The results are partially summarized in Table IX. Only the major features will be discussed here.

The ionic strength dependence down to ~ 0.01–0.035 M is small but, for reduction by both cytochromes, consistent with the interaction of their positively-charged heme-edge faces with a slightly negative charge on HiPIP. The greater sensitivity in this region of the oxidation by cytochrome c_2^{III} is not surprising in view of the increase in positive and negative charge on each reactant by one unit each. At $I = 0.001$ M, a significantly more rapid rate develops for reduction by cytochrome c^{II} consistent with charges more distant from the interaction site(s?) being deshielded by the decreased ionic strength. This increased rate appears attributable to slight possible decreases in $\Delta H^{\ddagger}$ and $\Delta S_{\ddagger}$ barriers. It seems interesting that the bacterial cytochrome appears to be a slightly more rapid reductant than the mammalian one in spite of a 0.06 V decrease in driving force.

The physiological reactions of HiPIP have not been clearly defined. Kennel et al. (1972) report that HiPIP mediates the rereduction by artificial donors of the bound high potential cytochrome c_{556} in *Chromatium* subchromatophore particles. Implications of the involvement of HiPIP in light-driven electron transport in *Chromatium vinosum* have

TABLE IX

SUMMARY OF KINETIC PARAMETERS FOR CYTOCHROME c REACTION WITH HiPIP

Conditions	pH	I	k (M^{-1} sec^{-1})	$\Delta H^{\ddagger}$	$\Delta S^{\ddagger}$
Reduction by horse heart cytochrome c^{II} (Sigma type VI)[a]					
[cyt c^{II}] > 10 [HiPIP$_{ox}$]	6.7–7.3	0.10	1.16 × 10^5	6.0	−15
[HiPIP$_{ox}$] > 10 [cyt c^{II}]	6.7–7.3	0.10	1.44 × 10^5	6.0	−15
[cyt c^{II}] > 10 [HiPIP$_{ox}$]	7.0	0.010	1.74 × 10^5	(5.9?)	(−14?)
[cyt c^{II}] > 10 [HiPIP$_{ox}$]	7.0	0.0050	3.12 × 10^5	—	—
[cyt c^{II}] > 10 [HiPIP$_{ox}$]	6.7–7.3	0.0010	8.40 × 10^5	5.6	−13
[HiPIP$_{ox}$] > 10 [cyt c^{II}]	6.7–7.3	0.0010	7.92 × 10^5	6.0	−12
[cyt c^{II}] > 10 [HiPIP$_{ox}$]	8.0	0.10	1.12 × 10^5	—	—
Reduction by *Rhodospirillum rubrum* cytochrome c_2^{II} [b]					
[HiPIP$_{ox}$] > [cyt c_2^{II}]	7.0	0.135	1.8 × 10^5	—	—
[HiPIP$_{ox}$] > [cyt c_2^{II}]	7.0	0.035	2.4 × 10^5	—	—
Oxidation by *Rhodospirillum rubrum* cytochrome c_2^{III} [b]					
[HiPIP$_{red}$] > [cyt c_2^{III}]	7.0	0.135	3.3 × 10^4	—	—
[HiPIP$_{ox}$] > [cyt c_2^{II}]	7.0	0.035	7.5 × 10^4	—	—

[a] Work of Charles Jacks at San Diego State (1976). Medium maintained with Tris–Tris HCl. Parentheses indicate estimates from other values.

[b] Mizrahi et al. (1976). Medium maintained with 20 mM potassium phosphate buffer and appropriate concentrations of NaCl.

been reported (Dutton and Leigh, 1973; Evans et al., 1974). However, HiPIP has recently been isolated from a denitrifying bacteria (T. E. Meyer, quoted in Mizrahi et al., 1976), and HiPIP-like sites seem to be present elsewhere in nonphotosynthetic systems. We suspect that such systems function widely as electron transfer catalysts between larger proteins much the same as the c cytochromes, ferredoxins, and probably, the rubredoxins.

In view of their size and the charge dependence described above, the reactants of Table IX might be expected to simulate the physiological reactions of HiPIP more closely than with simple reagents. It can easily be calculated that the cross-reactions are at least a factor of 10 faster than might be expected on the basis of Eq. (1) and the HiPIP self-exchange $\Delta G^{\ddagger}$ indicated by simple reagents. In order to explore the possible reasons for this, we have used data from or related to Tables II, VII, and IX to calculate the parameters of Table X, with the previous reservations regarding the media still in effect. While the results should proba-

TABLE X
Possible Parameters for the Cytochrome c^{II}–HiPIP Reaction[a]

System	$\Delta H°$	$\Delta S°$	$\Delta G°$	$\Delta H^{\ddagger}$	$\Delta S^{\ddagger}$	$\Delta G^{\ddagger}$
$I = 0.010\ M$						
cyt c^{II} to cyt c^{III}	16.8	36	6.0	12.4	−5	14.0
HiPIP$_{ox}$ to HiPIP$_{red}$	−22.8	−50	−8.0	[5.4]	[−9]	[7.4]
cyt c^{II} + HiPIP$_{ox}$ to cyt c^{III} + HiPIP$_{red}$	−6.0	−14	−2.0	(5.9)	(−14)	9.7
$I = 0.10\ M$						
cyt c^{II} to cyt c^{III}	14.4	28	6.0	9.4	−13	13.6
HiPIP$_{ox}$ to HiPIP$_{red}$	(−22.8)	(−50)	−8.0	[11.0]	[5.0]	[9.6]
cyt c^{II} + HiPIP$_{ox}$ to cyt c^{III} + HiPIP$_{red}$	(−8.4)	(−22)	−2.0	6.0	−15	10.6

[a] As for footnote a in Table IV, with data obtained or approximated as being appropriate at the respective ionic strengths.

bly not be taken literally, they seem to place the cytochrome c^{II} oxidation by HiPIP$_{ox}$ intermediate between the two extremes mentioned before or possibly in the high enthalpy, low entropy barrier mode. Most significant are the much lower values calculated for $\Delta G^{\ddagger}_{HH}$ than with the simple reactants (Tables VII and VIII). Thus, it appears that physiological reactants may well be able to react more rapidly with HiPIP, even in the absence of substantial specific ionic interactions, than any of the simple reactants [except possibly Co(phen)$_3^{3+}$ with a slightly lower calculated $\Delta G^{\ddagger}_{HH}$] appear to indicate.

The consideration of three mechanisms closely related to the three idealized behaviors of HiPIP discussed above has been offered by Mizrahi et al. (1976) for electron transfer between cytochrome c_2 and HiPIP. Our activation parameters for reduction by cytochrome c^{II} (Table IX) suggest at least one direction in which attempts can be made to better characterize the details of reaction. To the extent that the calculated HiPIP self-exchange activation parameters of Tables VII, VIII, and X can be qualitatively relied upon to differentiate between mechanisms, it appears that HiPIP reduction by cytochrome c^{II}, at least, is different in some significant way from reduction by Fe(CN)$_6^{4-}$ [and possibly Ru(NH$_3$)$_6^{2+}$, given the uncertainties present]. A very tentative proposition can be offered that cytochrome c may be more efficient than these "hard-sphere" reagents in effecting either a "conformational unwrapping" of the buried cluster during activation or a phenoxy radical pathway in ways that might be more closely simulated by the HiPIP$_{red}$

oxidation by hydrophobic Co(phen)$_3^{3+}$.* In any event, the contrast between the calculated parameters for different reaction systems is provocative. The extent to which these contrasts are not understood is indicative in some ways of how far we are from a real understanding of these behaviors.

Clearly, much more work and appropriate parameters are necessary in order to explore this question in detail. However, it appears to this author as being one of the most intriguing problems ever confronted in oxidation–reduction chemistry.

2. Other Iron–Sulfur Proteins

We are not familiar with any other kinetic studies of protein–protein reactions involving structurally characterized iron–sulfur proteins. The field is obviously ripe for further exploration.

IV. SURVEY OF THE REDOX BEHAVIOR OF STRUCTURALLY LESS CHARACTERIZED IRON–SULFUR PROTEINS

The possible roles of aromatic residues in the redox mechanisms of iron–sulfur proteins discussed above can be more closely scrutinized on the basis of recent results with the clostridial-type ferredoxins, one of which has been structurally characterized (Adman et al., 1973). Such a role was suggested by Packer et al. (1972) on the basis of ^{13}C resonance studies of *Clostridium acidi-urici* ferredoxin, which contains two Fe$_4$S$_4$(SR)$_4$ redox centers and two apparently abutant tyrosines.

Some questions regarding such a role were raised by the absence of one of these aromatic residues in the ferredoxin from *Clostridium M-E* (Tanaka et al., 1974) and by the full electron-carrier activity of the *C. acidi-urici* ferredoxin in two biological assays when one of the tyrosine residues had been replaced by Leu (Lode et al., 1974a). Conclusions regarding the latter behavior were dependent, however, on the remaining Tyr 30 being kinetically innocuous. A more precise definition of the role of such tyrosines seems to be afforded by the retention of full electron-transfer activity when Tyr 2 is replaced by Leu in an aromatic-free *C. M-E* ferredoxin (Lode et al., 1974b).

Thus, the role of the tyrosines in these proteins appear to be indirect, possibly along two of the lines previously mentioned, (a) and (b), and

---

*The seemingly less reactive reagents may have to await the activation of a nonadiabatic pathway if they lack the capacity to compensate for a potentially more rapid but structurally more demanding reaction mode of HiPIP.

not likely to include the third possibility, (c), of a tunneling role. (Some analogy might be found in the retention of pseudophysiological activity by the cytochrome c of *Peptococcus denitrificans*, in which such a substitution at Tyr 74 has occurred naturally, relative to the activity of those c cytochromes retaining that aromatic residue (Dickerson and Timkovich, 1975).) It is premature to extend a dismissal of such direct roles universally to proteins such as HiPIP, which differ by ~16 kcal mole^{-1} in ground-state redox potential. Such direct roles would seem, however, to require positive experimental support before they can be accepted on the basis of proximity of redox site and aromatic residue alone.

Also of potential relevance to the redox behavior of tetrameric iron–sulfur clusters is the report by Phillips *et al.* (1974) that contact-shifted resonances observed with a partially reduced protein bearing only one of these clusters (Stombaugh *et al.*, 1973) are compatible with a "slow" rate of electron exchange. This was found to be suggestive that the "fast" electron exchange characteristics previously observed with proteins bearing two such clusters might be due to intramolecular transfer between them. The interpretation of these magnetic resonance experiments is beyond the present competence of this author to evaluate, but is closely related to the interpretation of other magnetic resonance studies described in Section III,A,3 for the latter class of proteins for which the issue of intramolecular transfer seemed unresolved.

Mention should be made of the role that the pseudomonal rubredoxin seems relatively well defined to fulfill in hydroxylation processes. References to important experiments regarding this role from Coon's laboratory are available elsewhere (e.g., Orme-Johnson, 1973; Bennett, 1973). An especially interesting report has appeared implicating a small terminal subsegment of the putidaredoxin amino acid sequence as being important for the kinetic behavior of this protein physiologically (Sligar *et al.*, 1974). Cleavage of the tryptophan and glutamine residues from the carboxy terminus of putidaredoxin is therein reported to have inappreciable effects on the reduction potential or spectroscopic integrity of the redox site, but to diminish the binding constant to cytochrome $P450_{cam}$, the rate of its putidaredoxin-stimulated decay rate when oxygenated, and the specific activity of the complete hydroxylase system, all by a factor of ~50. While the net changes in ΔG° or $\Delta G^\ddagger$ attendant to such cleavage is seemingly small, ~2 kcal mole^{-1}, the inclusion of these terminal residues in the native protein emphasizes the probable importance of fine-tuning such energy characteristics in order to achieve temporal compatibility with related physiological components. Electrochemical studies of putidaredoxin and its selenium analogue have also been reported (Wilson *et al.*, 1973).

Restrictions of space and time prevent a detailed consideration of a variety of other systems in which iron–sulfur proteins function. Therefore, only an arbitrarily selected set of recent references indicative of this variety will be given.

The role of iron–sulfur proteins in the metabolism of formate (e.g., Thauer et al., 1975) and dinitrogen (e.g., M. N. Walker and Mortenson, 1974; Evans and Albrecht, 1974) are reviewed in Chapters 2 and 5 of this volume. We shall simply mention the effect of MgATP on the accessibility of the iron centers in clostridial azoferredoxin (G. A. Waker and Mortenson, 1974) and the intriguing model studies of nitrogenase active sites by Schrauzer's group, most of which have previously been referenced (Bennett, 1973).

Consideration of the studies of iron–sulfur proteins in photosynthetic systems is deserving of separate report. We shall cite here several recent articles (Malkin et al., 1974; Malkin and Aparicio, 1975; Evans and Cammack, 1975; Bengis and Nelson, 1975) that should provide access to the literature. Also, EPR studies of an oxidized clostridial-type ferredoxin are suggestive to Sweeney et al. (1974) of a superoxidized iron–sulfur cluster related presumably to that of $HiPIP_{ox}$.

The catalytic properties of a hydrogenase from *Chromatium* (MW ~100,000) having two subunits, including an Fe_4S_4 protein of the ferredoxin class, have been studied (Gitlitz and Krasna, 1975). The heterolytic cleavage of H_2 to yield a free proton and an enzyme-bound hydride ion was considered. The hydrogenase was found to be incapable of catalyzing the reduction by H_2 of pyridine nucleotides or ferredoxin. Quite recent hydrogenase studies by Erbes et al. (1975) have suggested the binding of gaseous ligands by one or two equivalent Fe_4S_4 clusters.

Some kinetic characteristics of an adenyl sulfate reductase from *Desulforibris vulgaris* have been reported (Bramlett and Peck, 1975). Finally, an adrenodoxin reductase from bovine adrenal cortex has been studied (Foster and Wilson, 1975) and a revised reduction potential for adrenodoxin reported (Huang and Kimura, 1973).

ACKNOWLEDGMENTS

The author wishes to thank M. Cusanovich, E. Deutsch, and H. Gray for providing preprints of their manuscripts. We also wish to acknowledge helpful comments from them and from A. Budgor, B. Chance, M. Delbrück, J. Hurst, R. Linck, S. Lippard, R. Marcus, C. Mead, P. Schmidt, N. Sutin, and H. Taube regarding aspects of the material discussed herein. The viewpoints expressed are, of course, entirely the author's own responsibility. We appreciate the opportunity to examine protein models generously provided by Joe Kraut and Dick Dickerson. We have enjoyed the collaboration with Bob Bartsch, Martin Kamen, Walt Lovenberg, and Terry Meyer in our own protein studies. Publication of the chapter would not have been possible without the excellent cooperation provided by Valeta Drown, who expertly typed the manuscript. Finally, the support provided by the National Science Foundation and the donors of the Petroleum Research Fund, administered by the American Chemical Society, for part of our own experimental studies is acknowledged with gratitude.

REFERENCES

Adman, E. T., Sieker, L. C., and Jensen, L. H. (1973). *J. Biol. Chem.* **248**, 3987.
Asher, L. E., and Deutsch, E. (1972). *Inorg. Chem.* **11**, 2927.
Asher, L. E., and Deutsch, E. (1976). *Inorg. Chem.* **15**, 1531.
Baker, B. R., Basolo, F., and Neumann, H. M. (1959). *J. Phys. Chem.* **63**, 371.
Beattie, J. K., Sutin, N., Turner, D. H., and Flynn, G. W. (1973). *J. Am. Chem. Soc.* **95**, 2052.
Beinert, H., Ackrell, B. A. C., Kearney, E. B., and Singer, T. P. (1975). *Eur. J. Biochem.* **54**, 185.
Bengis, C., and Nelson, N. (1975). *J. Biol. Chem.* **250**, 2783.
Bennett, L. E. (1973). *Prog. Inorg. Chem.* **18**, 1.
Bennett, L. E., and Taube, H. (1968). *Inorg. Chem.* **7**, 254.
Bennett, L. E., Jacks, C. A., Ewall, R. X., Calvert, J., and Raymond, W. N. (1974). *Abstr., 167th Meet., Am. Chem. Soc.* INOR 193.
Benson, S. W. (1960). "The Foundations of Chemical Kinetics," p. 405. McGraw-Hill, New York.
Bergman, I. (1954). *Trans. Faraday Soc.* **50**, 829.
Bramlett, R. N., and Peck, H. D., Jr. (1975). *J. Biol. Chem.* **250**, 2979.
Bray, R. C. (1975). *In* "The Enzymes," 3rd ed. (P. D. Boyer, ed.), Vol. 12, p. 300. Academic Press, New York.
Bruice, T. C., Maskiewicz, R., and Job, R. (1975). *Proc. Natl. Acad. Sci. U.S.A.* **72**, 231.
Burwell, R. L., Jr., and Pearson, R. G. (1966). *J. Phys. Chem.* **70**, 300.
Cammack, R. (1973). *Biochem. Biophys. Res. Commun.* **54**, 548.
Campion, R. J., Deck, C. F., King, P., Jr., and Wahl, A. C. (1967). *Inorg. Chem.* **6**, 672.
Carter, C. W., Jr., Kraut, J., Freer, S. T., Alden, R. A., Sieker, L. C., Adman, E., and Jensen, L. H. (1972). *Proc. Natl. Acad. Sci. U.S.A.* **69**, 3526.
Carter, C. W., Jr., Kraut, J., Freer, S. T. Xuong, Ng. H., Alden, R. A., and Bartsch, R. G. (1974a). *J. Biol. Chem.* **249**, 4212.
Carter, C. W., Jr., Kraut, J., Freer, S. T., and Alden, R. A. (1974b). *J. Biol. Chem.* **249**, 6339.
Caughey, W. S. (1971). *Adv. Chem. Ser.* **100**, 248.
Chance, B. (1972). *FEBS Lett.* **23**, 3.
Churchill, M. R., and Wormald, J. (1971). *Inorg. Chem.* **10**, 1778.
Creutz, C., and Sutin, N. (1975). *Proc. Nat. Acad. Sci. U.S.A.* **72**, 2858.
DeVault, D., and Chance, B. (1966). *Biophys. J.* **6**, 825.
Dewar, M. J. S., Hashmall, J. A., and Venier, C. G. (1968). *J. Am. Chem. Soc.* **90**, 1953.

Dewar, M. J. S., Hashmall, J. A., and Trinajstić, N. (1970). *J. Am. Chem. Soc.* **92**, 5555.
Dickerson, R. E., and Timkovich, R. (1975). *In* "The Enzymes" (P. D. Boyer, ed.), 3rd ed., Vol. 11, p. 397. Academic Press, New York.
Dickson, D. P. E., and Cammack, R. (1974). *Biochem. J.* **143**, 763.
Dogonadze, R. R., Ulstrup, J., and Kharkats, Y. I. (1973). *J. Theor. Biol.* **40**, 259 and 279.
Dutton, D. L., and Leigh, J. S. (1973). *Biochim. Biophys. Acta* **314**, 178.
Eaton, W. A., and Lovenberg, W. (1970). *J. Am. Chem. Soc.* **92**, 7195.
Eaton, W. A., George, P., and Hanania, G. I. H. (1967). *J. Phys. Chem.* **71**, 2016.
Elder, R. C., Florian, L. R., Lake, R. E., and Yacnych, A. M. (1973). *Inorg. Chem.* **12**, 2690.
Erbes, D. L., Burris, R. H., and Orme-Johnson, W. H. (1975). *Proc. Nat. Acad. Sci. U.S.A.* **72**, 4795.
Evans, M. C. W., and Albrecht, S. L. (1974). *Biochem. Biophys. Res. Commun.* **61**, 1187.
Evans, M. C. W., and Cammack, R. (1975). *Biochem. Biophys. Res. Commun.* **63**, 187.
Evans, M. C. W., Lord, A. V., and Reeves, S. G. (1974). *Biochem. J.* **138**, 177.
Ewall, R. X., and Bennett, L. E. (1974). *J. Am. Chem. Soc.* **96**, 940.
Farina, R., and Wilkins, R. G. (1968). *Inorg. Chem.* **7**, 514.
Foster, R. P., and Wilson, L. D. (1975). *Biochemistry* **14**, 1477.
Frost, A. A., and Pearson, R. G. (1961). "Kinetics and Mechanism," 2nd ed., p. 211. Wiley, New York.
George, P., Hanania, G. I. H., and Irvine, D. H. (1959). *J. Chem. Soc.* p. 2548.
George, P., Eaton, W. A., and Trachtmann, M. (1968). *Fed. Proc., Fed. Am. Soc. Exp. Biol.* **27**, 526.
Gillard, R. D., and Maskill, R. (1968). *Chem. Commun.* p. 160.
Gitlitz, P. H., and Krasna, A. I. (1975). *Biochemistry* **14**, 2561.
Grimes, C. J., Piszkiewicz, D., and Fleischer, E. B. (1974). *Proc. Natl. Acad. Sci. U.S.A.* **71**, 1408.
Grossman, B., and Wilkins, R. G. (1967). *J. Am. Chem. Soc.* **89**, 4230.
Gupta, R. K. (1973). *Biochim. Biophys. Acta* **292**, 291.
Gupta, R. K., Koenig, S. H., and Redfield, A. G. (1972). *J. Magn. Reson.* **7**, 66.
Halpern, J., and Orgel, L. E. (1960). *Discuss. Faraday Soc.* **29**, 7.
Hanania, G. I. H., Irvine, D. H., Eaton, W. A., and George, P. (1967). *J. Phys. Chem.* **71**, 2022.
Hart, E. J., and Anbar, M. (1970). "The Hydrated Electron." Wiley (Interscience), New York.
Hodges, H. L., Holwerda, R. A., and Gray, H. B. (1974). *J. Am. Chem. Soc.* **96**, 3132.
Hopfield, J. J. (1974). *Proc. Natl. Acad. Sci. U.S.A.* **71**, 3640.
Huang, J. J., and Kimura, T. (1973). *Biochemistry* **12**, 406.
Hurst, J. K., and Lane, R. H. (1973). *J. Am. Chem. Soc.* **95**, 1703.
Isied, S. S., and Taube, H. (1973). *J. Am. Chem. Soc.* **95**, 8198.
Jacks, C. A. (1976). Ph.D. Thesis, San Diego State University, San Diego, California.
Jacks, C. A., and Bennett, L. E. (1974). *Inorg. Chem.* **13**, 2035.
Jacks, C. A., Bennett, L. E., Raymond, W. N., and Lovenberg, W. (1974). *Proc. Natl. Acad. Sci. U.S.A.* **71**, 1118.
Jensen, L. H. (1974). *Annu. Rev. Biochem.* **43**, 461.
Kassner, R. J. (1972). *Proc. Natl. Acad. Sci. U.S.A.* **69**, 2263.

Kassner, R. J. (1973). *J. Am. Chem. Soc.* **95,** 2674.
Kassner, R. J., and Yang, W. (1973). *Biochem. J.* **133,** 283.
Kemeny, G., and Mahanti, S. D. (1975). *Proc. Natl. Acad. Sci. U.S.A.* **72,** 999.
Kennel, S. J., Bartsch, R. G., and Kamen, M. D. (1972). *Biophys. J.* **12,** 882.
Kistenmacher, T. J., and Stucky, G. D. (1968). *Inorg. Chem.* **7,** 2150.
Krishnamurty, K. V., and Wahl, A. C. (1958). *J. Am. Chem. Soc.* **80,** 5921.
Krupka, R. M., Kaplan, H., and Laidler, K. J. (1966). *Trans. Faraday Soc.* **62,** 2754.
Kurtin, S. L., McGill, T. C., and Mead, C. A. (1971). *Phys. Rev. B* **3,** 3368.
Lambeth, D. O., and Palmer, G. (1973). *J. Biol. Chem.* **248,** 6095.
Lane, R. H., Sedor, F. A., Gilroy, M. J., Eisenhardt, P. F., Bennett, J. P., Jr., Ewall, R. X., and Bennett, L. E. (1976a). *Inorg. Chem.* **15,** in press.
Lane, R. H., Sedor, F. A., Gilroy, M. J., and Bennett, L. E. (1976b). *Inorg. Chem.* **15,** in press.
Lane, R. W., Ibers, J. A., Frankel, R. B., and Holm, R. H. (1975). *Proc. Nat. Acad. Sci. U.S.A.* **72,** 2868.
Lauher, J. W., and Ibers, J. A. (1975). *Inorg. Chem.* **14,** 348.
Lode, E. T., Murray, C. L., Sweeney, W. V., and Rabinowitz, J. C. (1974a). *Proc. Natl. Acad. Sci. U.S.A.* **71,** 1361.
Lode, E. T., Murray, C. L., and Rabinowitz, J. C. (1974b). *Biochem. Biophys. Res. Commun.* **61,** 163.
Long, T. V., Loehr, T. M., Alkins, J. R., and Lovenberg, W. (1971). *J. Am. Chem. Soc.* **93,** 1809.
Lovenberg, W., and Sobel, B. E. (1965). *Proc. Natl. Acad. Sci. U.S.A.* **54,** 193.
Lovenberg, W., and Williams, W. M. (1969). *Biochemistry* **8,** 141.
McArdle, J. V., Gray, H. B., Creutz, C., and Sutin, N. (1974). *J. Am. Chem. Soc.* **96,** 5737.
McDonald, P. D., and Hamilton, G. A. (1973). *Oxid. Org. Chem.* **5B,** 104.
McElroy, J. D., Mauzerall, D. C., and Feher, G. (1974). *Biochim. Biophys. Acta* **333,** 261.
Malkin, R., and Aparicio, P. J. (1975). *Biochem. Biophys. Res. Commun.* **63,** 1157.
Malkin, R., Aparicio, P. J., and Arnon, D. I. (1974). *Proc. Natl. Acad. Sci. U.S.A.* **71,** 2362.
Marcus, R. A. (1956). *J. Chem. Phys.* **24,** 966.
Marcus, R. A. (1964). *Annu. Rev. Phys. Chem.* **15,** 155.
Marcus, R. A., and Sutin, N. (1975). *Inorg. Chem.* **14,** 213.
Margalit, R., and Schejter, A. (1973). *Eur. J. Biochem.* **32,** 492.
Mathews, R., Charlton, S., Sands, R. H., and Palmer, G. (1974). *J. Biol. Chem.* **249,** 4326.
Meyer, T. J., and Taube, H. (1968). *Inorg. Chem.* **7,** 2369.
Miller, J. R. (1975). *Science* **189,** 221.
Miller, W. G., and Cusanovich, M. A. (1975). *Biophys. Struct. Mech.* **1,** 97.
Mizrahi, I. A., Wood, F. E., and Cusanovich, M. A. (1976). *Biochemistry* **15,** 343.
Morton, R. A., Overnell, J., and Harbury, H. A. (1970). *J. Biol. Chem.* **245,** 4653.
O'Reilly, J. E. (1973). *Biochim. Biophys. Acta* **292,** 509.
Orme-Johnson, W. H. (1973). *Annu. Rev. Biochem.* **42,** 159.
Packer, E. L., Sternlicht, H., and Rabinowitz, J. C. (1972). *Proc. Natl. Acad. Sci. U.S.A.* **69,** 3278.
Palmer, G. (1975). *In* "The Enzymes," 3rd ed. (P. O. Boyer, ed.), Vol. 12, p. 1. Academic Press, New York.
Palmer, G., and Brintzinger, H. (1972). *In* "Electron and Coupled Energy Transfer

in Biological Systems" (T. E. King and M. Klingenberg, eds.), Vol. 1, Part B, p. 379. Dekker, New York.

Peisach, J., Blumberg, W. E., Bode, E. T., and Coon, M. J. (1971). *J. Biol. Chem.* **246,** 5877.

Peover, M. E. (1971). *In* "Reactions of Molecules at Electrodes" (N. S. Hush, ed.), p. 278. Wiley (Interscience), New York.

Phillips, W. D., McDonald, C. C., Stombaugh, N. A., and Orme-Johnson, W. H. (1974). *Proc. Natl. Acad. Sci. U.S.A.* **71,** 140.

Rapp, D. (1971). "Quantum Mechanics," pp. 150, 161, 189, and 194. Holt, New York.

Rawlings, J., Wherland, S., and Gray, H. B. (1976). *J. Am. Chem. Soc.* **98,** 2177.

Reynolds, W. L., and Lumry, R. W. (1966). "Mechanisms of Electron Transfer." Ronald Press, New York.

Riley, P. E., and Seff, K. (1972). *Inorg. Chem.* **11,** 2993.

Schubert, M. P. (1933). *J. Am. Chem. Soc.* **55,** 3336.

Schulman, R. G. Eisenberger, P., Blumberg, W. E., and Stombaugh, N. A. (1975). *Proc. Nat. Acad. Sci. U.S.A.* **72,** 4003.

Schwarzenbach, G., and Heller, J. (1951). *Helv. Chim. Acta* **34,** 576.

Sligar, S. G., Debrunner, P. G., Lipscomb, J. D., Namtvedt, M. J., and Gunsalus, I. C. (1974). *Proc. Natl. Acad. Sci. U.S.A.* **71,** 3906.

Snow, J. T., Finley, J. W., and Friedman, M. (1975). *Biochem. Biophys. Res. Commun.* **64,** 441.

Stasiw, R., and Wilkins, R. G. (1969). *Inorg. Chem.* **8,** 157.

Stein, C. A., Ellis, P. E., Elder, R. C., and Deutsch, E. (1976). *Inorg. Chem.* **15,** 1618.

Stellwagen, E., and Cass, R. D. (1975). *J. Biol. Chem.* **250,** 2095.

Stellwagen, E., and Shulman, R. G. (1973). *J. Mol. Biol.* **80,** 559.

Stewart, R. (1964). "Oxidation Mechanisms," p. 84. Benjamin, New York.

Stombaugh, N. A., Burris, R. H., and Orme-Johnson, W. H. (1973). *J. Biol. Chem.* **248,** 7951.

Stynes, H. C., and Ibers, J. A. (1971). *Inorg. Chem.* **10,** 2304.

Sutin, N. (1973). *In* "Inorganic Biochemistry" (G. Eichorn, ed.), Vol. 2, p. 611. Am. Elsevier, New York.

Sweeney, W. V., Bearden, A. J., and Rabinowitz, J. C. (1974). *Biochem. Biophys. Res. Commun.* **59,** 188.

Tabner, B. J., and Yandle, J. R. (1968). *J. Chem. Soc. A* p. 381.

Tabner, B. J., and Yandle, J. R. (1971). *In* "Reactions of Molecules at Electrodes" (N. S. Hush, ed.), p. 283. Wiley (Interscience), New York.

Tanaka, M., Haniu, M., Yasunobu, K. T., Jones, J. B., and Stadtmann, T. C. (1974). *Biochemistry* **13,** 5284.

Thauer, R. K., Käufer, B., and Fuchs, G. (1975). *Eur. J. Biochem.* **55,** 111.

Walker, G. A., and Mortenson, L. E. (1974). *Biochemistry* **13,** 2382.

Walker, M. N., and Mortenson, L. E. (1974). *J. Biol. Chem.* **249,** 6356.

Warner, L. G., Ottersen, T., and Seff, K. (1974). *Inorg. Chem.* **13,** 1904 and 2819.

Watenpaugh, K. O., Sieker, L. C., Herriott, J. R., and Jensen, L. H. (1973). *Acta Crystallogr., Sect. B* **29,** 943.

Weschler, C. J., and Deutsch, E. (1973). *Inorg. Chem.* **12,** 2682.

Weschler, C. J., and Deutsch, E. (1976). *Inorg. Chem.* **15,** 139.

Weschler, C. J., Sullivan, J. C., and Deutsch, E. (1973). *J. Am. Chem. Soc.* **95,** 2720.

Weschler, C. J., Sullivan, J. C., and Deutsch, E. (1974). *Inorg. Chem.* **13,** 2360.

Wilkins, R. G., and Yelin, R. E. (1968). *Inorg. Chem.* **7,** 2667.

Wilson, G. S., Tsibris, J. C. M., and Gunsalus, I. C. (1973). *J. Biol. Chem.* **248,** 6059.
Woods, M., Sullivan, J. C., and Deutsch, E. (1975). *J. Chem. Soc. Chem. Commun.* p. 749.
Woods, M.. Karbwang, J., Sullivan, J. C., and Deutsch, E. (1976). *Inorg. Chem.* **15,** 1678.
Young, R. C., Meyer, T. J., and Whitten, D. G. (1975). *J. Am. Chem. Soc.* **97,** 4781.
Yu, M. W., and Fritchie, C. J., Jr. (1965). *J. Biol. Chem.* **250,** 946.

CHAPTER 10

Recent Mössbauer Results of Some Iron–Sulfur Proteins and Model Complexes

P. G. DEBRUNNER, E. MÜNCK, L. QUE, and
CHARLES E. SCHULZ

I. Introduction.. 381
II. Rubredoxin... 388
   A. Ferric Rubredoxin.. 389
   B. Ferrous Rubredoxin...................................... 393
   C. Discussion.. 398
III. The MoFe Protein of Nitrogenase............................ 400
   A. Introduction.. 400
   B. Results... 402
   C. Discussion.. 409
IV. Model Compounds for Rubredoxin and 4 Fe–4 S* Clusters........ 411
   References... 415

I. INTRODUCTION

Our present understanding of the active centers found in iron–sulfur proteins is largely based on physical methods of investigation, in particular X-ray diffraction, optical, magnetic resonance, and Mössbauer spectroscopy. This chapter will concentrate on ^{57}Fe Mössbauer spectroscopy and demonstrate its application to two systems for which it proved to be particularly useful. Comparisons will also be drawn with the spectra of the recently synthesized model complexes. Other aspects of the proteins studied here have been covered in earlier chapters of these volumes. For rubredoxin from *Clostridium pasteurianum*, discussed in Section II, the reader should consult Chapter 3 and 4 by Eaton and Lovenberg and by Jensen, respectively, in Volume II. The MoFe protein from the nitrogen fixing system of *Azotobacter vinelandii*, described in Section III, is covered by Orme-Johnson and Davis in Chapter 2. In Chapter 7, Holm discusses model complexes more generally.

Mössbauer spectroscopy is sensitive to ^{57}Fe only, but it allows a quantitative determination of all the iron present in a sample regardless of its charge and spin state. Under favorable conditions, it has high resolution and provides a detailed characterization of the electronic ground state of the iron. The reader interested in a more systematic discussion of Mössbauer spectroscopy and its application to iron–sulfur proteins is referred to Chapter 6 of Volume II and Chapter 8.

The two proteins examined here differ greatly in complexity. Rubredoxin from *Clostridium pasteurianum* contains a single iron atom per molecule and is the simplest prototype of an iron–sulfur protein. The MoFe protein from the nitrogen-fixing system of *Azotobacter vinelandii*, on the other hand, contains approximately twenty iron atoms, an equal number of acid-labile sulfur, and possibly two molybdenum atoms. Our understanding of nitrogenase is still rudimentary. Chemical methods and EPR provided some first insight into the nature and function of the metal centers, but only in combination with Mössbauer spectroscopy was it possible to differentiate between various iron–sulfur clusters in the protein (Münck et al., 1975). By judicious choice of temperature and magnetic fields, up to four spectral components have been identified. Here, the good resolution and the ability to quantitate all the ^{57}Fe present in the protein are responsible for the success of the Mössbauer method.

In the case of rubredoxin, our knowledge is much more advanced. The structure of the protein is known to 1.5 Å resolution (Watenpaugh et al., 1973), and many spectral data have been reported (Eaton and Lovenberg, Chapter 3, Volume II). As will be shown, the Mössbauer spectra are particularly rich in information (Rao et al., 1972), and they can be interpreted quantitatively in terms of a spin Hamiltonian formalism. The parameters derived from such an analysis characterize the electronic ground state of the iron, and they can be compared with theoretical models, in particular with molecular orbital calculations of the active center (Loew and Lo, 1974a,b; Loew et al., 1974a,b; Norman and Jackels, 1975). Thus, the Mössbauer data will not only enable us to refine the calculations, but they will also provide a basis for comparison with recently synthesized model complexes (Lane et al., 1975) and with other iron–sulfur proteins.

Since Mössbauer spectroscopy was introduced already in Chapter 6 of Volume II, we will just briefly review the possibilities and limitations of the method and summarize, for easy reference, the relations needed for the interpretation of the spectra. More details can be found in the books by Goldanskii and Herber (1968) and by Cohen (1976).

The Mössbauer effect is observable with relatively few isotopes, of which ^{57}Fe is by far the most important biologically. The natural abundance of this isotope is only 2.2%, and since measurements with less than one micromole of ^{57}Fe are usually impractical, the proteins have to be enriched either by chemical exchange or by growing cultures on enriched media. Since the Mössbauer effect can be observed in solid samples only, the proteins have to be studied under nonphysiological conditions, either lyophilized or in frozen solutions. The latter are less likely to produce artifacts and are therefore preferable.

The Mössbauer effect is, as the synonymous expression "nuclear γ-ray resonance" indicates, a resonance phenomenon between the ground state of a nucleus and its first excited state. In ^{57}Fe, the quantum of energy exchanged is a γ ray of energy $E\gamma$ = 14.4 keV and half-width Γ = 4.5 10^{-9} eV. The resonance energy spectrum is scanned by Doppler shifting the source of radiation relative to the absorber by an amount $\delta E = (v/c)E\gamma$, where v is the source velocity and c is the speed of light. Usually, the transmission of a beam of γ rays through an absorber is recorded as a function of Doppler shift, and for convenience, the energy is indicated in velocity units. The minimum observable linewidth is 2Γ, where Γ is the Heisenberg half-width quoted above; 2Γ corresponds to a Doppler velocity of v = 0.196 mm/second. Practical spectrometers do not quite reach this theoretical limit, but minimum widths of 0.24 mm/second are routinely achieved.

The resolution of 2Γ ~ 10^{-8} eV is adequate to observe the hyperfine interaction of the ^{57}Fe nucleus with the atomic electrons, typically of the order of 5.10^{-7} eV. It is this hyperfine splitting that makes the Mössbauer effect an important spectroscopic tool, since it allows one to probe the internal electric and magnetic fields acting on the nucleus. The internal fields are mainly due to the electrons in the unfilled 3d shell, and the observed splittings yield information about the electronic ground state of the iron.

Three different types of hyperfine interactions can be observed in Mössbauer spectroscopy:

i. The isomer shift δ, an electric monopole interaction that is proportional to the electron density at the nucleus and that manifests itself as an overall shift of the absorption pattern relative to zero velocity

ii. The electric quadrupole interaction $\mathcal{H}_Q$ proportional to the electric-field gradient at the nucleus, which, taken alone, gives rise to just two lines

iii. The magnetic dipole interaction, which produces a six-line spectrum in the simplest case

TABLE I

Typical Mössbauer Parameters for High-Spin Ferrous and Ferric Iron in Four- and Sixfold Coordination to Oxygen or Sulfur[a]

Complex		δ_{Fe} (mm/second)[b]	$|\Delta E_Q|$ (mm/second)	$-A$ (MHz)
Fe^{3+}–O	(6)	0.35–0.60	$\lesssim 0.9$	26–29
	(4)	0.3–0.4	$\lesssim 0.9$	22–28
Fe^{3+}–S	(6)	0.2–0.45	$\lesssim 0.8$	18–26
	(4)	0.15–0.25	$\lesssim 0.5$	12–27
Fe^{2+}–O	(6)	0.9–1.3	$\lesssim 4$	17–38
	(4)	0.8–0.95	$\lesssim 2.5$	—
Fe^{2+}–S	(6)	0.8–1.0	—	22
	(4)	0.5–0.7	$\lesssim 3.2$	10–26[c]

[a] Goldanskii and Herber (1968), Chapter 3.
[b] Reiff (1973).
[c] Münck et al. (1972).

In general, all these terms are present simultaneously, and the resulting hyperfine patterns are complex and must be computer analyzed. Before we discuss the formalism used to parametrize the experimental spectra, we will consider the electronic ground state of the iron atom to be expected in the iron–sulfur proteins.

As far as we know, all iron–sulfur proteins have the iron coordinated to four sulfur atoms arranged in a distorted tetrahedron. In the terminology of crystal field theory, tetrahedral coordination leads to a weak ligand field, and the iron is consequently in the high-spin state. Thus, ferric iron Fe^{3+}, $(3d)^5$, has a spin of $S = \frac{5}{2}$ and ferrous iron Fe^{2+}, $(3d)^6$, has a spin of $S = 2$. In the sulfur-bridged complexes found in the 2 Fe–2 S* and 4 Fe–4 S* proteins, the iron atoms couple antiferromagnetically to the smallest allowed spin value, which is zero or $\frac{1}{2}$,† depending on the charge state of the iron ions involved. High-spin ferrous and ferric ions show characteristic features in their Mössbauer spectra, and an assignment of charge and spin state is usually possible on the basis of the spectra alone. The same statement holds for the spin-coupled clusters, as will be shown in Section III. Here we just point out that high-spin ferrous iron has a more positive isomer shift than ferric iron; furthermore, the shift is smaller for tetrahedral coordination than for an octahedral one, and smaller for sulfur than for oxygen ligands (Table I).

To extract information from the Mössbauer data, we need a theory that allows us to parametrize the spectra so that we can reproduce or

---

† The coupling scheme leading to the $S = \frac{3}{2}$ system observed in the MoFe protein of nitrogenase (Section III) has not been established yet.

simulate them using a small set of parameters. The formalism adopted is based on the spin Hamiltonian $\mathcal{H}_S$ familiar from EPR experiments.

The eigenstates and eigenvalues of a complex with spin S in a magnetic field **H** are given by the spin Hamiltonian

$$\mathcal{H}_S = D[S_z^2 - \tfrac{1}{3}S(S+1)] + E(S_x^2 - S_y^2) + \beta \mathbf{H} \cdot \tilde{\mathbf{g}} \cdot \mathbf{S} \quad (1a)$$

Here, the first two terms represent the zero-field or fine-structure splitting and S_x, S_y, and S_z are the components of the electron spin operator in a suitably chosen coordinate system fixed within a molecule. The last term is the electronic Zeeman interaction that has to be specified, in general, by a g tensor $\tilde{\mathbf{g}}$.

Oosterhuis (1974) found it necessary to augment the spin Hamiltonian, Eq. (1a), by a fine-structure term of fourth order, $\mathcal{H}_{S4}$, in order to describe six-coordinated high-spin ferric complexes of low symmetry.

$$\mathcal{H}_{S4} = D\tfrac{1}{6}\mu[S_{x'}^4 + S_{y'}^4 + S_{z'}^4 - \tfrac{1}{5}S(S+1)(3S^2 + 3S - 1)] \quad (1b)$$

The principal axes x', y', z' of this term need not coincide with the principal axes x, y, z of the quadratic term in Eq. (1a). In the analysis of rubredoxin (Section II), we will use the augmented spin Hamiltonian Eqs. (1a) and (1b). Not surprisingly, the extra parameter μ of Eq. (1b) allows one to find better simulations of the Mössbauer spectra than are possible with the parameters D and E only.

It should be noted that the sum of Eqs. (1a) and (1b) is still not the most general spin Hamiltonian possible for spins $S \geq 2$. Small axial terms of fourth order in S may be added to the fine structure, and the Zeeman interaction may contain cubic spin terms. We will ignore these small corrections, however.

For high-spin ferric iron, $S = \tfrac{5}{2}$, the zero-field terms split the sixfold degenerate free ion ^6S ground state into three Kramers doublets with energy separations determined by D, E, and μ, if the latter is included. The g tensor can normally be replaced by the free-spin value $g_0 = 2.0023$; deviations from isotropy of the order of 1% have been found in cases of low symmetry (Schneider et al., 1968). EPR transitions can be observed, in principle, between the two states of each Kramers doublet; from the resonance spectrum, it is then possible to determine the fine structure parameters D and E. It is convenient for the analysis of EPR data to describe each Kramers doublet as a fictitious-spin doublet with $S' = \tfrac{1}{2}$ because the (fictitious) spin Hamiltonian has the simpler form

$$\mathcal{H}_{S'} = \beta \mathbf{H} \cdot \tilde{\mathbf{g}}_i' \cdot \mathbf{S}' \qquad S' = \tfrac{1}{2}, \quad i = 1, 2, 3 \quad (1c)$$

Equations (1a) and (1c) are made equivalent by proper choice of the g' tensors $\tilde{\mathbf{g}}_i'$, for each of the three Kramers doublets $i = 1, 2, 3$ (Wickman *et al.*, 1966). Usually, the *g* tensors are determined directly from EPR experiments, and the value of E/D in Eq. (1a) is adjusted to best reproduce the data.

The spin Hamiltonian, Eq. (1a), also applies to high-spin ferrous iron, $S = 2$. The resulting spin eigenstates, however, are fundamentally different from those of the ferric ion since they are singlets rather than Kramers doublets.† The spin expectation values $\langle \mathbf{S}_k \rangle$ and the magnetic moments $\langle \mathbf{\mu}_k \rangle = \beta \tilde{\mathbf{g}} \langle \mathbf{S}_k \rangle$ therefore vanish for each eigenstate, $k = 1, \ldots, 5$, unless a strong magnetic field is applied. The field may be of external origin or may be the exchange field in antiferromagnetically ordered compounds. More generally, in a crystal field of low symmetry, any system with an odd number of electrons, i.e., half-integer spin, will give rise to Kramers doublets with observable magnetic properties such as magnetic hyperfine splitting in Mössbauer spectroscopy and magnetic resonance transitions in EPR experiments. Systems with an even number of electrons and integer spin, however, such as high-spin ferrous iron, generally show no EPR signals, and magnetic hyperfine splitting is observed in strong fields only.

We next have to consider the interactions of the ^{57}Fe nucleus with its environment that give rise to the splittings observed in the Mössbauer spectra. ^{57}Fe has spin $I_g = \tfrac{1}{2}$ in the ground state and $I_e = \tfrac{3}{2}$ in the excited state. Both states have magnetic moments, with g factors $g_g = 0.18$ and $g_e = -0.103$, respectively, and the excited state has, in addition, a quadrupole moment, $Q \cong 0.18$ barn. If $\mathcal{H}_M$ and $\mathcal{H}_Q$ stand for the magnetic and electric quadrupole interactions of the ^{57}Fe nucleus, respectively, the total system, nucleus plus electron shell, is described by the Hamiltonian

$$\mathcal{H} = \mathcal{H}_S + \mathcal{H}_M + \mathcal{H}_Q \tag{2}$$

with $\mathcal{H}_S$ given by Eq. (1a). The magnetic interaction can be written as

$$\mathcal{H}_M = \mathbf{S} \cdot \tilde{\mathbf{A}} \cdot \mathbf{I} - \beta_n g_n \mathbf{H} \cdot \mathbf{I} \tag{3a}$$

The first term of Eq. (3a) represents the magnetic hyperfine coupling of the nuclear spin $\mathbf{I}$ with the electron spin $\mathbf{S}$ and is specified by the tensor $\tilde{\mathbf{A}}$. The second term is the Zeeman interaction of the nuclear magnetic moment $\mathbf{\mu}_n = \beta_n g_n \mathbf{I}$ with an external field $\mathbf{H}$. Again, Eq. (3a) is the lowest approximation only; other, smaller terms may exist but have been ignored. In general, the hyperfine tensor $\tilde{\mathbf{A}}$ consists of three parts, (i) the isotropic and negative Fermi contact term, which is always present and usually dominates, and the anisotropic parts due (ii) to the orbital and

---

† For $E = 0$, the levels split into a singlet and two doublets, but since the latter are not Kramers doublets, the argument still holds.

(iii) to the dipolar interaction. The last two terms vanish for states with zero orbital angular momentum L such as the ^6S high-spin ferric ion but are important for the ^5D configuration of the ferrous ion. Typically, A/h is equal to -30 MHz for ionic Fe^{3+} compounds; in the more covalent iron–sulfur complexes, it is reduced to $\sim 70\%$ of this value.† As mentioned before, the three Kramers doublets of high-spin ferric iron are frequently described by a fictitious-spin formalism with $S' = \frac{1}{2}$, Eq. (1c). In this approach, the magnetic hyperfine interaction, which takes the form $A_0 \mathbf{S} \cdot \mathbf{I}$ in the spin $S = \frac{5}{2}$ representation, has to be changed to $\mathbf{S'} \cdot \tilde{\mathbf{A}}_i' \cdot \mathbf{I}$, $i = 1, 2, 3$, where the components of the tensors $\tilde{\mathbf{A}}_i'$ are proportional to those of the tensors $\tilde{\mathbf{g}}_i'$ (Wickman et al., 1966). Specifically, we have

$$A_x'/g_x' = A_y'/g_y' = A_z'/g_z' \tag{4}$$

for each of the Kramers doublets. The nuclear Zeeman interaction $\beta_n g_n \mathbf{H} \cdot \mathbf{I}$ is independent of the representation chosen for the electron spin.

In an external magnetic field H such that $H \gg |A|/(g_n \beta_n) \sim 10$ G the electronic Zeeman term $\beta \mathbf{H} \cdot \tilde{\mathbf{g}} \cdot \mathbf{S}$ is much larger than the nuclear hyperfine coupling $\mathbf{S} \cdot \tilde{\mathbf{A}} \cdot \mathbf{I}$, and the spin expectation values $\langle \mathbf{S} \rangle$ are found to a good approximation by diagonalizing Eq. (1a). We can then define an internal magnetic field $\mathbf{H}_{int}$,

$$\mathbf{H}_{int} = -\langle \mathbf{S} \rangle \tilde{\mathbf{A}}/(g_n \beta_n) \tag{5}$$

and can write Eq. (3a) in the useful form

$$\mathcal{H}_M = -\beta_n g_n (\mathbf{H}_{int} + \mathbf{H}) \cdot \mathbf{I} \tag{3b}$$

The important difference between Kramers and non-Kramers states should be stressed again. The latter, exemplified by Fe^{2+}, $S = 2$, show no magnetic splitting unless a strong polarizing field is present. Kramers doublets, on the other hand, frequently show magnetic splitting even in zero external field. Indeed, the small fields due to the nuclear magnetic moments are sufficient to lift the degeneracy of the doublets, and complicated Mössbauer spectra may result. Since no such spectra will be shown, we do not elaborate on this point.

The relations given so far assume slow spin relaxation and are applicable at sufficiently low temperatures. In general, more than one spin state may be populated at a given temperature, each contributing a component spectrum with an intensity given by the Boltzmann factor of the corresponding level. In the fast relaxation limit, on the other hand, the

† In spin-coupled clusters of the 2 Fe–2 S* or 4 Fe–4 S* type, the effective or "coupled" A tensors differ from the intrinsic, uncoupled A tensors considered here (Münck et al., 1972).

spin expectation value $\langle \mathbf{S} \rangle$ of Eq. (5) has to be replaced by a thermal average over all spin states, and only a single spectrum is observed.

The electric quadrupole interaction $\mathcal{H}_Q$ finally is given by

$$\mathcal{H}_Q = \tfrac{1}{12}eQV_{\zeta\zeta}[3I_\zeta^2 - I(I+1) + \eta(I_\xi^2 - I_\eta^2)] \tag{6a}$$
$$\eta = (V_{\xi\xi} - V_{\eta\eta})/V_{\zeta\zeta} \tag{6b}$$
$$V_{\xi\xi} + V_{\eta\eta} + V_{\zeta\zeta} = 0 \tag{6c}$$

Here, V_{ii}, $i = \xi, \eta, \zeta$, are the components of the quadrupole tensor, i.e., the negative of the electric-field gradient. Its principal axes frame, ξ, η, ζ, need not coincide with the axis x, y, z, of the zero-field splitting. With the convention $|V_{\xi\xi}| < |V_{\eta\eta}| < |V_{\zeta\zeta}|$, the asymmetry parameter η, Eq. (6b), is limited to the values $0 \leq \eta \leq 1$. In the absence of magnetic interactions, a quadrupole splitting $\Delta E_Q = \tfrac{1}{2}eQV_{\zeta\zeta}(1 + \tfrac{1}{3}\eta^2)^{1/2}$ is observed. Typically, ΔE_Q for high-spin ferric iron is less than 1 mm/second, since the $S = \tfrac{5}{2}$ state derives from a crystal-field configuration $(t_{2g})^3(e_g)^2$ with $\Delta E_Q = 0$. High-spin ferrous iron, on the other hand, has an extra 3d electron and shows quadrupole splittings of 2–3 mm/second approaching the value $\Delta E_Q \simeq 4.5$ mm/second for a free ferrous ion (Ingalls, 1969).

Equations (1–6) provide a theoretical framework for the interpretation of the Mössbauer spectra to be presented in Sections II–IV. All the information about the electronic ground state of the iron is contained in the tensor quantities D, E, $\tilde{\mathbf{g}}$, $\tilde{\mathbf{A}}$, and V_{ii}, expressions for which can be found in a number of references (Cohen, 1976; Goldanskii and Herber, 1968). Table I indicates characteristic values for the Mössbauer parameters δ, ΔE_Q, and A of high-spin ferrous and ferric iron. On the basis of these empirical values, it is usually possible to identify the charge states involved.

For any given set of parameters D, E, $\tilde{\mathbf{g}}$, $\tilde{\mathbf{A}}$, and V_{ii}, the Mössbauer spectra can be computed by well-known methods. The reverse process of parametrizing a given spectrum, however, is far from trivial. A trial and error approach is usually employed, i.e., spectra are simulated with an assumed set of parameters and the latter are varied until the resulting spectra agree with the data. The uniqueness of the solution can be checked by simulating spectra measured under different experimental conditions.

II. RUBREDOXIN

Eaton and Lovenberg gave a detailed account of the physicochemical properties of rubredoxin (Rd) in Chapter 3, Volume II, and in Chapter 4 of the same volume, Jensen discussed the three-dimensional structure of oxidized Rd, which is known to 1.5 Å resolution. Below, we report Mössbauer studies on chemically reconstituted Rd (Lovenberg and Williams, 1969) from *Clostridium pasteurianum*.†

A. Ferric Rubredoxin

The first Mössbauer spectra of Rd were published by Phillips *et al.* (1970). In a lyophilized sample these authors observed a quadrupole doublet with a splitting of $\Delta E_Q = 0.78 \pm 0.02$ mm/second at 77°K and partially resolved magnetic splitting at 4.2°K. Rao *et al.* (1972) studied frozen solutions of Rd from *Chloropseudomonas ethylica* and *C. pasteurianum* and found broad, indistinct spectra at liquid nitrogen temperature that changed to well-resolved six-line patterns at 4.2°K. Figure 1 shows the Mössbauer spectra of *C. pasteurianum* Rd in frozen solution measured at 1.5°, 10°, and 20°K in a weak magnetic field perpendicular to the direction of the γ rays. At 1.5°K, Fig. 1a, we observe essentially the same simple pattern as that reported by Rao *et al.* (1972), but at the higher temperatures, Figs. 1b and c, additional absorption appears between the inner four lines, at −6.5 mm/second and at +6.8 mm/second. A careful analysis shows that the spectra consist of a superposition of three well-defined component spectra, Fig. 2, each of which can be associated with one of the three Kramers doublets of the $S = \frac{5}{2}$ high-spin ferric iron. At the lowest temperatures, only the ground doublet is populated, Fig. 1a, and since its spectrum is a simple six-line pattern, the Mössbauer parameters are readily estimated. A comparison of Figs. 1b and 1c shows that the line on the far left arises from the highest doublet and is part of another six-line spectrum, the remaining lines of which largely overlap with the ground state spectrum. The middle doublet, finally, accounts for much of the intensity between the inner four lines (−3 mm/second $< v <$ 3 mm/second). It has considerably smaller overall splitting than the other component spectra and includes broad bands as opposed to the six sharp lines of the other two doublets.

Having identified the component spectra arising from the three Kramers doublets, we now apply the spin Hamiltonian formalism, Eqs. (1–6), to parametrize the spectra. Lang *et al.* (1971) and Oosterhuis (1974) observed composite spectra of the same type in six- or seven-coordinated high-spin ferric iron, and much of the discussion given here follows theirs or the earlier work by Wickman *et al.* (1966). The analysis of the spectra is greatly facilitated by the fact that EPR data are available. All the rubredoxins studied so far show a strong EPR signal

† Enriched Rd was provided by Dr. Walter Lovenberg, National Institutes of Health.

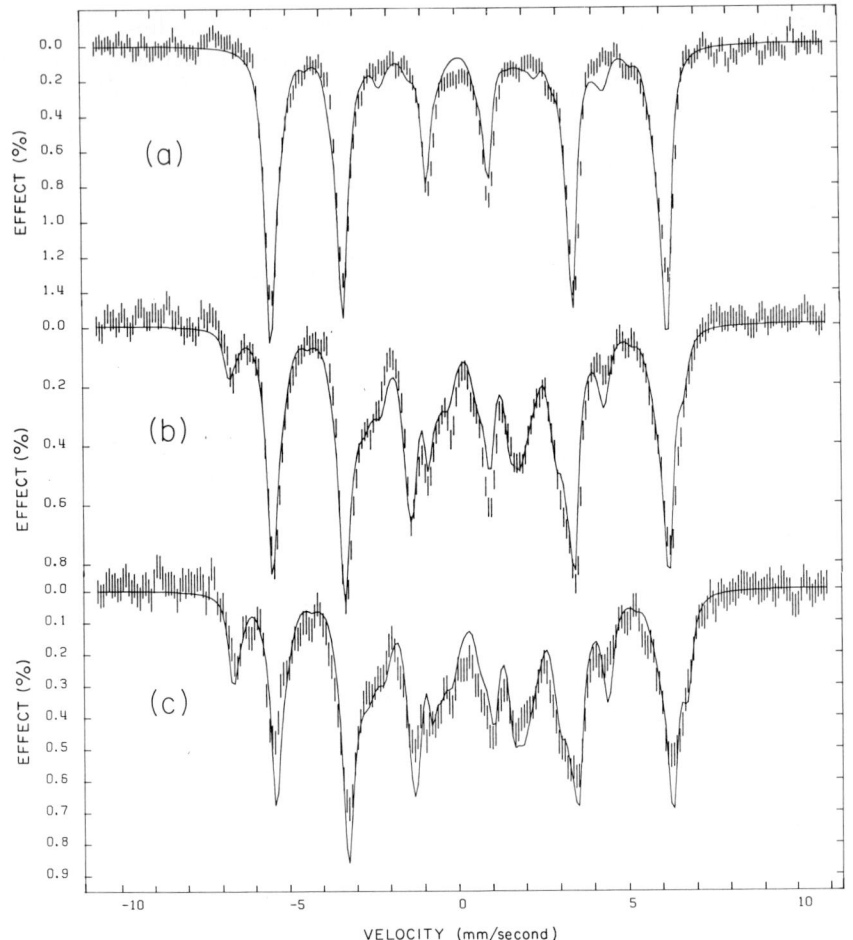

Fig. 1. Mössbauer spectra of a frozen solution of oxidized Rd from *C. pasteurianum* in a magnetic field of 1.3 kG perpendicular to the γ rays (a) at 1.5°K, (b) at 10°K, and (c) at 20°K. The solid lines show simulations obtained by the method described in Section II,A. The parameters used and their estimated uncertainties are as follows: $D = 2.5 \pm 0.5°$K, $E = 0.43 \pm 0.03°$K, $\mu = 0.3 \pm 0.1$; $\Delta E_Q = -0.5 \pm 0.1$ mm/second, $\eta = 0.2 \pm 0.1$, $A_x = -22.7 \pm 1.5$ MHz, $A_y = -21.5 \pm 0.4$ MHz, $A_z = -23.5 \pm 0.4$ MHz; $\Gamma = 0.3$ mm/second (full width at half-maximum, FWHM). The g tensors calculated with these parameters are $\tilde{g}_1' = (1.10, 9.35, 0.75)$, $\tilde{g}_2' = (3.98, 4.09, 4.55)$, $\tilde{g}_3' = (0.93, 0.73, 9.80)$.

near $g = 4.3$ and a single derivative peak near $g = 9.4$. Peisach *et al.* (1971) pointed out that the spin Hamiltonian Eq. (1a) very naturally explains these results if the fine structure term, $D[S_z^2 - \frac{1}{3}S(S+1) + E/D(S_x^2 - S_y^2)]$, has high rhombicity, i.e., for $E/D \gtrsim \frac{1}{3}$. The energies

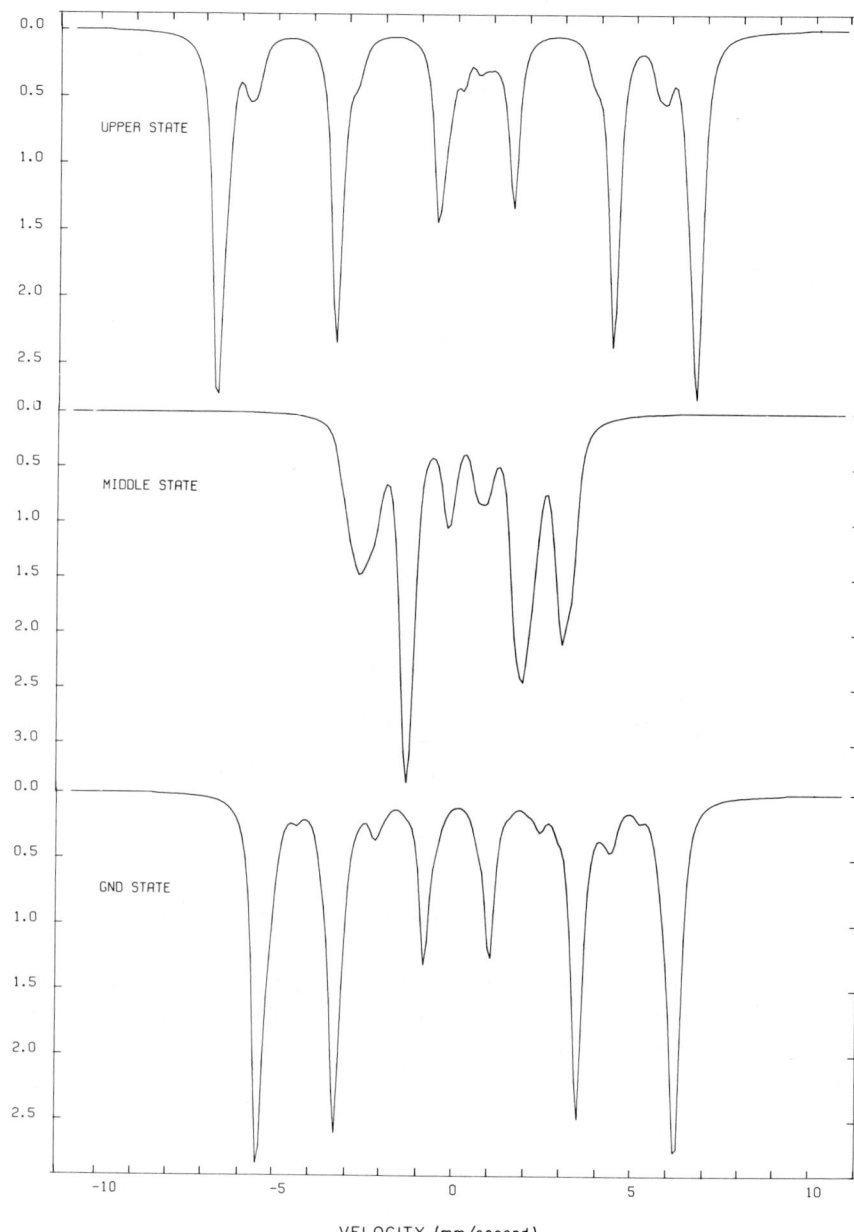

Fig. 2. Component spectra of the three Kramers doublets of oxidized Rd. An average over 15 × 15 different orientations of the molecule with respect to the applied field is shown calculated with the parameters of Fig. 1.

of the Kramers doublets then approach $E_2 = 4D\sqrt{7}/3$ and $E_3 = 2E_2$ relative to the ground state and the g tensors become $\tilde{g}_1' = (0.86, 9.68, 0.61)$, $\tilde{g}_2' = (4.29, 4.29, 4.29)$, and $\tilde{g}_3' = (0.86, 0.61, 9.68)$ in the fictitious spin representation, Eq. (1c) with $S' = \frac{1}{2}$. For the Mössbauer spectra, the strong anisotropy of $\tilde{g}_1'$ and $\tilde{g}_3'$ has important consequences. In the lowest and highest doublets, the electron spin tends to line up along the y and z axis, respectively, no matter how the molecule is oriented relative to the magnetic field. In the middle doublet, with its isotropic g value of 4.3, on the other hand, the electron spin takes the direction of the externally applied field.

The model just described provides a good approximation of the experimental facts. The strong EPR signal near $g = 4.3$ can be assigned to the middle doublet, whereas the $g \sim 9.4$ signal must be associated with the lowest one. By varying the ratio E/D, the energies of the doublets and their g tensors can be varied continuously between $E_2 = D$, $E_3 - 3D$, $\tilde{g}_1' = (6, 6, 2)$, $\tilde{g}_2' = (0, 0, 6)$, $\tilde{g}_3' = (0, 0, 10)$ for $E = 0$ and the values given above (Wickman et al., 1966).

For Rd from *Pseudomonas oleovorans* Peisach et al. (1971) not only measured $\tilde{g}_1'$ and $\tilde{g}_2'$, but they also determined the energy E_2 of the second Kramers doublet from the temperature dependence of the EPR intensity. Their experimental data, $\tilde{g}_1' = (1.22, 9.42, 0.90)$, $\tilde{g}_2' = (4.31, 4.0, 4.7)$, $E_2 = 7.7°K$, are best reproduced with $E/D = 0.28$, $D = 2.54°K$, leading to predicted values of $\tilde{g}_1' = (1.22, 9.52, 0.74)$, $\tilde{g}_2' = (4.20, 3.97, 4.58)$. Clearly, the agreement between theory and experiment is reasonable but not perfect, and other terms in the spin Hamiltonian are called for. Oosterhuis (1974) studied EPR and Mössbauer spectra of high-spin ferric iron in siderochromes and found that fourth-order terms, Eq. (1b), are needed to explain the data. We will adopt his model and let the energies E_2, E_3 and the g tensors $\tilde{g}_1'$, $\tilde{g}_2'$, and $\tilde{g}_3'$ be functions of the three fine structure parameters D, E, and μ, all referred to the same axes x, y, and z.

To analyze the Mössbauer spectra, we start with the lowest doublet, Fig. 1a. The sharpness of the lines indicates that the internal field H_{int} is practically the same for each molecule. The fact that the direction of the externally applied field does not affect the line intensities very much (cf. Fig. 1a with Fig. 5b of Rao et al., 1972) implies that the ground doublet has an easy axis of magnetization. This must clearly be the direction of the $g = 9.4$ component, i.e., the y axis of the fine structure term $D[S_z^2 - (35/12) + (E/D)(S_x^2 - S_y^2)]$. The quadrupole interaction, which is small in comparison with the magnetic hyperfine term, shifts the inner four lines relative to the outer two lines by an amount proportional to V_{yy}, the component of the quadrupole tensor along the prevailing internal field, $H_{\text{int}} \cong -A'_{1,y}\langle S_y' \rangle/(g_n\beta_n)$. Inspection of Fig. 1a yields the values $H_{\text{int}} = 365$ kG, $eQV_{yy}/2 \cong 0.28$ mm/second.

We next turn our attention to the third Kramers doublet, which has a g tensor even more anisotropic than the lowest doublet. The g tensor of this state has not been observed directly, but Peisach et al. (1971) predict values of $\tilde{g}_3' = (0.65, 0.41, 9.77)$ for Rd from P. oleovorans. Thus, the internal field is expected to point along the z axis, and since $g'_{3,z}$ is 3% larger than $g'_{1,y}$, Eq. (4) predicts an internal field and therefore an overall splitting that is 3% larger than for the ground doublet. Spectral decomposition shows, however, that the splitting is 13% larger for the upper doublet and we conclude that the A' tensors are not proportional to the g' tensors. We therefore drop the assumption leading to Eq. (4) that the hyperfine interaction in the $S = \frac{5}{2}$ representation is given by a constant, A_0, and will use a tensor $\tilde{A}$ instead.† Interestingly, in the third Kramers doublet, the quadrupole splitting, which is essentially given by V_{zz}, is negative and larger than in the ground doublet, $eQV_{zz}/2 = -0.5$ mm/second.

The model that we finally adopted to simulate the Mössbauer spectra is based on the spin Hamiltonian Eqs. (1a) and (1b) in the spin $S = \frac{5}{2}$ representation. The three parameters D, E/D, and μ specify the fine-structure splitting, and g is taken to be isotropic, $g_0 = 2$. The magnetic and electric hyperfine interactions are specified by the tensors $\tilde{A}$ and $\{V_{ii}\}$, for which we obtained estimates from an inspection of the data. To minimize the number of adjustable parameters, all tensors are assumed to have the same principal axes. The simulations obtained with this model are quite satisfactory; the solid lines in Fig. 1 represent the spectra calculated with the parameters listed in the figure caption, and Fig. 2 shows the component spectra of the individual Kramers doublets for comparison. It should be pointed out that not all the parameters can be determined unambiguously from the present data. This is particularly true for the fine structure parameters E/D and μ of Eqs. (1a) and (1b). Finite values of μ considerably improve the simulations, but reasonable solutions can be found for a range of E/D and μ. It is hoped that further experiments will remove this ambiguity.

B. Ferrous Rubredoxin

The susceptibility and Mössbauer data of Phillips et al. (1970) showed conclusively that the iron in reduced Rd is in the high-spin ferrous state with spin $S = 2$. Rao et al. (1972) observed well-resolved hyperfine splittings in Mössbauer spectra measured in strong magnetic fields. Their data suggested the feasibility of a quantitative analysis based on the

---

† Strictly speaking, either the assumption $\tilde{g} = g_0 = 2.0$ or $\tilde{A} = A_0$ in the $S = \frac{5}{2}$ representation has to be dropped. Since $\tilde{g}_3'$ has not been measured, we assume $g'_{3,z} \leq 10$ and are then forced to let $\tilde{A}$ be a tensor.

spin Hamiltonian discussed in Section I. We will show below that acceptable simulations can indeed be generated for the data taken at 4.2°K in different magnetic fields if slow spin relaxation is assumed; at higher temperatures, however, the simple model appears to break down as the spin relaxation rate becomes comparable with the nuclear Larmor frequencies. Since the number of parameters is large even in the slow relaxation model, and since little is known about their order of magnitude, it is difficult to assess the uniqueness of the solution.

Figure 3 shows the Mössbauer spectra of reduced Rd from *Clostri-*

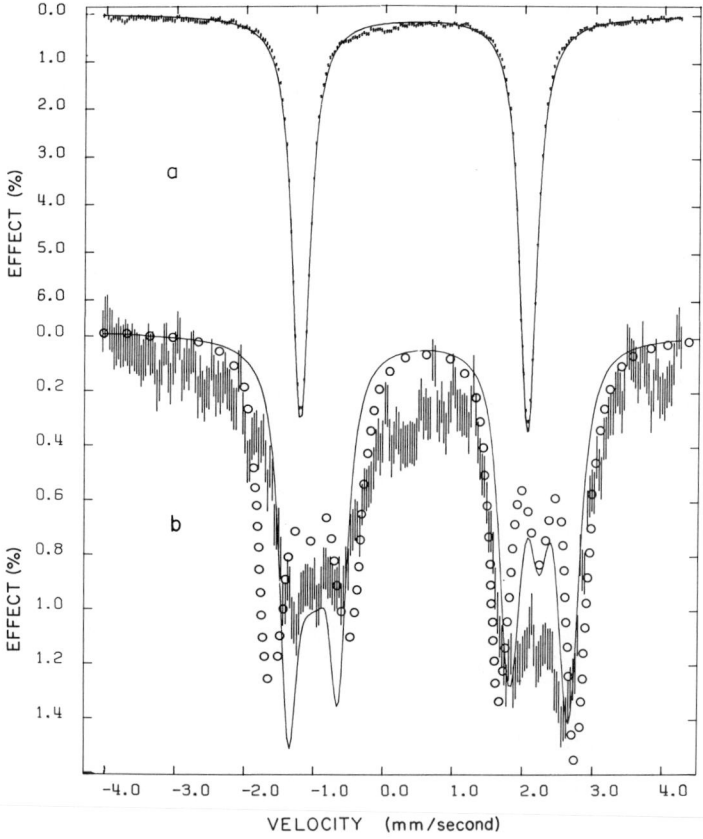

Fig. 3. Mössbauer spectra of reduced Rd from *C. pasteurianum* (a) at 4.2°K in zero field and (b) at 175°K in a field of 43 kG parallel to the γ rays. The solid line in (a) is a least squares fit of two Lorentzians of FWHM $\Gamma = 0.28$ mm/second, isomer shift $\delta_{Fe} = 0.70 \pm 0.02$ mm/second, and quadrupole splitting $\Delta E_Q = -3.25 \pm 0.01$ mm/second. The solid line is a simulation based on the parameters listed in Fig. 4 assuming fast relaxation. The dotted line in (b) shows a simulation for a diamagnetic complex with ΔE_Q as listed under (a) and with $\eta = 0.6$.

dium pasteurianum under different experimental conditions. At 4.2°K, a quadrupole doublet with a splitting of $|\Delta E_Q| = 3.25$ mm/second and an isomer shift of $\delta_{Fe} = 0.70$ mm/second is observed (Fig. 3a). The lines are only slightly wider (0.28 mm/second FWHM) than the minimum observable linewidth (0.24 mm/second FWHM), which indicates that the sample is quite homogeneous. As pointed out by Rao *et al.* (1972), the quadrupole splitting decreases very little with temperature.† Figure 3b shows a spectrum measured at 175°K in a parallel magnetic field of 43 kG. Both lines are considerably broadened by the magnetic interaction and have some structure. For diamagnetic complexes, it is possible to derive the sign and the asymmetry parameter of the quadrupole interaction from such spectra (Collins and Travis, 1967); for Rd, however, the situation is complicated by the paramagnetism of the ferrous ion, which gives rise to an internal magnetic field that opposes the applied field. It is most likely, though, that the quadrupole interaction is negative and that the asymmetry parameter is nonzero.‡ Furthermore, a comparison of the data with a spectrum calculated for a diamagnetic complex having the quadrupole splitting of Rd, $\Delta E_Q = -3.22$ mm/second and $\eta = 0.6$ (dotted line in Fig. 3b) shows that the internal field $\mathbf{H}_{int} = -\langle S \rangle \, \tilde{\mathbf{A}}/g_n\beta_n$, Eq. (5), must be of the order of -10 kG. Assuming that the magnetic susceptibility of Rd at 175°K is isotropic and obeys Curie's law, we estimate $\langle S \rangle$ to be -0.07 and find a rough average for the hyperfine interaction of $\langle A \rangle \approx -20$ MHz. This oversimplified model may give an adequate explanation for the reduced broadening of the higher energy line, but it certainly does not explain the peculiar wings of the lower energy peak. The solid line in Fig. 3b is a simulation based on the spin Hamiltonian parameters to be discussed below. It reproduces the higher energy peak reasonably well but not the peak to the left.

In an external field applied at helium temperature, Fig. 4, the two lines of the quadrupole doublet split into a characteristic pattern. From the higher-energy transition, which originates from the $I_z = \pm \frac{1}{2}$ nuclear level if the quadrupole splitting is negative, two lines of equal intensity emerge that are displaced by roughly ± 1.8 mm/second from the zero field position, leaving a weak unshifted line. The lower energy transition, on the other hand, splits into two relatively narrow doublets with an overall splitting approximately half that of the higher energy transition. Two facts are particularly noteworthy in these spectra; (i) the magnetic splitting saturates for relatively modest applied fields, and (ii) the spec-

---

† $|\Delta E_Q(T)|$ in millimeters per second: 3.25 (4.2°K), 3.28 (40°K), 3.23 (150°K), 3.21 (200°K). Errors ±0.01 mm/second.

‡ For $\eta = 1$, the distinction between positive and negative quadrupole interaction is meaningless, since $V_{xx} = 0$, $V_{zz} = -V_{yy}$.

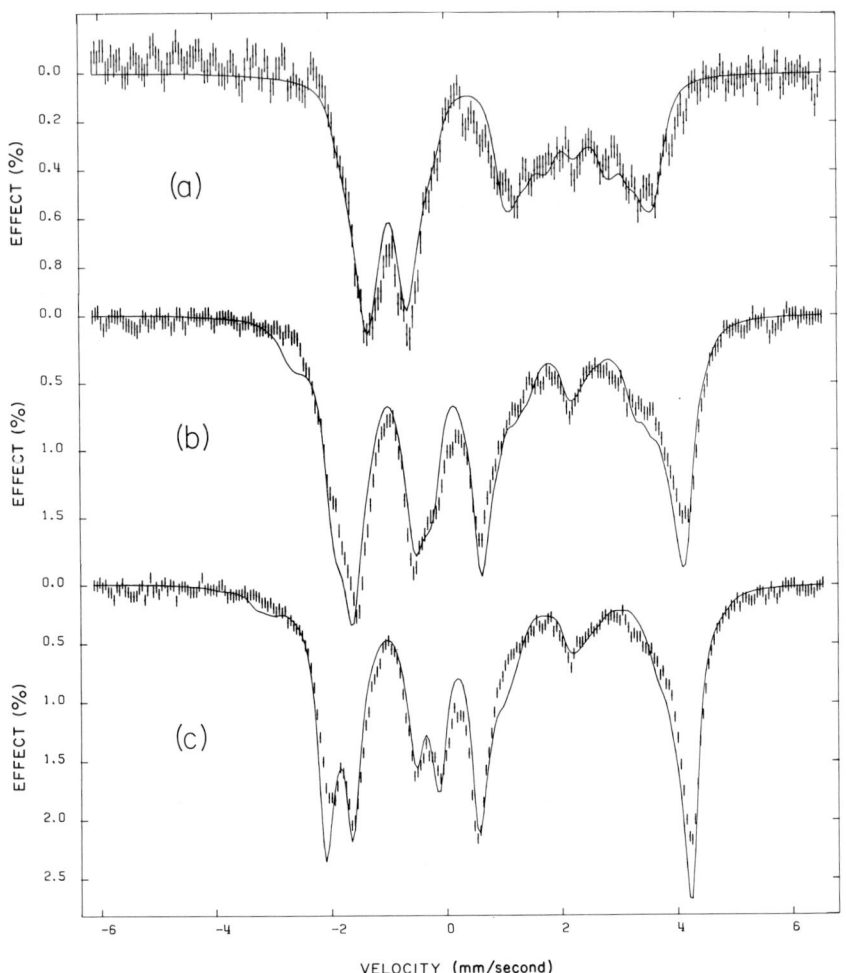

Fig. 4. Mössbauer spectra of reduced Rd from *C. pasteurianum* at 4.2°K in parallel magnetic fields of (a) 6.6 kG, (b) 15 kG, and (c) 24 kG. The solid lines represent the simulations discussed in Section II,B and were calculated in the slow relaxation limit with the following parameters: $D = 12°$K, $E = 1.4°$K, $\mu = 0.33$; $\Delta E_Q = -3.25$ mm/second, $\eta = 0.65$; $A_x = -17.4$ MHz, $A_y = -12.1$ MHz, $A_z = -38.5$ MHz.

trum consists of lines rather than broad bands in spite of the fact that the applied field has a different orientation for each molecule relative to the quadrupole tensor, which is fixed in the molecule. This implies that the induced internal field $\mathbf{H}_{int}$, Eq. (5), must have a fixed direction with respect to the molecular axis. It is indeed possible to approximate the line pattern of Fig. 4c using $\Delta E_Q = -3.25$ mm/second, $0.5 < \eta < 1$, and assuming an internal field of 125 kG along the y axis of the quadrupole tensor† following an analysis by Kündig (1967). The fact that the spectra measured by Rao et al. (1972) in perpendicular fields are very similar to those obtained in parallel fields, Fig. 4, points in the same direction; ferrous Rd must have an easy axis of magnetization. The question that then arises is how can such a situation be obtained in the spin Hamiltonian formalism, Eq. (1).

The simplest model assumes a quadratic fine-structure term with maximum rhombicity, i.e., $E/D = \frac{1}{3}$. The first two terms of Eq. (1a) then reduce to the expression $(2D/3)(S_z^2 - S_y^2)$. A change in sign of the parameter D is obviously equivalent to an interchange of the y and z axes. For $D > 0$, the y direction is the easy axis of magnetization, whereas for $D < 0$, it is the z direction. The five spin states have energies $\pm(4/\sqrt{3})D$, $\pm 2D$, 0, so that the splitting between the lowest two states is approximately $0.3D$. The spin expectation values along the easy axis of magnetization depend on the parameter $y = \beta g_y H_y / D$, and in lowest approximation they are the opposite of each other for the two lowest states $\langle S_y \rangle = \pm ay/(1 + by^2)^{1/2}$, where a and b are constants. In the limit of slow spin relaxation, each state contributes its own spectrum weighted with the appropriate Boltzmann factor. At helium temperature, the inequality $kT \ll 2D$ applies, and only the lowest two states are populated. Since they have internal fields of essentially equal magnitude but opposite sign, their spectra are practically identical. This simple model correctly predicts the easy axis of magnetization and, with a few modifications, it quantitatively accounts for the saturation. The condition $E/D = \frac{1}{3}$ can be relaxed and a fourth-order term, Eq. (1b), can be added so that all three fine structure parameters D, E, and μ are adjustable.

The solid lines in Fig. 4 show the simulations obtained using the spin Hamiltonian formalism, Eqs. (1-6), in the limit of slow relaxation. For simplicity, all interactions are taken to have the same principal axes, the orientation of which relative to the Rd molecule is not known. To further reduce the number of parameters the g tensor was expressed in terms of the fine structure splitting D and E, $g_x = 2 - 2(D - E)/\lambda$, $g_y = 2 - 2(D + E)/\lambda$, $g_z = 2$, with $\lambda = -100$ cm^{-1}, a result based on second-order perturbation theory for $D > 0$. The simulations are quite satisfactory, but the parameters derived, which are listed in the figure legend, should be considered as only preliminary. In principle, the same parameters should also reproduce the spectra measured at higher temper-

† Another possibility is to assume positive quadrupole interaction, $\eta \sim 0.6$, and an internal field along the z axis of the quadrupole tensor. In both cases, H_{int} is along the positive component of V_{ii}, Eq. (6).

atures in an applied field, e.g., Fig. 3b, which are more sensitive to the higher states of the spin quintet. Such an analysis has not been completed yet; it is difficult, because the spin relaxation rate appears to be changing from the slow limit at 4.2 to the fast limit at higher temperature. Further work on this problem is in progress.

C. Discussion

The data presented in Figs. 1, 3, and 4 demonstrate that Rd offers very favorable conditions for the determination by Mössbauer spectroscopy of a complete set of hyperfine and fine structure parameters, which in turn serve to characterize the electronic ground state of the iron. The success of the spectral simulations in Figs. 1 and 4 shows that the spin Hamiltonian formalism provides an adequate theoretical framework for the parametrization of the data. Here, we will compare the results obtained for Rd with data from other iron–sulfur proteins, and we briefly discuss the pertinent model calculations.

As far as ferric Rd is concerned, two of the findings are particularly noteworthy. One is the fact that the energy ratio E_3/E_2 of the third and second Kramers doublets relative to the ground doublet turned out to be $E_3/E_2 = 3.3 \pm 0.5$. In the model of Peisach et al. (1971) for the characteristic EPR signal at $g = 4.3$, the fine structure term in Eq. (1a) is assumed to have large rhombicity, $E/D \lesssim \frac{1}{3}$. This assumption leads to an energy ratio $E_3/E_2 \approx 2$, a value that is not compatible with our result. As Oosterhuis (1974) pointed out, inclusion of a fine structure term of fourth order, Eq. (1b), allows one to vary the energy ratio E_3/E_2 while maintaining the $g = 4.3$ signal. In fact, the second Kramers doublet has an isotropic g value of $g = 30/7 \cong 4.3$ as long as E/D and μ satisfy the linear relation $4\mu = 3 - 9E/D$ (Oosterhuis, 1974). Thus, a single fourth-order term, Eq. (1b), with the same principal axes as the second-order term, Eq. (1a), provides adequate flexibility to reproduce the Mössbauer data. To further elucidate this question, a combined analysis of the Mössbauer and EPR spectra is needed.

The second point of interest is the fact that the magnetic hyperfine interaction is anisotropic in the $S = \frac{5}{2}$ representation. This means that it contains components other than the Fermi contact term, such as orbital or spin dipolar contributions. It has to be pointed out, though, that the A values derived from Mössbauer experiments depend on the fine structure and Zeeman interactions chosen for the data analysis; more definitive numbers could be obtained if the g' tensors were better known or directly from an ENDOR experiment. The anisotropy of the A tensor derived from Fig. 1, $\tilde{A} = (-22.7 \pm 1.5, -21.5 \pm 0.4, -23.5 \pm 0.4)$

MHz is not without precedent. Fritz et al. (1971) reported ENDOR measurements on reduced 2 Fe–2 S* proteins, and from their data the intrinsic hyperfine tensors† of the high-spin ferric iron are found to be (-24, -21.4, -18.5) MHz for adrenodoxin and putidaredoxin, and (-21.9, -21.9, -18.4) MHz for spinach ferredoxin. The average hyperfine interaction, $\langle A \rangle = -22.6$ MHz is slightly larger in Rd than in the 2 Fe–2 S* proteins, for which a mean value of -21 MHz is found, but it is considerably smaller than the -30 MHz typically observed for highly ionic Fe^{3+}.

Considering the anisotropy of $\tilde{A}$, one might expect the Zeeman interaction to be anisotropic, too. In the $S = \frac{5}{2}$ representation, then, $\tilde{g}$ in Eq. (1a) would differ from the spin-only value, $g_0 = 2.0023$, as has been inferred for high-spin ferric iron in FeS_2Se_2 clusters (Schneider et al., 1968). Such an interpretation of the experimentally determined g' tensors is model dependent, however, and for Rd we do not have enough data to draw any meaningful conclusions.

In the analysis of oxidized Rd, we could draw heavily on the knowledge obtained from EPR experiments. In the case of reduced Rd, on the other hand, no such information is available and all the parameters, D, E, $\tilde{g}$, $\tilde{A}$, and (V_{ii}) must be derived from Mössbauer data. This is a nontrivial problem and requires much additional work. The simulations obtained for the low-temperature data, Fig. 4, are encouraging, but the parameters used should be considered tentative. It is possible, nevertheless, to draw some general conclusions from the limited data available.

The fine-structure or zero-field splitting in ferrous Rd is large, and it must contain a sizable rhombic component. As in the case of ferric Rd, inclusion of a fourth-order term, Eq. (1b), produces adequate simulations for a range of values of E/D and μ. The hyperfine interaction quoted in Fig. 4 is comparable in size with the intrinsic A tensors‡ of the ferrous iron in the reduced 2 Fe–2 S* proteins adrenodoxin (-13, -18, -26) MHz (Fritz et al., 1971), putidaredoxin (-11, -16, -26) MHz (Münck et al., 1972), and spinach ferredoxin (-8, -13, -26) MHz (Dunham et al., 1971). The quadrupole splitting, finally, $|\Delta E_Q| = 3.25$ mm/second at 4.2°K, is 20% larger than for the ferrous iron in putidaredoxin, $|\Delta E_Q| = 2.7$ mm/second (Münck et al., 1972).

A number of molecular orbital calculations relating to Rd have appeared using different methods of approximation. The more recent SCF-$X\alpha$-SW scheme is capable of producing realistic, spin-polarized wave-

† The intrinsic hyperfine tensor is $\frac{3}{7}$ of the measured value on account of spin coupling (Münck et al., 1972). The negative sign of $\tilde{A}$ follows from Mössbauer measurements in applied fields (Dunham et al., 1971; Münck et al., 1972).

‡ Because of spin coupling, the intrinsic A tensor is $-\frac{3}{4}$ times the measured value.

functions, but it has been limited so far to the highly symmetric tetrahedral clusters FeS_4^{6-} (Vaughan et al., 1974) and FeS_4^{5-} (Norman and Jackels, 1975) and to $Fe(SH)_4^-$ and $Fe(SCH_3)_4^-$ in D_{2d} symmetry (Norman and Jackels, 1975). Earlier iterative extended Hückel calculations of $Fe(SH)_4^-$ (Loew et al., 1974a,b; Loew and Lo, 1974b) and $Fe(SH)_4^{2-}$ (Loew and Lo, 1974a) used the bond lengths and angles determined by X-ray diffraction for oxidized Rd (Watenpaugh et al., 1973). In this last series of papers, many observable parameters have been calculated, and a meaningful comparison with the experimental data is possible. Specifically, for the ferric complex (Loew et al., 1974a,b; Loew and Lo, 1974b), the calculated g' tensors reproduce the measured values quite well, and the A tensor is predicted to have an anisotropy of $\pm 11.5\%$, or roughly twice the observed value. According to the model the largest and smallest components of the A tensor are found along the direction of the largest g' values in the ground doublet and the highest doublet, respectively, and the same is true for the experimental data. The magnitude and orientation of the predicted quadrupole tensor does not correlate as well with the measured values.

The calculations for the ferrous complex (Loew and Lo, 1974a) were based on the geometry of oxidized Rd, a procedure of questionable validity. The electric quadrupole and magnetic hyperfine interaction were calculated, but the agreement with experiments is not very good. Shulman et al. (1975) recently pointed out that the average iron–sulfur bond length increases by 0.05 Å upon reduction, and based on their data for Rd from *Peptococcus aerogenes,* they questioned the sizable differences in bond length reported for Rd from *C. pasteurianum* (Watenpaugh et al., 1973). Whenever definitive coordinates become available, the model calculations can be made more meaningful, in particular, for reduced Rd. We hope it will then be possible to correlate the experimental data with the model and to understand the peculiarities of the electronic ground state in Rd.

III. THE MoFe PROTEIN OF NITROGENASE

A. Introduction

The properties of the nitrogenase system have been described in detail by Orme-Johnson and Davis in Chapter 2 of this volume. Therefore, we can restrict ourselves here to those aspects of nitrogenase that are pertinent to the Mössbauer investigations.

Two Mössbauer studies of nitrogenase components have been reported. Smith and Lang (1974) have examined the system from *Klebsiella pneumoniae*, while Münck *et al.* (1975) have studied the MoFe protein from *Azotobacter vinelandii*. Considering the complexities of the proteins, it is gratifying to note that the intricate Mössbauer spectra of the proteins from both species are identical in all details. Thus, it appears that the art of preparing these enzymes has progressed to the point that reliable and reproducible results can be obtained by spectroscopic methods.

The Mössbauer methodology employed in the studies of the proteins from *Klebsiella* and *Azotobacter* are essentially the same. In one aspect, however, the investigations differ appreciably. Münck *et al.* (1975) closely tied the Mössbauer studies to a parallel EPR investigation performed on identical samples. This procedure yielded more information than was obtainable by either technique alone and allowed the characterization of the EPR active centers of the MoFe protein in great detail.

Here, we essentially follow the work of Münck *et al.* (1975) and use their nomenclature. The work of Smith and Lang will be referred to in the discussion. The primary objective of the Mössbauer study by Münck *et al.* (1975) was to establish whether iron (or molybdenum) atoms are associated with the EPR active centers. First, we briefly discuss how an unambiguous assignment can be accomplished.

In Mössbauer spectroscopy of ^{57}Fe-containing proteins, one generally encounters two types of spectra—quadrupole doublets and magnetic spectra. For iron in a paramagnetic complex with half-integer electronic spin S, one can observe, at low temperatures, a magnetically split Mössbauer spectrum associated with a Kramers doublet (oxidized rubredoxin, reduced plant-type and bacterial-type ferredoxins, and oxidized HiPIP are examples). Depending on the magnetic anisotropy, we can distinguish two types of spectra. For systems that are magnetically isotropic or moderately anisotropic, the intensities of the absorption lines depend quite strongly on the direction of an applied field $H \gg$ 10 G relative to the Mössbauer radiation. On the other hand, for magnetically anisotropic systems with $g_1 \simeq g_2 \ll g_3$, the spectra are quite insensitive to the direction of the applied field. For such doublets, it is generally difficult to observe an EPR signal, since the EPR transition amplitudes, which are proportional to the matrix elements of the perpendicular spin components S_1 and S_2 are quite small. This situation frequently occurs for Kramers doublets that are members of a spin multiplet, i.e., for systems with $S \geq \frac{3}{2}$. The ground doublet of oxidized rubredoxin, for instance, exhibits such an anisotropy and hence a field-independent Mössbauer spectrum.

In contrast, for iron complexes with zero or integer spin (non-Kramers systems) $\langle S \rangle$ and therefore $\mathbf{H}_{int}$, Eqs. (5) and (3b), are zero unless a strong magnetic field is applied (cf. Section I). Consequently, the Mössbauer spectrum will have no magnetic features and only a quadrupole doublet can be observed. Non-Kramers systems are in general not amenable to EPR spectroscopy, and the Mössbauer effect becomes a particularly useful tool capable of probing the environments of all iron atoms (reduced rubredoxin, the oxidized plant- and bacterial-type ferredoxins, and reduced HiPIP are non-Kramers systems). For compounds with integral spin the Mössbauer effect can serve two functions. It can elicit information about iron that is EPR silent in all steps of a catalytic cycle, or it can tell us about the state of the iron that becomes EPR silent as a consequence of a redox reaction. All situations described in the last two paragraphs are encountered in the MoFe protein.

The EPR properties of the MoFe protein have been discussed at length in Sections II,E and IV,C of Chapter 2 of this volume. We can summarize the results as follows. The observed g values at 4.32, 3.65, and 2.01 result from an $S = \frac{3}{2}$ spin system described by the spin Hamiltonian

$$\mathcal{K}(S = \tfrac{3}{2}) = D[S_z^2 - \tfrac{5}{4} + (E/D)(S_x^2 - S_y^2)] + g_0\beta \mathbf{S} \cdot \mathbf{H}$$

The first term splits the quartet into two Kramers doublets, separated in energy by $\Delta = 2D[1 + 3(E/D)^2]^{1/2}$ (see Fig. 3 of Chapter 2 in this volume). The values $g_0 = 2.0$ and $E/D = 0.055$ yield $g_x = 3.66$, $g_y = 4.32$, and $g_z = 1.98$ in reasonable agreement with the observed g values for the MoFe protein from *A. vinelandii*. The EPR signal results from the ground doublet, which is depopulated as the temperature is raised. From the temperature dependence of the EPR signal, the zero-field splitting was determined to be $\Delta/k = 15°K$. Furthermore, quantitation of the EPR signal yielded 2.1 spins per molecule, i.e., two $S = \frac{3}{2}$ centers are present per MoFe protein. We would like to emphasize here that Mössbauer studies of the same samples played an important role in corroborating the interpretation of the EPR results given here. For one thing, it was clear from the Mössbauer measurements that the EPR signal resulted from the lowest doublet and that the second doublet had to be more than 8°K above the ground doublet, i.e., $\Delta/k \geq 8°K$. Second, since the EPR active species was distinguishable in the Mössbauer spectra from all the other iron atoms, it was possible to verify that it was present in maximum concentration under the conditions of the EPR experiments. How valuable the EPR results were for the Mössbauer study will become evident in the following presentation of the Mössbauer results.

B. Results

Mössbauer spectra of the MoFe protein from *A. vinelandii* are shown in Figs. 5 and 6. The spectra shown in Fig. 5 were taken at 1.5°K in

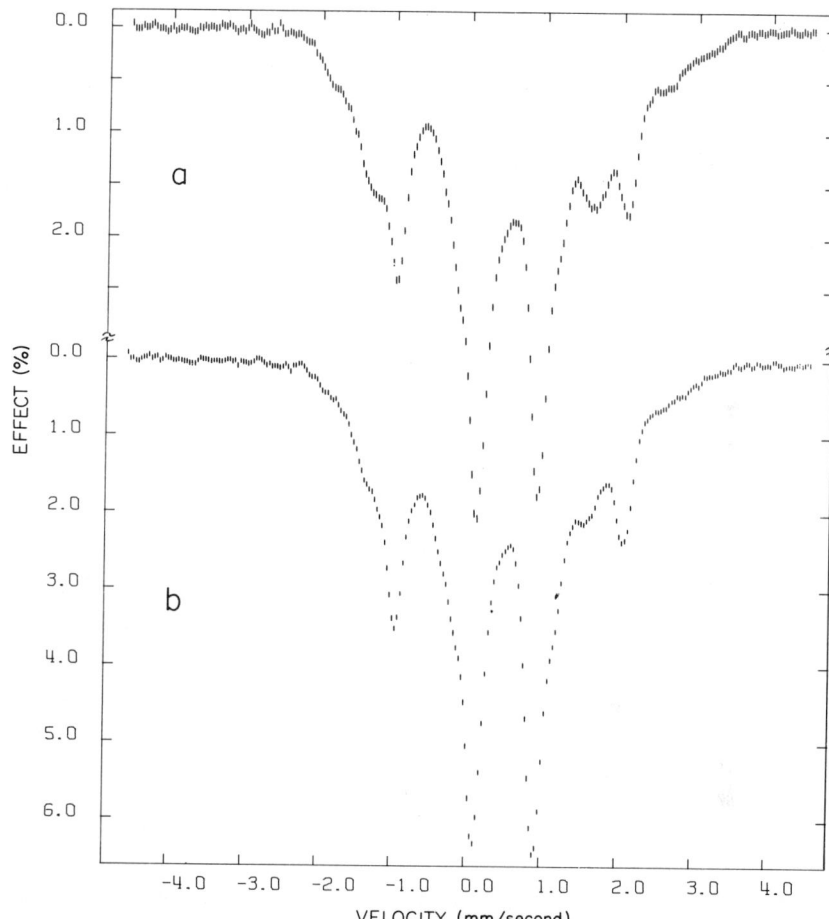

Fig. 5. Mössbauer spectra of the native MoFe protein from *A. vinelandii* taken at 1.5°K in a magnetic field of 360 G applied (a) parallel and (b) perpendicular to the Mössbauer radiation. The scales in Figs. 5 and 6 are plotted relative to a ^{57}Co(Rh) source, kept at room temperature. (From Münck *et al.*, 1975.)

transverse and parallel magnetic fields of 360 G. The spectrum in Fig. 6 was taken at 30°K. Münck *et al.* (1975) performed a spectral decomposition of the data and discussed the nature of these spectra in detail. In the following we briefly describe their results.

The solid line in Fig. 6 is the result of least-squares fitting four quadrupole doublets to the data. The area under each doublet reflects the number of iron atoms that contribute to a spectral component; determination of the area thus provides a quantitation of the number of iron

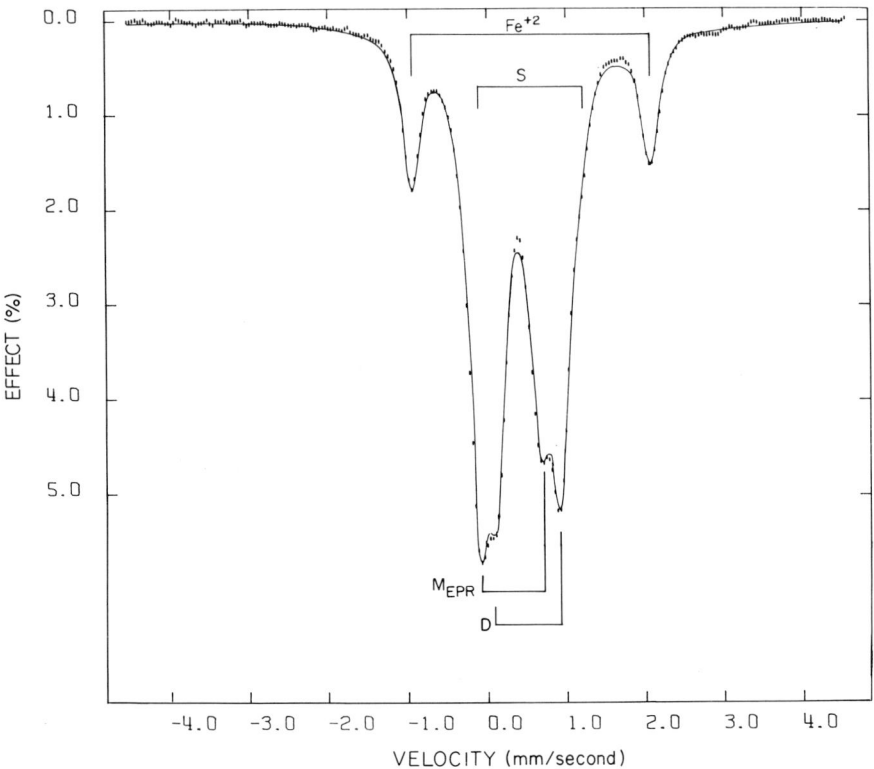

Fig. 6. Mössbauer spectrum of the native MoFe protein taken at 30°K. The solid line is the result of least squares fitting four quadrupole doublets to the data. (From Münck *et al.*, 1975.)

atoms in different environments. Three of these doublets are present also in the 1.5°K data. The fourth doublet (labeled M_{EPR}) changes, at lower temperatures, into a spectrum exhibiting paramagnetic hyperfine structure; this spectral component represents about seven iron atoms associated with the two $S = \frac{3}{2}$ centers.

One of the doublets, labeled Fe^{2+}, represents 14% (about three atoms) of the total amount of iron present. Its isomer shift $\delta = 0.69$ mm/second at 4.2°K (relative to Fe metal) and its quadrupole splitting $\Delta E_Q = 3.05$ mm/second imply high-spin ferrous iron. The isomer shift compares favorably with the distinctive isomer shifts associated with ferrous iron tetrahedrally coordinated to sulfur, e.g., reduced rubredoxin from *C. pasteurianum* ($\delta = 0.70$ mm/second at 4.2°K and $\Delta E_Q = 3.25$ mm/second, see Section II) and the Fe^{2+} component in reduced plant-type ferredoxins

($\delta = 0.58$ mm/second and $\Delta E_Q = 2.7$ mm/second at $4.2°K$ for putidaredoxin; see Münck et al., 1972).

A second doublet appears in the center of the spectra. It is best recognized in Fig. 5; this doublet, with $\delta = 0.64$ mm/second and $\Delta E_Q = 0.81$ mm/second is labeled D (see Fig. 6) (our ignorance about the nature of the iron centers presently does not allow a more informative labeling of the spectral components). Spectral component D represents 42% (about nine atoms) of the total iron present.

There is a third quadrupole doublet in the spectra displayed in Figs. 5 and 6. This component (labeled S) accounts for one iron atom with Mössbauer parameters $\Delta E_Q \simeq 1.4$ mm/second and $\delta \simeq 0.6$ mm/second. It could possibly be an impurity, although this species can be discerned also in the spectra from *K. pneumoniae*; however, Smith and Lang (1974) do not mention S explicitly.

As mentioned, spectral component M_{EPR} shows a magnetic pattern for temperatures below $8°K$. As the temperature is raised the magnetic hyperfine interaction averages out so that only the quadrupole interaction is observed. This temperature behavior parallels that of the observed EPR signal of the protein. At $1.5°K$, the spin–lattice relaxation rate is slow compared to the nuclear precession frequency ($\simeq 10$ MHz), and magnetic Mössbauer spectra have to be observed under these conditions. This is the case as can be seen from the spectra in Fig. 7 which resulted when spectral component D, Fe^{2+}, and S were subtracted from the spectra of Fig. 5. At about $16°K$, the electronic spin relaxation time becomes so short ($\simeq 10^{-10}$ sec) that the EPR lines are appreciably broadened. Such a short relaxation time leads to the collapse of the magnetic hyperfine splitting in the Mössbauer spectrum. Indeed, M_{EPR} is collapsed at $16°K$ into a quadrupole doublet. This doublet with Mössbauer parameters $\Delta E_Q = 0.76$ mm/second and $\delta = 0.40$ mm/second contains 38.5% of the total absorption. This amounts to about seven iron atoms and allows a further conclusion regarding the $S = \frac{3}{2}$ centers. Since about seven iron atoms contribute to M_{EPR} and the EPR spectrum quantitates to two spins per molecule, the iron atoms must be spin-coupled; otherwise the EPR signal should quantitate to about seven spins per molecule. Spin-coupling is further borne out when the magnitudes of the observed hyperfine fields are examined. While monoatomic Fe^{3+} typically exhibits splittings ranging from 10 to 17 mm/second, the patterns observed in Fig. 7 are typical of 4 Fe–4 S* clusters. The reduced magnetic splittings in these clusters result from a delocalization of the electronic spin over all four iron atoms. Thus, it would appear that the MoFe protein presents a new kind of 4 Fe–4 S* cluster, with spin-coupling resulting in a net spin of $S = \frac{3}{2}$.

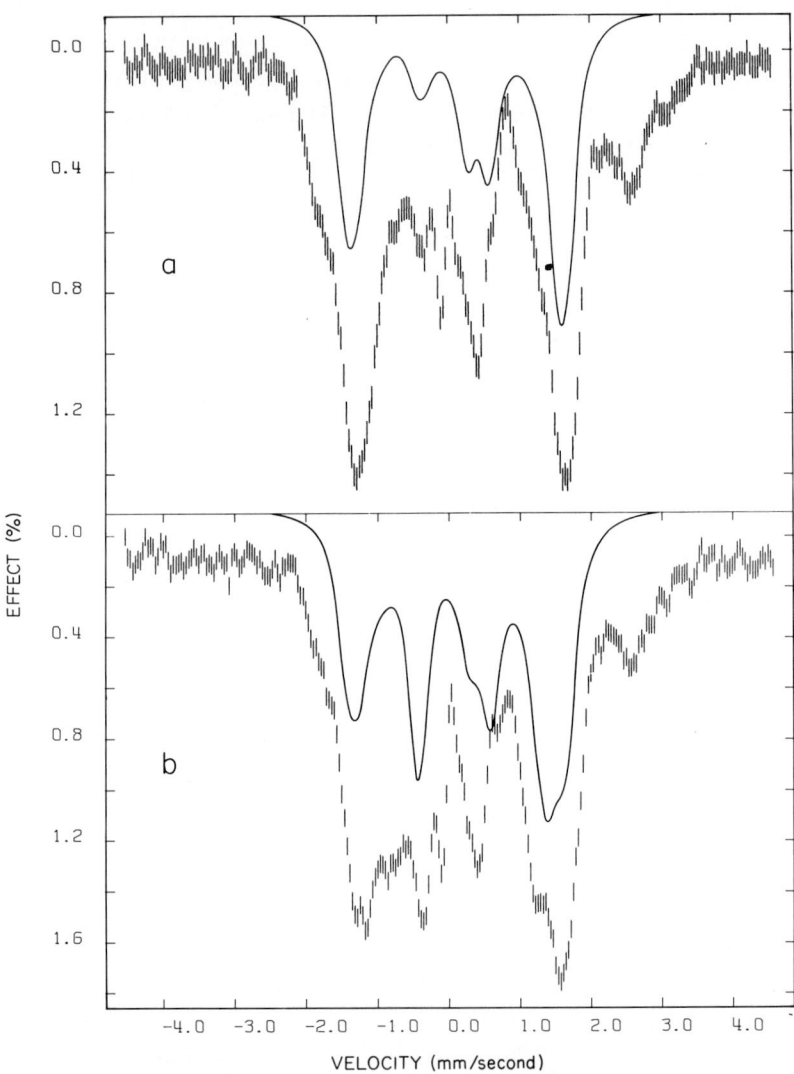

Fig. 7. Low temperature spectra of spectral component M_{EPR} in (a) parallel and (b) perpendicular applied field. Spectra were obtained from those shown in Fig. 5 by removing components D, Fe^{2+}, and S using the results from the least squares fit in Fig. 6. The displayed spectra contain at least two distinct components. For one of them, a computer simulation was attempted. The second magnetic spectrum shows distinct features around −2 mm/second and for velocities >2 mm/second. (From Münck et al., 1975.)

At 30°K, M_{EPR} exhibits a sharp quadrupole doublet. Thus, all iron atoms contained in M_{EPR} exhibit the same ΔE_Q and δ (within rather narrow margins). Quite often such information is used to claim that these iron atoms reside in identical environments. Such conclusions are not justified, since isomer shift and quadrupole splitting data alone (especially for small ΔE_Q) are too insensitive. A case in point is oxidized HiPIP. While all four iron atoms give rise to one quadrupole doublet at 77°K, at least two different environments can be distinguished in the magnetic hyperfine spectrum taken at 4.2°K. The same is true here for spectral component M_{EPR}. At least two subcomponents can be discerned in the spectra shown in Fig. 7. This further supports our conclusions regarding the nature of the $S = \frac{3}{2}$ centers. The presence of a half-integer electronic spin implies an odd number of electrons in each cluster. If we have four irons per cluster, they must be inequivalent.

One of the magnetic components of M_{EPR} is identified rather easily. It corresponds closely to the computed curve in Fig. 7. The other component has broader features but can be discerned at -2 mm/second and for velocities $v > 2$ mm/second. The computed curve in Fig. 7 was simulated for the ground doublet of an $S = \frac{3}{2}$ system, using the experimental g values of the MoFe protein, an isotropic magnetic hyperfine coupling constant of 12.1 MHz, and the experimentally determined quadrupole splitting. As can be seen from Fig. 7, the computed spectrum agrees quite well with the experimental data; specifically, the intensity changes between parallel and transverse magnetic field are reproduced quite well. Magnetically split Mössbauer spectra with absorption intensities sensitive to the direction of the applied magnetic field quite generally imply that an EPR signal can be observed under suitable conditions. Thus, there is no doubt that M_{EPR} indeed is associated with the observed EPR spectrum. The second magnetic spectrum is too diffuse to permit a meaningful computer simulation at this time.

One crucial experiment can further establish the assignment of the magnetic M_{EPR} spectrum to the $S = \frac{3}{2}$ centers. As discussed by Orme-Johnson and Davis in Chapter 2 of this volume, the EPR signal decreases by a factor ten in the fixing mixture, i.e., when MgATP, a proper reductant, and the Fe protein are added to the MoFe protein. If the disappearance of the EPR signal implies that the $S = \frac{3}{2}$ centers have each been transformed by a one-electron reduction process, then the two iron clusters will have integer, possibly zero, electronic spin, and the associated Mössbauer spectrum will exhibit quadrupole interactions only. This is indeed observed.

A Mössbauer spectrum of the MoFe protein in the fixing mixture, taken at 4.2°K, is shown in Fig. 8. The magnetic component M_{EPR} has

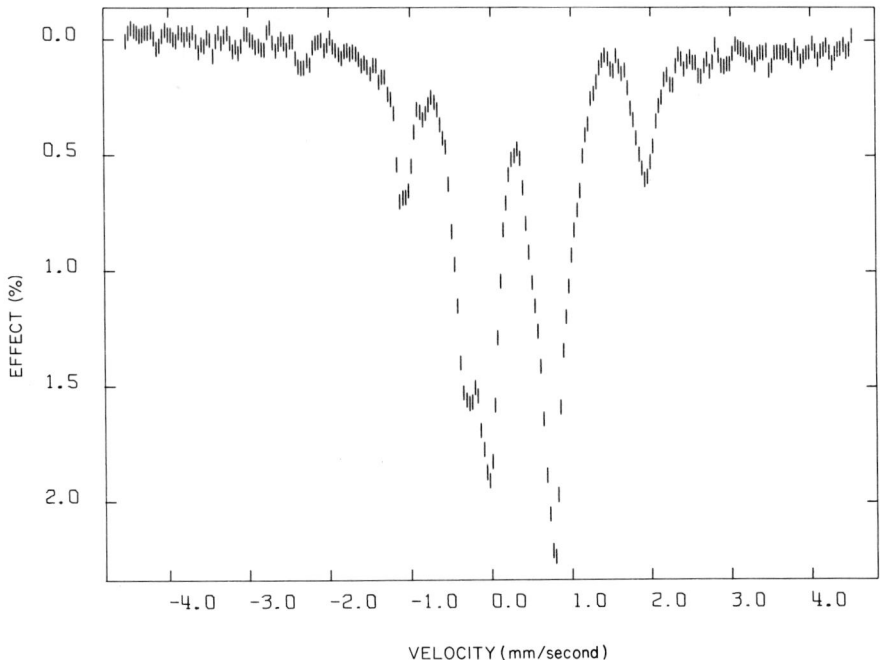

Fig. 8. Mössbauer spectrum of the ^{57}Fe enriched MoFe protein, MgATP, and dithionite (fixing mixture). The spectrum was taken at 4.2°K; the velocity scale is plotted relative to ^{57}Co(Rh), kept at 4.2°K. (From Münck et al., 1975.)

vanished, and it appears as a quadrupole doublet. Furthermore, the spectral components D, Fe^{2+}, and S present themselves in the same spectroscopic state as in the native protein. Therefore, these components can be removed by subtraction from the spectrum in Fig. 8 using the results of the least-squares fit applied earlier to the spectra of the native sample, Fig. 6. The remainder should be M_{EPR} as it appears in the fixing mixture. After subtraction, we are left with a well-defined absorption at -0.25 mm/second and a broad absorption band at 0.8 mm/second. This observation suggests that M_{EPR} appears in the fixing mixture as a superposition of quadrupole doublets with splittings ranging from 0.95 to 1.3 mm/second. The isomer shift has increased slightly, suggesting that each $S = \frac{3}{2}$ cluster has been *reduced* by one electron.

So far we have described the results of Münck et al. (1975). They are in essential agreement with the work by Smith and Lang (1974) on the *K. pneumoniae* protein. The latter authors associated M_{EPR} correctly with the observed EPR signals, but they did not interpret it as a spin $S = \frac{3}{2}$ center. Table III in Chapter 2 of this volume shows that both groups are in good agreement regarding the identification and quantitation of the various iron components.

Smith and Lang (1974) also studied a higher oxidation state of the MoFe protein. They showed that upon oxidation with Lauth's violet, all iron components change their spectroscopic state. In particular, Smith and Lang associate a quadrupole doublet with Mössbauer parameters $\Delta E_Q = 0.9$ mm/second and $\delta = 0.35$ mm/second with a higher oxidation state of M_{EPR}. Again, the spectra of the oxidized forms of the MoFe protein from *Klebsiella* and *Azotobacter* are almost identical. Thus, it appears that the $S = \frac{3}{2}$ clusters are capable of assuming at least three different oxidation states. We will discuss the Lauth's violet form of the protein in more detail in the following section.

C. Discussion

From EPR and Mössbauer studies a reasonably consistent picture has emerged regarding the $S = \frac{3}{2}$ centers. Certainly the Mössbauer effect has contributed more information than could be hoped for considering the complexities of the protein. Below we present some speculations about the nature of the $S = \frac{3}{2}$ center that should stimulate further research.

Smith and Lang (1974) have noted that the isomer shifts, the quadrupole splittings, and the magnetic interactions for the $S = \frac{3}{2}$ centers exhibit a certain resemblance to the high-potential Fe protein from *Chromatium*. We would like to pursue this suggestion a bit further. Surprisingly, the magnetic behavior of the $S = \frac{3}{2}$ centers is well described by an isotropic g value, although the Zeeman interaction of a spin $\frac{3}{2}$ system is expected to be anisotropic. The oxidized state of HiPIP is generally considered to consist of one ferrous and three ferric atoms. If we assume the same model for the $S = \frac{3}{2}$ centers, we can argue that spin coupling of three high-spin ferric irons to one high-spin ferrous iron should result in a fairly isotropic g tensor, since each of the constituents has a fairly isotropic $\tilde{g}$. Indeed, the Zeeman term is $g_0 \beta \mathbf{H} \cdot \mathbf{S}$ for high-spin ferric compounds with $g_0 = 2$, $S = \frac{5}{2}$. High-spin ferrous iron in tetrahedral environments has rather isotropic g values, also. Dunham *et al.* (1971) have estimated g values of 2.12, 2.07, and 2.00 for the high-spin ferrous iron component of spinach ferredoxin.

In addition, in such a model we might expect that those iron atoms in the cluster that derive from a high-spin ferric configuration will show rather isotropic magnetic hyperfine interactions. This could explain why the theoretical Mössbauer spectrum computed with an isotropic coupling constant in Fig. 7 matches one of the two magnetic subspectra so well.

The question then arises as to whether the two magnetic subspecies shown in Fig. 7 have an area ratio of 3:1. The data are not incompatible with such a ratio; our understanding of these spectra, however, is not adequate to draw firm conclusions. Interestingly, the second magnetic subspectrum of M_{EPR} has features characteristic of a rather anisotropic A tensor as one might expect from a ferrous iron atom.

The suggestion that the spin $S = \frac{3}{2}$ clusters in their EPR-active state correspond to oxidized HiPIP has another interesting consequence; oxidation by Lauth's violet of the $S = \frac{3}{2}$ cluster should produce a state analogous to "superoxidized" HiPIP. To our knowledge, there is no evidence for this oxidation state in any other protein or in a synthetic 4 Fe–4 S* cluster.

Spectral components D and Fe^{2+} are quite distinct from M_{EPR}. These components are observed both in the native protein and under fixing conditions. The data show clearly that D and Fe^{2+} are physically separated from the $S = \frac{3}{2}$ clusters. D and Fe^{2+} do not undergo spectroscopic changes when passing from one enzymatically relevant state to another. We do not believe, however, that they are inert ingredients of the protein but only that they remain reduced in the native enzyme and in the fixing mixture.

The Mössbauer parameters of component Fe^{2+} are quite close to those of reduced rubredoxin. A deeper probing into the nature of Fe^{2+}, however, gave some interesting results. Both Smith and Lang (1974) and Münck et al. (1975) noted that components D and Fe^{2+} behave diamagnetically $(S = 0)$ in the presence of a strong applied magnetic field. These findings show clearly that component Fe^{2+} behaves quite unlike rubredoxin when the magnetic properties are examined. Smith and Lang have suggested the possibility that species Fe^{2+} consists of a pair of high-spin ferrous iron atoms spin-coupled to zero net spin. Münck et al. (1975) noticed that spectral components D and Fe^{2+} occur in the ratio 3:1 and they suggested that they might reside in yet another type of 4 Fe–4 S* cluster. Spin-coupling then could result in diamagnetism. These suggestions are difficult to test in the native enzyme and in the fixing mixture, since both D and Fe^{2+} appear as simple quadrupole doublets. If these atoms could be reversibly oxidized to paramagnetic states, more definite information could be obtained. The Lauth's violet-oxidized form of the MoFe protein seems to offer this possibility.

We have already mentioned that M_{EPR} appears as a quadrupole doublet (i.e., as an integer or zero spin species) in the Lauth's violet-oxidized form of the MoFe protein. The remainder of the spectrum consists of a magnetic component that is a complex mixture of magnetic species (see Fig. 3 of Smith and Lang, 1974). It is very likely that this magnetic spectrum (labeled M1 by Smith and Lang) reflects component D in a higher oxidation state. Whether component Fe^{2+} is also contained in M1 can only be established by a detailed analysis. Moreover, it must be shown that the oxidation of D is reversible. Some indirect information in this regard is available. Smith et al. (1972) have reported that (1) the EPR signal of the native enzyme is fully recovered upon reduction and (2) that full catalytic activity is retained. These observations suggest that not only M_{EPR} but also D can be oxidized reversibly.

At the present time it is not clear whether those spectroscopic states observed in the Lauth's violet oxidized protein are enzymatically relevant. Nevertheless, a thorough study might give some valuable clues to the nature of spectroscopic components D and Fe^{2+}. It is worthwhile to note that component M1 observed by Smith and Lang (1974) represents species with unpaired electronic spins. The Mössbauer absorption at 4.2°K implies that the electronic ground state associated with M1 is a Kramers doublet with g values such that $g_1 \cong g_2 \ll g_3$. Unfortunately, the EPR transition probability in such a doublet is extremely small. Very anisotropic doublets, however, are generally associated with multiplets $S \geq \frac{3}{2}$. Therefore, it should be possible to detect an EPR signal associated with one of the Kramers doublets belonging to this multiplet. Although it has been reported by Smith et al. (1972) that the Lauth's violet state of the MoFe protein does not exhibit an EPR signal, this state should be reexamined in the light of the Mössbauer results.

IV. MODEL COMPOUNDS FOR RUBREDOXIN AND 4 Fe–4 S° CLUSTERS

The iron–sulfur complexes synthesized by Holm and his co-workers present an unusual opportunity to study the active sites of iron–sulfur proteins outside their protein environments and to correlate these findings with protein structure and function. Holm has discussed these model complexes in Chapter 7 of this volume, and we will limit our treatment to those aspects that are relevant to our Mössbauer discussions.

The complexes with the formulation $[Fe_4S_4(SR)_4]^{n-}$ serve as synthetic analogues to bacterial ferredoxins (Fd) and high-potential iron proteins (HiPIP) (Averill et al., 1973). Four oxidation states for these cluster compounds have been attained electrochemically (DePamphilis et al., 1974) and their correspondences to actual protein states are as follows:

$[Fe_4S_4(SR)_4]^{1-} \leftrightarrow [Fe_4S_4(SR)_4]^{2-} \leftrightarrow [Fe_4S_4(SR)_4]^{3-} \leftrightarrow [Fe_4S_4(SR)_4]^{4-}$
$Fd_{s\text{-}ox} \quad\leftrightarrow\quad Fd_{ox} \quad\leftrightarrow\quad Fd_{red}$
$HiPIP_{ox} \quad\leftrightarrow\quad HiPIP_{red} \quad\leftrightarrow\quad HiPIP_{s\text{-}red}$

Only two states, the dianion and the trianion, have been synthesized chemically so far, and only these two have been studied by Mössbauer spectroscopy (Frankel et al., 1975).

The dianions, corresponding to Fd_{ox} and $HiPIP_{red}$, exhibit Mössbauer spectra consisting of quadrupole doublets with splittings of $\Delta E_Q \cong 1.2$ mm/second centered near 0.34 mm/second (Table II). These parameters are identical for polycrystalline samples and frozen solutions. The isomer shifts are somewhat smaller than the protein values and probably reflect the difference in their respective environments.

Holm et al. (1974) have studied the dianion with R = benzyl in strong applied magnetic fields. The results confirm the diamagnetism of the compound. The principal component of the electric quadrupole tensor was found to be positive, and this result is in essential agreement with those found for reduced HiPIP and oxidized Fd (Dickson et al., 1974; Thompson et al., 1974). We would like to comment on one aspect. Holm et al. (1974) state that the results for the model complex "provide evidence that all iron sites are structurally and electronically equivalent." Certainly, the gross features of all four irons appear to be the same. However, the theoretical curve for the 80 kG spectrum in Fig. 1 of Holm et al. (1974) does not match the experimental data very well. We consider it likely that at least two different values for the asymmetry param-

TABLE II
MÖSSBAUER PARAMETERS OF MODEL COMPLEXES AND IRON–SULFUR PROTEINS

	T (°K)	ΔE_Q (mm/second)	δ (mm/second)	Reference[a]
$(Et_4N)_2[Fe_4S_4(SCH_2Ph)_4]$	77	1.26	0.34	1
	4.2	1.26	0.34	1
$(n-Bu_4N)_2[Fe_4S_4(SPh)_4]$	4.2	1.10	0.35	1
$(Me_4N)_2[Fe_4S_4(S-t-Bu)_4]$	4.2	1.10	0.35	1
$(Ph_4As)_2[Fe_4S_4(SEt)_4]$	4.2	1.17	0.34	1
Chromatium $HiPIP_{ox}$	77	0.80	0.31	2
Chromatium $HiPIP_{red}$	77	1.13	0.42	2
C. pasteurianum Fd_{ox}	77	0.91	0.43	3
	4.2	1.08	0.44	3
$[Fe_4S_4(SCH_2Ph)_4]^{3-}$	77	1.15	0.46	1
$[Fe_4S_4(SPh)_4]^{3-}$	77	1.08	0.45	1
C. pasteurianum Fd_{red}	77	1.25	0.57	3
	4.2	1.54	0.58	3
$(Et_4N)[Fe(S_2-o-xyl)_2]$	77	0.57	0.13	4
C. pasteurianum Rd_{ox}	4.2	0.5	0.25	5

[a] References: (1) Frankel et al. (1975); (2) Dickson et al. (1974); (3) Thompson et al. (1974); (4) Lane et al. (1975); (5) Section II.

eter η must be used to describe the four iron atoms of the cluster. In addition, it would be worthwhile to analyze the high-field data of reduced HiPIP (Dickson et al., 1974) and oxidized Fd (Thompson et al., 1974). It appears that the 60 kG spectra for both proteins are very similar, but lacking a detailed analysis, no statement can be made about the equivalence of the four iron atoms in the respective clusters.

The trianions are paramagnetic ($S = \frac{1}{2}$) like their protein analogues Fd_{red} and $HiPIP_{s-red}$. The model complexes exhibit a quadrupole doublet at 77°K with $\Delta E_Q = 1.10$ mm/second and $\delta = 0.45$ mm/second (Frankel et al., 1974), which, again, are somewhat smaller than the respective protein values. The observed increase in isomer shift indicates that the electron is added to an orbital with essentially metal character. To observe magnetic hyperfine interactions at low temperatures and weak magnetic fields, the complex has to be dissolved in a proper solvent to provide magnetic dilution. Figure 9 shows a spectrum of $[Fe_4S_4(SCH_2Ph)_4]^{3-}$ in DMF/THF; the data were taken at 4.2°K in a parallel field of 1000 G. The observed magnetic splittings are quite similar to those found in Fd_{red} (Thompson et al., 1974) and $HiPIP_{s-red}$ (Dickson and Cammack, 1974).

It is interesting to note that the synthetic analogues have more negative redox potentials than the proteins (DePamphilis et al., 1974) and that wrapping a polypeptide containing four cysteines around the 4 Fe-4 S* core moves the redox potential closer to those of the proteins (Que

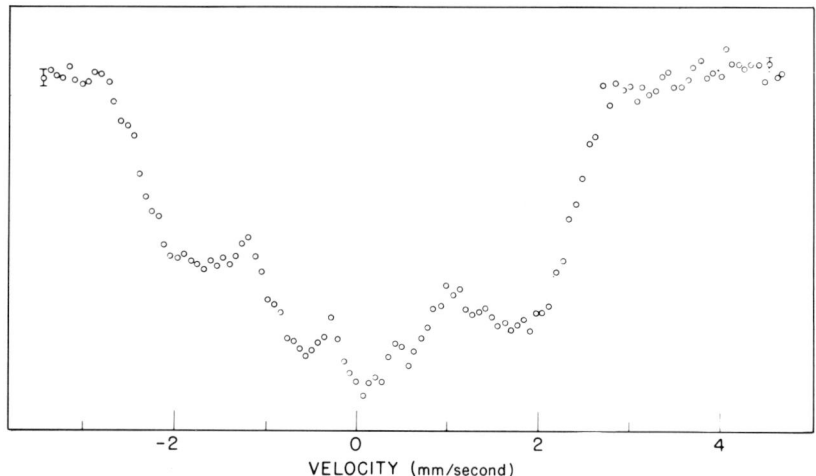

Fig. 9. Mössbauer spectrum of $[Fe_4S_4(SCH_2Ph_4)]^{3-}$ in DMF/THF at 4.2°K with a 1 kG parallel field. Absorption due to residual dianion is observed at 0.7 and −0.4 mm/second. (From Frankel et al., 1974.)

et al., 1974). No Mössbauer data were reported for the latter, but it would certainly be very interesting to look for correlations between the redox potential and the Mössbauer parameters, in particular the isomer shift.

The mononuclear complex $(Et_4N)[Fe(S_2\text{-}o\text{-}xyl)_2]$ mimics the iron coordination in rubredoxin (Rd). The iron environment in the protein however, consists of four iron–mercaptide bonds, one of which is about 0.2 Å shorter than the others (Watenpaugh *et al.*, 1973). The complex, in an essentially unconstrained environment, has four Fe–S bonds of roughly equal length and approaches T_d Fe–S_4 microsymmetry (Lane *et al.*, 1975). The compound was investigated with Mössbauer spectroscopy on crystalline material and on concentrated frozen solutions in DMF and DMF/CH_2Cl_2 (Lane *et al.*, 1975). The quadrupole splitting $\Delta E_Q = 0.57$ mm/second is comparable with that found in rubredoxin from *C. pasteurianum* (see Section II) and the isomer shift $\delta = 0.13$ mm/second is somewhat smaller than that of Rd. As in oxidized rubredoxin, the iron in this complex has a ^6A ground state, which is split by spin–orbit coupling into three Kramers doublets. Magnetic Mössbauer spectra for the three doublets are, in principle, observable at low temperatures. The magnetically concentrated materials did not allow the observation of magnetic spectra at 4.2°K in small applied magnetic fields, since spin–spin relaxation effects average out the magnetic hyperfine interaction. When a strong external field is applied, however, the levels are sufficiently split to yield a sizable population difference at 4.2°K. This allows for a finite value of a thermally averaged spin $\langle S \rangle_{th}$ (in the fast relaxation limit, the spin expectation values of the individual spin levels of the sextet are thermally averaged); i.e., the term $A_0 \langle S \rangle_{th} \cdot \mathbf{I}$ is observable and a proper analysis yields A_0. The complex was investigated in an applied field of 80 kG at 4.2°K, and an internal magnetic field of -380 kG was obtained, in good agreement with rubredoxin.

In conclusion, the iron–sulfur complexes discussed above serve as excellent synthetic analogues for the active sites of iron–sulfur proteins, and the investigation of their properties has augmented our understanding of these proteins. The Mössbauer studies carried out so far underscore the utility of this spectroscopic method and should stimulate further work on the nature of these complexes. To name a few of the questions that deserve further scrutiny, we mention the inequivalence of the iron atoms in the 4 Fe–4 S* clusters and computer simulation of the spectra observed for the trianionic cluster. Another topic of considerable interest is the electronic structure of the iron in the tetramercaptide complex as compared with that in rubredoxin. The deviations from tetrahedral symmetry are much larger in the latter and give rise to a highly anisotropic ground state of the iron, as shown in Section II. The same type of information can be obtained for the tetramercaptide complex if magnetically dilute samples are studied. Thus, further Mössbauer investigations should contribute greatly to our present rudimentary understanding of these iron–sulfur complexes.

NOTE ADDED IN PROOF. The iron centers of the MoFe protein have been characterized in more detail in cooperation with Dr. W. H. Orme-Johnson. Although data evaluations are still in progress, some important facts have emerged. The Wisconsin group has performed titration studies of the MoFe protein with Lauth's violet. Removal of about six electrons yields a Mössbauer spectrum like that shown in Fig. 3 of Smith and Lang (1974). More importantly, two electrons can be removed before the EPR signal of the $S = \frac{3}{2}$ centers is affected. After removal of two electrons about 70% of spectroscopic components D and Fe^{2+} have vanished, and the magnetic Mössbauer spectrum M1 of Smith and Lang (1974) has appeared. Thus the removal of two electrons transforms D and Fe^{2+} (about eight iron atoms) into M1. Hence D and Fe^{2+} reside in spin-coupled clusters. Our preliminary analysis supports the earlier suggestion that the clusters are of the 4 Fe–4 S* class consisting of three iron atoms of D-type and one atom of Fe^{2+}-type. Moreover, these data show unambiguously that M_{EPR} appears as a quadrupole doublet ($\Delta E_Q \approx 0.9$ mm/second, $\delta = 0.35$ mm/second) in the Lauth's violet oxidized form. This establishes that the $S = \frac{3}{2}$ centers can exist in three distinct oxidation states.

Kostikas et al. (1976) recently reported Mössbauer studies of structural analogs of reduced rubredoxin. The low-temperature, high-field spectra of tetrakisthiophenolatoiron(II) are very similar to those of reduced rubredoxin, in fact, the parameters used for the simulations in Fig. 4 reproduce the spectra of the model complex adequately. The structure of the analog has been determined by Dr. D. Coucouvanis and co-workers.

REFERENCES

Averill, B. A., Herskovitz, T., Holm, R. H., and Ibers, J. A. (1973). *J. Am. Chem. Soc.* **95**, 3523.

Cohen, R. L. (1976). "Applications of Mössbauer Spectroscopy," Vol. 1. Academic Press, New York.

Collins, R. L., and Travis, J. C. (1967). *In* "Mössbauer Effect Methodology" (I. J. Gruverman, ed.), Vol. 3, p. 123. Plenum, New York.

DePamphilis, B. V., Averill, B. A., Herskovitz, T., Que, L., Jr., and Holm, R. H. (1974). *J. Am. Chem. Soc.* **96**, 4159.

Dickson, D. P. E., and Cammack, R. (1974). *Biochem. J.* **143**, 763.

Dickson, D. P. E., Johnson, C. E., Cammack, R., Evans, M. C. W., Hall, D. O., and Rao, K. K. (1974). *Biochem. J.* **139**, 105.

Dunham, W. R., Bearden, A. J., Salmeen, I. T., Palmer, G., Sands, R. H., Orme-Johnson, W. H., and Beinert, H. (1971). *Biochim. Biophys. Acta* **253**, 134.

Frankel, R. B., Herskovitz, T., Averill, B. A., Holm, R. H., Krusic, P. J., and Phillips, W. D. (1974). *Biochem. Biophys. Res. Commun.* **58**, 974.

Frankel, R. B., Averill, B. A., and Holm, R. H. (1975) *J. Phys. (Paris)* **35**, Colloq. C6, 107.

Fritz, J., Anderson, R., Fee, J., Palmer, G., Sands, R. H., Orme-Johnson, W. H., Beinert, H. Tsibris, J. C. M., and Gunsalus, I. C. (1971). *Biochim. Biophys. Acta* **253**, 110.

Goldanskii, V. I., and Herber, R. H. eds. (1968). "Chemical Applications of Mössbauer Spectroscopy." Academic Press, New York.

Holm, R. H., Averill, B. A., Herskovitz, T., Frankel, R. B., Gray, H. B., Siiman, O., and Grunthamer, F. J. (1974). *J. Am. Chem. Soc.* **96**, 2644.

Ingalls, R. (1969). *Phys. Rev.* **188**, 1045.

Kostikas, A., Petrouleas, V., Simopoulos, A., Coucouvanis, D., and Holah, D. G. (1976). *Chem. Phys. Lett.* **38**, 582.

Kündig, W. (1967). *Nucl. Inst. Methods* **48**, 219.

Lane, R. W., Ibers, J. A., Frankel, R. B., and Holm, R. H. (1975). *Proc. Natl. Acad. Sci. U.S.A.* **72**, 2868.

Lang, G., Aasa, R., Garbett, K., and Williams, R. J. P. (1971). *J. Chem. Phys.* **55**, 4539.

Loew, G. H., and Lo, D. Y. (1974a). *Theor. Chim. Acta* **32**, 217.

Loew, G. H., and Lo, D. (1974b). *Theor. Chim. Acta* **33**, 137.

Loew, G. H., Chadwick, M., and Steinberg, D. A. (1974a). *Theor. Chim. Acta* **33**, 125.

Loew, G. H., Chadwick, M., and Lo, D. (1974b). *Theor. Chim. Acta* **33**, 147.

Lovenberg, W., and Williams, W. M. (1969). *Biochemistry* **8**, 141.

Münck, E., Debrunner, P. G., Tsibris, J. C. M., and Gunsalus, I. C. (1972). *Biochemistry* **11**, 855.

Münck, E., Rhodes, H., Orme-Johnson, W. H., Davis, L. C., Brill, W. J., and Shah, V. K. (1975). *Biochim. Biophys. Acta* **400**, 32.

Norman, J. G., Jr., and Jackels, S. C. (1975). *J. Am. Chem. Soc.* **97**, 3833.

Oosterhuis, W. T. (1974). *Struct. Bonding (Berlin)* **20**, 59.

Peisach, J., Blumberg, W. E., Lode, E. T., and Coon, M. J. (1971). *J. Biol. Chem.* **246**, 5877.

Phillips, W. D., Poe, M., Weiher, J. F., McDonald, C. C., and Lovenberg, W. (1970). *Nature (London)* **227**, 574.

Que, L., Jr., Anglin, J. R., Bobrik, M. A., Davison, A., and Holm, R. H. (1974). *J. Am. Chem. Soc.* **96**, 6042.

Rao, K. K., Evans, M. C. W., Cammack, R., Hall, D. O., Thompson, C. L., Jackson, P. J., and Johnson, C. E. (1972). *Biochem. J.* **129**, 1063.

Reiff, W. M. (1973). *In* "Mössbauer Effect Methodology" (I. J. Gruverman and C. W. Seidel, eds.), Vol. 8, p. 89. Plenum, New York.

Schneider, Y., Dischler, B., and Räuber, A. (1968). *J. Phys. Chem. Solids* **29**, 451.

Shulman, R. G., Eisenberger, P., Blumberg, W. E., and Stombaugh, N. A. (1975). *Proc. Natl. Acad. Sci. U.S.A.* **72**, 4003.

Smith, B. E., and Lang, G. (1974). *Biochem. J.* **137**, 169.

Smith, B. E., Lowe, D. J., and Bray, R. C. (1972). *Biochem. J.* **130**, 641.

Thompson, C. L., Johnson, C. E., Dickson, D. P. E., Cammack, R., Hall, D. O., Weser, U., and Rao, K. K. (1974). *Biochem. J.* **139**, 97.
Vaughan, D. J., Tossell, J. A., and Johnson, K. H. (1974). *Geochim. Cosmochim. Acta* **38**, 993.
Watenpaugh, K. D., Sieker, L. C., Herriot, J. R., and Jensen, L. H. (1973). *Acta Crystallogr., Sect. B* **29**, 943.
Wickman, H. H., Klein, M. P., and Shirley, D. (1966). *Phys. Rev.* **152**, 345.

Author Index

Numbers in italics refer to the pages on which the complete references are listed.

A

Aasa, R., 31, *58*, 389, *416*
Abel, E. A., 214, *272*
Abragam, A., 290, *327*
Ackers, G. K., 18, *60*
Ackrell, B. A. C., 65, 71, 74, 75, 76, 78, 79, 85, 93, *97, 98, 100*, 351, *375*
Adams, D. M., 248, *272*
Adman, E. A., 208, 211, 218, 223, 260, *273*
Adman, E. T., 75, *97*, 109, *118*, 158, 159, 161, 162, 163, 164, 165, 166, 167, 168, 170, 174, 175, 176, 177, 180, 181, 185, 187, 188, *202, 203*, 208, 218, 233, *273, 280*, 286, *327, 328*, 357, 369, 372, *375*
Adams, J. M., 126, *151*
Adler, E., 148, *151*
Ahrland, S., 131, *151*
Albracht, S. P. J., 67, 68, 70, 85, 87, 91, 95, *97*
Albrecht, S. L., 29, *59*, 374, *376*
Alden, R. A., 75, 89, *97, 98*, 158, 159, 160, 161, 162, 163, 164, 166, 167, 168, 170, 171, 172, 173, 174, 176, 177, 178, 179, 182, 184, 185, 186, 187, 188, 191, 194, 195, 198, *202, 203*, 206, 208, 209, 218, 222, 223, 228, 233, 255, 269, *273, 275, 279*, 286, 307, *328*, 356, 357, 364, 368, *375*
Ali, A., 213, *273*
Alkins, J. R., 248, *277, 352, 377*
Allen, E., 110, *119*
Alvarez, A., 147, *154*
Anbar, M., 361, *376*
Anderson, R. E., 192, *202*, 235, 255, *273*, 308, 324, *327*, 399, *416*
Andreesen, J. R., 132, 139, 140, 141, 146, 149, *151, 153*
Andreoli, A. J., 129, *155*
Andrew, I. G., 137, *151*
Anger, G., 192, *202*, 255, *273*, 308, 324, *327*

Anglin, J. R., 213, 219, 230, 234, 238, 242, 245, 260, 261, 271, *273, 279*, 413, *416*
Antanaitis, B. C., 178, 192, 198, *202, 203*, 248, 249, 255, *273, 280*, 322, *327*
Antonini, E., 206, *273*
Aparicio, P. J., 93, *98*, 208, *277*, 374, *377*
Appleby, C. A., 41, *58*
Argos, P., *278*
Armstrong, J. J., 106, *118*
Arnon, D. I., 107, 112, *119*, 208, *277*, 308, *327*, 374, 377
Aronson, A. I., 117, *118*
Aronson, J. N., 126, *151*
Asakura, T., 63, 66, 71, 75, 81, 82, *99*
Asher, L. E., 346, 349, *375*
Atherton, N. M., 299, *327*
Ausubel, F., 2, 4, 5, 8, 9, 11, 12, *14*
Averill, B. A., 27, 28, *58*, 160, 164, 166, 168, 171, 182, 192, 196, *202, 203*, 218, 219, 223, 225, 226, 228, 229, 230, 231, 232, 234, 236, 237, 239, 240, 242, 245, 246, 249, 252, 253, 256, 257, 268, *273, 274, 275, 276*, 286, 287, 308, 315, 321, *327, 328, 329*, 411, 412, 413, *415, 416*
Azoulay, E., 147, *151*

B

Bachofen, R., 308, *327*
Bäckström, D., 82, 87, *97*
Baginsky, M. L., 78, *97*
Bailey, C. B., 126, *151*
Baker, B. R., 343, *375*
Balch, A. L., 215, 218, 256, *273, 274*
Bale, J. R., 27, *58*, 262, *273*
Ball, E. G., 124, *155*
Ballou, D. P., 96, *97*
Barber, M. J., 302, *328*
Barker, H. A., 127, 135, *151*
Baron, J., 209, 229, *275*
Bartels, K., 194, *203*

Bartholomaus, R. C., 108, *119*
Bartsch, R. G., 75, 93, *97*, 158, 159, 173, 175, 176, 177, 178, 179, 187, 192, *202*, *203*, 223, 242, 251, 255, *273*, *279*, 286, 302, 310, 324, *328*, *329*, 356, 357, 364, 369, *375*, *377*
Basolo, F., 343, *375*
Bauchop, T., 137, *151*
Baugh, R. F., 66, 82, 95, *97*
Bearden, A. J., 73, *93*, *98*, *99*, 209, 232, 235, 250, 251, 255, 269, *273*, *274*, *280*, 301, 302, 304, 310, *328*, *329*, 374, *378*, 399, 409, *416*
Beattie, J. K., 350, *375*
Beck, G., 111, *118*
Beck, J.-P., 111, *118*
Beck, W., 149, *151*
Beckwith, J. R. 105, *118*
Beinert, H., 20, 27, *58*, *59*, 63, 65, 66, 67, 69, 70, 71, 72, 73, 74, 75, 76, 78, 80, 82, 83, 84, 85, 89, 90, 91, 92, 93, 94, 95, 96, *97*, *98*, *99*, *100*, 108, *119*, 218, 250, 251, *274*, *281*, 301, 302, 304, 308, *328*, *329*, 351, *375*, 399, 409, *416*
Bekes P., 127, *151*
Bell, C. M., 212, *273*
Bender, R. A., 3, 5, 7, 13, *14*
Benedict, C. R., 150, *151*
Benfield, G. L., 294, 302, *329*
Bengis, C., 374, *375*
Bennett, L. E., 265, 267, *273*, *276*, 332, 335, 336, 337, 338, 340, 343, 344, 345, 346, 347, 349, 350, 351, 352, 353, 354, 355, 356, 358, 364, 369, 373, 374, *375*, *376*, *377*
Bennett, J. P., Jr., 345, 346, 349, *377*
Benson, A., 106, 114, 116, *119*
Benson, S. W., 338, *375*
Berden, J. A., 90, *100*
Bergensen, F. J., 35, 36, 44, *58*
Bergman, I., 361, *375*
Bernal, I., 215, 216, 256, *273*, *274*, *275*
Bernhardt, F. H., 272, *273*
Besch, P. K., 112, *119*
Birktoft, J. J., 194, *203*
Blaschkewski, H. P., 123, 127, 141, 142, *153*
Blaylock, B. A., 127, *151*
Bleaney, B., 290, *327*
Blomstrom, D. C., 162, *202*, 310, *327*

Blumberg, W. E., 225, 249, 250, *273*, *278*, *280*, 299, *328*, 352, 356, *378*, 392, 393, 398, 400, *416*
Bobrik, M. A., 160, 168, 171, 197, *202*, *203*, 213, 223, 226, 230, 234, 238, 242, 245, 257, 258, 260, 261, 271, *273*, *279*, 413, *416*
Bode, E. T., 356, *378*
Bogner, L., 250, *274*, 321, *328*
Boime, I., 111, *118*
Boiwe, T., 206, *274*
Bolle, A., 110, *119*
Bollinger, R. E., 109, *119*
Bolton, J. R., 93, *98*
Boon, J. W., 217, *273*
Boras, R. C., *59*
Boulter, D., 148, *154*
Bovarnick, M., 147, *151*
Bradshaw, W. H., 135, 146, *151*
Brady, F. D., 129, *152*
Bramlett, R. N., 374, *375*
Brandén, I., 206, *274*
Brantner, R. V., 75, 82, 93, *97*
Branzoli, U., 150, *151*
Bray, R. C., 20, 28, 29, 50, 54, *58*, *60*, 96, *100*, 302, *328*, *329*, 347, 351, *375*, 411, *416*
Breitenbach, M., 249, 251, *275*, 308, 310, 324, *328*
Brenchley, J. E., 3, 4, 5, 7, *13*, *14*
Brill, W. J., 2, *14*, 16, 29, 30, 31, 32, 35, 43, 44, 54, 57, *58*, *59*, *60*, 121, 135, 136, 146, *151*, 255, *278*, 327, *329*, 382, 401, 403, 404, 407, 408, *416*
Brintzinger, H., 209, 241, 242, *273*, *278*, 287, 307, *328*, *329*, 332, *377*
Brodie A. M., 256, *273*
Brodrick, J. W., 113, 114, 117, *118*
Brostigen, G., 217, *273*
Brown, A., 139, 140, 146, *151*
Brown, G. M., 126, 128, *151*, *156*
Brown, J. R., 106, *118*
Brown, L. P., 112, *119*
Bruice, T. C., 107, *118*, 160, 175, 178, *202*, *203*, 258, 260, 264, 270, *273*, *276*, 347, 349, *375*
Brunori, M., 206, *273*
Bryant, M. P., 123, 136, 146, *151*, *154*, 155

AUTHOR INDEX 421

Buchanan, B. B., 107, 114, *118, 119*
Bui, P. T., 34, *58*
Bulen, W. A., 35, 43, 46, 47, *58, 59, 60*
Burg, A. W., 126, 128, *151*
Burge, W. D., 147, *153*
Burns, R. C., 16, 18, 21, 43, 46, 47, *58, 59*
Burris, R. H., 16, 17, 25, 27, 31, 34, 35, 36, 37, 38, 39, 43, 44, 45, 47, 49, 50, 51, 53, 55, 56, 57, *58, 59, 60,* 63, *98,* 209, 229, 230, 242, 249, 262, *273, 274, 280,* 306, *330,* 373, 374, *376, 378*
Burwell, R. L., Jr., 366, *375*
Butt, V. S., 125, 148, *153*

C

Caldwell, R. A., 268, *281*
Calvert, J., 358, *375*
Cambier, H. W., 19, *59*
Cammack, R., 93, *98,* 106, 107, 113, *118,* 163, 164, *202,* 207, 209, 211, 230, 233, 241, 242, 249, 250, 251, 252, 255, 269, 270, 272, *273, 274, 275, 276, 278, 279, 280,* 285, 287, 288, 292, 295, 299, 300, 301, 302, 303, 307, 308, 309, 310 311, 312, 314, 319, 321, 324, *327, 328, 329,* 357, *376,* 382, 389, 392, 393, 395, 397, 412, 413, *415, 416, 417*
Campion, R. J., 343, 364, 366, *375*
Cannella, C., 110, *118*
Cannon, F. C., 2, *14*
Carter, C. W., Jr., 75, 89, *97, 98,* 158, 159, 160, 161, 162, 163, 164, 166, 167, 168, 170, 171, 172, 173, 175, 177, 178, 179, 182, 184, 185, 186, 187, 188, 191, 195, 198, 202, *202, 203,* 208, 218, 222, 223, 228, 233, 255, 269, *273, 275,* 286, 307, *328,* 356, 357, 358, 364, 368, *375*
Cass, R. D., 343, *378*
Cauquis, G., 231, *274*
Caughey, W. S., 363, *375*
Cavallini, D., 110, *118*
Cerdonio, M., 234, 236, *274*
Chadwick, M., *277,* 382, 400, *416*
Chambers, D. A., 110, *119*
Chan, I., 2, 9, *14*
Chance, B., 63, 71, 75, 81, 82, 90, *97, 98, 99,* 358, *375*
Chandler, J. L. R., 129, *155*

Chandra, S., 216, *273*
Chang, H. C., 150, *153*
Chappel, J. B., 90, *98*
Charlton, S., 63, *99,* 249, *278,* 308, 324, *329,* 369, *377*
Chase, T., 123, 127, *151*
Chatt, J., 131, *151*
Cheeseman, P., 136, *151*
Cheldelin, V. H., 125, *155*
Chen, C. H., 18, 45, 55, *59*
Chen, J. S., 262, 263, 272, *275, 278*
Chowdhury, A., *152*
Christensen, J., 111, 112, *118*
Chu, C. T.-W., 216, 217, 253, 255, *275*
Churchill, M. R., 215, 221, 226, *274,* 352, *375*
Clare, M., 249, *274*
Clegg, R. A., 92, *98*
Cleland, W. W., 42, 46, *58, 59*
Cobley, J. G., 92, *97, 98*
Coffman, R. E., 236, *280*
Cohen, R. L., 382, 388, *415*
Cole, J. A., 123, 147, *151*
Cole, R. D., 109, *119*
Coleman, J. M., 216, *274*
Coles, C. J., 78, *97*
Coll, R. J., 212, *277*
Collins, R. L., 395, *415*
Cone, J. E., 126, 128, *151, 154*
Connelly, N. G., 215, *274*
Coon, M. J., 229, 250, *277, 278,* 356, *378,* 392, 393, 398, *416*
Cooper, A., 206, *274*
Cooper, T. G., 150, *151*
Coppens, P., 217, *277*
Cotton, S. A., 214, 249, *274*
Coucouvanis, D., 214, 215, 216, 220, 231, 242, 256, *274, 276,* 415, *416*
Cox, E. G., 215, *280*
Crane, F. L., 82, 96, *97, 98*
Cremona, T., 66, 94, *97, 99*
Crespi, H. L., 235, 255, *273*
Cretney, W. C., 28, 31, *59*
Creutz, C., 343, 351, *375, 377*
Crichton, R. R., 206, *274*
Crosse, B. C., 214, *272*
Cusanovich, M. A., 106, 107, *118, 278,* 302, 310, *329,* 343, 364, 365, 366, 370, 371, *377*

D

Dahl, L. F., 215, 216, 217, 253, 255, *274, 275, 280, 281*
Dalton, H., 16, 23, *58,* 268, *274*
Darnall, D. W., 102, *118*
Davenport, H. E., 301, *328*
Davies, N. R., 131, *151*
Davis, B. R., 216, *273*
Davis, K. A., 66, 78, 95, *97, 98*
Davis, L. C., 27, 28, 29, 30, 31, 32, 34, 35, 36, 40, 41, 42, 43, 45, 47, 48, 51, 53, 54, 57, *58, 59, 60,* 255, *278,* 327, *329,* 382, 401, 403, 404, 406, 407, 408, *416*
Davison, A., 213, 219, 230, 234, 238, 242, 245, 247, 260, 261, 271, *273, 274, 279,* 413, *416*
Davison, D. C., 148, *151*
Dawes, E. A., 137, *151, 155*
Dearman, H. H., 212, 219, *280*
Debrunner, P., 229, 230, 251, *275,* 278, 301, 302, *329,* 373, *378,* 384, 387, 399, 405, *416*
Deck, C. F., 343, 364, 366, *375*
Decker, K., 128, 130, 137, 138, *151, 152, 155*
De Klerk, H., 75, 93, *97,* 175, 178, 179, *202,* 242, *274*
DeKok, A., 41, *59*
DeKok, J., 76, 96, *97*
De Leo, A. B., 3, 4, 5, 7, 12, 13, *14*
De Luca, H. F., 93, *99*
Delwiche, C. C., 147, *153*
de Médicis, R., 217, *274*
De Moss, J. A., 147, *153*
Denmark, S. E., 160, *203,* 219, 222, 225, 226, 227, 229, 230, 232, 233, 242, 257, 258, *278*
De Pamphilis, B. V., 160, 182, 196, *202, 203,* 219, 222, 225, 226, 227, 229, 230, 231, 232, 233, 234, 242, 253, 257, 258, 268, *274, 276, 278,* 411, 413, *415*
DerVartaniam, D. V., 73, 75, 82, 92, *97, 100*
Dethlefsen, U., 213, *279*
D'Eustachio, A. J. D., 140, *153*
Deutsch, E., 346, 348, 349, 350, *375, 378, 379*
De Vault, D., 358, *375*
Dewar, M. J. S., 361, *375, 376*

Dickerson, R. E., 206, *274, 280,* 334, 358, 361, 369, 373, *376*
Dickson, D. P. E., 209, 230, 232, 250, 251, 252, 255, *274, 278, 280,* 287, 306, 307, 308, 310, 311, 312, 314, 319, 321, 324, *328, 329, 330,* 357, *376,* 412, 413, *415, 416, 417*
Dietz, R. E., 247, *278*
Dilworth, M. J., 17, *58*
Dischler, B., 385, 399, *416*
Dixon, R. A., 2, *14*
Doemeny, P. A., 219, *279*
Dogonadze, R. R., 363, *376*
Donohue, J., 185, *202*
Dooijewaard, G., 67, 68, 70, *97*
Douglas, J. E., 256, *273*
Douglas, M. W., 123, 147, *151*
Downie, J. A., 92, *98*
Dowsing, R. D., 299, *328*
Dreyer, W. J., 34, *59*
Dubourdieu, M., 106, *118*
Dukes, G. R., 258, 259, *274*
Dunham, W. R., 65, 89, 95, *99,* 209, 233, 234, 235, 236, 250, 251, 253, 255, 266, 269, *273, 274, 278, 279,* 301, 302, 304, *328,* 399, 409, *416*
Dus, K., 75, 93, *97,* 175, 178, 179, *202,* 211, 242, *274, 280*
Dutton, D. L., 370, *376*
Dutton, P. L., 69, 81, 90, 91, *97, 98, 99*

E

Eady, R. R., 16, 18, 21, 25, 29, 36, 46, 48, 52, 55, 56, *58, 59, 60*
Easterday, R. L., 150, *153*
Eaton, W. A., 239, 242, 247, *274, 276,* 341, 352, 366, *376*
Edenharder, R., 127, 142, *153*
Edmondson, D. E., 106, 107, *118,* 122, 150, *153*
Ehrenberg, A., 82, 87, *97,* 192, *202,* 255, *273,* 308, 324, *327*
Eicher, H., 249, 250, 251, *274, 275,* 308, 310, 321, 324, *328*
Eisenberg, D., 206, *274*
Eisenberg, R., 214, 215, 216, *274, 279*
Eisenberger, P., 225, *280,* 352, *378,* 400, *416*
Eisenhardt, P., 345, 346, 349, *377*

AUTHOR INDEX

Eisenstein, K. K., 308, *328*
Eklund, H., 206, *274*
Elder, R. C., 345, 350, *376, 378*
El Ghazzawi, E., 132, 140, 141, 146, *151*
Eliezer, Z., 250, *279*
Ellis, P. E., 350, *378*
Elowe, D., 66, *98*
Elsden, S. R., 128, *153*
Elstner, E. F., 125, 126, 128, 129, *151*, 294, 302, *329*
Emiliani, E., 127, *151*
Enoch, H. G., 132, 147, *152*
Epstein, E. F., 215, *274*
Erbes, D. L., 27, *58, 59*, 63, *98*, 262, *274*, 374, *376*
Erecínsko, M., 71, 80, 90, *97, 98*
Estabrook, R. W., 209, 229, 241, 242, *275, 278*, 307, *329*
Evans, H. J., 16, 18, 21, 47, *59*
Evans, M. C. W., 29, 33, 34, *59*, 93, *98*, 251, 255, *274, 275, 279*, 299, 300, 301, 302, 307, 309, 310, 314, 319, 321, 324, *328, 329*, 370, 374, *376*, 382, 389, 392, 393, 395, 397, 412, 413, *416*
Ewall, R. X., 340, 345, 346, 358, *375, 376, 377*

F

Fahrenholz, F., 213, *273*
Fan, A., 250, *279*
Farina, R., 343, *376*
Federici, G., 110, *119*
Fee, J. A., 108, *118*, 218, 234, 235, 242, 247, *274, 275, 278*, 304, *328*, 399, *416*
Feher, G., 358, *377*
Fehlhammer, W. P., 149, *151*, 255, *280*
Feigelson, P., 129, *152*
Ferguson, G., 215, 218, 256, *275*
Figueroa, I., 147, *154*
Finazzi,-Agro, A., 110, *118*
Finley, J. W., 349, *378*
Fisher, R. J., 147, *152*
Flahaut, J., 214, *275*
Flatmark, T., 83, *98*
Fleet, M. E., 217, *275*
Fleisher, E. B., 347, *376*
Florian, L. R., 345, *376*
Flossdorf, J., 126, 147, 149, 150, *155*
Flynn, G. W., 350, *375*

Foner, S., 234, 235, 240, *275*
Forget, P., 149, *152*
Foster, R. P., 374, *376*
Foust, G. P., 247, *278*
Fox, J. L., 106, *118*
Fox, R., 302, *328*
Frankel R. B., 160, 166, 192, *203*, 219, 221, 225, 226, 227, 229, 230, 234, 235, 240, 242, 243, 250, 251, 252, 256, 257, *275, 276, 277, 278*, 287, 308, 315, 321, *328, 329*, 350, 352, *377*, 382, 412, 413, 414, *416*
Frazier, W. A., 19, 59
Freer, S. A., 165, 166, *203*
Freer, S. T., 75, 89, *97, 98*, 158, 159, 160, 161, 162, 163, 164, 165, 166, 167, 168, 170, 171, 172, 173, 174, 176, 177, 178, 179, 182, 184, 185, 186, 187, 188, 191, 195, 198, *202, 203*, 206, 208, 209, 218, 222, 223, 228, 233, 255, 269, *273, 275, 279*, 286, 307, *328*, 356, 357, 364, 368, *375*
Friedman, M., 349, *378*
Fritchie, C. J., Jr., 216, 228, *275, 279*, 351, *379*
Fritz, J., 304, *328*, 399, *416*
Frost, A. A., 366, *376*
Fry, K. T., 66, *98*
Fuchs, G., 130, 131, 132, 133, 134, 139, 140, 146, 150, *155*, 374, *378*
Fujii, T., 124, 147, *152*
Fujita, T., 147, *152*
Fukuyama, T., 147, 149, *152*

G

Gall, R. S., 216, 217, 253, 255, *275*
Garbett, K., 299, *327*
Ganapathy, K., 20, *58*
Garbett, K., 389, *416*
Garfinkel, D., 2, *14*
Garing, J. C., 213, *273*
Garland, P. B., 92, *98*
George, P., 341, 343, 352, 366, *376*
Gersoncle, K., 249, 250, 251, *274, 275*, 308, 310, 321, 324, *328*
Gerwing, J., 111, 112, *118*
Gesteland, R. F., 110, *119*
Geyda, J. P., 292, *328*
Ghambeer, R. K., 139, *155*
Ghazarian, J. G., 93, *99*
Gholson, R. K., 129, *155*

Gibson, J. F., 209, 249, 285, 287, 292, 294, 299, *274, 275, 328, 329, 330*
Gillard, R. D., 299, 349, *327, 376*
Gillum, W. O., 219, 234, 235, 240, 262, 263, *275, 276*
Gilmour, A. D., 212, *275*
Gilroy, M. J., 345, 346, 247, 349, *377*
Ginsburg, A., 2, 3, *14*
Gitlitz, P. H., 374, *376*
Glickson, J. D., 237, *275, 279*
Gold, L. M., 110, 111, *119*
Goldanskii, V. I., 382, 384, 388, *416*
Goldberg, R. B., 2, 4, 5, 8, 9, 11, 12, *14*
Goldberger, R. F., 2, *14*
Good, M. L., 216, *273, 275*
Gook, K. A., 18, 21, 25, 29, 36, 48, *59*
Gordon, J. K., 2, *14*
Gottschalk, G., 132, 140, 141, 146, *151, 152*
Gray, H. B., 160, 166, 192, *203*, 234, 236, 242, 245, 251, 252, *274, 276, 279*, 321, *329*, 342, 343, 364, 365, 366, 367, *376, 377, 378*, 412, *416*
Gray, T. A., 92, *98*
Graziani, M. T., 110, *118*
Grey, I. E., 250, *279*
Grimes, C. J., 347, *376*
Grimmelickhuijzen, C. J. P., *98*
Gröbner, P., 123, 127, 141, 142, *153*
Grossman, B., 342, *376*
Grossman, S., 92, *97, 98*
Grunthaner, F. J., 160, 166, 192, *203*, 245, 251, 252, 321, *329*, 412, *416*
Guerrieri, F., 81, *99*
Gunsalus, I. C., 108, *119*, 209, 211, 218, 229, 230, 251, *275, 278, 280, 281*, 301, 302, 304, *328, 329*, 373, *378, 379*, 384, 387, 399, 405, *416*
Gupta, M. P., 212, *278*
Gupta, R. K., 340, 341, 344, *376*
Gutpa, V. D., 214, *278*
Gurney, E., 1, 2, *14*
Guroff, G., 126, 128, *151, 154*
Gustafsson, I., 82, 87, *97*
Gutman, M., 73, 90, 91, *98*

H

Haaker, H., 41, *59*
Hadfield, K. L., 46, *59*

Hall, D. O., 33, 34, *59*, 192, *202*, 207, 209, 211, 233, 241, 249, 250, 251, 252, 255, 272, *273, 274, 275, 276, 279, 280*, 285, 287, 288, 292, 294, 295, 299, 300, 301, 302, 303, 306, 307, 308, 310, 311, 312, 314, 319, 321, 324, *327, 328, 329, 330*, 382, 389, 392, 393, 395, 397, 412, 413, *416, 417*
Hall, S. R., 217, *275*
Halliwell, B., 125, 148, *152, 153*
Halpern, J., 363, *376*
Ham, N. S., *280*
Hamilton, G. A., 358, *377*
Hamilton, W. C., 215, *275*
Hamilton, W. D., 31, 36, 37, 49, 50, *59*
Hanania, G. I. H., 343, 352, 366, *376*
Haniu, M., 181, 182, 183, 184, *203*, 211, *280*, 372, *378*
Hannaway, C., 215, 218, *275*
Hansen, R. E., 63, 65, 67, 69, 70, 72, 75, 80, 89, 94, *96, 97, 99*, 108, *119*, 218, *281*
Hanstein, W. G., 66, 73, 94, 95, *98*
Harbury, H. L., 269, *275*, 343, *377*
Hardy, R. W. F., 16, 18, 21, 43, 46, 47, *58, 59*, 106, *118*, 140, *153*
Harmon, H. J., 96, *98*
Harold, F. M., 136, *152*
Harrington, A. A., 124, *152*
Harrison, J. E., 123, 143, *154*
Harrison, P. M., 206, *275*
Hart, E. J., 361, *376*
Hartzell, C. R., 20, *58*, 69, 96, *97, 98*
Haselkorn, R., *118*
Hashmall, J. A., 361, *375, 376*
Haslett, B. G., 106, 107, 113, *118*
Hatefi, Y., 63, 65, 66, 73, 75, 78, 89, 94, 95, *97, 98, 99*
Havelka, M. D., 16, *59*
Hayaishi, O., 126, 127, 128, *152, 154, 155*
Hayaishi, T., 128, *152*
Healy, P. C., 215, *275*
Heidrich, H.-G., 86, 87, 95, *97*
Heinen, W., 27, *59*
Heller, J., 342, *378*
Henninger, H., 137, *155*
Henzl, M. T., 27, 28, 44, 47, 53, *58*
Herber, 382, 384, 388, *416*
Herriott, J. R., 158, 187, *204*, 208, 217, 221, *275, 281*, 352, *378*, 382, 400, 411, *417*
Herskovitz, T., 160, 164, 166, 168, 171, 182, 192, 196, 198, *202, 203*, 218, 219, 223, 225,

226, 228, 229, 230, 231, 232, 233, 234, 235, 236, 237, 239, 240, 242, 245, 246, 248, 249, 251, 252, 253, 256, 257, 268, *273, 274, 275, 276, 280,* 286, 287, 308, 315, 321, *327, 328, 329,* 411, 412, 413, 415, 416
Hersott, J. R., 225, *279*
Heupel, A., 125, 126, 128, 129, *151*
Hill, H. A. O., 249, *274*
Hill, R., 301, *328*
Hirai, H., 212, 219, *280*
Hirose, M., 34, *59*
Hoare, D. S., 147, *154,* 206, *275*
Hodges, H. L., 342, *376*
Höpner, T., 121, 127, 142, 143, 144, 145, 147, 149, 150, *152, 153, 155*
Hoffström, L., 82, 87, *97*
Holah, D. G., 220, 242, 256, *274, 276,* 415, *416*
Hollander, F. J., 231, 256, *274, 276*
Holm, R. H., 27, *60,* 107, *119,* 160, 162, 164, 166, 169, 170, 171, 182, 192, 196, 197, 198, *202, 203, 204,* 207, 213, 218, 219, 221, 223, 225, 226, 227, 228, 229, 230, 231, 232, 233, 234, 235, 236, 237, 238, 239, 240, 241, 242, 243, 244, 245, 246, 248, 249, 250, 251, 252, 253, 254, 256, 257, 258, 259, 260, 261, 262, 263, 264, 265, 267, 268, 270, 271, *273, 274, 275, 276, 277, 278, 279, 280,* 286, 287, 308, 315, 321, *327, 328, 329,* 350, 352, *377,* 382, 411, 412, 413, 414, *415, 416*
Holsten, R. D., 18, 21, *58, 59*
Holwerda, 342, *376*
Holzer, H., 2, *14*
Hong, J.-S., 107, 108, 109, 111, 112, *118,* 218, 242, *276*
Hopfield, J. J., 359, *376*
Horrocks, W. D., 237, *276*
Horrocks, W. D., Jr., 237, *277*
Hoskins, B. F., 217, *276*
Howard, R. L., 16, 18, 21, 47, *59,* 66, *98*
Hoy, T. G., 206, *275*
Hu, W., 215, *276*
Huang, J. J., 235, 241, 248, *276,* 374, *376*
Huang, T. C., 18, 19, 20, 21, 23, 28, *59*
Huber, R., 194, *203,*
Huennekens, F. M., 127, 129, *152, 154,* 212, 253, 254, *281*
Hummel, J. P., 34, *59*
Hunt, S. V., 128, 137, *151*
Hurst, J. K., 347, *376*

Hwang, J. C., 45, *59*
Hyndman, L. A., 43, *59*

I

Ibers, J. A., 160, 166, 168, 171, *202, 203,* 215, 218, 219, 221, 223, 225, 226, 227, 228, 229, 230, 232, 233, 234, 236, 242, 243, 250, 251, 253, 256, 257, 258, 265, 266, 267, *273, 276, 277, 278, 279, 280,* 286, *327,* 334, 350, 352, 355, *377, 378,* 382, 411, 412, 414, *416*
Iizuka, T., 234, 235, *278*
Ingalls, R., 388, *416*
Ingledew, W. J., 63, 71, 85, 89, 90, *98, 99*
Irvine, D. H., 343, *376*
Ishimura, Y., 272, *278*
Isied, S. S., 347, *376*
Irvine, D. H., 366, *376*
Islam, K. M. S., 215, 218, *275*
Israel, D. W., 16, 18, 21, 47, *59*
Itagaki, E., 147, *152*

J

Jacks, C. A., *276,* 336, 337, 338, 352, 353, 354, 355, 358, *375, 376*
Jackels, S. C., 253, *278,* 400, *416*
Jackson, E. K., *59*
Jackson, P. J., 250, 251, *279,* 299, 300, 319, *329,* 382, 389, 392, 393, 395, 397, *416*
Jakoby, W. B., 125, 126, *152*
James, E., 40, *59*
Jayne, J., 212, *277*
Jeannin, S., 212, *276*
Jeannin, Y., 212, *276*
Jeng, D., 109, *118*
Jensen, L. H., 75, *97,* 109, *118,* 158, 159, 161, 162, 163, 164, 165, 166, 167, 168, 170, 171, 174, 175, 176, 177, 180, 181, 185, 187, 188, *202, 203, 204,* 208, 211, 217, 218, 221, 223, 226, 233, 269, *273, 275, 276, 280, 281,* 286, 307, *327, 328,* 352, 357, 369, 372, *375, 376, 378,* 382, 400, 414, *417*
Job, R. C., 107, *118,* 160, *202, 203,* 258, 260, 264, 270, *273, 276,* 347, 349, *375*
Johansson, G., 216, *276*
Johnson, C. E., 33, 34, *59,* 209, 230, 249, 250, 251, 252, *273, 274, 275, 278, 279,*

286, 287, 288, 294, 296, 299, 300, 301, 302, 303, 304, 305, 306, 307, 308, 310, 311, 312, 314, 319, 321, 324, *328*, *329*, *330*, 382, 389, 392, 393, 395, 397, 412, 413, *416*, *417*

Johnson, K. H., 162, 170, 198, *204*, 253, 254, *276*, *280*, *281*, 400, *417*

Johnson, P. A., 122, 124, 143, 145, 147, *152*

Johnston, H. M., 2, *14*

Jones, J. B., 181, 182, 183, 184, *203*, 372, *378*

Jones-Mortimer, M. C., 143, 145, 147, *152*

Jørgensen, C. K., 214, *276*

Joyner, A. E., *151*

Jungerman, K., 123, 127, 128, 130, 131, 132, 134, 136, 137, 138, 139, 142, 146, 150, *151*, *154*, *155*, *156*

K

Käufer, B., 131, 132, 133, 134, 137, 140, 146, 150, *155*, 374, *378*

Kagamiyama, H., 211, *276*

Kallai, O. B., 206, *274*, *280*

Kallio, R. E., 124, *152*

Kalvius, M. G., 249, 251, *275*, 308, 310, 324, *328*

Kamen, M. D., 175, 178, 180, 200, *203*, 206, *279*, 369, *377*

Kaneda, T., 143, *152*

Kano, Y., 34, *59*

Kaplan, H., 366, *377*

Kassner, R. J., 268, 269, *276*, 334, *376*

Kato, N., 124, 126, 147, *153*

Katz, N., 130, 146, *152*

Kavedia, C. V., 212, *278*

Kearney, E. B., 65, 71, 74, 75, 76, 78, 79, 85, 93, *97*, *98*, *100*, 351, *375*

Kearny, J. J., 135, 146, *153*

Keech, D. B., 125, *154*

Keilin, D., 61, *98*

Kelly, B., 112, *118*

Kemeny, G., 339, *377*

Kennedy, C. K., 2, *14*

Kennedy, I. R., 40, *59*, 268, *281*

Kennel, S. J., 369, *377*

Kent, S. S., 128, *153*

Kester, W. R., *278*

Ketchum, P. A., 19, *59*

Kharkats, Y. I., 363, *376*

Kiefer, G., 219, *279*

Kimura, T., 209, 211, 218, 235, 241, 242, 247, 248, *274*, *276*, *280*, 374, *376*

King, P., Jr., 343, 364, 366, *375*

King, T. E., 63, 66, 71, 73, 74, 75, 76, 77, 79, 80, 82, 83, 95, *97*, *98*, *99*, *100*

Kirchniawy, H., 127, 128, 130, 131, 137, 142, 146, *152*, *155*

Kiso, Y., 213, *281*

Kistenmacher, T. J., 266, *276*, 352, *377*

Kjekshus, A., 217, *273*, *276*

Klein, M. P., 299, *330*, 385, 386, 392, *417*

Kleiner, D., 18, *59*

Klotz, I. M., 102, *118*

Knappe, J., 123, 127, 141, 142, 143, 145, 147, 149, *152*, *153*, *156*

Knight, E., 140, *153*, 310, *327*

Knight, E., Jr., 106, *118*, 162, *202*

Knowles, P. F., 302, *329*

Knox, W. E., 126, *154*

Koch, S., 265, *276*

Koenig, S. H., 340, 341, *376*

Koivusalo, M., 124, 126, *155*

Kolthoff, I. M., 211, 212, *277*, *280*

Kornberg, H. L., 128, *153*

Kosaka, A., *155*

Kostikas, A., 415, *416*

Krakow, G., 148, *153*

Krampitz, L. O., *154*

Krasna, A. I., 374, *376*

Kraut, J., 75, 89, *97*, *98*, 158, 159, 160, 161, 162, 163, 164, 165, 166, 167, 168, 170, 171, 172, 173, 174, 177, 178, 179, 182, 184, 185, 186, 187, 188, 191, 194, 195, 198, *202*, *203*, 206, 208, 209, 218, 222, 223, 228, 233, 255, 269, *273*, *275*, *279*, 286, 307, *328*, 356, 357, 358, 364, 368, *375*

Krishnamurty, K. V., 335, *377*

Kröger, A., 147, *153*

Krupka, R. M., 366, *377*

Krusic, P. J., 160, *203*, 246, 251, *275*, 287, 308, 315, 321, *328*, 413, *416*

Kubas, G. J., 214, 217, 249, *281*

Kündig, W., 397, *416*

Kukla, D., 194, *203*

Kula, M., 126, 147, 149, 150, *155*

Kunishima, M., 212, 219, *280*

Kurahashi, K., 102, *118*
Kurtin, S. L., 358, *377*
Kutzbach, C., 122, 125, *153*
Kuwana, T., 69, *98*

L

Lachenal, D., 231, *274*
Laemmli, U. K., 111, *118*
Laidler, K. J., 366, *377*
Laishley, E. J., 130, *153*, 163, *203*
Lake, R. E., 345, *376*
La Mar, G. N., 239, 240, *276, 277*
Lambeth, D. O., 368, *377*
Lamfrom, H., 110, *119*
Lamobowitz, A. M., 90, *100*
Lane, M. D., 150, *153*
Lane, R. H., 345, 346, 347, 349, 350, 352, *376, 377*
Lane, R. W., 219, 221, 225, 226, 229, 230, 234, 242, 243, 250, 251, 267, *277*, 382, 412, 414, *416*
Lang, G., 28, 29, 32, 37, *60*, 327, *330*, 389, 400, 405, 408, 409, 410, 411, 414, *416*
Lappin, A. G., 211, *277*
Lauher, J. W., 266, *277*, 352, 355, *377*
Lavinge, G., 212, *276*
Lawford, H. G., 92, *98*
Lecomte, J. R., 35, *58*
Leder, P., 111, *118*, 126, *153*
Lee, I. Y., 81, *98*
Leek, A. E., 125, 148, *153*
LeGall, J., 106, *118*, 140, *153*, 305, *330*
Leigh, J. S., 71, 72, 75, 82, *99*, 370, *376*
Leigh, J. S., Jr., 71, 80, *98*
Leimenstoll, G., 130, 138, 146, *152*
Leipoldt, J. G., 217, *277*
Le Minor, L., 123, *154*
Lengyel, P., 103, *118*
Lester, R. L., 132, 147, *152*
Leussing, D. L., 211, 212, 219, *277*
Levin, E., 136, *152*
Levine, M., 206, *278*
Levine, R. P., 106, *118*
Levy, J. G., 112, *118*
Li, L. F., 139, 140, 146, *153*
Lillehoj, E. B., 127, *153*
Lim, J., 63, 71, 74, 75, 76, 77, 80, 83, *99*
Lin, P. M., 130, *153*
Lindmark, D. G., 127, 142, *153*

Lindsay, J. G., 81, 90, *98, 99*
Lindstrom, E. S., 147, *156*
Linke, H. A. B., 141, *153*
Linnane, A. W., 147, *153*
Lipmann, F., 102, 103, *118*, 126, 129, *153*
Lippard, S. J., 214, 215, 216, 267, 269, *274, 276, 277*
Lipscomb, J. D., 209, 229, 230, *275*, 373, *378*
Lipscomb, W. N., 206, 216, *276, 277, 279*
Little, H. N., 148, *154*
Little, R. G., 267, *277*
Livingston, D. M., 126, *153*
Livingstone, S. E., 214, *277*
Ljones, T., 17, 25, 26, 31, 36, 37, 47, 49, 50, 51, 52, *59, 60*
Ljungdahl, L. G., 123, 132, 135, 138, 139, 140, 146, 148, 149, *151, 153, 155*
Llinás, M., 207, *277*
Lo, D. Y., 253, *277*, 382, 400, *416*
Lode, E. T., 109, *118*, 182, 184, *203*, 229, 250, *277, 278*, 372, *377*, 392, 393, 398, *416*
Loehr, T. M., 248, *277*, 352, *377*
Loew, G. H., 253, *277*, 382, 400, *416*
Löw, H., 88, *98*
Loo, A., 2, 6, *14*
Long, T. V., 248, *277*, 352, *377*
Lord, A. V., 93, *98*, 307, 310, *328*, 370, *376*
Lorusso, M., 81, *99*
Lovenberg, W., 27, *59*, 104, 114, *118, 203*, 207, 209, 217, 233, 234, 241, 242, 247, 248, 250, 251, 272, *274, 275, 276, 277, 279*, 299, 308, 324, *329*, 352, 353, 354, 355, *376, 377*, 389, 393, *416*
Low, K. B., 8, 12, *14*
Low, W., 247, *277*
Lowe, D. J., 28, 29, 37, 50, *60*, 96, *100*, 302, *328, 329*, 411, *416*
Lucas-Lenard, J., 103, *118*, 126, 129, *153*
Ludowieg, J., 148, *153*
Lugay, J. C., 126, *151*
Lumry, R. W., 265, *279*, 334, 335, 338, 339, 359, 361, 363, *378*
Lusty, C. J., 66, *98*
Lynden-Bell, R. M., 302, *329*

M

McBride, B. C., 123, 136, *151, 153*
Macara, I. G., 206, *275*

McArdle, J. V., 343, *377*
McAuley, A., 211, 212, *275, 277*
McAuliffe, C. A., 211, 214, *277*
McCarthy, J., 209, 229, *275*
McCarthy, K. F., 104, *118*
McCormick, N. G., 123, 127, *153*
McDonald, C. C., 27, *59,* 192, *203,* 234, 237, 241, 250, 251, 255, *275, 277, 279,* 299, 308, 324, *329,* 373, *378,* 389, 393, *416*
McDonald, P. D., 358, *377*
McElroy, J. D., 358, *377*
McGill, T. C., 358, *377*
MacGillaury, C. H., 217, *273*
Machinist, J. M., 66, *98*
McKenzie, E. D., 212, *273*
Mackey, L. N., 69, *98*
McKnight, G. S., 111, *119*
Macnicol, P. K., 41, *58*
Madanski, C., 19, *59*
Magasanik, B., 3, 4, 5, 7, 12, 13, *13, 14*
Mahanti, S. D., 339, *377*
Mahler, H. R., 66, *98*
Makashima, S., 212, 219, *280*
Malavolta, E., 147, *153*
Malkin, R., 93, *98, 99,* 108, *118,* 208, 218, 269, *273, 277,* 374, *377*
Marcus, R. A., 333, 335, 338, 339, 343, 359, 360, 362, *377*
Margalit, R., 341, 344, *377*
Margoliash, E., 206, *274*
Marks, R. H. L., 269, *275*
Marshall, W., 296, 300, *329*
Martin, R. B., 212, *280*
Martin, R. L., 215, 231, 235, *277*
Marty, B., 147, *151*
Maruyama, H., 150, *153*
Maskiewitz, R., 160, 175, 178, *202, 203,* 264, *273,* 347, 349, *375*
Maskill, R., 349, *376*
Mason, H. S., 20, *58*
Mason, J. I., 209, 229, *275*
Mason, R., 207, 209, 230, 249, *278, 281,* 299, 306, *327, 330*
Massey, V., 106, *118,* 122, 125, 150, *151, 153*
Matarese, R., 110, *119*
Mather, J. H., 148, *153*
Mathews, B. W., 206, *278*
Mathews, F. S., 206, *278*

Mathews, M. B., 148, *154*
Mathews, R., 63, *99,* 249, *278,* 308, 324, *329, 377*
Mathur, H. B., 212, *278*
Matsubara, H., *276,* 307, *329*
Mauzerall, D. C., 358, *377*
Mayer, A., 249, 251, *275,* 308, 310, 324, *328*
Mayerle, J. J., 160, *203,* 219, 222, 225, 226, 227, 229, 230, 232, 233, 234, 237, 239, 240, 242, 251, 257, 258, *276, 278*
Mayhew, S. G., 106, *118,* 247, *248,* 299, *327*
Mayr, M., 78, *97, 98*
Mazelis, M., 148, *154*
Mead, C. A., 358, *377*
Meeks, J. R., 229, 230, *275*
Mehler, A. H., 126, 128, *154, 155*
Mehrotra, R. C., 214, *278*
Meyer, T. E., 175, 178, 180, 200, *203*
Meyer, T. J., 215, 218, 256, *275,* 334, 338, 339, 351, 354, *377, 379*
Midwinter, G. G., 127, *154*
Miller, J. R., 358, *377*
Miller, W. G., 343, *377*
Mitchell, B., 111, 112, *118*
Mitchell, P., 92, *99*
Mitchel, P. C. H., 235, *278*
Mislan, J. P., 212, *277*
Mizrahi, F. A., *278*
Mizrahi, I. A., 364, 365, 366, 370, 371, *377*
Möckel, W., 127, 142, *153*
Moleski, C., 235, *278*
Moll, B., 106, *118*
Morandi, C., 2, 4, 5, 8, 9, 11, 12, *14*
Morgan, T. V., 75, 82, *97*
Morris, J. A., 23, 40, *58, 59*
Morris, J. G., 137, *151*
Morrison, J. F., 40, 42, *59*
Mortenson, L. E., 16, 17, 18, 19, 20, 21, 23, 25, 28, 34, 36, 38, 39, 40, 47, 51, 53, 55, 56, *58, 60,* 107, 109, *118, 119,* 130, 146, *154,* 198, *203,* 242, 244, 248, 249, 262, 263, 264, 271, 272, *275, 278, 279, 280,* 374, *378*
Morton, R. A., 343, *377*
Moss, T. H., 178, 192, 198, *202, 203,* 235, 255, *273, 278, 280,* 302, 310, 322, *327, 329*

AUTHOR INDEX

Moustafa, E., 40, *59*
Mower, H., 106, 114, 116, *119*
Müller, U., 143, 144, 145, 147, 149, 150, *152, 155*
Münck, E., 29, 30, 31, 32, *59*, 229, 230, *275, 278,* 301, 302, 327, *329,* 382, 384, 387, 399, 401, 403, 404, 405, 406, 407, 408, *416*
Muetterties, E. L., 256, *280*
Mukai, K., 248, 280
Mukherjee, A. D., 217, *276*
Muller, J. L. M., 76, 96, *97*
Mullinger, R. N., 192, *202,* 209, 230, 255, *273, 278,* 292, 306, 307, 308, 310, 311, 312, 319, *327, 328, 329*
Multani, J. S., 28, 31, *59*
Munkres, K. D., 115, *119*
Murphy, J. C., 132, 147, *155*
Murray, C. L., 109, *118,* 182, 184, *203,* 372, *377*
Murray, K. S., 212, *278*
Murray, S. G., 211, 214, *277*

N

Nagatani, H., 4, *14*
Nakashima, T., 106, 114, 116, *119*
Nakayama, H., 127, *154*
Namtvedt, M. J., 373, *378*
Nantvedt, M. J., 218, *281*
Nakos, G., 130, *154*
Nason, A., 19, *59,* 147, 148, *154*
Neilands, J. B., 108, *118*
Nepokroeff, C., 117, *118*
Nelson, N., 374, *375*
Nelson, S. M., 149, *154*
Neumann, H. M., 343, *375*
Neurauter, C., 139, 140, 146, *151*
Newburgh, R. W., 125, *155*
Newman, D. J., 305, *330*
Newman, L., 212, 219, *277*
Newman, P. J., 212, *278*
Nicholas, D. J. D., 27, *59*
Nicholls, P., 125, *154*
Nicholson, D. C., 217, *276*
Niedermaier, S., 147, *153*
Nitz, R. M., 111, 112, *118*
Nomura, M., 103, *118*
Nordstrom, B., 206, *274*
Norman, J. G., Jr., 162, 170, 198, *204,* 253, 254, *278, 281,* 400, *416*

Normore, W. M., 148, *153*
Nozaki, M., 272, *278*

O

O'Farrell, P. Z., 111, *119*
Ogata, K., 124, 126, 147, *153*
Ohlsson, I., 206, *274*
Ohmura, E., 126, 128, *154*
Ohnishi, T., 63, 66, 67, 68, 70, 71, 72, 73, 74, 75, 76, 77, 80, 81, 82, 85, 89, 90, 91, 92, *98, 99*
Ohrloff, C., 138, *152, 155*
O'Kelley, J. C., 147, *154*
Oosterhuis, W. T., 389, 392, 398, *416*
Orbach, R., 292, *329*
Ordal, E. J., 123, 127, 147, 149, *152, 153*
O'Reilly, J. E., 366, *377*
Orgel, L. E., 363, *376*
Orme-Johnson, N. R., 63, 65, 67, 69, 70, 72, 75, 80, 89, 93, 94, *98, 99*
Orme-Johnson, W. H., 26, 27, 28, 29, 30, 31, 32, 34, 35, 36, 37, 39, 40, 41, 42, 44, 47, 49, 50, 51, 53, *58, 59, 60,* 63, 65, 75, 89, *98, 99,* 108, *119,* 159, 163, 200, *203,* 207, 209, 218, 229, 230, 233, 234, 235, 242, 249, 250, 251, 255, 262, 272, *273, 274, 278, 279, 280, 281,* 301, 302, 304, 308, 327, *328, 329,* 332, 372, 373, 374, *376, 377, 378,* 382, 399, 401, 403, 404, 406, 407, 408, 409, *416*
Oro, J., 125, 148, *154*
Orton, J., 212, *273*
Ottersen, T., 349, *378*
Overnell, J., 343, *377*

P

Pachowsky, H., 272, *273*
Packer, E. L., 182, *203,* 297, *329,* 372, *377*
Page, F. M., 212, *278*
Palmer, G., 20, 25, 27, 28, 31, 36, 38, 51, 55, *59, 60,* 63, 66, *97, 99,* 108, *118,* 122, *154,* 209, 218, 234, 235, 236, 237, 241, 242, 247, 249, 250, 251, *273, 274, 275, 278, 279,* 285, 287, 301, 302, 304, 307, 324, *328, 329,* 332, 347, 368, 369, *377,* 399, 409, *416*
Palmer, G. A., 96, *97*
Pannan, C. D., 217, *276*

Paolella, P., 127, 142, *153*
Papa, S., 81, *99*
Papaefthymiou, G. C., 219, 229, 242, 251, 267, *277*
Pappalardo, P., 247, *278*
Parak, F., 249, 250, 251, *274, 275*, 308, 310, 321, 324, *328*
Paris, C. G., 3, 5, 7, *14*
Parker, D. A., 235, *278*
Parker, D. J., 138, 140, *155*
Pasek, E. A., 231, *278*
Patterson, G. S., 231, *279*
Pauling, L., 169, *203*
Peacock, D., 148, *154*
Pearson, R. G., 131, *154*, 366, *375, 376*
Peck, H. D., Jr., 130, 146, *153, 154*, 163, *203*, 305, *330*, 374, *375*
Pedelty, R., 231, *276*
Pederson, J. I., 93, *99*
Peel, J. L., 299, *327*
Peisach, J., 249, 250, *273, 278*, 356, *378*, 392, 393, 398, *416*
Peover, M. E., 361, *378*
Perrin, D. D., 134, *154*
Petering, D., 235, 247, *278*
Petersson, L., 192, *202*, 255, *273*, 308, 324, *327*
Petrouleas, V., 415, *416*
Phillips, W. D., 27, *59*, 160, 162, 164, 166, 192, *202, 203*, 218, 219, 227, 229, 234, 236, 237, 239, 240, 241, 246, 249, 250, 251, 255, *275, 276, 277, 278, 279*, 287, 299, 308, 315, 321, 324, *328, 329*, 373, *378*, 389, 393, 413, *416*
Pichinoty, F., 123, *154*
Piéchaud, M., 123, *154*
Pignolet, L. H., 231, *279*
Pine, M. J., 123, *154*
Pinsent, J., 132, 147, *154*
Piszkiewicz, D., 347, *376*
Plowman, J., 126, 128, *151, 154*
Poe, M., 192, *203*, 234, 236, 237, 250, 251, 255, *275, 279*, 299, 324, *329*, 389, 393, *416*
Pollick, P. J., 216, *274*
Postgate, J. R., 2, *14*, 16, 17, 18, 21, 25, 29, 36, 48, *58, 59*, 209, 230, 249, *281*, 306, *330*
Poston, J. M., 140, *154*
Prewitt, C. T., 216, *279*
Prince, R. C., 81, *99*

Prival, M. J., 3, 4, 5, 7, 13, *14*
Puig, J., 123, *154*
Purvis, J., 209, 229, *275*

Q

Quadri, S. M. H., 147, *154*
Que, L., Jr., 27, *60*, 107, *119*, 160, 168, 171, 182, 196, 197, *202, 203*, 213, 219, 223, 226, 229, 230, 231, 232, 233, 234, 238, 242, 244, 245, 253, 257, 258, 260, 261, 262, 263, 264, 268, 271, *273, 274, 279*, 411, 413, *415, 416*
Quayle, J. R., 122, 123, 124, 125, 143, 145, 147, 150, *152, 154*
Quiocho, F. A., 206, *279*
Quispel, A., 16, 17, *60*

R

Rabinowitz, J. C., 107, 108, 109, 111, 112, 113, 114, 115, 117, *118, 119*, 123, 127, *151*, 182, 184, *203*, 208, 218, 232, 242, 269, *276, 277, 280*, 297, *329*, 372, 374, *377, 378*
Racker, E., 72, 82, *99*, 124, *154*
Rader, J. I., 127, 129, *154*
Rae, K. K., 215, *279*
Räuber, A., 385, 399, *416*
Ragan, C. I., 72, 82, *99*
Rall, S. C., 109, *119*
Rao, K. K., 192, *202*, 207, 209, 211, 230, 233, 241, 249, 250, 251, 252, 255, 272, *273,, 274, 275, 276, 278, 279, 280*, 285, 287, 288, 295, 299, 300, 301, 302, 303, 308, 310, 312, 314, 319, 321, 324, *327, 328, 329, 330*, 382, 389, 392, 393, 395, 397, 412, 413, *416, 417*
Rappoport, D. A., 125, 148, *154*
Rasse, D., 286, *329*
Rawlings, J., 234, 236, 242, *274, 279*, 364, 365, 366, 367, *378*
Raymond, W. N., *276*, 352, 353, 354, 355, 358, *375, 376*
Reddy, C. A., 136, 146, *154*
Redfield, A. G., 340, 341, *376*
Reeder, D. J., 135, 146, *151*
Reeves, S. G., 307, 310, *328*, 370, *376*
Reger, D. L., 247, *274*
Reiff, W. M., 250, 256, *275, 279*, 384, *416*

AUTHOR INDEX

Reynolds, W. L., 265, *279,* 334, 335, 338, 339, 359, 361, 363, *378*
Rhoads, R. E., 111, *119*
Rhodes, H., 29, 30, 31, 32, *59,* 255, *278,* 327, *329,* 382, 401, 403, 404, 406, 407, 408, *416*
Ribbons, D. W., 123, 143, *154*
Richards, R., 249, *274*
Richert, D. A., 125, 135, 148, *155*
Riebeling, V., 136, *154*
Riederer-Henderson, M. A., 146, *154*
Rieske, J. S., 63, 80, 81, *99*
Riley, P. E., 349, *378*
Rimpler, M., 213, *279*
Rivera-Ortiz, J., 45, 56, 57, *60*
Rivest, R., 216, *280*
Robertson, G. B., 215, 231, *270*
Robertson, J. H., 215, *280*
Robertus, J. D., 194, *203*
Robinson, I. M., *151*
Rocha, V., 129, *155*
Rohde, N. M., 215, 231, *277*
Rose, Z. B., 124, *154*
Rosser, R. J., 129, *155*
Rossmann, M. G., 199, 201, *203*
Rothman-Denes, L. B., *118*
Roxbury, J. M., 143, *152*
Rühlmann, A., 194, *203*
Ruiz-Herrera, J., 147, *154*
Rupprecht, E., 128, 130, 131, 137, 138, 146, *152, 155*
Ruschig, U., 143, 144, 145, 147, 149, 150, *152, 155*
Rush, J. D., 209, 230, *278, 329*
Russell, A. S., 16, 18, 21, 47, *59*
Ruzicka, F. J., 65, 67, 71, 72, 75, 82, 83, 84, 85, 89, 93, 95, *97, 98, 99*

S

Sagers, R. D., 135, 146, *153*
Sahm, H., 124, 126, 147, 149, *155*
Salemme, F. R., 206, *279*
Salmeen, I., 234, 237, 250, 251, *274, 279,* 301, 302, 304, *328,* 399, 409, *416*
Salser, W., 110, *119*
Samson, L., 206, *274*
Sands, R. H., 32, *59,* 63, 65, 73, 89, 95, *97, 99,* 207, 209, 233, 234, 235, 236, 249, 250, 251, 253, 255, 266, *273, 274, 278, 279,* 285, 287, 301, 302, 304, 308, 324, *328, 329,* 369, *377,* 399, 409, *416*
San Pietro, A., 66, *98,* 237, *279,* 302, 310, *329*
Sasaki, R. M., 307, *329*
Sato, R., *152*
Saxton, R. E., 129, *155*
Sayers, D. E., 225, *279*
Schacht, J., 127, 142, *153*
Schaupp, A., 139, 140, 146, *151*
Schejter, A., 341, 344, *377*
Schepler, K. L., 65, 89, 95, *99*
Scherer, P., 137, *155*
Schimkat, M., 147, *153*
Schimke, R. T., 111, *119*
Schlaak, H. E., 249, 251, *275,* 308, 310, 324, *328*
Schlegal, H. G., 140, *151*
Schmidt, W., 128, 137, *152*
Schmitt, T., 123, 127, 141, 142, *153*
Schneider, Y., 385, 399, *416*
Schnitker, U., 132, 134, 139, 146, *155*
Schöberl, A., 213, *279*
Schön, G., 123, 127, 142, *152*
Schonbaum, G. R., 125, *154*
Schrauzer, G. N., 219, *279, 280*
Schubert, M. P., 212, *279,* 349, *378*
Schütte, H., 124, 126, 147, 149, 150, *155*
Schulman, M., 125, 135, 138, 139, 140, 148, *155*
Schulman, R. G., 352, *378*
Schultz, A. J., 215, 216, *279*
Schunn, R. A., 216, *279*
Schutt, H., 2, *14*
Schwager, P., 194, *203*
Schwartz, G., 117, *119*
Schwarz, K., 131, *155*
Schwarzenbach, G., 149, *155,* 342, *378*
Schweiger, M., 110, *119*
Schweet, R., 110, *119*
Sedor, F. A., 345, 346, 347, 349, *377*
Seff, K., 349, *378*
Seidman, C., 2, *14*
Senior, P. J., 3, 11, *14*
Shah, V. K., 2, *14,* 28, 29, 30, 31, 32, 35, 36, 37, 43, 44, 45, 47, 48, 49, 50, 53, 54, 57, *58, 59, 60,* 255, *278,* 327, *329,* 382, 401, 403, 404, 406, 407, 408, *416*
Shanmugam, K. T., 2, 4, 5, 6, 8, 9, 11, 12, *14,* 16, *60,* 107, *119*

Sheraga, H. A., 286, *329*
Shethna, Y. I., 306, *330*
Shimazono, H., 127, *155*
Shimizu, M., 4, *14*
Shimoyama, M., 129, *155*
Shirai, N., 213, *281*
Shiraishi, S., 63, 71, 85, *99*
Shirley, D. A., 299, *330*, 392, *417*
Shulman, R. G., 225, *280*, 343, 378, 400, *416*
Shum, A. C., 132, 147, *155*
Siegel, L. M., 108, *119*
Sieker, L. C., 75, *97*, 109, *118*, 158, 159, 161, 162, 163, 164, 165, 166, 167, 168, 170, 171, 176, 177, 180, 181, 187, 188, *202*, *203*, *204*, 208, 211, 217, 218, 221, 223, 226, 233, 269, *273*, *275*, *280*, *281*, 286, 307, *327*, *328*, 352, 357, 369, 372, *375*, *378*, 382, 400, *416*
Siiman, O., 160, 166, 192, *203*, 242, 245, 251, 252, *276*, 321, *329*, 412, *416*
Siliva, C. K., 93, *98*
Silverstein, R., 35, 43, 47, 57, *60*
Simon, G. L., 255, *280*
Simopoulos, A., 209, 230, *278*, 306, 307, 308, 310, 311, 312, 319, *329*, 415, *416*
Simpson, E. R., 209, 229, *275*
Singer, T. P., 65, 66, 67, 71, 72, 73, 74, 75, 76, 78, 82, 85, 90, 91, 92, 93, 94, 95, *97*, *98*, *99*, *100*, 351, *375*
Skelley, D. S., 112, *119*
Skyrme, J., 92, *98*
Slack, G. N., *280*
Slater, E. C., 76, 81, 90, 91, 96, *97*, *98*, *100*
Slater, J. C., 253, *280*
Sletten, D., 242, *274*
Sletten, K., 75, 93, *97*, 175, 178, 179, *202*
Sligar, S. G., 373, *378*
Sly, W. S., 125, *155*
Smith, B. E., 18, 21, 25, 28, 29, 32, 36, 37, 48, 50, *59*, *60*, 96, *100*, 327, *330*, 400, 405, 408, 409, 410, 411, 414, *416*
Smith, F. G., 127, *153*
Smith, R. V., *59*, 299, *329*
Smith, S. S., 106, *118*
Snow, J. T., 349, *378*
Snow, M. R., 215, *280*
Sobel, B. E., 242, *277*, 352, *377*
Söderlund, G., 206, *274*

Söll, D., 103, *118*
Spiro, T. G., 198, *203*, 248, 249, *280*
Sreenivasaya, M., 148, *151*
Stadtherr, L. G., 212, *280*
Stadtman, E. R., 2, 3, *14*, 125, 140, *154*, *155*
Stadtman, T. C., 123, 127, 136, 146, *151*, *155*, 181, 182, 183, 184, *203*, 372, *378*
Stallcup, M. R., 115, 117, *119*
Stangroom, J. E., 299, *327*
Stasiw, R., 366, *378*
Staudinger, H., 272, *273*
Stein, C. A., 350, *378*
Steinberg, D. A., 253, *277*, 382, 400, *416*
Steiner, D. F., 103, *119*
Steinfink, H., 250, *279*
Stellwagen, E., 343, *378*
Stephenson, M. P., 137, *155*
Stern, E. A., 225, *279*
Sternlicht, H., 182, *203*, 297, *329*, 372, *377*
Stewart, J. M., 217, *275*
Stewart, R., 358, *378*
St. John, R. T., 2, *14*
Stokstad, E. L. R., 122, 125, *153*
Stombaugh, N. A., 31, 39, *60*, 209, 225, 230, 242, 249, 251, *279*, *280*, 306, *330*, 352, 373, *378*, 400, *416*
Strahs, G., 160, *203*
Straub, D. R., 231, *278*
Streicher, S. L., 1, 2, 3, 4, 5, 7, 8, 9, 11, *14*, 16, *60*
Stricks, W., 212, *280*
Strittmatter, P., 124, *155*
Stucky, G. D., 266, *276*, 352, *377*
Studier, F. W., 111, *119*
Stynes, H. C., 267, *280*, 334, *378*
Sugiura, Y., 212, 213, 219, *280*
Suhadolnik, R. J., 1, *151*
Sukhani, D., 214, *278*
Sullivan, J. C., 348, 349, 350, *378*, *379*
Sundquist, J. E., 31, 39, *60*
Surzycki, S. J., 106, *118*
Sutin, N., 265, 267, *280*, 332, 333, 338, 339, 343, 350, 351, 363, *375*, *377*, *378*
Suzuki, K., 209, 229, *275*
Swank, R. T., 115, *119*
Swanson, S., 206, *280*
Sweeney, W. V., 109, *118*, 182, 184, *203*, 232, 236, 269, *280*, 372, 374, *378*
Switkes, E. S., 247, *274*

T

Tabner, B. J., 361, *378*
Tabor, H., 128, *152, 155*
Tagawa, K., 112, *119*
Tager, H. S., 103, *119*
Takano, T., 206, *274, 280*
Takeda, M., 126, *155*
Tamaoki, T., 124, 126, 147, *153*
Tanaka, M., 106, 114, 116, *119*, 181, 182, 183, 184, *203*, 207, 210, 211, 212, 213, 219, *280, 281*, 372, *378*
Tang, S. C., 265, *276*
Tang, S.-P., 248, 249, *280*
Tang, S.-P. W., 198, *203*
Tani, Y., 124, 126, 147, *153*
Tano, K., 219, *279, 280*
Tasaki, A., 235, *276*
Taube, H., 334, 338, 339, 343, 347, 354, *375, 376, 377*
Taylor, D., 215, 231, *277*
Taylor, G. A., 125, *154*
Taylor, J. E., 212, 219, *280*
Taylor, W. E., 229, *275*
Tchen, T. T., 150, *151*
Tebbe, F. N., 256, *280*
Tedro, S., 175, 178, 179, 180, 200, *203*
Telfer, A., *59*, 93, *98*
Terzis, A., 216, *280*
Thauer, R. K., 127, 128, 130, 131, 132, 133, 134, 135, 136, 137, 138, 139, 140, 142, 146, 150, *151, 152, 154, 155*, 374, *378*
Thomas, J. T., 215, *280*
Thompson, C. L., 251, 252, *279, 280*, 287, 299, 300, 310, 319, *329, 330*, 382, 389, 392, 393, 395, 397, 412, 413, *416, 417*
Thomson, A. J., 253, *280*, 321, *330*
Thorneley, R. N. F., 38, 41, 46, 48, 50, 51, 52, 55, 56, *60*
Thornley, J. H. M., 209, *275*, 287, 292, *328, 330*
Timkovich, R., 334, 358, 361, 369, 373, *376*
Tischer, T. N., 211, 212, *277*
Tisdale, H. D., 78, *97*
Tissieres, A., 103, *118*
Tkacheva, Z. G., 147, *155*
Tomati, U., 110, *119*
Tomita, A., 212, 219, *280*

Tomkins, G. M., 111, *118*
Toms-Wood, A., 136, *151*
Tonomura, K., 124, 147, *152*
Tosi, L., 148, *153*
Tossell, J. A., 400, *417*
Trachtmann, M., 341, *376*
Trakatellis, A. C., 117, *119*
Trautwein, A., 121, 143, 144, 145, 147, 149, *152*
Travis, J. C., 163, *203*, 305, *330*, 395, *415*
Treichel, P. M., 216, *281*
Trinajstić, N., 361, *376*
Trinh-Toan, 255, *280*
Tsai, R. L., 218, *281*
Tsibris, J. C. M., 108, *119*, 158, *204*, 209, 218, 233, 235, 251, *278, 280, 281*, 301, 302, 304, *328, 329*, 373, *379*, 384, 387, 399, 405, *416*
Tso, M.-Y. W., 17, 18, 21, 25, 31, 34, 35, 36, 37, 38, 47, 49, 50, *59, 60*
Tsuchiya, D. K., 307, *329*
Tsunoda, J. N., 181, *204*
Tubb, R. S., 2, *14*
Turner, D. H., 350, *375*
Turner, G. I., 35, 36, 41, 44, *58*
Tyler, B. M., 3, 4, 5, 7, *14*
Tzeng, S. F., *151, 155*

U

Udaka, S., 148, *153*
Ulstrup, J., 363, *376*
Uotila, L., 124, 126, *155*

V

Vahrenkamp, H., 214, *281*
Vaisey, E. B., 125, *155*
Valentine, R. C., 1, 2, 4, 6, 9, 11, *14*, 16, *60*, 163, *204*, 229, *281*
Vallee, B. L., 134, *155*, 206, 207, 225, *281*
Vallin, I., 88, *98*
Vandecasteele, J. F., 43, 47, *60*
Van Gelder, B. F., 20, *58*
Vanngard, T., 31, *58*
Vaughan, D. J., 400, *417*
Veeger, C., 41, *59*, 73, *100*
Venkatachalam, C. M., 173, *204*
Venier, C. G., 361, *375*
Vennesland, B., 148, *153, 154*

Vergamini, P. J., 214, 217, 249, *281*
Vetter, H., 127, 142, *153*
Vetter, J., *156*
Villarejo, M., 110, *119*

W

Wacker, W. E. C., 134, *155*, 206, 207, 225, *281*
Wadzinski, A. M., 123, 143, *154*
Wagner, C., 126, *151*
Wagner, F., 124, 147, *155*
Wahl, A. C., 335, 343, 364, 366, *375*, 377
Walker, F. A., 239, *276*
Walker, G. A., 39, 51, 55, *60*, 374, *378*
Walker, M. N., 374, *378*
Wang, J., 212, 219, *280*
Wang, J. H., 308, *328*
Wang, R.-H., 234, 236, *274*
Ward, F. B., 123, 147, *151*
Ward, J. C., 214, *281*
Ward, M. A., 23, *58*
Warme, P. K., 286, *329*
Warner, L. G., 349, *378*
Watari, H., 235, *276*
Watenpaugh, K. D., 158, 159, 164, 172, 175, 177, 185, 187, *202*, *204*, 208, 211, 217, 221, 223, 226, 269, *273*, *281*, 352, *378*, 382, 400, 414, *417*
Wearer, L. H., 206, *278*
Webster, R. E., 126, *155*
Weger, M., 247, *277*
Wegham, H. J., 90, *100*
Wei, C. H., 215, 216, *274*, *281*
Weiher, J. F., 160, 162, 164, 166, *202*, *203*, 218, 219, 227, 229, 231, 234, 236, 250, 251, *276*, *278*, *279*, 299, 310, 324, *327*, *329*, 389, 393, *416*
Weinstein, B., 213, *273*
Weitzman, P. D. J., 268, *281*
Wenning, J., 137, *155*
Weschler, C. J., 346, 348, 349, 350, *378*
Weser, U., 250, 251, 252, *280*, 287, 310, 319, 321, *328*, *330*, 412, 413, *417*
Westley, J., 110, *119*
Whalley, F. R., 292, *328*
Wharton, D. C., 20, *58*
Whatley, F. R., 106, 107, 113, *118*, *275*, 285, 287, 292, 301, *328*, *330*

Wherland, S., 364, 365, 366, 367, *378*
White, A. H., 215, *275*
Whiteley, H. R., 123, 127, *153*, *156*, 181, *204*
Whitten, D. G., 351, *379*
Wickman, H. H., 299, *330*, 385, 386, 392, *417*
Wikström, M. K. F., 69, 90, 91, *100*
Wilkes, G. R., 216, *281*
Wilkins, C. J., 256, *273*
Wilkins, R. G., 342, 343, 366, *376*, *378*
Williams, R. J. P., 207, 225, 267, *281*, 389, *416*
Williams, W. M., 242, *277*, 352, *377*
Willison, K. R., 38, 41, 46, 50, *60*
Willnow, P., 143, 144, 145, 147, 149, 150, *152*, *155*
Wilson, D. F., 69, 71, 75, 81, 82, 90, 91, *97*, *98*, *99*, *281*
Wilson, G. S., 373, *379*
Wilson, L. D., 374, *376*
Wilson, P. W., 27, 43, *59*, 147, *152*,
Winter, D. B., 63, 71, 74, 75, 76, 77, 80, 82, *99*
Wohlrab, H., 63, 66, 75, *99*
Wohlueter, R. M., 2, *14*
Wojcicki, A., 216, *274*
Wolf, W. A., 126, 128, *156*
Wolfe, R. S., 121, 123, 135, 136, 146, *151*, *153*, *155*, *156*
Wolfenden, R. V., 199, *204*
Wolin, E. A., 121, 123, 135, 136, 146, *151*
Wolin, M. J., 123, 136, 146, *151*, *154*
Wong, K. Y., 111, *118*
Wood, F. E., *278*, 364, 365, 366, 370, 371, *377*
Wood, H. G., 123, 135, 138, 139, 140, 146, 150, *151*, *153*, *155*
Wood, N. P., 127, 142, *153*, *156*
Woods, M., 348, 349, *379*
Woody, R. W., 158, *204*, 209, 233, *280*
Wormald, J., 215, 221, 226, *274*, 352, *375*
Wrigley, C. W., 147, *153*

X

Xuong, Ng. H., 158, 159, 160, 161, 162, 170, 173, 176, 177, 178, 187, *202*, 206, 218, 223, *273*, *279*, 286, *328*, 356, 357, 364, *375*

Y

Yacnych, A. M., 345, *376*
Yagi, T., 146, *156*
Yajima, H., 213, *281*
Yan, J. F., 212, 219, *280*
Yandle, J. R., 361, *378*
Yang, C. S., 212, 253, 254, 268, *281*
Yang, C. Y., 162, 170, 198, *204*
Yang, W., *276*, 334, *377*
Yasunobu, K., 181, 182, 183, 184, *203*
Yasunobu, K. T., 106, 114, 116, *119*, 181, *204*, 207, 210, 211, 213, *280, 281*, 372, *378*
Yates, M. G., 46, 48, 52, 56, *60*
Yelin, R. E., 342, 343, *378*
Yoch, D. C., 137, 146, 147, *156*, 163, *204*, 229, 269, *280, 281*
Yonetani, T., 63, 66, 71, 75, 81, *99*, 234, 235, *278*
You, K. S., 66, 95, *98*

Young, R. C., 351, *379*
Yu, C. A., 79, *100*
Yu, L., 79, *100*
Yu, M. W., 351, *379*

Z

Zaugg, W. S., 63, 80, 81, *99*
Zelitch, I., 125, 129, *156*
Zeppezauer, E., 206, *274*
Zeylemaker, W. P., 73, *100*
Zgorzalla, W., 249, 251, *275*, 308, 310, 324, *328*
Zimmerman, J. K., 18, *60*
Zipser, D., 105, *118*
Zubay, G., 110, *119*
Zubieta, J. A., 207, 209, 214, 216, 230, 249, 268, *274, 278, 281*, 306, *330*
Zumft, W. G., 16, 17, 18, 19, 20, 21, 23, 25, 28, 31, 36, 38, 47, 51, 53, 56, *59, 60*

Subject Index

A

Acetylene substrate for nitrogenase, 45–46
Acid solvolysis, 264–265
Active center
  core extrusion, 262
  magnetic resonance, 283–300
  Mössbauer chemical shift, 297–298
  spectroscopy, 289–298
Active site, 205–281, 283–330
Adenosine triphosphate, see ATP
Adrenal ferredoxin, see Adrenodoxin
Adrenodoxin, 229, 301, 319
Adenylation enzyme cascade, 2
Aldehyde carboxylate oxidoreductase, 251
Alcohol dehydrogenase, 125
  hepatic, 206, 348
Aldehyde oxidase, 122, 125
Alkylthiolate, 257
Aminoacylferredoxin, 109
Antimycin A, 80
Apoferredoxin, 107–110, 113–117
Arylthiolate, 257–260
ATP, 17, 42, 47, 90, 91
  binding, 37, 44, 45
  hydrolysis, 46
  and nitrogenase, 43
  protonation, 40
Azotobacter chroococcum nitrogenase, 48
Azotobacter vinelandii
  ferredoxin, 22, 29, 36, 41, 43, 45, 84, 269, 382, 401–404, 409
  MoFe protein, 30, 32
  nitrogenase, 255

B

Bacillus polymyxa ferredoxin, 29, 137, 147, 163, 209, 234, 249

Bacillus stearothermophilus ferredoxin, 209, 305–325
Bacteriophage T4, 110
  lysozyme, 110
Benzene, 361
Beta vulgaris, 148
Bijvoet difference density map, 160, 161
Bohr magneton, 291

C

Candida boidinii, 147, 149
Candida utilis, 66
  NADH dehydrogenase, 92
Carbon dioxide oxidoreductase, 129–134
  function, 130–131
  inactivation, 134
  inhibitor, reversible, 131, 134
  metals, 132
  properties, 131
Carbon metabolism, bacterial of single C compounds, 143
Carboxypeptidase A, bovine, 206
Catalase, 125
Center 5, 80–82
Chlamydomonas reinhardi, 106
Chloropseudomonas ethylica, 299
Chromium(III) substitution chemistry, 346–347
Clostridium acidi-urici, 108, 109, 112–114, 121, 127, 134–136, 146, 184, 232, 234, 247, 297, 372
Clostridium butyricum, 123, 138, 142
Clostridium cylindrisporum, 125, 127, 135, 146
Clostridium formicoaceticum, 124, 136, 146
  CO_2 reduced to formate, 140–141

SUBJECT INDEX

Clostridium kluyveri monocarboxylic acid cycle, 137–138
Clostridium M-E, 184, 372
Clostridium pasteurianum, 16, 22, 25–29, 37, 39, 42, 45, 46, 50, 104, 106, 109, 111–117, 124, 129–134, 138–140, 146, 158, 208, 210, 213, 221, 225, 234, 248, 263, 271, 286, 299, 301, 306–310, 316, 319, 321, 324, 325, 352, 368, 382, 389, 390, 394–396, 400, 412, *see also* Ferredoxin; Rubredoxin
Clostridium thermoaceticum, 123, 124, 136, 146, 149
CO_2 reduced to formate, 138–140
Cluster-binding cavity of Fe_4S_4* center, 172–187
  changes, 193–195
  conformational, 195–196
Coenzyme A transferase, 125
Contact shift equation, 239
Coordination unit, biological
  mononuclear, 206
  polynuclear, 206
Creatine kinase, 40, 41
Crucifera, 148
Cubane, inorganic, 350
Cyanide, 134, 150
Cysteinyl derivatives, 268
Cytochrome c, 61, 62, 102, 206, 340–344, 355, 369–373

D

Degeneracy, orbital, 254
Desulfovibrio desulfuricans, 209
Desulfovibrio gigas, 106, 163, 305
Desulfovibrio vulgaris, 106, 112, 146
Disulfide, 349
Dithionite
  oxidation, 42–44
  titration with, 77

E

Edman degradation, 109
Electron spin relaxation, 290–293
Electron transfer, 231–232, 347
  biological, 267
  mitochondrial, 61–100
  in nitrogenase components, 43–54

Electron tunneling, 358, 362, 363
Encephalomyocarditis virus RNA, 111
Entatic state hypothesis, 207
Equisetum protein, 211
Escherichia coli, 3, 105, 123, 134, 142, 147, 149
  episome F'33, 10
  induction in, 105
  lactose utilization, 105
  nitrate reductase, 149
Escherichia coli–Klebsiella pneumoniae hybrid, 10
ETF dehydrogenase, 82–84
Euglena gracilis, 126, 294
Extrusion reaction, 260–264

F

FDH, *see* Formate dehydrogenase
Ferritin, 206
Ferredoxin, 101, 102, 106–117, 138–141, 223, 285–288, 368, 369, 372
  adrenal, *see* Adrenodoxin
  amino acid composition, 114, 116
  analogues, 411–415
  assay, 112–113
  bacterial, *see* individual species of bacteria
  bound, discovered (1971), 93
  clostridial, *see* Clostridium
  common ancestor, 200–202
  cooperativity, 199–200
  4 Fe, 310–313
  immunology, 111–112
  plant type, *see* *Spinacia oleraceae* ferredoxin
  synthesis, regulation of, 106–107
  and visible light, 106–107
  x-ray analysis, 157–204
Ferredoxin-NADP reductase, 112
Ferredoxin: CO_2 oxidoreductase, reduced, 129–134
Flavin, 350
Flavodoxin, 106, 140
Flavoprotein, 83
Formaldehyde dehydrogenase, 124
Formate
  CO_2 reduction to, 138–141
  and ferredoxin, 129–142
  in metabolism, 127–129

oxidation to CO_2 by cell-free
  bacterial extract, 130
Formate dehydrogenase, 122, 124, 129–137, 143–150
  in cell-free extract, 136–137
  comparisons, 149–150
  ferredoxin-dependent, 129–137
  in formate metabolism, 143–150
  function, 135, 143
  inactivation, 145
  inhibition, 145
  kinetics, 144–145
  properties, 135, 143–144, 146–148
Formate kinase, 126
Formyl-FH_4 dehydrogenase, 122, 125
Formylmethionine, protein synthesis and, 129
Formyl tetrahydrofolate synthetase, 127
N-Formylaspartate deformylase, 126
S-Formylglutathione hydrolase, 126
N-Formylmethionine deformylase, 126
Fourier difference map, 188–192, 195–196
Franck–Condon
  activation, 334, 359, 360
  barrier, 347
  rearrangement, 196, 359
  relaxation rate, 360

G

β-Glactosidase, 110
Glutamate dehydrogenase, 5–11
Glutamate synthetase, 5–7
GTP-8-formylhydrolase, 126, 128

H

Hammet constant, 232
Hemoglobin, 206
  synthesis by lysate of rabbit reticulocyte, 110
Hepatoma cell of rat, 111
High potential iron protein (HiPIP), *Chromatium vinosum* strain D, 22, 29, 76, 82, 89, 94, 158, 159, 162, 172–175, 178, 179, 208, 210, 223, 232, 234, 242, 248, 250, 285, 286, 305–310, 313–320, 322–324, 340–343, 356–372, 409
  analogues, 411–415
  crystals, 188–189

Fe_4S_4* center
  active, 160–164
  analogous synthetic, 164–172
  protein-bound, 165–172
  synthetic, 164–172
  preparation, 188–189
  reduced versus oxidized, 187–196
  stabilization, 188–189
  three-state hypothesis, 160–164
  x-ray analysis, 157–204
Histidase, 3, 4
Histidine utilization gene, see *Hut* system
Holoferredoxin, see Ferredoxin
Hueckel molecular orbital calculation, 253
Hut system, 3, 7
  operon, regulation of, 7
Hydrogen bond, 184–187, 193, 202
Hyperfine interaction, 383–384, 399

I

Imidazole, 361
Indole, 361
Inner sphere reactions, 345–347
Insulin, 102
Iron
  electron spin relaxation, 292–293
  equivalence, 317–322
  hyperfine interaction, magnetic, 293–296
    transferred, 296–297
  magnetic moment, 290–292
  mixed valence states discovered, 166
  oxidation level in Fe_4–S_4* centers, 166
  valence, 317–322
    mixed states discovered, 166
Iron complex
  active site analogues, 217–228
    structure, 217–228
    synthesis, 218–220
  binuclear, 215–216
  expanded lattice, 217
  1 Fe analogue, 219–220, 222, 225, 227
  2 Fe analogue, 219, 227
  4 Fe analogue, 218–219, 227–228
  mononuclear, 215
  with sulfide, 211–214
  tetranuclear, 216–217
  with thiol, 211–214
  trinuclear, 216

SUBJECT INDEX 439

Iron protein, 220–228, *see also*
    Nitrogenase
Iron–sulfur analogues, 205–281
    active site, 207–211
        core extrusion, 260–264
        structure, 208–209
    chelate, 210–211
    chemical reactivity, 256–265
        acid solvolysis, 264–265
        active site core extrusion, 260–264
        protonation, 264–265
        thiolate ligand substitution, 257–260
    coordination units, 214–217
        binuclear, 215–216
        extended lattice, 217
        mononuclear, 215
        tetranuclear, 216–217
        trinuclear, 216
    electronic features, 233–256
        absorption spectra, 241–247
        EPR spectra, 249–250
        magnetic susceptibility, 233–236
        Mössbauer spectra, 250–253
        nonanalogue complexes, 255–256
        proton magnetic resonance, 236–241
        resonance Raman spectra, 247–249
        theoretical electronic structural
            models, 253–255
    model peptides of, 213
    properties, physical, 228–256
        electron transfer, 231–232
        and ligand structure, 232
        oxidation-reduction level, 228–233
    redox centers, units as, 265–271
        and ligand structure, 268–271
        structure, 265–267
    synthetic, of the active site, 205–281
Iron–sulfur proteins
        spectroscopy, 289–298
    angles in cluster, 168–169
    biosynthesis, 101–119
    center of electron-dense atoms, 96,
        160–164
        binding cavities, 172–187, 193–195
        as chromophore, large, 172
        coordination geometry, 184–187
        and energy conservation, 89–93
        geometry, 167, 170
            coordination, 184–187
            uncertainty of, 165

    interatomic distances, 168–169, 191
    of mitochondrion, 66–89
    models for, 411–415
    NH—S hydrogen bond, 184–187
    resonances, related, 88–89
    and side chain, 177–184
        aliphatic, 180, 181
        aromatic, 182–184
        hydrophobic, 177
    spin state, 170–171
    structure, changes of, 189–193
    symmetry, 171–172, 192–193
    unassigned, 82–88
        center, 5, 82–84
        ferredoxin-type, 86–88
        HiPIP, 84–86
classification, 285
in formate metabolism, 121–156
Mössbauer results, 381–417
nitrogenase, see Nitrogenase
oxidation level, 232–233
oxidation-reduction potentials, list of,
    230
redox chemistry, *see* Redox chemistry
redox mechanisms, 331–379
Rieske's, 80, 81
rubredoxin, *see* Rubredoxin of
    Clostridium pasteurianum
structure, review of, 158–159

J

Jahn–Teller distortion, 254, 255

K

Klebsiella aerogenes histidase,
    biosynthesis of, 4
Klebsiella pneumoniae, 2–4, 22, 29, 48,
    137, 146, 400, 401, 408, 409
    Asm^- mutant, 5
    glutamate auxotroph, 5
    glutamate dehydrogenase, 5, 6
    histidine utilization (*hut*) gene, 3
    MoFe protein, 32
    nitrogenase-depressed mutants, 1–14
Klebsiella pneumoniae–Escherichia coli
    hybrid, 10
Kramers
    doublet, 385, 386, 391

state, 387
theorem, 289
Kynurenine formamidase, 126

L

Ligand
  field splitting, 291
  structure and redox potential, 268–271
  substitution reaction, 261
Lysozyme of bacteriophage T4, 110

M

Magnesium adenosine triphosphate (MgATP)
  binding, 34–40
  EPR spectrum, 36–40
  and Fe protein, 34–43, 50–52, 55
  hydrolysis, 39
  Michaelis constant, 34–36
  reaction rate, MgADP, 40–43
Magnetic hyperfine coupling, 289, 293–296, 300
Marcus cross-reaction
  enthalpy, 339
  entropy, 339
Marcus free energy correlation, 336–339
Marcus model of reaction between two proteins, 359
Marcus relationships, 335
Mercaptoacetate in ligand, 211
2, 3-Mercaptopropan-1-ol, 212
Metal, 19–25
  analysis, difficulties of, 20
Metalloprotein
  models, 339–345
  redox behavior, 339–345
Methane bacteria, 123, 126, see also individual species
Methanobacillus omelianskii, 121, 136, 146
  S strain, 136
Methanobacter ruminantium, 146
Methanobacterium vannielii, 136
Methanococcus capsulata, 143
Methanococcus vannielii, 146
Methanomonas methanooxidans, 143
Methanosarcina barkeri, 127
Micrococcus aerogenes, 109

Micrococcus lactilyticus, 123, 127, 368
Mitochondrion
  electron paramagnetic resonance spectrum, 64, 75
  electron transfer system, 61–100
  Fe–S center, 66–89
  inner membrane, 86
  outer membrane Fe–S protein, 86–88, 95
Molecule, paramagnetic, 237
Molybdenum–iron protein, see Nitrogenase
Moment, magnetic of iron, 290–292
Monocarboxylic acid cycle, 137–138
Mössbauer
  chemical shift, 297–298
  nucleus, 233
  spectroscopy, 250–253
    description of the method, 383–388
    hyperfine interactions, 383–384
Myoglobin, 206

N

NADH-dehydrogenase, 66–73
Nitrobacter sp., 90
Nitrobacter agilis, 147
Nitrogen fixation genes, 1–2
  chronology of discoveries, 2
  construction of strains which export much NH_4, 2
  episomes, 2
  mapping, 2
  transfer to *E. coli*, 2
Nitrogenase, 15–60, 400–411
  absorbance, 49–54
  acetylene as substrate for, 45–48
  and carbon dioxide, 54
  composition of protein components, 17–34
  EPR properties, 27–32
  Fe–S centers, 26–27
    displacement, 27
    identification, 27
  metal composition, 19–25
  MoFe protein, 400–411
  Mössbauer spectrum, 32–34
  spectra
    Mössbauer, 32–34
    ultraviolet, 25–26
    visible, 25–26

SUBJECT INDEX

subunit structure, 17–19
electron transfer sequence, 43–54
  steady-state evidence for, 43–46
  energy of activation, 46–49
  EPR, 49–54
Fe content, 21, 22, 26, 33
Fe protein containing MgATP, 39, 55
  interaction with MgATP, 34–43
    binding studies, 34–36
    EPR spectrum, 36–40
    Michaelis constant, 34–36
    reaction rate controlled by
      MgATP/MgADP ratio, 40–43
    redox potential, 36–40
  hypotheses, 55–57
mapping regulatory genes, 8–11
Mo content, 22, 23
MoFe protein, 19, 25–26, 55–57
molecular weight of components, 18, 22
oxygen lability, 20
purification, 49
as a reductive dephosphorylation system, 57
S content, 22
Nitrogenase depressed mutants of
  Klebsiella pneumoniae, 4–11
  isolation, 4–5
  properties, biochemical, 5–8

O

Outer sphere, 345
  cross-reactions between redox couples
    Marcus free energy correlation, 336–339
    Marcus relationships, 335
Ovalbumin, 111
Oxalate decarboxylase, 127

P

Parsley ferredoxin, 110
Penicillamine, 212
Peptide deformylase, 126
Peptococcus aerogenes ferredoxin, 158, 159, 162, 163, 173, 175–177, 182, 184, 200, 208, 210–213, 218, 223, 225, 286, 305, 307, 369, 400

Peptococcus denitrificans, 373
Peptostreptococcus elsdenii, 106
Phaseolus aureus, 148
Phaseolus vulgaris, 106, 107
Phenol, 361
Phenoxide oxidation, 358
Photosynthesis, 351
Pisum sp., 148
PMR spectrum, 238, 239
Polypeptide
  biosynthesis, 103
  elongation, 103
  folding of chain, 172–177
  intiation, 103
  mRNA, 104
  polyribosome, 104
  secondary structure, 185–187
  side chains, aromatic amino acid, 182–184
    nonpolar amino acid, 177–182
  termination, 103
  tRNA, 104
Polyribosome, 104, 111, 114
Protein
  biosynthesis, bacterial, 103–105
    major pathway of, 102
  C state center, 322
  crystallography of structure, 165–166
  1 Fe center, 298–300
  2 Fe–2 S* centers, 288, 301–305
  4 Fe–4 S* centers, 305–326
  Mössbauer data, 302–305, 310–315
  reconstitution, 262
  spin coupling, 322–326
  structure refinement technique, 165–166, 189–192
Proteus rettgeri, 147
Proton
  activation, 17
  magnetic resonance, 236–241
Protonation, 264–265
Pseudomonas AM1, 122, 126, 143, 147
Pseudomonas methanica, 143
Pseudomonas oleovorans, 229, 250, 392
Pseudomonas oxalaticus, 121, 125, 128, 143, 144, 147, 149, 150
  formate dehydrogenase, 143–149
    function, 143
    inactivation, 145
    inhibitor, 145

kinetic properties, 145
properties
　　kinetic, 144–145
　　molecular, 143–144
Pseudomonas putida, 272
　　ferredoxin, *see* Putidaredoxin
Purine as energy source, 135
Putidaredoxin, 223, 285, 301
Pyruvate-formate lyase, 127
　　activation, 141–142
　　ferredoxin-dependent, 141–142

Q

Quadrupole interaction, electric, 298, 388
Quadrupole splitting, 300, 317, 318

R

Redox chemistry
　　developments, recent, 332–351
　　inner sphere rearrangement term, 334
　　outer sphere rearrangement term, 335
　　outer sphere reactions, 332–345
Resonance, nuclear gamma-ray, *see* Mössbauer effect
Rhizobium japonicum, 22, 29, 41
Rhodanese, 110
Rhodopseudomanas gelatinosa, 178, 179
Rhodopseudomonas palustris, 147
Rhodopseudomonas spheroides, 81
Rhodospirillum rubrum, 106, 107, 123, 340, 369, 370
Ribonucleic acid, *see* RNA
Rieske's Fe–S protein, 80, 81
RNA
　　messenger, 104–105
　　transfer, 104
Rotenone, 87
Rubredoxin of *Clostridium pasteurianum*, 104, 158, 207–210, 242, 251, 298–300, 317, 319, 350, 352–356, 373, 388–400, 414
　　electron transfer, 353
　　　　kinetics, 353, 354
　　as electron transfer protein- extremely reactive, 355
　　ferric, 389–393
　　ferrous, 393–398
　　first isolated (1965), 352

models for, 411–415
x-ray absorption spectroscopy, 352
Ruthenium, 351

S

S organism, 146
Saccharomyces cerevisiae, 76, 77, 147
Sarcina ventriculi, 137
Scenedesmus ferredoxin, 303
Site, active, *see* Active site
Spectrophotometry, stopped flow, 258
Spin coupling, 322–326
Spin Hamiltonian, 290–299, 385, 386, 389, 393, 397
Spinacia oleraceae ferredoxin, 148, 210, 234, 287, 304, 319, 368
Spirulina maxima, 26, 244
Streptococcus faecalis, 123
Structure refinement technique in crystallography, 165–166, 189–192
Succinate dehydrogenase, 73–80
　　deactivation, 78
　　electron acceptor, artificial, 73
　　Fe–S center, 73–77
　　HiPIP, 76, 85
　　oxidation-reduction potential, 71
　　thermodynamic features, 77
　　turnover, 77
Succinate-ubiquinone reductase, 74
Sulfide, 211–214
Sulfinic acid, 349
Sulfite reductase, 108
Sulfur ligand covalency, 297
Synechoccus lividas, 235
Synthetic analogue approach, 227

T

Taft constant, 232
Thiocapsa pfennigii, 178, 179
Thiol, 211–214
　　oxidation pathway to disulfides, 349
Thiolate, 346–347, 350
　　ligand substitution, 257–260
Three-state hypothesis for HiPIP, 163–164, 233, 286
　　summarized, 196–202
Titration
　　coulometric, 69

SUBJECT INDEX

direct, 69
oxidoreductive, 69
Tunneling, quantum-mechanical, 358, 362, 363
Tysosine 19, 182, 183, 195, 357, 372
  electron transfer dynamics, 358
  oxidizability, 361
Tyrosine aminotransferase, 111

U

Ubiquinol-cytochrome c reductase, 80–82
Ubisemiquinone, 88–89

V

Vibrio succinogenes, 147

X

X-ray diffraction, 284
Xanthomonas pruni, 126, 128
Xanthine oxidase, 63, 285, 302, 348

Z

Zeeman splitting, 295

Molecular Biology

An International Series of Monographs and Textbooks

Editors

BERNARD HORECKER

Roche Institute of Molecular Biology
Nutley, New Jersey

NATHAN O. KAPLAN

Department of Chemistry
University of California
At San Diego
La Jolla, California

JULIUS MARMUR

Department of Biochemistry
Albert Einstein College of Medicine
Yeshiva University
Bronx, New York

HAROLD A. SCHERAGA

Department of Chemistry
Cornell University
Ithaca, New York

HAROLD A. SCHERAGA. Protein Structure. 1961

STUART A. RICE AND MITSURU NAGASAWA. Polyelectrolyte Solutions: A Theoretical Introduction, *with a contribution by Herbert Morawetz*. 1961

SIDNEY UDENFRIEND. Fluorescence Assay in Biology and Medicine. Volume I—1962. Volume II—1969

J. HERBERT TAYLOR (Editor). Molecular Genetics. Part I—1963. Part II—1967

ARTHUR VEIS. The Macromolecular Chemistry of Gelatin. 1964

M. JOLY. A Physico-chemical Approach to the Denaturation of Proteins. 1965

SYDNEY J. LEACH (Editor). Physical Principles and Techniques of Protein Chemistry. Part A—1969. Part B—1970. Part C—1973

KENDRIC C. SMITH AND PHILIP C. HANAWALT. Molecular Photobiology: Inactivation and Recovery. 1969

RONALD BENTLEY. Molecular Asymmetry in Biology. Volume I—1969. Volume II—1970

JACINTO STEINHARDT AND JACQUELINE A. REYNOLDS. Multiple Equilibria in Protein. 1969

DOUGLAS POLAND AND HAROLD A. SCHERAGA. Theory of Helix-Coil Transitions in Biopolymers. 1970

JOHN R. CANN. Interacting Macromolecules: The Theory and Practice of Their Electrophoresis, Ultracentrifugation, and Chromatography. 1970

WALTER W. WAINIO. The Mammalian Mitochondrial Respiratory Chain. 1970

LAWRENCE I. ROTHFIELD (Editor). Structure and Function of Biological Membranes. 1971

ALAN G. WALTON AND JOHN BLACKWELL. Biopolymers. 1973

WALTER LOVENBERG (Editor). Iron-Sulfur Proteins. Volume I, Biological Properties—1973. Volume II, Molecular Properties—1973. Volume III, Structure and Metabolic Mechanisms—1977

A. J. HOPFINGER. Conformational Properties of Macromolecules. 1973

R. D. B. FRASER AND T. P. MACRAE. Conformation in Fibrous Proteins. 1973

OSAMU HAYAISHI (Editor). Molecular Mechanisms of Oxygen Activation. 1974

FUMIO OOSAWA AND SHO ASAKURA. Thermodynamics of the Polymerization of Protein. 1975

LAWRENCE J. BERLINER (Editor). Spin Labeling: Theory and Applications. 1976

T. BLUNDELL AND L. JOHNSON. Protein Crystallography. 1976

in preparation

HERBERT WEISSBACH AND SIDNEY PESTKA (Editors). Molecular Mechanisms of Protein Biosynthesis

QP
552
I7
L69
v.3

OCT 20 1977